6th International Conference on Transportation and Traffic Engineering (ICTTE 2017)

MATEC Web of Conferences Volume 124 (2017)

Hong Kong, China
1 - 3 July 2017

Editors:

M. Figueira
R. Montemanni
G. Lodewijks
J. Rzesny

ISBN: 978-1-5108-4883-2

Printed from e-media with permission by:

Curran Associates, Inc.
57 Morehouse Lane
Red Hook, NY 12571

Some format issues inherent in the e-media version may also appear in this print version.

Printed by Curran Associates, Inc. (2017)

For additional information, please contact EDP Sciences – Web of Conferences at the address below.

EDP Sciences – Web of Conferences
17, Avenue du Hoggar
Parc d'Activité de Courtabœuf
BP 112
F-91944 Les Ulis Cedex A
France

Phone: +33 (0) 1 69 18 75 75
Fax: +33 (0) 1 69 28 84 91

contact-edps@webofconferences.org

Additional copies of this publication are available from:

Curran Associates, Inc.
57 Morehouse Lane
Red Hook, NY 12571 USA
Phone: 845-758-0400
Fax: 845-758-2633
Email: curran@proceedings.com
Web: www.proceedings.com

TABLE OF CONTENTS

PREFACE

2017 6th International Conference on Transportation and Traffic Engineering (ICTTE 2017) will be held in Hong Kong during July 1-3, 2017. This conference covered recent trends and progresses made in the field of transportation and traffic engineering. The technical objective of this conference is to address key issues associated with science and technology in this rapidly evolving field of research and to promote contact between basic researchers and technological needs for real industrial applications.

The activities of the conference included keynote speeches and author presentations in sessions. The invited talks were focused on the potential application of transportation and traffic engineering in various fields of application.

ICTTE 2017 has received a total submission of 102 full length articles. And 48 manuscripts were selected after peer review and were recommended for publication in this volume, which are addressing diverse topics related to recent trends in transportation and traffic engineering, including modeling and analysis of traffic system, traffic facilities construction and operation system, traffic planning and management, traffic safety and risk assessment, urban logistics and information system, traffic and environment, vehicle design and manufacturing engineering, industrial production and management.

We hope the volume will offer the readers a good opportunity to access the state-of -the art developments and future directions of researches on transportation and traffic engineering.

We're truly grateful to all the members of the conference technical program committee, the reviewers who spared their valuable time, for their advice which have certainly helped to improve the quality, accuracy, relevance and the sincere efforts to maintain the quality of each paper selected for this conference program and volume for publication. And thank every author and participant who has contributed to the success of the conference.

Prof. Monteiro Figueira, University Lusofona, Portugal

June 26, 2017

edp sciences

MATEC Web of Conferences

Statement of Peer review

In submitting conference proceedings to MATEC Web of Conferences, I certify to the Publisher that I adhere to the **Policy on Publishing Integrity** of the journal in order to safeguard good scientific practice in publishing.

1. All articles have been subjected to peer review administered by the proceedings editors.
2. Reviews have been conducted by expert referees, who have been requested to provide unbiased and constructive comments aimed, whenever possible, at improving the work.
3. Proceedings editors have taken all reasonable steps to ensure the quality of the materials they publish and their decision to accept or reject a paper for publication has been based only on the merits of the work and the relevance to the journal.

Title, date and place of the conference

6th INTERNATIONAL CONFERENCE ON TRANSPORTATION AND TRAFFIC ENGINEERING

JULY 1-3 2017 HONG KONG

Proceedings editor(s):

PROF. MONTEIRO FIGUEIRA - LUSOFONA UNIV. - PORTUGAL

PROF. ROBERTO MONTEMANNI

PROF. GABRIEL LODEWIJKS

PROF. JAGIENKA RZESNY

Date and editor's signature

10th SEPT 2017

MATEC Web of Conferences **124**, 01001 (2017) DOI:10.1051/matecconf/201712401001
ICTTE 2017

Established Method of High-Speed Railway Survey Control Network

H.Y. Wang , F.Y. Yang and J.F. Gao

[1] University of Science and Technology Liaoning College of Civil Engineering, Anshan, China
[2] China Railw ay 19th Bureau Group 2nd co.,LTD

Abstract. It is very important of high-speed railway construction surveying and mapping. Especially the establishment of construction survey control network and searching aiming at layout, measurement, data processing of adjustment calculation of the horizontal control network and the elevation control network. The most important part is the establishment of CP II and CPIII control network. The elevation control mainly adopts second order leveling. Horizontal and elevation accuracy must to meet the requirements of high speed railway construction survey.

1 Introduction

With the progress of technology, high-speed railway has obtained a rapid growth. The new-built speed of 250~300km per hour and the old line of 200km per hour are regard as high-speed railway in Western Europe. The international railway agreement which was signed in Geneva in 1985 by the United Nations Economic Commission has stipulated that the speed of new special of passenger train high-speed railway is 300km per hour and the speed of new mixed type of passenger and freight trains high-speed railway is 250km per hour. In order to achieve the high speed condition for railway trains, the precision of high-speed railway track must be kept in the millimeter range. However, the traditional railway engineering measurement technology cannot meet the requirement of high-speed railway construction. The basis for the high-speed railway precise engineering surveying is setting up horizontal and elevat-ion control network at all levels. Under the control of all levels of precision measurement control network, we can achieve accurate construction and the accuracy of track laying to ensure high speed and safe driving of passenger train. In this paper, the construction technology of high-speed railway construction survey control network is illustrated by a total length of 24.265km high-speed railway construction.

2 Horizontal control network

WGS-84 coordinate system is adopted in the horizontal coordinate system. According to the principle of grading network, we establish the horizontal control network in four grades. Along the routes, the country's high level horizontal control points are rare and the accuracy is also poor. Therefore, we establish the first GPS framework as the benchmark of horizontal coordinate on the whole line (CP0). The second level is basic horizontal control network (CPI). The third level is line control network (CPII). The four levels is tracking laying and foundation piles control network (CP III) [1-5].

After the exploration design units surveys CP0、CPI、CPII、and elevation control network the construction organization carry on making piles and repeating survey. The CP III control points should be measured by construction organization [9-14].

2.1 Technical requirements of plane control network at all levels

On the basis of CPII, the construction control network adopts the method of free station intersection. During the construction phase, connecting the CPII and the CP III control network is the main task. CP0、CPI、CPII and CP III network requirements are shown in table 1、table2 and table3.

Table1. Layout requirements of horizontal control network at all levels

Levels of Control Network	Methods of Measurement	Grades of Measurement	Point Spacing	Remarks
CP 0	GPS	National B -Level	30~50km	
CPI	GPS	B-Level	≥1000m	≤4km A Pair of Point

	GPS	C-Level		
CPII	Traverses	Third Level	800～1000m	
CP III	Linear-angular Intersection Network Network	—	50～70m A Pair of Point	Point to Point Distance 10～20m

Table2. Accuracy Indices of GPS Measurement

Levels of Control Network	Mean Error of Baseline	Relative Mean Error of the Weakest Line
CP0	≤0.7"	—
CPI	≤1.3"	1/170000
CPII	≤1.7"	1/100000

Table3. Technical requirements for traverse survey

Levels of Control	Echo Length (km)	Length (m)	Mean Error of Distance Measurement (mm)	Mean Error of Angle Measurement (″)	Mean Error of Coordinate in Adjacent Points (mm)	Relative Mean Error of the Total Length of the Traverse	Misclosure Limitation of Azimuth (″)	Corresponding Traverse Grades
CP II	≤5	400~800	3	1.8	8	1/55000	$\pm3.6\sqrt{n}$	Third Level

Note: n is the number of angle of traverse ring

2.2 The horizontal control network shape and the connection measurement methods of CP III

In order to meet the requirements of CP III measurement, we should carry on a comprehensive repetition measurement and maintenance of basic plane control network CPI, line control network CPII and elevation control network before CP III measurement. Furthermore, we also need to recover the horizontal elevation points which are lost and destroyed.

On the basis of basic plan control network CPI, line control network CPII and elevation control network, CP III measurement can be laid and measured by construction organization according to the needs of underline engineering and rail construction. As shown in figure1 ▲ is CPI, ▲ is CPII, ⊗ is CP III. The number of 1, 2, 3……14、15、16 in the figure are just the code in this article and not the real CP III point [6-8]. The connecting measurement method: For example, after the instrument set up, centered, and leveled over at station 2, observing CPI、CPII, using liner-angular intersection we can calculate the coordinate of the station. Then observing CP III again, we can get the coordinate of CP III, too.

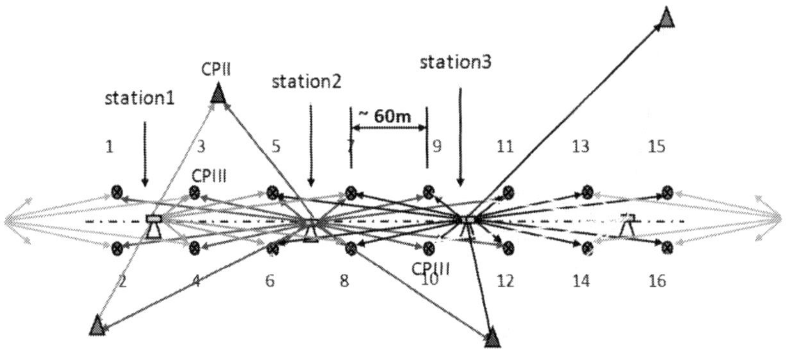

Figure 1. Connecting Measurement Methods of CP III Horizontal Control Network

2.3 The observation instrument and limit standard of CP Ⅲ horizontal network

2.3.1 Measuring Instrument

According to the estimation of the position and the relative position accuracy of the CP Ⅲ horizontal network, the nominal accuracy of the total station can be used to carry out the standard network observation is mean error of horizontal direction observation in one round ≤±1.0″ , mean error of distance measurement ≤±(1mm+2ppm),in addition, the total station should have a motor drive, automatic collimation and data recording function. Other accuracy of the total station cannot meet the accuracy requirements of the horizontal network measurement.

2.3.2 Station Limit

The method for measuring horizontal direction of CP Ⅲ plane network is full circle direction observation method which should control the main accuracy indices including the difference value return to zero in semi-observation , the difference between direction value the same direction returning zero and 2c of the mutual difference value in different value on same direction. Accuracy indices are shown in table 4.

Table 4. The accuracy indices of full circle direction observation

control network name	instrument	set	the difference value return to zero in semi-observation	2c of the mutual difference	the difference between direction value the same direction returning zero
CP Ⅲ	0.5″	2	6″	9″	6″
	1″	3	6″	9″	6″

2.4 The adjustment methods of CP Ⅲ horizontal control network

Adjustment should not reduce the actual observation accuracy of CP Ⅲ network, and ensure the smoothness of the track laying construction. Adjustment mainly has the following three calculating models: free network adjustment and correction traditional restriction network adjustment and leveling algorithm of CP Ⅲ .

3 The elevation control network

1985 National Elevation datum is adopted in the elevation system(1)The elevation control network shall be measured according to the required precision of the second-order leveling and the leveling base control network should be measured once on the whole line. When connecting to another railway, to make it easier to determine the relationship between the two railway elevation systems, we always need to carry on connecting measure with bench marks of the other railway. (2)Each bench mark should be set at 2km.Besides, according to the actual situation the key project area(such as bridges, long tunnels and special roadbed structure) shall be added. Bench marks which away from the centre line should be between 50~150m can use a same pile with the plane control points, or be set alone.(3)Leveling method should be adopted in elevation control.

3.1 Measurement of the elevation control network

The second-order leveling is adopted in the whole line and the leveling route is laid along the line in principle. Each bench mark should be set at abound 2km and each deeply buried bench mark should be set at abound 20km. According to the actual situation the key project area(such as bridges, long tunnels and special roadbed structure) shall be added. Bench marks which away from the centre line should be between 50~150m need to use a same point with CP Ⅱ and CP Ⅲ as far as possible. The location of the bench marks should be safe and quiet. Besides, the places which is on the soli solid, convenient conversation and long term preservation are also take into account. Line bench marks should bury stone by rule and finish the description of station. The buried standard of deeply buried bench mark should in accordance with the situation of stratum along the line. The point can be buried to supporting layer or by specially designed according to the actual situation leveling method should be adopted in elevation control and digital level is a good choice. In the process of observation, each section shall be an even number. After leveling operation, every leveling route should calculate accident mean error MΔ according to the different values of height difference between segments when the number of leveling network rings more than20, we should calculate closure error Mw according the number of rings yet. MΔ and Mw should meet the stipulate, otherwise the routes which have big closure error should be re-measured.

Calculation Formula:

$$M_{\Delta} = \sqrt{\frac{1}{4n}[\frac{\Delta\Delta}{L}]} \quad (1)$$

$$M_{W} = \sqrt{\frac{1}{N}[\frac{WW}{L}]} \quad (2)$$

In the Formula: Δ ——Difference Values of Height Difference Between Segments (mm);

L ——Measuring Length or Loop Length (km);

n ——the Number of Segments;

W ——Closing Error of Rings (mm);

N ——the Number of Rings

3.2 Establish the foundation piles elevation control network

The points of the fourth horizontal control network (CPI) are the common points of the horizontal and the elevation control network. Elevation measurement of the CP III control point should be carried out after the completion of the CP III horizontal measurement and in accordance with the requirements of precision leveling measurement, starting and closing at second order basic bench mark. It is suggested that the elevation of the control points should be measured by the digital level and use tightly adjusted.

3.3 Method of Observation and Adjustment

1) As shown in figure 2 and figure 3, the observation lines can be seen that whether direct or reversed, all the bench mark of CP III can be linked in the routes.

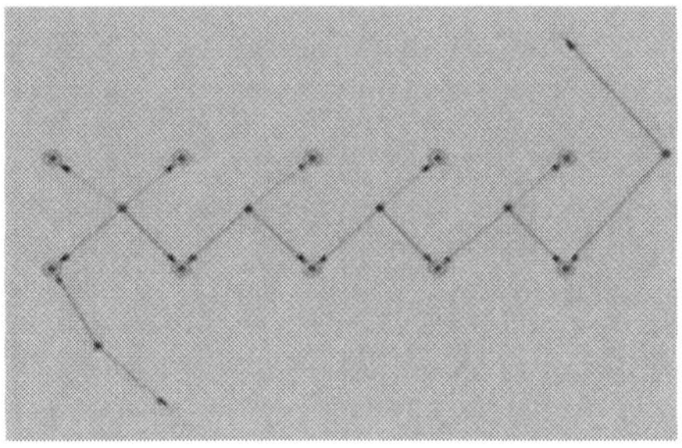

Figure 2. Diagrammatic sketch of the direct leveling route

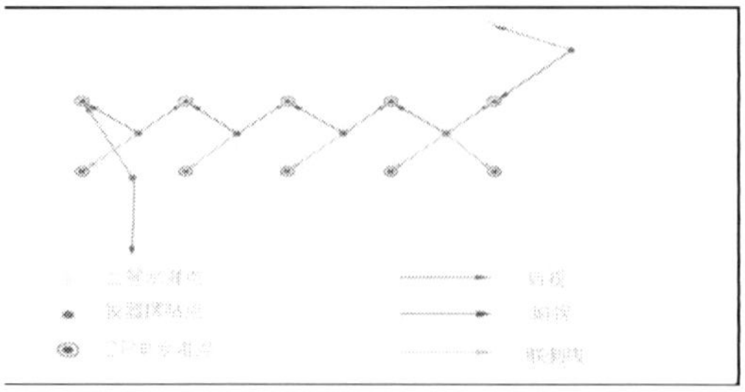

Figure 3. Diagrammatic sketch of the reversed leveling route

2) Each measurement stations are set between the four points and do not have to move the instrument, so as to improve the efficiency of field observation.

3) The length of the horizontal line between the points in the leveling line is also approximately equal. In order to ensure the correctness of CP III leveling measurement, the method of independent round-trip observation should be adopted to check the level line.

4) The method of indirect adjustment is adopted by the elevation control network adjustment. In addition, restriction adjustment is adopted by connecting to two order leveling point around survey area.

4 Conclusion

According to the above control measurement, after the adjustment, the precision index of high-speed railway control measurement is that the correction value of the horizontal distance should be≤±3"(refer to the station to the CP III point), the mean square error of unit weight of horizontal direction should be≤±2.5",the relative error of the relative position of CP III point should≤±1mm and the elevation accuracy of the whole station should meet the requirement of the second order leveling.

References

1. LI Shuliang,,LIU Chenglong, NI Xiantao. Study on the surveying method of CP III vertical control network in high-speed railway[J]. Science of Surveying and Mapping, 2011, 36(1): 45-47.
2. Liu Chenglong, Yang Xuefeng, Lu Jiankang, He Bo. Construction and adjustment method for CP III trigonometric leveling network of high-speed railway[J]. Journal of southwest jiaotong university, 2011, 46(3): 434-439.
3. Hao Yadong, Li Jufang, Liu Li. High-speed rail ballastless track layout of control network and precision control[J]. Bulletin of Surveying and Mapping, 2011,(8): 51-53.

DOI:10.1051/matecconf/201712401001

4. Zhou Lingyan, Liu Chenglong, Gao Hongtao, Nie Huxiao. Application to a new method of control network inside tunnels in hi-speed railways[J]. Journal of geomatics science and technology, 2014, 31(6): 570-575.
5. Wang Lei, Liu Chenlong, Yang Xuefeng, Chen Haijun. Three-Dimensional adjustment calculation for free station and its precision evaluation in high-speed railway[J]. 2011, 28(4): 258-261.
6. Yan G F, Gao S, Zhang J H, et al. Research on Rigorous Adjustment Lapping between Adjacent of CP III Plane Network[J]. Journal of the China Railway Society, 2014.
7. Ren X C, Zhou D W. Research and Application on the Lapping Methods between Adjacent Sections for CP III Track Control Network for High-speed Railway[J]. Journal of Railway Engineering Society, 2014.
8. Puziewicz J, Matusiak-Ma †F k M, Ntaflos T, et al. Subcontinental lithospheric mantle beneath Central Europe[J]. International Journal of Earth Sciences, 2015, 104(8):1-12.
9. Yang Feng yun Application of GPS Technology for Water Pipeline network in a big city Beijing The Publishing House of Surveying and Mapping 2002.11 P35—36
10. Langley R.B. GLONASS: Review and Update, GPS World, 1997(7), 46—51
11. Zanaoa N. Preliminary Evaluation of the Russian GLONASS System as a Potential Geodetic Tool, Journal of Geodesy, 1998(72), 356—363
12. Gourevitch, S., 1994. Inplications of Z Technology for Civilian Positioning. Proceedings of the ION-GPS-94 Meeting, September 1994, Salt Lake City, Utah, pp. 249-255.
13. Ham,S., 1997.Quality Control Issues Relating to Ambiguity Resolution for Real- Time GPS kinematic Positioning ,Journal of Geodesy, 71(6), pp. 351-361.
14. Lorenz et al, US patent Number 5, 134, 407 Magellan Corporation, FGCS Test Report, January 2000

Multi-Path Signal Synchronization Model with Phase Ratio and Dispersion Constraints

Tzu-Jung Huang, Hsun-Jung Cho, Chien-Chia Liäm Huang

Department of Transportation and Logistics Management, National ChiaoTung University, Hsinchu, Taiwan 30010 ROC

Abstract. Proper designs of offsets forsignalized intersections in an urban arterial system can help improve the level of service (LoS) significantly. The MAXBAND model has been the most widely used approach for this purpose. In this paper, a MAXBAND model in which the lengths of phases in a cycle are variable is proposed. In other words, the lengths of phases in a cycle, as well as the cycle itself, are no longer fixed.Besides, theincorporation oftraffic dispersion module in the proposed model allows that the travel times from upstream intersection to the downstream one are taken into account.Real-world application on an arterial system in Taiwanfor evaluating the performance of the proposed model is conducted to validate the methodology. Actual traffic flow data are collected through on-site experiments. Results suggest that the improvements are around 44%, on average, in terms of total delay of the entire network.

1 Introduction

An effective signal control strategyis crucial in improving the LoS for coordinated signalized intersections. By adjusting the offsets of the coordinated intersections in an arterial system, the control delays and the stopping times of vehicles can be well controlled, thereby providing better LoS. The MAXBAND model, proposed by Morgan and Little1, is one effective approach for this purpose. The concept of the MAXBAND model is to maximizethe total progression bandwidth of a two-way coordinated signalized system by adjusting the offsets at each intersection over atime interval.

The very first MAXBAND model was proposed by Little et al.2 in 1966. The model encompassestime constraintsdescribing timerelationship between the up- and down-stream intersections. Chang et al.3proposed the MAXBAND-86 that extendsLittle's model to solve multi-arterial closed network problems. Yet, neithermodel has considered the lead-lag left-turning phases and the queuing clearance times.Gartner et al.4solved the problem by developing a multiband model thatincorporates the left-turn phase sequence optimization and initial queue clearance times.Themajor contribution of multiband model resides in the consideration of the flow volume in the model.

Based on the multiband model, several extensions have been proposed thereafter. Stamatiadis et al.[1]67proposed the Multiband-96 model, extending the multiband model to arterial networks. Lu et al.8extended the multiband model by incorporating the travel time constraints with traffic dispersion module9. Zhang et al.10improved Gartner's model by relaxing the limitation of symmetrical progression. However, all these models above do not consider the turning flow from the minor legs crossing the arterial. In this case, the efficiency of the arterial systemmay increase, whereas that of the entire network may not necessarily. Most recently, Yang et al.11proposed a multi-path progression model featuring the optimization of the phase sequences and the determination of critical paths in the associated network. This model can handle the intersections with high turning volumes on the minor legs by adding a new path through the arterial.

A major restriction in Yang's model resides in that the length of the common cycle, as well as that of the phases, for the coordinated intersections are fixed.Such a restriction may possibly lead to unsatisfactorylocal solutions. Therefore, based upon this observation, an extension of their model by relaxing such restrictionsis proposed so that the lengthsof phases in a cycle, as well as the cycle itself, become variable. Such flexibilities can certainly lead to better solutions, compared to the fixed case. Besides, theincorporation ofRobertson's traffic dispersion module9allows oneto take into accountthe travel times from the upstream intersection to the downstream one, thereby providing more realistic results.Consequently, the resultant model becomes a mixed-integer quadratically constrained programming (MIQCP) problem which is NP-hard 1213in essence. We then exploit Gurobi solver14to search for an approximate feasible solution. Plus,empirical experiments are conducted for evaluating our methodology, and the results suggest that the performance of our model comes with an improvement being around 44% on average, in terms of total control delay per minute.

This is how the rest of the paper is organized. In Section 2, we first review of base model proposed by Yang et al.11, and then introduce the revised model thereafter. In Section 3, we elaborate on how the experiments as well as the numerical simulations are performed. Conclusions will be given in Section 4.

2 Model Specifications

In this section, the review of the multi-path progression model proposed by Yang et al.11is given. Thereafter, tworevised models, one by relaxing the phase ratios in the model, and the other by adding the dispersion module to the model, are introduced.

2.1. Multi-path progression model - A review

Variables to be used in the multi-path progression modelare given inTable 1.

Table 1. Notations

$P(\bar{P})$	path sets of outbound(inbound) paths
$b_p(b_{\bar{p}})$	total bandwidths of the outbound (inbound) path p ($\bar{p}$)
$\varphi_p(\varphi_{\bar{p}})$	weight for outbound(inbound) path $p(\bar{p})$.
$\beta_{p,l,i}(\beta_{\bar{p},l,i})$	binary, equals 1 if path $p(\bar{p})$ is on a green light in phase l at intersection i
$w_{p,i}(w_{\bar{p}i})$	the time from the right(left) side of red of path p ($\bar{p}$)at intersection i to the progression line
θ_i	offset of intersection$i$
$r_{p,i}(\bar{r}_{p,i})$	left- (right-)end red time of the band in a cycle
$\tau_{p,i}(\tau_{\bar{p},i})$	queue clearance time in the path$p$at intersection$i$
$t_i(\bar{t}_i)$	travel time from intersection i to $i+1$ for outbound(inbound) link
$m_{p,i}$	integer, loop variable
$x_{l,m,i}$	binary, equals 1 if phase l precedes phase m at intersection i
$\phi_{l,i}$	ratio of phase $l$to a cycle at intersection$i$

The multi-path progression model proposed by Yang et al. 11is given below

$$Max \sum_p (\varphi_p b_p) + \sum_{\bar{p}} (\varphi_{\bar{p}} b_{\bar{p}}) \quad (1)$$

s.t.

$$w_{p,i} + b_p \le \sum_l \beta_{p,l,i}\phi_{l,i} \quad (2)$$

$$w_{\bar{p},i} + b_{\bar{p}} \le \sum_l \beta_{\bar{p},l,i}\phi_{l,i} \quad (3)$$

$$\theta_i + r_{p,i} + w_{p,i} + t_i + m_{p,i} = \theta_{i+1} + r_{p,i+1} + w_{p,i+1} + \tau_{p,i+1} + m_{p,i+1} \quad (4)$$

$$-\theta_i + \bar{r}_{\bar{p},i} + w_{\bar{p},i} - \tau_{\bar{p},i} + \bar{t}_i + m_{\bar{p},i} = -\theta_{i+1} + \bar{r}_{\bar{p},i+1} + w_{\bar{p},i+1} + m_{\bar{p},i+1} \quad (5)$$

$$r_{p,i} \le \sum_l \beta_{p,m,i}x_{l,m,i}\phi_{l,i} + M(1-\beta_{p,m,i}) \quad \forall m \quad (6)$$

$$\bar{r}_{p,i} \le \sum_l \beta_{p,m,i}x_{m,l,i}\phi_{l,i} + M(1-\beta_{p,m,i}) \quad \forall m \quad (7)$$

$$r_{p,i} + \bar{r}_{p,i} + \sum_l \beta_{p,l,i}\phi_{l,i} = 1 \quad (8)$$

$$r_{\bar{p},i} \le \sum_l \beta_{\bar{p},m,i}x_{l,m,i}\phi_{l,i} + M(1-\beta_{\bar{p},m,i}) \quad \forall m \quad (9)$$

$$\bar{r}_{\bar{p},i} \le \sum_l \beta_{\bar{p},m,i}x_{m,l,i}\phi_{l,i} + M(1-\beta_{\bar{p},m,i}) \quad \forall m \quad (10)$$

$$r_{\bar{p},i} + \bar{r}_{\bar{p},i} + \sum_l \beta_{\bar{p},l,i}\phi_{l,i} = 1 \quad (11)$$

$$x_{l,l,i} = 0, \ \forall l \quad (12)$$

$$x_{l,m,i} + x_{m,l,i} = 1, \ \forall l \ne m; \ \forall i \quad (13)$$

$$x_{l,n,i} \ge x_{l,m,i} + x_{m,n,i} - 1, \ \forall l \ne m \ne n; \ \forall i \quad (14)$$

$$x_{l,n,i} + x_{n,m,i} = 1, \forall l \ne m \ne n; \ \forall i \quad (15)$$

$$x_{l,m,i} = 1, \quad \forall l \ne m; \ \forall i \quad (16)$$

$$x_{l,m,i} - x_{l',m',i+1} = 0, \quad \forall i \quad (17)$$

The decision variables include bandwidth $b_p(b_{\bar{p}})$, the interference variable $w_{p,i}(w_{\bar{p},i})$, the phase ratio $\phi_{l,i}$, the left(right) side red time $r_{p,i}(\bar{r}_{p,i})$, the offset θ_i, the loop variable $m_{p,i}$, and the phase relative position variable $x_{l,m,i}$.

The parameters include the weight of bandwidth $\varphi_p(\varphi_{\bar{p}})$, the available phase parameter $\beta_{p,l,i}(\beta_{\bar{p},l,i})$, the travel time $t_i(\bar{t}_i)$, and the queue clearance time $\tau_{p,i}(\tau_{\bar{p},i})$. M is a penalty parameter.

The objective function (1) is to maximize the sum of bandwidths of outbound (P) and inbound paths($\bar{P}$). The weights $\varphi_p(\varphi_{\bar{p}})$can be adjusted in accordance with the volumes of the associated paths. Here we have $\varphi_p = \varphi_{\bar{p}} = 1$.

Constraints (2) and (3) are interference constraints. They guarantee that the progression band of path $p(\bar{p})$ will only use the available green time at intersection i. Constraints (4) and (5) are loop constraints, representing the relationship of the intersection i and $i + 1$ while they are synchronized.

Constraints (6)-(17) are phase sequenceconstraints at each intersection so that the green band competition amongdifferent paths are controlled. Specifically, Constraints (6)-(11) are the boundary conditions for the left/right side red times of progression band. Constraints (12)-(17) ensure that the resulting phase sequence is feasible.

2.2. Revised model 1: Making $\phi_{l,i}$ a variable

In the progression model introduced in Section 2, the phase ratio $\phi_{l,i}$ is a constant. Here this restriction is relaxed and $\phi_{l,i}$is now made a variable. By doing so, constraints associated with $\phi_{l,i}$ become nonlinear in essence. The problem then becomes a nonlinearly constrained mixed integer optimization problem, or more specifically, a mixed integer quadratically constrained programming (MIQCP) problem. Besides,constraints

(15)-(17) will not be used, similar to the manipulation in Yang et al.11

Along with the relaxation onthe phase ratio, the restriction that the cycle length is also fixed is also relaxed. To do this, we addconstraint (18) to ensure that the cycle length is now varying butbounded.

$$\frac{1}{e} \le Z \le \frac{1}{f} \quad (18)$$

where Z is the reciprocal of the cycle length, and e (f) is the upper (lower) bound of cycle length.Based on (18), the travel time from intersection i to intersection i+1 can then be formulated as

$$t_i = \frac{d_i}{v_i} * Z \quad (19)$$

$$\bar{t_i} = \frac{\bar{d_i}}{\bar{v}_i} * Z \quad (20)$$

where d_i $(\bar{d}_i)$ is the distance from outbound (inbound) intersection i to $i+1$; and v_i $(\bar{v}_i)$is the free flow speed of the link.

The sum of the phase lengthsatany intersection must equal the cycle length; in other words, the total ratio of phases at an intersection must equal 1. Thus,

$$\sum_{l} \phi_{l,i} = 1, \forall i \quad (21)$$

To ensure that the resulting MIQCP is solvable,according to13, the model must come withlower boundson the phase ratio variable $\phi_{l,i}$, to begiven in (22). To correctly set the bounds for the ratio variable, we resort to the capacity equation discussed in Roess et al.15, given by

$$c_q = s_q * (\frac{g_q}{C})$$

wheres_q is the saturation flow rate for lane or lane group q, $g_q$is the effective green time for lane or lane group q, and C is signal cycle length. This equation can be rewritten as$\frac{c_q}{s_q} = \frac{g_q}{C}$. Since the critical volume of lane V_q^c will never be bigger than the capacity c_q ,we then have$\frac{g_q}{C} = \frac{c_q}{s_q} \ge \frac{V_q^c}{s_q}$. Here,the ratio$\frac{g_q}{C}$ can be considered as aproportion of the phase length to the cycle length, which isalso the ratio of phase(ϕ_q). Consequently, we have

$$\frac{g_q}{C} = \phi_q = \frac{c_q}{s_q} \ge \frac{V_q^c}{s_q}.$$

Therefore, the lower bounds for the phase ratio variable $\phi_{l,i}$are given below

$$\begin{aligned} \phi_{l,i} &\ge \frac{V_{l,i}^c}{s}, \; if \textstyle\sum_l \frac{V_{l,i}^c}{s} < 0.9; \\ \phi_{l,i} &\ge \frac{V_{l,i}^c}{s} * (\frac{1}{\sum_l \frac{V_{l,i}^c}{s} + 0.1}), \; otherwise. \end{aligned} \quad (22)$$

The threshold 0.9 is set to evade over-saturation, so that the remaining one-tenth of the entire cycle can be used as further adjustments, if needed.

Therefore, the revisedmodel, termed as RMPPM1, is given as follows:

(RMPPM1)$Max\ \sum_p(\varphi_p b_p) + \sum_{\bar{p}}(\varphi_{\bar{p}} b_{\bar{p}})$

s.t.

Eqs. (2) – (14), (18) – (22)

2.3. Revised model 2: adding traffic dispersion

In RMPPM1, the parameter t_i is mainly defined by the ratio of distance between an intersection to thedownstream$d_i$to the free flow speed of the linkv_i. In the development of MAXBAND model, there aretwo methods to processv_i. One is to make it a constant, and the other is to set a boundary for it, see LP2 in Little et al.2. Both methods have their disadvantages. For the first method, the real free flow speed might never be reached even when traffic is moderate to dense; and the second method requires new variables, thereby introducing more complexities to the problem.Based upon this, Lu et al.8proposed a model that combined the multiband model proposed by Gartner et al.4and the traffic flow dispersion model proposed by Robertson9, so that the calculation of the link travel timecan be more realistic.

In this study, we encompass the suggestion by considering the traffic dispersion module in the proposed RMPPM1, named RMPPM2. The travel time constraints(19)-(20)arethen replacedby(23)-(24)

$$t_i = \sum_{T_i = Time_{fast}}^{Time_{slow}} T_i F(1-F)^{T_i - Time_{fast}} * Z \quad (23)$$

where $F = \frac{1}{1+0.35T_i}$, $Time_{fast} = \frac{d_i}{V_{fast}}$, $Time_{slow} = \frac{d_i}{V_{slow}}$.

$$\bar{t_i} = \sum_{T_i = \overline{Time}_{fast}}^{\overline{Time}_{slow}} T_i F(1-F)^{T_i - \overline{Time}_{fast}} * Z \quad (24)$$

where $F = \frac{1}{1+0.35T_i}$, $\overline{Time}_{fast} = \frac{\bar{d_i}}{\overline{V}_{fast}}$, $\overline{Time}_{slow} = \frac{\bar{d_i}}{\overline{V}_{slow}}$. $Time_{fast}$ $(\overline{Time}_{fast})$ is thetravel time of fastest vehicle between intersections$i$ and i+1 in an outbound(inbound) path, while $Time_{slow}$ $(\overline{Time}_{slow})$ is thetravel time of slowest vehicle between intersection $i$and i+1 in an outbound(inbound)path.

Consequently, the RMPPM2 is written as the follows:

RMPPM2: $Max\ \sum_p(\varphi_p b_p) + \sum_{\bar{p}}(\varphi_{\bar{p}} b_{\bar{p}})$

s.t.

Eqs. (2) – (14), (18), (21) – (24)

3 NUMERICAL EXPERIMENTS

3.1 Empirical study

Areal-world applicationis conducted on the proposed models. The actual flow data of an arterial system,containing a highway ramp and surface signalized intersections,in Chubei City in Taiwan are analysed. Figure 1shows the topology for the underlying arterial system. The traffic data were collected during 17:00 p.m. – 21:00 p.m. in April 13th, 2013. Fivecritical paths are chose for evaluation in accordance with the traffic data. Path 1 contains the traffic flow from the northbound off-ramp through the major arterial; paths 2 and 3 pass through the major arterial in opposite directions; and paths 4 and 5 are from minor legs of the surface signalized intersections to the major arterial. We mark the three intersections, from right to left, as 1, 2 and 3, respectively.Also, paths 1 and 2 are outbound paths, and paths 3, 4, and 5 are inbound paths.

3.2. Numerical implementations

For numerical implementations, both problem solving and traffic simulations for performance evaluations are carried out. For problem solving, a C++ program is coded for connecting theGurobisolver for obtaining a feasible timing plan for the proposed models introduced in Section 2. Obtained signal timing plansthen serve as the input to the traffic software integrated system (TSIS) for random simulations. A total of 500 independent replications are performed and output the associated performance measures.

On-going timing plans as well as those suggested in Yang et al.11are used for comparisons.Designs of phasesat each intersection of the two plansaregiven inTable 2and Table 3.

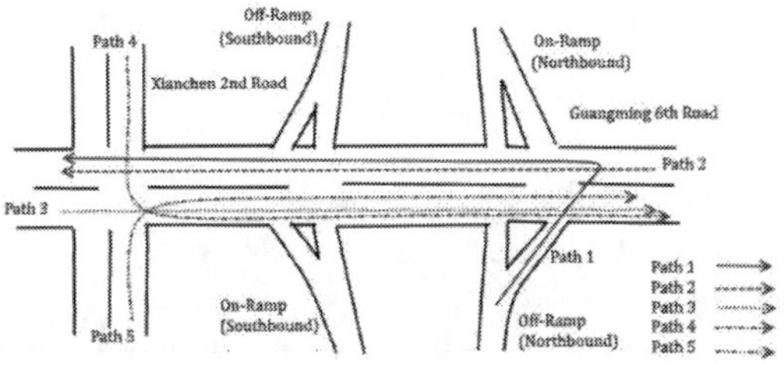

Figure 1. Chubei Network

Table 2. On-going Timing Plan Design

On-going plan	Phase 1	Phase 2	Phase 3	Phase 4
Intersection 1				
Intersection 2				
Intersection 3	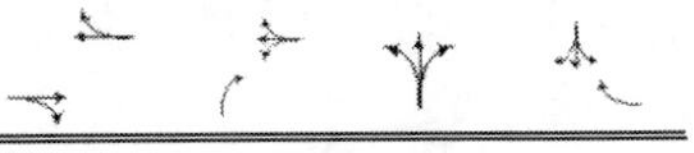			

Table 3. Yang's11 Timing Plan Design

Yang's plan	Phase 1	Phase 2	Phase 3	Phase 4
Intersection 1				
Intersection 2				
Intersection 3	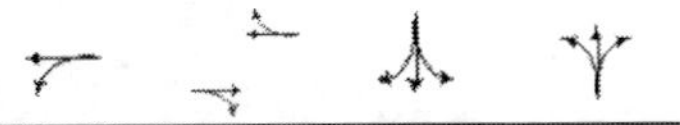			

Numerical results are summarized inTable 4andTable 5. The revised results (Plan b, Plan c, Plan y and Plan z) are derived from solving our models. Note that both the phase sequences and cycle lengths differ from the original timing plans (Plan a and Plan x).

From Table 4, we see that the cycle length suggested by the proposed model is now longer than the original timing plan. From the timing plan in Plan b, we observe that the lengths of the phase 1'sat intersections 1 and 2 have increased. In particular, this allows more traffic on the major arterial in outbound directions (path 1 and path 2). On the other hand, from the result of Plan c, the cycle length decreases in Plan a,while it increases in Plan b. Also, comparing Plan b and Plan c, the ratio of the phase 1 decreases while the ratio of phase 2 increases.Thissuggests that sacrificing the benefit of outbound pathscan improve that of the inbound paths.

From Table 5, as forthe plan suggested by Yang et al., we can see that comparing the original plan (Plan x), the cycle length decreases both in the results of RMPPM 1 (Plan y) and RMPPM 2 (Plan z). For intersection 3 with high right-turn traffic, the model result suggests a longer length of that phase so that the right-turn traffic can dissipate more therein.This can be observed from the results that the ratio of phase 4 at intersection 3 increases. The cycle length of Plan z is longer than Plan y, but all the phase ratios are closely the same.

3.3. Performance evaluations

To evaluatethe system performance under different signal timing plans, we adopt the total control delay per minute in the network as the performance measure.Figure 2 andFigure 3 show the results for the timing plan containing in Table 4 and Table 5, respectively.

Colored lines in Figure 2, from the top to the bottom,correspond tothe total control delay per minute of plans a, b, and c, respectively. It is apparent from Figure 2 that Plan c outperforms the rest planswith respect to the on-going timing plan. Compared to the original signal (Plan a), the improvement with Plan c is around 44.39% in terms of total control delay per minute of the network; and around 15.59% compared to plan b.It can be noticed that Plan c, as a matter of fact, comes with the shortest cycle length. We infer that while the cycle length becomes longer, the phase lengthsbecome longer. In this case, at the intersection 1, the left-turning flow from the westbound will form a queue if the red time is too long, preventing the eastbound left-turning traffic flow at intersection 2 from dissipating.

Lines on Figure 3 are the total control delay per minute of plans x, y, and z, respectively. Results show that Plan y is the best amongst the rest. Theimprovement is around 59.57% of delay compared to Plan x; and 13.12% of delay compared to Plan z.

We further compare the performance of each path under the Plan a to Plan c and plan y. This is because Plans c and y are respectively the best alternatives derived from solving the RMPPM1and RMPPM2. When we compare Plan c to Plan a, the improvements of the paths are around 20-52% on average, resulting basically from the improvements on the arterial paths (Path 2 and 3). In terms of delay, the improvements are around 37%-49%. Yet, when compared to Plan y, even though the performance of Path 2 and 3 has improved around 17%-

65%, the performance of the minor paths is less satisfactory, being around -7%-33%.

Table 4. Model Results For On-going Timing Plan

Plan Number		a			b			c		
Timing Plan		On-going			On-going(RMPPM 1)			On-going(RMPPM 2)		
	#	Offset	Phase Num	Phase Length	Offset	Phase Num	Phase Length	Offset	Phase Num	Phase Length
	1		1	56		2	61		1	55
Intersection1	2	33	2	71	73	1	87	62	3	39
	3		3	53		3	52		2	56
	1		1	56		3	52		1	52
Intersection2	2	3	2	69	0	1	70	0	2	58
	3		3	55		2	78		3	39
	1		1	40		4	57		1	34
	2		2	48		1	57		2	43
Intersection3		28			50			47		
	3		3	30		2	46		3	30
	4		4	62		3	40		4	43
Cycle Length		180			200			150		

Table 5. Model Results Timing Plan Suggested in Yang et al.11

Plan Number		x			y			z		
Timing Plan		Yang			Yang(RMPPM 1)			Yang(RMPPM 2)		
	#	Offset	Phase Num	Phase Length	Offset	Phase Num	Phase Length	Offset	Phase Num	Phase Length
	1		1	89		3	66		1	51
Intersection1	2	0	2	44	50	2	39	0	3	74
	3		3	47		1	46		2	44
	1		1	79		3	39		2	66
Intersection2	2	3	2	41	0	2	58	41	1	59
	3		3	60		1	52		3	44
	1		1	37		2	43		4	34
	2		2	47		1	34		2	48
Intersection3		17			54			28		
	3		3	68		3	43		1	39
	4		4	28		4	30		3	48
Cycle Length		180			150			169		

DOI:10.1051/matecconf/201712401002

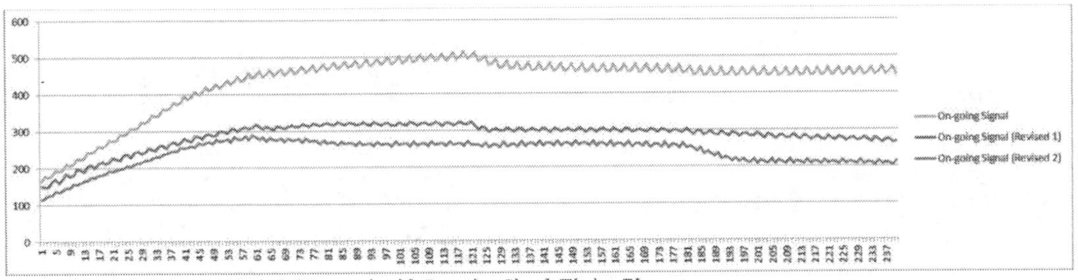

Figure 2. Delay of Control of Chubei Network with On-going Signal Timing Plan

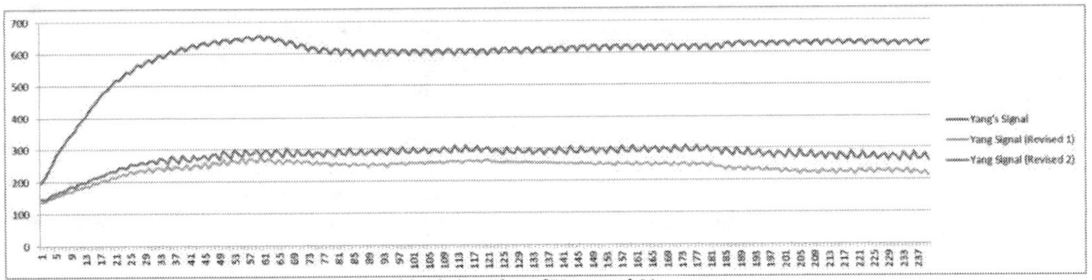

Figure 3. Delay of Control of Chubei Network with Timing Plan of Yang et al.11

Lastly, we compare Plan c to Plan y. From the perspective of path performance, Plan c is around 10% better than Plan y with respect to Paths 1 and 5. Yet, on the other hand, Plan y yields better results on Path 3. As to Paths 2 and 4, Plan y is better. Despite this fact, Plan c is the best alternative in terms of the aggregate network delay, possibly due to the lack of efficiency in the turning flow at intersection 3 under Plan y.

Conclusions

In this paper, extensions of the signal synchronization model proposed in Yang et al.11are given by letting the ratios of phases become variable, and by incorporating the traffic flow dispersion module for adjusting the travel times. Numerical evidences suggest a total improvement of around 44.39% and 59.57% with on-going timing plan and those suggested by Yang et al.11, respectively, in terms of the total control delay per minute of the entire network. We observe that in this experiment, the results of RMPPM 2 are not always better than that of RMPPM 1. This may be due to interference of the paths that we have not considered in the experiment - such as the path flow from the northbound of intersection 2 and the left-turning flow at intersection 2 - or the red time is too longer to remove the queue in each cycle and keep the progression.

Comparisons on the performance of each path under different plans are carried out as well. Results show that compared to the on-going timing plan (Plan a), with Plan c, the total control delay of each path can be improved around 20% to 52%; with Plan y, the total control delay of paths mainly in the major, such as path 2 and path 3, will be improved around 17% to 65%. But when it comes to paths through the branches, the results of our model have a smaller contribution for the total control delay of them, due to the difference of the signal design.

As of this moment, discussions of the signal synchronization problem still are limited by assuming that the design of the phases comprising a cycle is pre-defined. A potential future research direction resides in the optimal design of phases for signalized intersections. A proper design of phases can certain help improve the LoS of both the underlying signalized intersections and the entire network associated.

Acknowledgment

This study is supported by Ministry of Science and Technology (MOST), Taiwan, with Grant No. MOST-105C581; and Chunghwa Telecom Laboratories, Taiwan, with Grant No. 105R951, with Department of Transportation and Logistics Management, National ChiaoTung University.

References

1. Morgan, J. T. and Little, J. D. C., "Synchronizing Traffic Signals for Maximal Bandwidth," in Operations Research, **12 (6)**, pp. 896-912 (1964).
2. Little, J. D. C., "The Synchronization of Traffic Signal by Mixed-Integer Linear Programming," in Operations Research, **14 (4)**, pp. 568–594 (1966).
3. Chang, Cohen, Liu, Chaudhary, and Messer, "MAXBAND-86: program for optimizing left-turn phase sequence in multi-arterial closednetworks," in Transportation Research Record, **1181**, pp. 61–67 (1998).
4. Gartner, N. H., Assman, S. F., Lasaga, F., and Hou, D. L., "A Multi-Band Approach to Arterial Traffic Signal Optimization," in Transportation Research Part B, **25 (1)**, pp. 55-74 (1991).

5. Stamatiadis, C., Gartner, N.H., "MULTIBAND-96: a program for variable-bandwidth progression optimization of multiarterial traffic networks,"in TransportationResearch Record, **1554**, pp. 9–17 (1996).
6. Gartner, N.H. and C. Stamatiadis, "Arterial-Based Control of Traffic Flow in Urban Grid Networks," in Mathematical and Computer Modelling, **35**, pp. 657-671 (2002).
7. Gartner, N.H. and C. Stamatiadis, "Progression Optimization Featuring Arterial- and Route-Based Priority Signal Networks," in Journal of Intelligent Transportation Systems, **8**, pp. 77-86 (2004).
8. Lu, S., Liu, X., and Dai, S., "Revised MAXBAND Model for Bandwidth Optimization of Traffic Flow Dispersion,"in 2008 ISECS International Colloquium on Computing, Communication, Control, and Management, pp. 85–89 (2008).
9. Robertson, *"TRANSYT: a traffic network study tool, Road Research Laboratory Report"*, LR 253. Road Research Laboratory, Crowthorne (1969).
10. C. Zhang, Y.Xie, N. H. Gartner, C.Stamatiadis, and T.Arsava, "AM-Band: An Asymmetrical Multi-Band model for arterialtraffic signal coordination ," in Transportation Research Part C, **58**, pp. 515–531 (2015).
11. Yang, X., Cheng, Y., and Chang, G. L. "A multi-path progression model for synchronization of arterial traffic signals," in Transportation Research Part C: Emerging Technologies, **53**, pp. 93-111 (2015).
12. P. Pardalos, "Global optimization algorithms for linearly constrained indefinite quadratic problems," in Computers and Mathematics with Applications, **21**, pp. 87–97 (1991).
13. R. Jeroslow, "There cannot be any algorithm for integer programming withquadratic constraints," in Operations Research, **21**, pp. 221–224 (1973).
14. Optimization, G. (2012). "Gurobi optimizer reference manual", 2, 1-3.Retrieved Oct. 31th, 2016, from http://www.gurobi.com.
15. Roess, R. P., Prassas, E., andMcShane, W. R.,*"Traffic Engineering"*, Pearson Education Inc.(2004)

DOI:10.1051/matecconf/201712401003

Platoon Dispersion Analysis Based on Diffusion Theory

Mohamed Badhrudeen[1], Helen Thomas[2], Lelitha Devi Vanajakshi[3] and Anuj Sharma[4]

[1]*Project Officer, Department of Civil Engineering, Indian Institute of Technology Madras, Chennai, India.*
[2]*Former Graduate Student, Department of Civil Engineering, Indian Institute of Technology Madras, Chennai, India.*
[3]*Associate Professor, Department of Civil Engineering, Indian Institute of Technology Madras, Chennai, India.*
[4]*Associate Professor, Department of Civil, Construction, and Environmental Engineering, Iowa State University, Iowa, USA.*

Abstract: Urbanization and growing demand for travel, causes the traffic system to work ineffectively in most urban areas leading to traffic congestion. Many approaches have been adopted to address this problem, one among them being the signal co-ordination. This can be achieved if the platoon of vehicles that gets discharged at one signal gets green at consecutive signals with minimal delay. However, platoons tend to get dispersed as they travel and this dispersion phenomenon should be taken into account for effective signal coordination. Reported studies in this area are from the homogeneous and lane disciplined traffic conditions. This paper analyse the platoon dispersion characteristics under heterogeneous and lane-less traffic conditions. Out of the various modeling techniques reported, the approach based on diffusion theory is used in this study.The diffusion theory based models so far assumed thedata to follow normal distribution. However, in the present study, the data was found to follow lognormal distribution and hence the implementation was carried out using lognormal distribution. The parameters of lognormal distribution were calibrated for the study condition. For comparison purpose, normal distribution was also calibrated and the results were evaluated. It was found that model with log normal distribution performed better in all cases than the one with normal distribution.

1 Introduction

The rapid urbanisation of city areas all over the world is negativelyaffecting the urban mobility. One of the main consequences of urbanization is traffic congestion, where vehicles move at a slower speed leading to increased travel time. Thus,it is imperative to manage the traffic system efficiently using appropriate traffic management methods. A widely used technique to manage the movement of traffic in urban roads is signal control. Traffic signal controls the movement of traffic by sharing the available time separately for conflicting movements topreclude any potential conflicts. Further, it ensures the safety of the vehicles and pedestrians. Hence, awell-designed traffic signal is instrumental in keeping the traffic efficient and safe. In order to design a signal, one must consider different factors that influence the performance of traffic signal. One such factor is the knowledge of platoon dispersion.

Platoons are group of vehicles getting discharged together from a queue at the start of a green signal, which will be travelling together for a distance, before they get dispersed. The rate of dispersion depends on the speed and manoeuvrability of the vehicles present. This phenomenon is called platoon dispersion. Understanding the platoon behaviours will enable us to design and coordinate the successive signals.Coordinating the signals can be achieved if they are close enough for the platoon to reach the downstream signal without much dispersion. In such cases, the green time can be assured for the platoon at the downstream signal, making the total vehicle delay to be minimized.

There have been many studies that looked into the behaviour of platoon vehiclesand thedispersion characteristics of platoon [1-8]. However, most of the studies were reported from homogeneous and lane following traffic conditions. In countries like India, where the heterogeneous traffic consists of multiple classes of vehicles, and no lane-discipline, the platoon dispersion characteristics will be different and the above models may not perform well. Therefore, it warrants a need to study the platoon dispersion under heterogeneous and lane-less traffic conditions and is the aim of the present study. This study analyses the platoon dispersion characteristic under Indian traffic conditions. Data obtained from an urban arterial in Chennai, India, is used for validation.

2 Literature Review

There have been several studies starting from 1950's on platoon dispersion [2-10]. In general, the available platoon dispersion models can be classified in to three groups depending on the theories based on which the models were developed, and are: a) Kinematic wave theory based models, b) Recurrence theory based models, and c) Diffusion theory based models.

DOI:10.1051/matecconf/201712401003

Kinematic wave theory models was first proposed by Lighthill and Whithman[3], which assumed a functional relationship between flow (q), concentration (k), and distance (x) along the road[11]. Additionally, the shock waves experienced in the traffic is considered to be more "kinematic" than "dynamic". Geroliminis and Skabardonis [12] combined kinematic wave theory model with Markov theory to develop an analytical model that could be used to estimate platoon dispersion. Ng et al. [13] proposed an integration of Rakha vehicle dynamics model and LWR model to estimate the arrival profile of platoons on signalized intersections. Seddon [11] observed that application of Kinematic theory to platoon dispersion needs information about speed-density relationship, making it less suitable for real time applications.

One of the most popular recurrence theorybased model was proposed by Robertson [1]. In Robertson's model, the platoon dispersion was explained by flow profiles observed at different points. The downstream flow profiles are estimatedusing the upstream flow profile and the dispersion characteristics. Studies conducted by Lam [14] and Seddon [15] concluded that platoon data predicted by Robertson's model gave predictions closer to the actual values. Similarly, Tarnoff and Parsonson [16] reported good performance of Robertson's model.Fernandez et al.[17] pointed out that even with limitedfield data,Robertson's model has the ability to represent the reality at a reasonable level of accuracy.

The first platoon dispersion model based on diffusion theory was proposed by Pacey [2]. Pacey'smodel assumedthat the vehicles speed isconstant and distributed normally, which is critical to predict downstream flow profile [18].The assumption in Pacey's model makes it apt for application in steady state traffic condition [19]. Many other studies have used Pacey's model using the same assumption [6, 18, 20-22]. Pacey's model has broad applicability in modelling speed distribution, and was shownto be accurate under road conditions without any side interferences [6]. Wang et al.[23] proposed a non-transformation normal distribution based platoon dispersion model. Wei et al.[24]assumed that the platoon speed follows a truncated normal distribution, andthe expected number of vehicles at the downstream wascalculated. Further, differentstudiesvalidated the accuracy of the Pacey's model by assuming the platoon speeds followed truncated normal distributions [25], log normal distribution [26], truncated Gaussian mixture distribution [27], and truncated mixed phase distribution [28].Dhamaniya and Chandra [29] reported that speeds of vehicles following normal distribution may not be always true under mixed traffic.

Based on the above review, it is clear that not much research has been reportedon platoon dispersion characteristics under heterogeneous and lane-less traffic conditions. The applicability and accuracy of the existing models might not hold if the models were to be used under such condition [30]. The present study is one of the first such studies toanalysethe platoon dispersion process under heterogeneous and lane-less traffic condition based on diffusion theory models.

3 Data Collection

An urban road stretch in Chennai, India was chosen for this study. The stretch contains a T-intersection that connects Sardar Patel road and Rajiv Gandhi Salai. The distance of the whole stretch, starting from Madhya Kailash (MK), is nearly 2 km. At the other end of the stretch is the Tidel park intersection. Each direction has 3 lanes with lane width of 3.5m each. A schematic of the study stretch is shown in Figure 1.

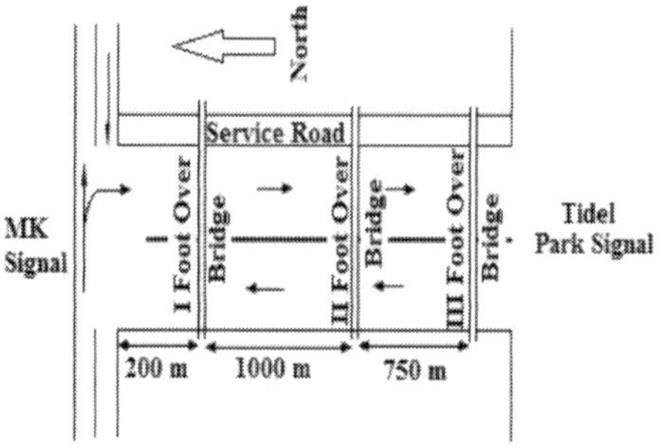

Figure 1. Schematic sketch of Study Site showing downstream data collection locations

Traffic flow observed on this stretch was around 100,000 vehicles per day. The vehicles flowing out from the Madhya Kailash intersection in the southbound direction were considered for analysis. Data were collected using video cameras installed at four locations, Madhya Kailash (MK), I Foot over bridge (I FOB), II Foot over bridge (II FOB), and III Foot over bridge (III FOB).Data included the type of vehicle that crosses the reference point, and time of crossing. Vehicles were classified into4 categories: Two-wheelers (2Ws), Three-wheelers (3Ws), Light Motor vehicles (LMVs) that included passenger cars, and light commercial vehicles, and Heavy Motor Vehicles (HMVs), which were mainly buses and trucks. Four days data were collected during three one-hourperiods: 6:30 to 7:30 a.m., 7:30 to 8:30 a.m., and 2:45 to 3:50 p.m. Data were collected during off-peak traffic condition to make the data extraction easy, since the data extraction was carried out manually from the recorded videos. Apart from the flow data at four locations, travel time data for the mentioned time periods were collected using Bluetooth sensor. Bluetooth sensors were placed during the same time as the videos being recorded at the four locations to obtain the travel time for each section. The penetration rate for Bluetooth sensors was found to be around 10%.

4 Data Analysis

To start with, the individual vehicle flow data collected were grouped into 5 seconds intervals for all locations and time intervals. The average composition of different classes of vehicles during the study duration was found to be 47%, 41%, 8% and 4% of 2Ws, LMVs, HMVs and

DOI:10.1051/matecconf/201712401003

3Ws respectively. It can be seen that 2Ws and LMVs constitute majority of the vehicles.Table 1shows the vehicle flows observed during different time intervals.

Table 1. Average flow at different locations during different times of day

Flow at different locations (vehicles/ hour)			
Location	Duration of day		
	6.30-7.30 a.m	7.30-8.30 a.m	2.45-3.50 p.m
MK	1418	1852	2627
I FOB	1695	2308	3474
II FOB	1672	2245	3239
III FOB	1320	1685	-

Additionally, a preliminary analysis was conducted on the travel time data collected using Bluetooth (BT) sensors to identify the distribution followed. For this, the IFOB to IIFOB section was selected. Bluetooth data at these two points were gathered and sorted. Bluetooth data mainly consists of two important parameters a) Mac ID, and b) Detected time. Mac ID is a unique alphanumeric sequence assigned to the BT enabled electronic devices. Sensor identifies such electronic devices and records the Mac ID and its time of detection. These data is then screened for outliers and they are compared from consecutive locations to calculate traveltime. An algorithm was written in Matlaband the travel time between IFOB to IIFOB was extracted.

Once the travel time data were extracted, different probability distributions namely, Normal, Weibull, Uniform, Log Normal, Logistic, and Exponential were tried.For each distribution, K-S test [31] was carried out to measure the goodness of fit. The significance level (α) was assumed to be 1%. The K-S test statistic (D) was found and compared with the critical value. Among the different distributions, Log normal was found to fit the travel time data best, followed by Normal distribution. Table 2shows the goodness of fit details. This finding was used in the next stage of analysis.

Table 2. Details of the K-S test on fitted distributions

	Sample size	Significance level (α)	Test Statistic (D)	Critical Value
Log Normal	170	1%	0.07	0.12
Normal	170	1%	0.11	0.12

5 Methodology

In this study, diffusion theory basedplatoon dispersion model [2], is used as the base model and is discussed below.

The rationale behind Pacey's theory is to make use of normal distribution of vehicle speeds to predict arrival of vehicles at downstream locations. It states that if speeds in traffic streams are normally distributed, the dispersion of corresponding platoons can be described by dispersion in speeds under the following assumptions:

i. Vehicles travel at constant speed.
ii. Vehicle speed is independent of platoon position.
iii. There is no overtaking.

Pacey derived travel time distribution of vehicles for a road segment from the assumption that the speed of the vehicles follows normal distribution. Once the travel time distribution is derived the vehicles in the downstream location was predicted. Seddon[18]discretizedPacey'scontinuous equation as the flow pattern will not be represented by a continuous curve. The discretized equation was obtained as

$$q'(t+T) = \sum q(t)p(T), \qquad (1)$$

where, $q(t)$isthe flow at upstream location at time t, and $q'(t+T)$ is the flow at downstream location at time $(t+T)$. $p(T)$ is the probability of travel time of T. Seddon obtained travel time distribution by assuming that vehicle speed follows normal distribution.However, the actual distribution of travel time need not always be normal and is explored in this study.

As presented in Table 2, the travel time data in the present study followed log normal distribution better than normal distribution. Hence, the above formulation was implemented using log normal distribution instead of normal distribution. Equation 2 shows the probability density function of log normal distribution.

$$p(T) = \frac{1}{(T)\sigma\sqrt{2\pi}} e^{-1/2\left(\frac{\ln(T)-\mu_{ln}}{\sigma_{ln}}\right)^2}, \qquad (2)$$

The parameters for the log-normal distribution are μ_{ln} and σ_{ln}, which are the mean and standard deviation respectively of the natural logarithm of travel time. Calibration of these parameters were done based onthe lowest RMSE between estimated and actual downstream flow. Different values of parameters were tried (μ_{ln}ranging from 1 to 7 in increments of 0.1 and σ_{ln}ranging from 0 to 5 in increments of 0.1). The time interval used was 5 seconds. The set of μ_{ln} and σ_{ln}values that resulted in lowest RMSE was taken as the best-fit values. The calibrated parameter values obtained are shown in Table 3.

Table 3. Calibrated values of parameters values obtained by assuming log-normal distribution of travel times

		MK to I	MK to II	MK to III-	I to II	I to III	II to III
6.30 - 7.30 a.m	μ_{ln}	3.60	4.40	4.80	4.30	4.20	3.60
	σ_{ln}	2.70	2.70	2.80	2.40	2.40	2.40

7.30 - 8.30 a.m.	μ_{ln}	3.80	4.60	4.80	4.20	4.90	4.00
	σ_{ln}	2.70	2.90	2.80	2.00	3.00	2.10
2.45 - 3.50 p.m.	μ_{ln}	4.60	5.40	-	4.70	-	-
	σ_{ln}	3.00	3.10	-	2.80	-	-

The performance of this model was compared with a model that assumed normal distribution for travel time. The parameters μ_n and σ_n, which are the mean and standard deviation of travel times respectively, are the parameters to be calibrated for this model. The same procedure was followed here also. In this case, μ_n ranging from 0 to 300 seconds (with 5 seconds increment) and $\sigma_n$ranging from 0 to 40 seconds (with increment of 0.1 seconds) were used.For each combination of μ_n and σ_n, the RMSE between estimated and actual downstream flow was calculated. The pair of μ_n and $\sigma_n$that gave the lowest RMSE was taken as the calibrated value. The parameter values obtained are shown in Table 4.

Table 4. Calibrated values of parameters values obtained assuming normal distribution of travel times

		MK to I	MK to II	MK to III	I to II	I to III	II to III
6.30 - 7.30 a.m.	μ_n	30.0	75	125	70	85	40
	σ_n	14.3	14.30	17.30	11.1	9.40	10.2
7.30 - 8.30 a.m.	μ_n	35.0	110	120	65	125	50
	σ_n	12.3	15.90	18.20	8.30	17.7	6.80
2.45 - 3.50 p.m.	μ_n	85.0	185	-	95	-	-
	σ_n	17.8	20.50	-	13.9	-	-

The calibrated values from Tables 3 and 4were used to predict the downstream flow. Error is calculated in terms of Root Mean Square Error (RMSE). Table 5 tabulates the RMSE observed for different road segments across different time periods.

Table 5. RMSE values obtained for Lognormal (LN) and Normal (N) models

		MK to I-FOB	MK to II-FOB	MK to III-FOB	I-FOB to II-FOB	I-FOB to III-FOB	II-FOB to III-FOB
6.30 - 7.30 a.m.	LN	2.77	3.1	2.56	2.55	2.55	3.26
	N	4.21	4.41	3.21	3.31	3.43	5.22
7.30 - 8.30 a.m.	LN	3.12	4.24	3.98	3.92	2.67	3.89
	N	5.21	6.21	4.99	5.88	4.12	5.11
2.45 - 3.50 p.m.	LN	5.1	4.22	-	3.89	-	-
	N	6.67	5.78	-	5.22	-	-

From Table 5, it can be seen that for all the cases, model that used log normal travel time distribution performed better than the one that used normal distribution.

5 Summary and Conclusion

The main objective of this study was to understand the nature of platoon dispersion in a lane less heterogeneous Indian traffic conditions. An urban arterial road stretch in Chennai, India was chosen as the study site. Data were collected by recording videos at four different points of the road. Additionally, travel time data were collected using Bluetooth sensors.Flow datawere extracted at the four locations for three time periods of the day at every 5 sec intervals. Travel time data were analysed to identify the probability distributions it follows using K-S test. Log normal distribution and Normal distribution were found to be bestfits. Methodology used in this study was based onSeddon's model [18]. However, instead of the normal distribution assumed in the original study, the present study used lognormal distribution.The parameters of lognormal distribution were calibrated for the present data. For comparison purpose, normal distribution was also calibrated. It was found that model with log normal distribution performed better in all cases than the one with normal distribution. Though the model employed in this study takes platoon dispersion into account for estimating vehicle arrival, it does not provide any information regarding how external and internal factors impact the dispersion itself. In addition, the methodology can be extended to analyze the arrival profile of different classes of vehicles.

Acknowledgements

The authors acknowledge the support for this study as part of the sub-project CIE/10-11/169/IITM/LELI under the Centre of Excellence in Urban Transport project funded by the Ministry of Urban Development, Government of India, through letter No. N-11025/30/2008-UCD and the Indo-US Science and Technology Forum through grant No. IUSSTF/JC-Intelligent Transportation Systems//95-2010/2011-12.

References

1. D.I. Robertson,Report No. LR253, *Road Research Laboratory*,(Berkshire, England, 1969).
2. G.M. Pacey, Research Note No. RN/2665/GMP, *Road Research Laboratory*, (Berkshire, England, 1956).
3. M.J. Lighthil, G.B. Witham, *Proc. of the Royal Society*,A229, 317-345 (1955).
4. B.J. Lewis, Highway Research Board BulletinNo. 178 (1958).
5. E.P. Graham, D.C. Chenu, Traffic Eng., .**32**, 11-13 (1962).
6. R. Herman, R.B. Potts, W.R. Rothery, Traffic Eng. Control,**5**, 9, 529-533 (1964).
7. J.M. Smelt, *Master of Science Thesis*, Monash University, Department of Civil Engineering, Clayton, Victoria, Australia, November 1983.
8. J.A. Hillier, R.W. Rothery, Transp. Sci. **1,** 2, 81–94 (1967).
9. Y. Lu, X. Yang, IEEE Transactions on Intelligent Trans. Sys., **15**, 1, 334 – 344 (2014).
10. A.H.F. Chow, S. Li, W.Y. Szeto, D.Z.W. Wang, Transportmetrica B. Trans. Dynamics, **3**, 3, 169 – 191 (2015).
11. P.A. Seddon, Traffic Eng. Control, **13,** 8, 332-336 (1971).
12. N. Geroliminis, A. Skabardonis, Transp. Res. Rec.**1934**, 116–124 (2005).
13. K.M. Ng, K. Salam, M.B.I. Reaz, *Proc. 5th Intl. Conf. On Intelligent Sys. Modelling and Simulation*, 490 – 495 (2014).
14. J.K. Lam,*Proc. of the Seventh Int. Symp. on Trans. and Traffic Theory*, 119–144 (1977).
15. P.A. Seddon, Traffic Eng. Control, **13**, 10, 442-444 (1972b)
16. P.J. Tarnoff, P.S. Parsonson, Rep. No.233, *National Cooperative Highway Research* Program, (1981).
17. R. Fernandez, E. Valenzuela, F. Casanello, C. Jorquera, Transp. Res. Part A,**40,** 5, 386-398 (2006).
18. P.A. Seddon, Traffic Eng. Control,**13,** 9, 388-390 (1972a).
19. C.Q. Liu, P.K. Yang, J. of Tongji University,**24,** 6, 636-641 (1996).
20. K.L. Bang, *M.S. thesis*, Ohio State University, Ohio, (1967).
21. M. Tracz, Traffic Eng. Control, **16,** 11, 490–492 (1975).
22. A. Polus, Transp., **8**, 2, 141–151 (1979).
23. D. H. Wang, F. Li, X.M. Song, J. Jilin Univ., **39**(4), 891–896 (2009).
24. M. Wei, W. Jin, L. Shen, J. Appl. Math., **2012**, Article ID 727839, 13 pages (2012).
25. M. Wei, B. Sun, W.Z.Jin, J.Guangxi Univ. (natural Science edition), (2013).
26. M. Wei, B. Sun, J. Beijing Univ. of Technology, (2013).
27. W. Wu, W. Jin, L. Shen, J. Appl. Math., **2013**, Article ID 480965, 9 pages (2013).
28. Y.S. Jiang, Z. Yao, X. Ding, X. Luo, *95th Annual Meeting on Trans. Res. Board*, 456 – 462 (2016).
29. A. Dhamaniya, S. Chandra, Int. J. Civ., Struc. Const., Arch., Eng.7, 11, (2013).
30. V.T. Arasan, S.H.Kashani, Transp. Res. Rec. No. **1852**, 175-182 (2003).
31. I.T. Young, J.Histochem.Cytochem.**25**, 7,935 – 941 (1977).

Establishment and Empirical Analysis of Evaluation Index System for Basic Public Service Level of Transportation

Jingling Jiang[1], Luyang Gong[2] and Yanhong Li[3]

[1, 2, 3] *China Academy of Transportation Sciences, Room 805, No.240, Chaoyang District, Beijing China*

Abstract. The establishment of evaluation index system of basic public service level of transportation is of great guiding significance to improving China's comprehensive transportation development, whereas it is lacking in a complete set of practical and standardized evaluation index system for basic public services of transportation. This paper analyzes the connotations and characteristics of basic public services of transportation to determine its scope and content at present and develop the rating index system and index calculation method that cover four aspects of urban public transportation, rural passenger transportation, rural highways and universal postal service of three categories. Through analysis and comparison of the comprehensive evaluation methods of multiple indexes and based on the structural characteristics of the basic public service evaluation indexes of transportation, the paper combines AHP with fuzzy evaluation method to build a quantitative model of multi-index comprehensive evaluation to provide comprehensive evaluation method for basic public service evaluation of transportation. In addition, it also takes Xiji County of Ningxia Hui Autonomous Region as example to evaluate the basic public service level of the transportation, and it can detect through evaluation problems and deficiencies existing in basic public services of its transportation, so as to take corresponding measures for those problems and deficiencies.

1 Introduction

The 18th National Congress of CPC explicitly put forward that equal access to basic public services should be generally achieved by 2020.As one of the important service industries for protection and improvement of people's livelihood, transportation shall promote the close correlation of the equal access to basic public services to the transportation industry. In addition, the current period and the period from now on are vital to fully upgrading the basic public service level of transportation. But the precondition for equal access to basic public services of transportation is that researches shall be conducted on the difference between basic public service levels of transportation in different areas.It requires researches on basic public service items of transportation at first and then the establishment of evaluation index system in which certain indexes and steps shall be taken to provide support and basis for evaluation of basic public service levels of transportation in different areas.

At present, the country is lacking in researches on evaluation indexes and evaluation methods for basic public services of the transportation. When it comes to index system and evaluation model, there is no reference for evaluation of basic public service level of transportation in the country. Development and equalization of basic public services of transportation in different areas cannot be described quantitatively, bringing promotion in equal access to basic public services of transportation certain obstacles. But some researches have been carried out in the country with regard to the evaluation indexes of basic public services. For instance, Li Jian took the second level indexes based on the first level index system established with 6 basic public service items to research the mechanism that how major indexes influence the supply of basic public services[1]. XuCuimei established the evaluation index system for equal access to basic public services in Hainan and reached the conclusion that equal access to residence is the biggest factor of comprehensive evaluation after analysis through entropy evaluation method[2]. Based on the actual investigation in 198 prefecture-level cities in China, Lin Yangyan et al took basic education, basic medical care and health, environmental protection and infrastructure as four evaluation indexes to figure out the Gini coefficient of the supply of basic public services in the country[3]. Ma Hao et al established the evaluation index system for equal access to public services based on empirical researches on representative provinces and cities of east, central, west regions and gave suggestions for promoting equal access[4]. Wang Xinmin et al established the multi-level evaluation index system for equal access to basic public services, then determined index weight with AHP and adopted the grey correlation comprehensive evaluation model to evaluation the equal access to basic public services in 31provincial areas throughout the country[5]. In addition, researches conducted with regard to evaluation of basic public

DOI:10.1051/matecconf/201712401004

services provide reference for researches on basic public services of transportation.

2 Understanding of basic pbulic service of transportation

2.1 Connotation and characteristics of basic public services of transportation

The State Council issued the Twelfth Five-year Plan for National Basic Public Service System in July 2012 to give the clear definition of basic public services in a formal document for the first time that basic public services refer to the public services that are established on some social common understanding, led and provided by the government and compatible with the economic and social development level, aiming at safeguarding the basic survival and development needs of all citizens. Citizens are entitled to basic public services while the government is liable for providing basic public services. Basic public services refer to the services that the government provides for all citizens mainly for free or public benefit to satisfy the most fundamental need of all citizens and all citizens have equal access to services. Compared to basic public services, non-basic public services refer to high level services the government provides for the society in order to give response to higher level public demand by exercising the public powers or resources in its hands. In addition, the public can voluntarily decide whether or not to enjoy the non-basic public services according to their own conditions and needs[6].

As clothes, food, shelter and travel are the most fundamental needs for survival, the ultimate goal of the transportation industry is to provide the economic society direct travel and transportation services. In the Twelfth Five-year Plan for National Basic Public Service System, the government explicitly pointed out the scope of basic public services, also including public services in transportation in close relation to people's living environment in a broad sense. The scope of basic public services covers the field of transportation, which does not mean that the scope and content of entire transportation service are included in the scope of basic public services. It is just like the nine-year compulsory education in the field of education that is included in the scope of basic public services, but higher education is outside the scope of basic public services. Public services of transportation can also be divided into basic public services and non-basic public services of transportation.

With a reference to the definition of basic public services in the Twelfth Five-year Plan for National Basic Public Service System, the basic public services of transportation refer to the public services led and provided by the government and compatible with the development level and stage of the economic society, aiming at safeguarding the basic travel demand of all citizens. The basic characteristics are as follows: First is being led by the government; second is being compatible with the development stage; third is safeguarding the basic travel demand and fourth is indicating the idea of comprehensive transportation.

2.2 Scope and content of basic public services of transportation

According to the Twelfth Five-year Plan for National Basic Public Service System, basic public services of transportation during the period of the "Twelfth Five-year Plan" include providing administrative villages with highways and regular passenger buses, full coverage of public transportation in urban built-up area, making each township have post office and village accessible by postal communication and so on. The content proposed of basic public services of transportation mainly focuses on the field of infrastructure, which relies mainly on the local government's comprehensive planning and overall arrangement based on the local natural, geographical and ecological environment and other actual conditions.

Based on the understanding of basic public services of transportation and strict definition of the development stage where the country's economic society is, basic public services of transportation can be divided into two parts, that is, transportation infrastructure and transportation services. It is known that infrastructure is the basic condition for public travel, and from the viewpoint of transportation infrastructure, it mainly includes the infrastructure in highway, waterway, postal service and other categories that are more popular at present, such as road highway infrastructure system, transport station of transportation and so on. If looking from a perspective of transportation services, some basic transportation services serving the general public are included in the scope of basic public services, such as public transportation, urban-rural regular bus line and so on in urban public transportation system, providing the general public convenient and low-cost travel services. As a result, the paper suggests that the current basic public service items of transportation shall include urban public transportation, rural passenger transportation, rural highway and universal postal services.

3 Establishment of evaluation index system for basic public service level of transportation

1 Selection principle of evaluation index

There are a lot of indexes involved in evaluation of basic public services of transportation, but be sure to select the indexes as few as possible for overall evaluation of the development level to reflect the most important and comprehensive information, making each index relatively independent, quantifiable and universal.

(1) Systematicness

Evaluation index system can fully reflect the development of basic public services of transportation as an organic whole. It is required to reflect function of basic public service items of transportation and the systematicness and coordination between items and the inside, otherwise the evaluation results are unreliable, which may even misguide the behaviors of the superior management department and the internal management.

(2) Scientificity

The indexes selected should be scientific and reasonable, and can reflect the complete picture and development level of basic public services of transportation truly and objectively. In addition to this, index concept shall be precise and scientific rather than ambiguous, the explanation shall be accurate without different meanings and the data source and calculating range shall be clear and stay the same[7].

(3) Objectivity

Make sure the evaluation indexes are selected in an objective and fair way, data source is reliable and accurate and the evaluation method is scientific.

(4) Operability

When it comes to index selection, be sure to obtain the statistics released by the statistical department on a regular basis as much as possible or obtain through sampling survey and typical survey. Those indexes selected shall be measurable and easy to be quantified.

(5) Comparability

It requires that the data used shall be comparable. The indexes selected shall have the uniform definition and measurement standard in each evaluation object to ensure evaluation and comparison are conducted on the same basis. For indexes that have unclear concept and cannot be measured, be sure to select no evaluation index as possible[8].

(6) Conciseness

Indexes shall be typical and representative because basic public services of transportation involve many factors and there are a huge number of indexes reflecting the current development. Under the premise of ensuring systematicness, be sure to reflect the most important characteristics with minimum indexes and try to select those representative comprehensive and major indexes as possible.

There are many types of indexes related to basic public services of transportation, but precise quantification does not mean accurate evaluation. For overall evaluation of the development level, be sure to select the indexes as few as possible to reflect the most important and comprehensive information, making each index relatively independent, quantifiable and universal. All in all, be sure to adhere to the unity of principles of systematicness, scientificity, objectivity, operability, comparability and conciseness when designing and screening the indexes.Principles of systematicness, scientificity and objectivity have a far-reaching significance for theoretical discussion on the evaluation index system of basic public services of transportation; while principles of operability, comparability and conciseness are conductive to the promotion and application of the index systemin actual evaluation.

3.2 Category of indexes

It is the government's job to provide basic public services. Firstly, the government shall provide basic public services in terms of policy, fund and so on, that is, use supply-side indexes to evaluate the government's supply of basic public services. However, the country has a vast territory with extremely unequal distribution of population and resources, so the government's supply cannot be the only one taken into account[9].For instance, population density and other factors have an important influence on this. Even though the same government supply, the service level may be different due to different population distribution and density. Therefore, the paper holds that the government's actual ability to provide basic public services needs to be evaluated under the constraint of the existing supply-side indexes, which reflect the basic public service level to which everyone has access, that is, indexes related to ability. Constrained by supply indexes and ability indexes, results shall be taken into account that the basic public service have achieved, such as the proportion of public travel by public transportation, accident rate and so on.

Based on the analysis above, the evaluation index system for basic public services of transportation shall be established from three aspects of supply, ability and result. When evaluating the basic public services of transportation of some area, be sure to establish the index system including the three indexes mentioned above and apply the method of giving different coefficient or weight to comprehensive evaluation.

3.3 Evaluation index

There are many factors that affect the basic public service level of transportation. But specific to three aspects of each service item of basic public services of transportation, there are a lot of evaluation indexes involving every aspect of economic and social life. In addition, it seems to be tedious and difficult to evaluate all indexes, so the paper selects the important indexes that are representative and available in data as evaluation indexes.

Based on the content and items of the basic public services of transportation analyzed and combined with the present situation of basic public service level of transportation, the paper follows the principle of operability and other index selection principles to classify the evaluation indexes for basic public services of transportation into 18 indexes, which can be summed up as four aspects, that is, urban public transportation, rural passenger transportation, rural highway and universal postal service.

Evaluation indexes for basic public services of transportation are shown in Table 1, including specific definitions of evaluation indexes and calculating methods.

DOI:10.1051/matecconf/201712401004

Table 1 Evaluation Indexes for Basic Public Services of Transportation

First level index	Second level index	Third level index	Index definition	Calculating method
Urban public transportation	Supply	Operating subsidy system for public buses and trolleybuses and arrival rate	The ratio of the actual subsidy for urban public transportation and total policy subsidy reasonably calculated during the statistical period. (Unit: %)	Arrival rate of operating subsidy for public transportation=Actual subsidy for urban public transportation/total policy subsidy reasonably calculated ×100%
	Ability	Public vehicle population per 10,000 people	Average operating public buses per 10,000 people calculated by urban population during the statistical period. (Unit: Standard vehicle/10,000 people)	Public vehicle population per 10,000 people=Total public vehicle population / urban population
		Coverage ratio of public transport stations within 500m	Ratio of cover area within a radius of 500m of the public transportation stations within the built-up area of the central urban area and the area of built-up area in the central urban area during the statistical period. (Unit: %)	Coverage ratio of public transport stations within 500m= Cover area within a radius of 500m of the public transportation stations within the built-up area of the central urban area / the area of built-up area in the central urban area ×100%
	Result	Passenger satisfaction rate of public transportation	Average score of effective questionnaires about passenger satisfaction of public transportation service quality during the statistical period. (Unit: %)	Passenger satisfaction rate of public transportation=Copies of satisfied questionnaires / total copies of questionnaires ×100%
Rural passenger transportation	Supply	Support policy level of rural passenger transportation	Support policies for rural passenger transportation issued by the local government or industry management department.	It can be divided into four aspects, subsidy policy for vehicle purchase, preferential policy for ticket price, subsidy policy for policy loss and subsidy policy for station construction, and relevant policies have been issued, that is, full score for the item. In addition, the policy document shall prevail.
	Ability	Proportion of incorporated villages accessible by passenger cars	Proportion of incorporated villages accessible by passenger buses of the total incorporated villages in the administrative region (Unit: %).	Proportion of incorporated villages accessible by passenger cars=Number of incorporated villages accessible by passenger cars / total number of incorporated villages ×100%
		Proportion of islands with a population of more than 500accessible by regular liners	Proportion of islands with a permanent resident population of more than 500 that are accessible by regular liners of the total in the administrative region (Unit: %).	Proportion of islands with a population of more than 500 accessible by regular liners / number of islands with a population of more than 500
	Result	Passenger capacity per capita	Ratio of road and waterway passenger capacity and the permanent resident population in the administrative region (Unit: number of times/ per capita per day).	Passenger capacity per capita=(Annual road passenger capacity + annual waterway passenger capacity)/(permanent resident population ×365)
		Death rate of traffic liability accidents per 10,000 vehicles in rural passenger transportation	Traffic death toll due to equal and above liabilities of corresponding mileage for rural passenger service vehicles during the statistical period. (Unit: people/million vehicle kilometers)	Death rate of traffic liability accidents per 10,000 vehicles in rural passenger transportation= Traffic death toll of rural passenger transportation / total operating mileage of rural passenger transportation
Rural highway	Supply	Support policy for rural highways	Fund support policies for rural highway construction, management and maintenance and implementation during the statistical period.	Including proportion of supporting funds for rural highway construction, expenditures for management agencies of rural highway incorporated in financial budget and implementation of funds for maintenance of rural highways. Calculate the policy score according to the proportion of supporting funds for construction, and

				calculate the scores of the two policies according to expenditures for management agencies of rural highway incorporated in financial budget and implementation of funds for maintenance of rural highways
	Ability	Rate of incorporated villages provided with hardened roads	Proportion of incorporated villages provided with hardened roads of the total incorporated villages in the administrative region (Unit: %).	Rate of incorporated villages accessible by highways=Number of incorporated villages provided with hardened roads / total number of incorporated villages ×100%
	Result	Proportion of rural highways with medium and above technical conditions	Proportion of rural highway mileage with medium and above technical conditions of the total rural highway mileage in the administrative region (Unit: %).	Proportion of rural highways with medium and above technical conditions=(High-quality+quality+medium -quality rural highway mileage) / total rural highway mileage ×100%
Universal postal service	Supply	Support policy for universal postal service	Support policies for infrastructure construction land, capital and so on and implementation of universal postal services during the statistical period	With relevant policies issued, full score for relevant polices, otherwise no score.
	Ability	Passing rate of facilities in post office branches	Ratio of number of postal office branches in the administrative region and the number of postal office branches to be designed according to the universal postal service standard (Unit: %).	Passing rate of facilities in post office branches=Number of postal office branches/ number of postal office branches to be designed ×100%
		Proportion of incorporated villages accessible by postal communication	Ratio of incorporated villages accessible by postal communication and total incorporated villages in the administrative region (Unit: %)	Proportion of incorporated villages accessible by postal communication=Number of incorporated villages accessible by postal communication / number of incorporated villages * 100%
	Result	Passing rate of postal services in terms of time limit	Passing situation of business hour, time of checking the postbox (mailbox) and time limit for the entire mail according to the requirements in Universal Postal Services(YZ/T 0129-2009) during the statistical period.	Calculate according to the requirements for business hours in different regions stipulated in Universal Postal Services (YZ/T 0129-2009), satisfactory ones can have the scores.

4 Evaluation modeling for basic public service level of transportation

4.1 Non-dimensionalization of evaluation index value

Calculate the evaluation value of each evaluation index involved in basic public services of transportation with the above-mentioned calculating method for each evaluation index and then adopt the calculating method of non-dimensionalization of evaluation index (qualitation and quantification) to non-dimensionalize the indexes.

4.2 Establishment of evaluation sets

Divide the evaluation sets into five different levels, that is, "excellent", "good", "average", "poor" and "extremely poor". Then constitute the domain of discourse of evaluation with these evaluation levels and adopt the statistical analysis technique to determine the standard evaluation value of each index based on experts' findings as shown in the table.

Table 2 Table of Basic Public Service Level Grading of Transportation

Comprehensive evaluation	Excellent	Good	Average	Poor	Extremely poor
Score criteria	100-80	80-60	60-40	40-20	20-0

4.3 Determine the degree of membership of evaluation indexes and calculate the weight of index values

Before comprehensive evaluation, be sure to determine the degree of membership of the evaluation index to each level selected in the first place r_{ijt}. Measure the frequency k_{ijt} of each index at each evaluation level separately according to the levels provided by experts involved in investigation according to evaluation sets in Table 2, and then figure out the single factor evaluation sets of each index is:

$$r_{ij} = \{(k_{ij1}/n),(k_{ij2}/n),(k_{ij3}/n),(k_{ij4}/n),(k_{ij5}/n)$$

, where n refers to the number of experts involved in the investigation.

Based on experts' findings, the statistical analysis technique shall be adopted to determine the standard evaluation value of each index and AHP to calculate the weight of each sub goal in comprehensive evaluation.

4.4 Implementation of multi-index fuzzy comprehensiveevaluation

Firstly is to build the fuzzy relation matrix according to the certain fuzzy membership and then make the final evaluation in the principle of maximum degree of membership.

Assuming the evaluation goal of the evaluation question as B (comprehensive evaluation value), relevant evaluation index () matrix is $Q = \{x_1, x_2, x_3, \cdots, y_1, y_2, y_3, \cdots, z_1, z_2, z_3, \cdots\}$ and relevant weight matrix is $W = \{w_1, w_2, \cdots, w_n\}$

$$B = Q \cdot W^T = \sum_{i=1}^{i} x_i w_i + \sum_{j=1}^{j} y_j w_j + \sum_{l=1}^{l} z_l w_l + \sum_{h=1}^{h} v_h w_h$$

(3-1)

Where: x_i, y_j, z_l, v_h ——Evaluation value of each constraint;

w_n ——Relevant weight of each constraint.

B refers to the final comprehensive evaluation results of basic public service level evaluation of transportation and the evaluation standard is shown in Table 2.

5 EMPIRICAL ANALYSIS

The paper takes Xiji County of the Ningxia Hui Autonomous Region (located in the concentrated poverty-stricken area around Liupan Mountain) as the typical example to evaluate the basic public service level of the transportation and the detailed evaluation results are shown in Table 3.

Table 3 Index Scores of Basic Public Service Level Evaluation of Transportation in Xiji County

Type of service	S.N.	Specific index	Score
Urban public transportation	1	Operating subsidy system for public buses and trolleybuses and arrival rate	50
	2	Public vehicle population per 10,000 people	60
	3	Coverage ratio of public transport stations within 500m	30
	4	Motorized share ratio of public transportation	20
	5	Passenger satisfaction rate of public transportation	50
Rural passenger transportation	6	Support policy level of rural passenger transportation	76
	7	Proportion of incorporated villages accessible by passenger cars	62
	8	Proportion of islands with a population of more than 500 accessible by regular liners	100
	9	Passenger capacity per capita	50
	10	Death rate of traffic liability accidents per 10,000 vehicles in rural passenger transportation	85
Rural	11	Support policy level of rural	75

highway		highways	
	12	Rate of incorporated villages provided with hardened roads	70
	13	Proportion of rural highways with medium and above technical conditions	70
	14	Treatment rate of hidden dangers in county and township roads	40
Universal postal service	15	Support policy for universal postal service	82
	16	Passing rate of facilities in post office branches	76
	17	Proportion of incorporated villages accessible by postal communication	100
	18	Passing rate of postal services in terms of time limit	75
Total:			48

Based on the score of each evaluation index, the total score of basic public service evacuation of transportation in Xiji County is 48 at a medium level, indicating the backward development. It can be seen from Table 3 that different types of basic public services of transportation differ largely in development level, but universal postal service has relatively favorable development level while urban public transportation, rural passenger transportation and rural highway have relatively lower development level. From the scores of specific indexes, it can be observed that Xiji County will increase efforts in operating subsidies for urban public transportation, increase the coverage of public transportation stations and increase the number of public buses and trolleybuses in the future to improve the motorized share ratio of public transportation and passenger satisfaction; raise the rate of incorporated villages accessible by passenger cars and lift the rural passenger capacity; increase efforts in treatment of hidden dangers in county and township roads and raise the conditions for road passage and safety level.

References

1. Li, J. Research on Evaluation Index System of Basic Public Services [J], Commercial Research, 2011, 409 (05): 48-56.
2. Xu, C. Research on Evaluation Index System of Equal Access to Basic Public Services—Take Hainan for Example [J], Research and Discussion, 2014(3): 48-52.
3. Lin, Y., Zhang, X., Li, Y. Equal Access to Basic Public Services: Index System, Comprehensive Evaluation and Status Analysis—Based on Empirical Research in 198 Prefecture-level Cities in the Country [J], Fujian Tribune. Humanity and Social Science Edition, 2014(6): 184-192.
4. Ma, H., Zeng, X. Establishment of Evaluation Index System with Equal Access to Basic Public Services in the Country—Based on Empirical Research in Representative Provinces in East, Central and West China [J], Jianghan Tribune, 2011(11): 23-25.
5. Wang, X. Nan, R. Establishment and Application of Evaluation System of Equal Access to Basic Public Services—Based on Empirical Research in 31 Provincial Areas in the Country [J], Soft Science, 2011, 25(7): 21-26.
6. An, T., Ren, Q. Equal Access to Public Services: Theories, Problems and Strategies [J]. Top HR, 2007 (08): 48-53.
7. Chen, H. Research on China's Basic Public Service System [J]. Scientific Socialism, 2007(3): 98-100.
8. Shao, L., Bai, C. Research on Method for Establishment of Systematic and Comprehensive Evaluation Index System [J]. Journal of Naval University of Engineering, 2008, 20(3): 48-52.
9. Zhao, Y., Li, F. Analysis on Factors with Influence on Regional Differences of Basic Public Services in China [J]. Academic Journal of Shanxi University of Finance and Economics,2009(8):15-22.

A cognitive and metacognitive approach for exploring link between urban transport environment and driver behaviour: case of roundabout crossed by tramway line

Fatiha Moutchou[1], and Abdelghani Cherkaoui[1]

[1]Research team "Engineering, management and optimization of Systems" (EMOSYS), Industrial Engineering Department, Mohammadia School of Engineering, Rabat, Morocco, BP 765 Rabat agdal, Code postal 10100, Morocco

Abstract. The proliferation of roundabouts and tramway lines are among the most spectacular transformations of Moroccan roads and landscapes in recent years. These changes raise serious concerns about the effects of these new urban and road developments on traffic, on drivers' behaviours, and thus on road safety. These issues bring up the question concerning the adequacy between the choices of road designers and the expectations of road users. Indeed, sustainable urban transport and well functioning road cannot be designed without a holistic understanding of the way drivers act in changing environment through which they move in time and space. Consequently, there is a strong need to further knowledge about practices and behaviours of road users. The present research investigates the cognitive and metacognitive strategies use by drivers in a complex technical system such as roundabout crossed by tramway line. In this article we propose an empirical model to explore the effect of urban transport environment on drivers' behaviours. The paper presents an exploratory study based on a data triangulation approach including database analysis, interview, observation, and think aloud protocol. The paper presents the main findings of the study and discusses their implications for road safety.

1 Introduction

The evolution of contemporary cities was characterized by changes in lifestyles and mobility practices. What is changing according to Nuvaloti [1], are the functions of the cities. Cities have become places where people go to work, consume, visit but less so to live.

In Morocco, due to the growth in population, the rapid urbanization, and the increase in daily mobility in the last decades, transport has become a highly strategic issue for urban areas. Indeed, if transport has shaped cities and guided their growth and their evolution, it constitutes today one of their major problems.

According to the World Bank assessment report on urban transport sector in Morocco, published in 2009 [2], «Public transport systems perform poorly and are not responsive to the changing demands of the population. The road infrastructure and the management of traffic are also lagging behind, and congestion is increasing fast ». So, solving problems of congestion or time lost in public transport call for a rethinking of mobility, particularly by redesigning road and expanding alternative transportation.

In Morocco, there has been much interest in tackling congestion and pollution in urban areas through the promotion of roundabouts and tramway lines. This choice comes often from the facility of turning back, the fluidity and accessibility that they bring. Indeed tram roundabouts are seen to be an effective tool in managing traffic flow in a safe and aesthetic manner.

But if these new infrastructure projects begin to develop today, their implementation in urban areas has lead to changes both in terms of urban shape, as on overall functioning of the city.

The roundabout crossed by a tramway is the intersection of the road and a tramway line. This refers to the definitions of 'the road', 'the railway', and their diversities. Tram-car interaction within the road network takes place in the context of existing road infrastructure, which has limited resource. But beyond this competition for space [3], the roundabout crossed by the tramway line is finally the interaction between two transport systems with contexts, constraints, and traffic rules specific to each of them.

Overall, it has to be stressed that is not only the built space, but also the lived and travelled space that must be taken into account to overcome major challenges that transportation and road safety present in urban area, and to make the system user friendly and well adapted to the qualifications and limitations of road users.

Research on road safety has long been focused on corrective approach after the occurrence of the accident. This approach has reached its limits. The current reflections are mostly focusing on issues relating to cohabitation and behaviours of urban road users.

'Traffic calming' techniques implemented in Europe, in the late 1970's [4], and the "vision zero approach" developed in Sweden [5], are a good illustration of measures which aim to resolve traffic and safety problems with preserving and enhancing neighbourhood liveability. These approaches, focusing on the relationship between space and users practices, also open up interesting perspectives on the responsibility for safety which must be shared between those who design, and those who use the road transport system.

All these foreign experiences are of great interest, not to copy ready-made solutions, but to draw lessons from other countries that are very useful for the formulation of strategies and models adapted to the Moroccan context.

In this perspective, the question of the relationship between driver and his environment serves as a guide to our reflection. This study is concerned with aspects of urban environment that influence driver's behaviours. The focus is on the cognitive and metacognitive strategies of drivers when crossing a roundabout with tramway line.

In this article, we first introduce models of driving behaviour used in our analysis. Then central concepts and issues related to cognition, and metacognition are highlighted. With this as a background, we propose a model to study the interaction tram-car driver. We present in this paper our experimental study to collect data on driver behaviour when crossing roundabout with tram and we conclude with the main findings of our research and their implications for road safety.

2 THEORETICAL FRAMEWORKS

2.1 urban environment complexity and drivers 'cognitive resources

It is opportune to start this introductory literature review by recalling that driver in urban environnement faces an overflow of information coming from (others users, neighbourhood, infrastructures design, roads signs...). Thus, it comes back to driver ability to find the needed information from the plethora of information available.

As Healey [6] states, the challenge in urban regions is that an urban region or a city is not a "thing", but an imagined phenomenon, a conception of a very complex set of overlapping and intersecting relations, understood in different ways by different people. City changes all the time and so do the mental representation of the city. This approach gained insight from the work of Kevin Lynch [7], a planner who argued that each individual holds a unique "image" of his or her city, a visual representation that guides through daily life and maps out meaning.

Because the driver reactions cannot be changed, great attention is paid nowadays to the meaning of the space the driver is moving in. This area of research and action is known as "positive guidance"[8], "self explaining road" [9], and "road readability" [10] all these terms have been applied to those road environment which are in line with the expectations of the road users, eliciting safe behaviour simply by design. A malfunctioning urban environment is therefore identified by the fact that drivers do not behave as intended. The need to understand more about driver behaviour in relation to urban environment has guided our research to an approach that takes into account the Moroccan road context and the driver behaviours in this context.

2.2 Role of cognition approach to identify drivers needs

There is an expanding research literature examining the effect of urban environment on drivers' behaviours [11-15]. The literature shows that in order to design a safe and well functioning road and urban development projects, we need to have a good understanding of how we operate as roads users. For this purpose, the cognitive approach is especially suitable for analysing human functions such as perception, mental representation, and decision making.

This study is based on two theoretical models that originate from cognitive psychology and are frequently mentioned in the literature on road user behaviour, namely Endsley's model of situation awareness, and the hierarchical structure of the driving task as described by Michon.

The concept of situation awareness proposed by Endsley [16] refers specifically to "the perception of elements in the environment within a volume of time and space, the comprehension of their meaning and projection of their status in the near future". Applying Endsley's model to the driving environment, the driver firstly perceives road situations. The driver then, has to understand what he has perceived by forming "a holistic picture" of the elements in the environment and their complex relationships. Finally, the driver projects his understanding into the near future, in order to take the most appropriate action. [17].

The concept of situation awareness focuses on the mental picture of the situation that people find themselves in, and how this picture can be distorted or improved by internal and external factors. These characteristics make it a suitable model for studying factors involved in unsafe or undesirable situations.

Also to better understand drivers' behaviours, the model of Michon [18] is a good starting point. Michon distinguishes three subtasks in driving with different temporal and cognitive requirements. The strategical level, related to planning trip (e.g., vehicle to use, route to take). The tactical level, that requires taking decisions about driving speed (e.g., passing a lead vehicle), and how to handle specific traffic situations such as crossing a crossroad. At the operational level, drivers exercise maneuvers allowing vehicle control (i.e., braking and steering).

Strategical level and tactical level have been found to be more cognitively demanding than operational level [19]. Indeed, at the strategic level, decisions are not really restricted by time, while at the tactical and even more so at the operational level, decisions are to be taken immediately. In this study we focus on the tactical level.

2.3. Metacognitive strategies in a driving context

The term "metacognition" is used to denote higher-order cognition about cognition, or "thinking about one's thinking" [20]. It refers to "the ability to understand and monitor one's own thoughts and the assumptions and implications of one's activities" [20, 21]. The difference between metacognition and cognition is the conscious, or controlled, aspect of processes. Whereas, cognitive skills are needed to perform a task, metacognitive skills are necessary to understand how it was performed. The assumption underlying this distinction is that metacognition helps to improve performance [22, 23]. The person's awareness of his strengths and limitations allows an optimal selection of the strategies to perform a task, as well as an assessment of the gap between his performance and his objectives in order to correctly manage risk resulting from task complexity. Thus metacognition could provide us cues to examine what drivers are thinking and how they are acting. Research on metacognition has largely focused on topics of learning and teaching. However, little research has attempted to extend metacognition to driving context.

In this study, we borrow from the work of Reulier [23], her description of the activities carried out by a reader to control his understanding, to illustrate our analysis of drivers' metacognitive difficulties.

The table 1 below presents our taxonomy of metacognitive difficulties in driving context.

Table 1. Metacognitive difficulties in a driving context

Metacognitive knowledge and strategies	**Metacognitive difficulties**
Knowledge about person/self	✓ Poor perception of one's own driving skills ✓ Excessive self-confidence ✓ Lack of self-confidence
Knowledge about task	✓ Lack of knowledge about task features
Knowledge about strategies	✓ Lack of flexibility in the use of strategies ✓ Lack of knowledge about strategies
Metacognitive skills for planning	✓ Difficulties in the route planning, choosing vehicle or remembering itinerary
Metacognitive skills for controlling and monitoring	✓ Lack of sensitivity to inconsistencies in road environment information
Metacognitive skills for regulating	✓ Failure of self-regulation

3 Proposal of a model to assess the interaction "tram-car driver"

Figure 1 contains the model we came up with to assess the interaction car-tram in roundabout. The articulation of models presented before seemed to us particularly interesting to fully understand human behaviour in typical driving situations such as roundabout crossed by tramway line. We also combine notions of spatial cognition to our analysis to suggest how cognitive mapping might be employed to help us better understand and predict driver behaviour.

The figure 1 below, represents a system composed of three spheres related to each other which are influenced by variables to maintain the equilibrium of the system. The successful adjustment or functional failures are regarded as the result of drivers' actions.

The three spheres can be described as follows:

Environment: the Environment to apprehend is considered as a system with its components and which is influenced by variables derived from Cauvin's work [24], such as places called landmarks or nodes (visibility, legibility, size, neighbourhood,) and links. (linearity, number of lanes, connections).

Driver: the driver is approached through the concept of "situation awareness" of Endsley [16]. The cognitive processes can be seen as a continuous cycle of perception, comprehension, projection and decision. In this study these processes are supposed to be influenced by the personal characteristics of the subject such as (metacognition, personality, expectation and motivation), and its status such as (age, gender and experience).

Action: the driver behaviour is studied through the actions (i.e. the driving task), divided by drawing inspiration from the model of Michon, in three level (strategic, tactic and operational). The drivers' actions are influenced by constraints related to activity goals (work, leisure...) as well as links with the space (traffic flow, mode of locomotion).

Thus, this framework allows generating and testing hypotheses about the specific causes of safe and unsafe driving behaviour.

DOI:10.1051/matecconf/201712401005

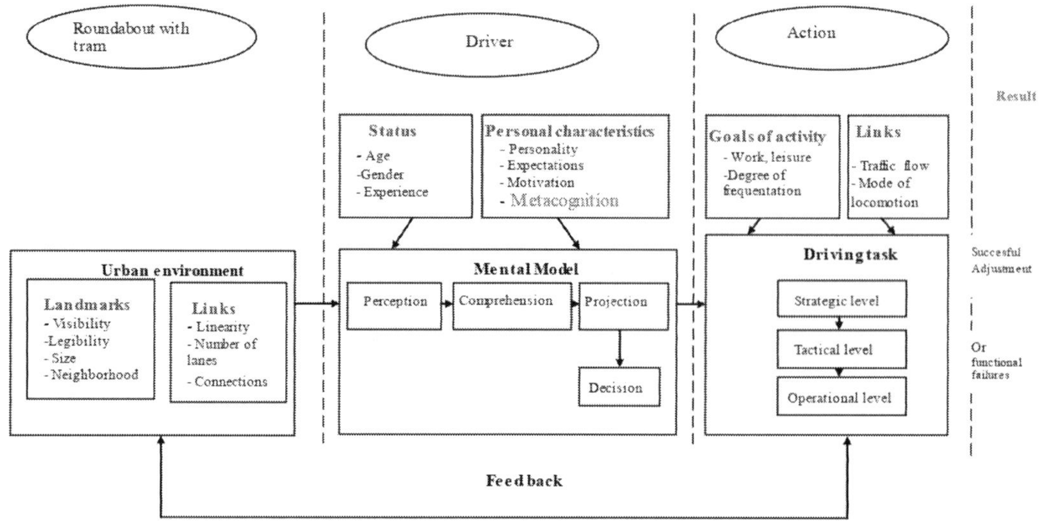

Figure 1. Framework for the tram-car driver interaction assessment.

4. Research Method

To validate our hypotheses, a particular attention is paid to the collection of data. For this purpose, our methodological proposal is based on the combination of quantitative methods (statistical analysis) and qualitative methods (related to drivers' expectations and their driving experiences). This requires implementation of several experimental protocols including accident database analysis, interview, observation, and think aloud protocol. The main advantage of this approach is that each event could be verified from different avenues.

4.1 Tram accident analysis

The Rabat Salé Tramway company (STRS) has collected data on accidents, between trams and others roads users, occurring in Rabat sale Tram network. Our analysis of the STRS accident database was carried out in order to have a global view on tram accidents, when, with whom and in which circumstances the accident happened.
The data used in our research covers the years 2011 to 2014. These four years, were chosen in order to have sufficient data to draw conclusions about a possible trend of evolution.

4.2 Roundabout visits

The choice of the roundabout studied named "IBN ROCHD Roundabout" was based on the number of accidents that happened there. Indeed there were 54 tram accidents between 2011 and 2014 at tram roundabouts. 17 of which were in "IBN ROCHD Roundabout".

Several visits, in vehicles and on foot, were carried out at the roundabout studied. This involved spending periods of time at the roundabout, observing users behaviours.

The observation was based on an analysis grid developed to carry out the assessment of factors that can affect the roundabout safety. The items included in this grid are elements of roundabout environment and users' behaviours which are inspired from several references studied [24] and detailed in our previous articles [3],[25].

4.3 Interviews and think aloud protocol

The experimentation was set up in two steps consisting of a verbal or think-aloud protocol and a retrospective interview.

- Verbal protocol: Twenty participants (eleven males and nine females) were part of the study. Participants were asked to carry out a think- aloud protocol by making a continuous dialogue of their internal thoughts related to their drive, and their progression through the roundabout studied. This involved drivers verbally expressing how they felt and making comments about their behaviours, especially as regards to the environmental conditions which influenced them and came into play in the various situations they encountered. An assessment of driver behaviours (speed, overtaking and driver's bodily expressions) was recorded on a grid developed beforehand.

- Retrospective interview: following the drive a retrospective interview was conducted. The participants were asked about their general impression and about particular points of the itinerary (signalling, visibility ....).

In order to deepen our analysis, verbalization and interviews of vehicle drivers were completed by the questionnaire of the tramway drivers, and interview with tram operators, to investigate users' behaviour from tram drivers' viewpoint.

Two journeys within a tram driver's cab were also performed. Tram Drivers were asked to think aloud during the drive, and verbal assessments of the driving situations were recorded.

4.4 Waiting time measures

Perceived waiting time at tram roundabouts differs from the real waiting time. For the purposes of this study, we measured the real value of the cycle of traffic lights at the roundabout studied to gain more knowledge concerning the role of tram in influencing driver's perception of waiting time in roundabouts with trams.

5. Findings

5.1 Design of the roundabout studied and visibility issues

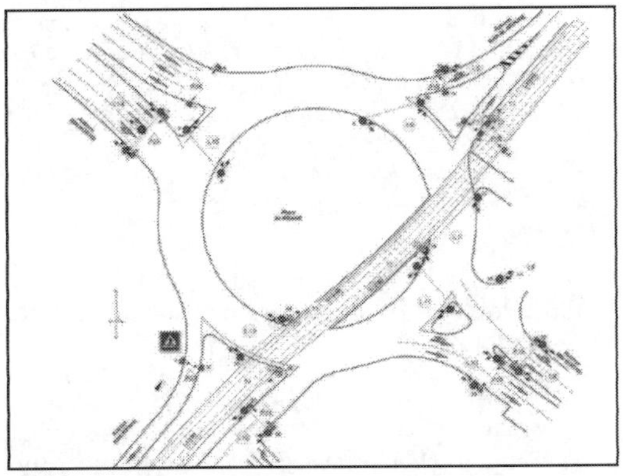

Figure 2. The layout of the signalized roundabout under study

"IBN ROCHD Roundabout" is a tram roundabout with four arms. The configuration is following: two lanes going into intersection and one separate lane for vehicles to turn right. Diagram of intersection geometry is in figure2.

Concerning the visibility of the roundabout studied, our analysis reveals the following shortcomings:

- At some locations there is considerable visual clutter from surrounding infrastructure, which can detract from the primary safety messages "give way to tramway" and "beware of tram";

-The curvature of the intersection creates a difficult angle from which to observe on-coming trams.

5.2 The results of the analysis of STRS accident Database

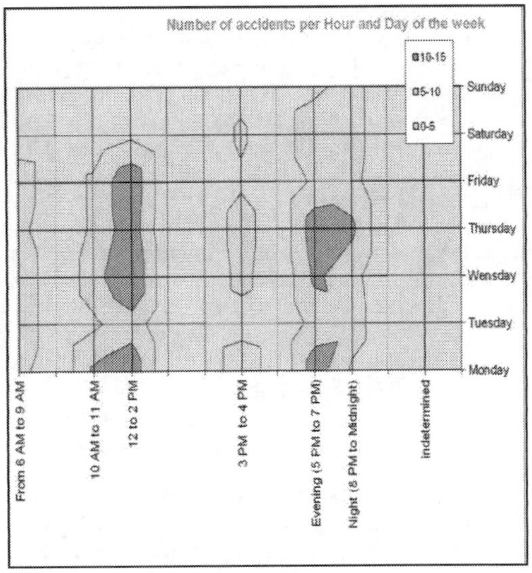

Figure 3. Number of accidents per hour and day of the week

This graph shows that the highest number of accidents occurs during peak hours of lunch break and at the evening from Monday to Friday with the greatest proportion of crashes on Wednesday and Friday for the time slot between 12 p.m to 2 p.m.

This confirms the assumptions of our model and can be explained by the motivations of the drivers (the need to arrive at destination at time), and high workload. These findings are also to be expected due to traffic flow:

- Home-Based work trips that are more concentrated during these time slots;
- Most of private schools in Morocco close on Wednesday afternoons, and several parents must return to the school to pick up their children;
- Friday experiences also a very significant traffic flow for the time of the prayer from 12 p.m to 2 p.m.

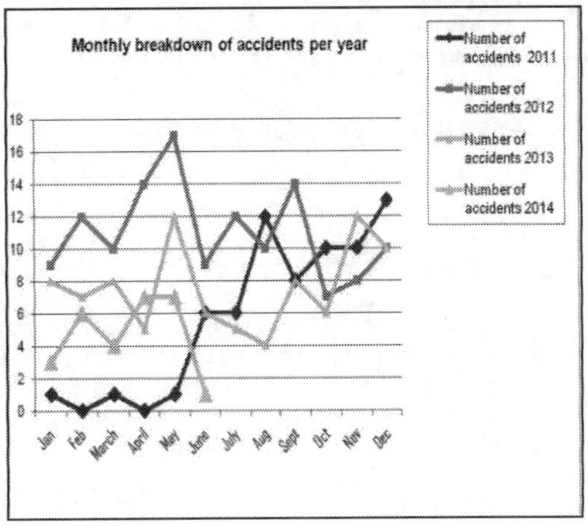

Figure 4. The monthly distribution of accidents

The larger proportion of tram accidents occurs during the winter months of September and December as shown in figure 4. This is most likely due to problems of reduced visibility due to fog and rain. Concerning the high rate of accidents in May, it may be linked to the organization of an international festival during this time of year in the cities of Rabat and Sale, Morocco. This festival is experiencing a strong affluence of non-resident visitors. This finding is consistent with the assumptions of our model and is assumed to stem from the lack of familiarity of non resident drivers with tram roundabouts. So, the roundabout with tram line should be designed to meet drivers' expectations so that drivers know where they are (landmarks), and how to navigate in it (self explaining urban development).

5.3 Metacognitive strategies of tram's drivers

Most Tramway drivers questioned seem to perceive accidents and incidents as the results of factors over which they have no control. Their awareness of risks helps them to be more careful and to question some aspects of their know-how in order to better regulate their activities. The following are two examples of what some participants stated:

"After this accident, i started to make better use of the rear-view mirrors and cameras".

"As *you see, the driver hesitates to cross. I start by warning of the tram's approach using the warning horn, but at the same time, i slow down a bit in case he would go through the red light".*

The qualitative data analysis suggests that tram drivers are likely to encounter unsafe situations and need to anticipate other road users 'actions in advance to avoid conflicts. So, drivers have to use their metacognitive strategies to evaluate, to examine alternative sources or use context clues. Monitoring strategies are then, strongly associated with the ability to assess situations, to judge, or to make good decisions. The likelihood of occurrence of an accident is therefore greater when there is a lack of knowledge about strategies or lack of self-regulation.

5.4 The drivers' ability to understand road signs

The roads signs are very visible in the roundabout studied. Despite this, the results of our study show that they do not attract driver's attention because of the width of the road and the multiple solicitations of attention. The question of perception of road signs raises also the question of understanding of their meaning. If drivers do not perceive the tram signs, it is probably because they do not understand them. Indeed when roads signs don't fulfil an information need of the driver, he has no need to read them and therefore often ignore them. Indeed, this has been pointed out by one of the participants:

"I focus my attention on the traffic lights. I do not notice the tram road signs. To be honest, i do not even look for them"

For many drivers there is no difference between all the tram signs. So, they cannot understand what is expected of them. It was also noted that tram warning signs were worded differently in Arabic and French. The sign "tram crossing the road ahead", is translated to "look out for trams". This discrepancy increases the ambiguity of the instruction.

These results confirm the assumptions of our model and can be explained by driver's situation awareness. Using their mental model, drivers try to perceive road signs that they believe are relevant while they just ignore those judged without useful information.

5.5 Perceived waiting time

According to the interviewees, the difficulties encountered in the roundabout studied are related to the cycle time of traffic signal.

If from our measures, the red light duration does not exceed 2 minutes even when the tramway is passing, drivers believe however that the average waiting time in the roundabout is about 15 minutes.

As Urry [26] points out, the classic view of travel time is basically built on the idea that "the time that is spent travelling is unproductive and wasted- dead time", and therefore a time to reduce.

The taxi drivers and the drivers with the experience of getting stuck seem to have the fear of congestion when the traffic is too crowded. Due to the number of stops experienced when queuing at the traffic light, drivers come to believe that a signal is dysfunctional, and they may try to run red light and to cross without the necessary caution. It was also stated that drivers alter their route choice to avoid getting stuck in the roundabout studied.

5.6 Effects of roundabout environment on driver behaviour

Many unfamiliar drivers with the area may not be expecting trams; the roundabout should provide information about trams using signage and surface delineation. From our analysis there is a number of design features at the roundabout studied that can play a role in unwanted events and contribute to unsafe conditions:

- stop line is not demarcated, either with a white line or appropriate pavement marking, to attract the driver's attention and to help drivers to identify safe stopping location;
- Parked cars before and after the tramway line may result in vehicle drivers slowing and being caught on the tracks while a tram approaches;
- The roundabout studied is a roundabout with traffic light; however signs giving priority to vehicles within the roundabout were maintained with traffic lights, which can lead to confusion and errors for unfamiliar drivers.

6. Discussion

The study has shown that our framework for the assessment of the interaction between tram and car driver has provided a good basis for the analysis of driver behaviour, and the causes of accidents in the roundabout studied. Indeed, results of our study have cast light on a number of problems associated with tram roundabouts and have explored some of the road design and human factors that contribute to the difficulties experienced by drivers at these locations. From our finding three key points must be highlighted:

- Cognitive and metacognitive processes

Metacognitive difficulties in crossing roundabouts with tram due to lack of strategies or lack of self regulation can place drivers at a greater risk of crash. Also no experienced or unfamiliar drivers with roundabout crossed by tram may be more prone to unsafe crossing.

It was acknowledged that most drivers will fail to read signs and consequently do not benefit from the information provided by these.

- Task challenges

The key challenges in crossing tram roundabouts is the goal of the activity (the need to arrive at work at time) which can place added stress and time pressure upon the driver and lead to risky driving in order to save time.

The traffic volume can add to the complexity of the task and may explain the distorted perceptions of waiting time in roundabouts with tram.

- Urban environmental factors

A number of characteristics of tram roundabouts are likely to exert some influence on driver behaviour such as: legibility and recognition of the traffic signs content.

Possible negative effects on driver safety from the roundabout design might occur from lack of pavement marking and the restriction of visibility

Since we cannot "redesign" the driver, one key finding of this study is that the urban environment must be designed to allow the driver to discern, to identify, and to choose easily in this environment the indices for the effective regulation of its activity

7. Conclusion

In a complex and multidimensional system such as tram roundabouts, forming an overall picture of the whole interactions and factors influencing drivers' behaviours is too challenging due to the need to take into account factors that are difficult to quantify.

The interest of this new approach is to contribute to a better understanding of the mechanisms of perception and understanding of the road environment that lead to deviation or successful adjustment.

Evidence has been found that road scene elements and traffic condition play a role in the difficulties that drivers encounter when crossing these intersections. This study has opened avenues of possibilities for further research in the area of urban design. Indeed these findings can help managers and road designers in their search for improvement of existing facilities, enhancement of those that seem relevant and finally designing of future infrastructure. Another future research opportunity is to investigate if these finding in roundabout can be generalized to the entire network by expanding our model in roundabout to other locations.

References

1. G. Nuvolati, Resident and Non Resident population: Types of conflicts. *P.Pucci and M.Colleni* (Eds.), Understanding mobilities for designing contemporary cities, Research for development. (2016)
2. World Bank. Morocco, Urban Transport Sector Development Policy Loan Project. Report Document AB4928,(2009).
3. F.Moutchou, A. Cherkaoui and E. El koursi, Human factors in the analysis of the Tram- car drivers at Intersections. Journal of civil engineering and architecture, Vol **8**, Serial No 77, April (2014).
4. K. Kjemtrup and L. Herrstedt, Speed management and traffic calming in urban areas in Europe: a historical view. *Accident Analysis & Prevention* **Vol.24**, No. 1, pp 57-65,(1992).
5. J. Whitelegg and G. Haq, Vision zero: adapting a target of zero for road traffic fatalities and serious injuries. Stockholm environment institute,(2006).
6. P. Healey, Urban complexity and spatial strategies: towards a relational planning for our times. *Taylor§ Francis*, (2007).
7. K. Lynch, The image of the city. Cambridge Massachussettes, *MIT Press*, (1960).
8. M. Millot, Analysis of the complex links between urban forms and road safety. Study report of CERTU (Centre for studies on road networks, transport, urban planning and public construction),(2004).
9. G.Weller and M.Dietz, Definition, comparison and evaluation of existing self-explaining road approaches in Europe. Deliverable N° 1, SWOV, (2010).
10. F. Saad, In depth analysis of interactions between drivers and the road environment. *4th Workshop of International Cooperation on Theories and Concepts in Traffic Safety (ICTCT)*, Vienna, pp. 65-79,(1991).
11. P. Salmon, M. Regan and I. Johnston, 2006, Human Error and Road Transport: Phase Two—A Framework for an Error Tolerant Road Transport System. Monash University Accident Research Centre report, (2006).
12. G. Helmers, Explanatory model for road user behaviour: Implications for the design of road and traffic environment. Nordic human factors guideline, (2014)
13. M. Millot, Urban growth and road safety: the complex influence of urban forms. Ph.D. Thesis, National Advanced Civil Engineering School (ENPC), Department of Transport, Paris, December.9, (2003). (In French)
14. D. Fleury, Categorization of road scenes and safety from accident: analysis to typical representations. Intellectica N°15, pp. 97-123, (1992).(In French)
15. R. Davidse, Assisting the older driver, intersection design and in car devices to improve the safety of the older driver. Swov, Nederland, (2007).
16. M. Endsley and D. Kaber, Team situation awareness for process control safety and performance. Process Safety Progress 17 (1) 43-48, (1998).
17. B.Bailly, Drivers situation awareness: fundamentals aspects, methods, and tool development for drivers' training. Ph .D. Thesis, Lyon 2 University, Specialty

cognitive psychology, December.13, (2004). (In French)
18. J.A. Michon, A critical view of driver behavior models: what do we know, what should we do?. In: *L. Evans, R.C. Schwing* (Eds.), Human Behavior and Traffic Safety, Plenum Press, New York, pp. 485-524, (1985).
19. G.M. Matthews, D.J. Bryant, R.Webb and J.L. Harbluk, 2001, Model for situation awareness and driving: Application to analysis and research for intelligent transportation systems. Transportation Research Record, (1779), (2001).
20. J.H. Flavell, Metacognition and cognitive monitoring— A new area of cognitive-developmental inquiry. American Psychologist, 34, 906– 91, (1979).
21. A.L. Brown, J.D. Bransford, R.A. Ferrara and J.C. Campione, Learning, remembering, and understanding. In J.H. Flavell, and E.M. Markman (Eds.), *Handbook of child psychology*: **Vol. 3**. Cognitive, (1983).
22. H. Patrick, Social self-regulation: Exploring the relations between children's social relationships, academic self-regulation, and school performance. Educational Psychologist, 32(4), 209–221, (1997).
23. J. Reulier, Verbal interactions between peers and development of metacognition in students with reading comprehension difficulties. Thesis, University of Quebec in Rimoski (2012). (In French)
24. An approach of intra-urban spatial cognition, http://cybergeo.revues.org (accessed Jan. 27, 1999).
25. F. Moutchou, A. Cherkaoui and E. El koursi, Factors influencing driver's behaviour at *intersections crossed by tram. Proceeding of the 3rd International conference on road and rail infrastructure, CETRA* 2014, Split, Croatia, pp 751-760, (2014).
26. J.Urry, Travelling times. European Journal of Communication Vol **21**(3): 357-372, (2006).

Construction of Visual Inspection Database for Catenary on High-speed Railways

Yan-guo Wang[1], Dapeng Xie[1], Haiying Wang[1], Xiangdong Li[2], Jing Shi[1], Hailang Li[1], Kexin Xiao[2] and Jincai Li[2]

[1]*Institute of Infrastructure Inspection, China Academy of Railway Sciences, Beijing 100081, China*
[2]*Beijing IMAP Technology Co., Ltd, Beijing 100081, China*

Abstract. With the rapid development of computer vision, techniques of machine vision and visual inspection have been applied into the inspection of catenary on high-speed railways. Visual inspection systems have been developed and super-high-resolution images are captured to check the status of catenary components. Automatic recognition of defects becomes very important since the number of images is too huge to be manually checked one by one. However, it is not easy for the development of recognition algorithms on catenary components. There are many types of defects to be checked on different kinds of catenary components, but the number of defect images is too small in real world. In this paper, a solution was proposed and implemented. An on-site data acquisition system was designed and developed, and different types of defects were manually made on different catenary components beforehand. Finally, a visual inspection database was successfully constructed, including plenty of different kinds of catenary components, different types of defects, in different inspection conditions. The visual inspection database will be of great use in the development and test of recognition algorithms for catenary.

1 Introduction

Catenary [1] is the one of the main infrastructures along high-speed railways which supplies power for the trains. Regular inspections have to be performed in order to guarantee the status of catenary. With the rapid development of computer vision [2, 3], techniques of machine vision and visual inspection have been applied into the inspection of catenary [4, 5]. Several visual inspection systems have been developed and super-high-resolution images are captured to check the status of catenary components [6-8].

However, the subsequent analysis on the visual inspection data is quite a difficult task. The number of images is too huge so that it will be very inefficient to be manually checked one by one. Automatic analysis is necessary but the development of the detection algorithms is not easy. There are so many kinds of catenary components to be checked, and hundreds of defect types to be automatically detected on different components. Development and test of recognition algorithms rely on a visual inspection database with image data of high quality, which should cover different catenary components and different types of defects. Unfortunately, defects on catenary components are rarely seen in real world and difficult to be collected, so that the number of defect images is too small for the development and test of recognition algorithms.

In this paper, a solution was proposed to construct the visual inspection database. The data acquisition was performed on a typical railway line which was under construction. An on-site data acquisition system was designed and developed, and different types of defects were manually made on different catenary components beforehand. Finally, a visual inspection database was successfully constructed, including plenty of different kinds of catenary components, different types of defects, in different inspection conditions.

The rest of the paper is organized as follows. The on-site data acquisition system is described in section 2. The arrangement of defects per-made and data acquisition is discussed in section 3. The final visual inspection database is reported in section 4. Then we draw the conclusions.

2 Data acquisition system

Catenary is composed of contact wire, cantilever, additional wire, and hanging post (in tunnels), which is illustrated in Fig. 1. There are many kinds of components in different regions of catenary, which vary greatly in size and location.

DOI:10.1051/mateconf/201712401006

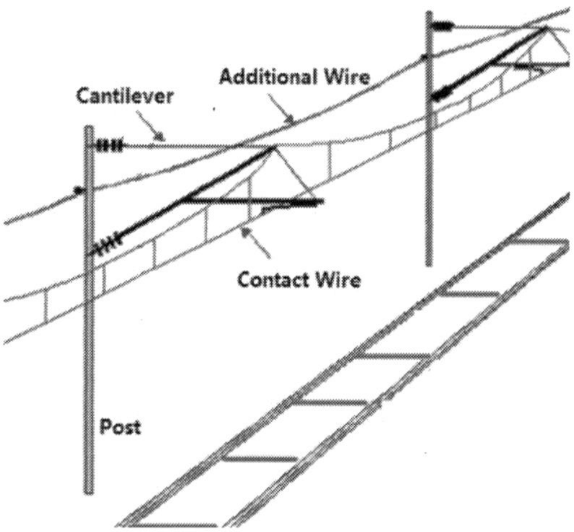

Figure 1. Structure of catenary

The data acquisition system was designed based on catenary maintenance ladders (Fig. 2), which can be operated on railway lines under construction.

Figure 2. Catenary maintenance ladder.

18 super-high-resolution cameras are used to capture detailed images of catenary components in different view directions, with an image resolution up to 29 megapixels. Fig. 3 illustrates the arrangement of the camera system.

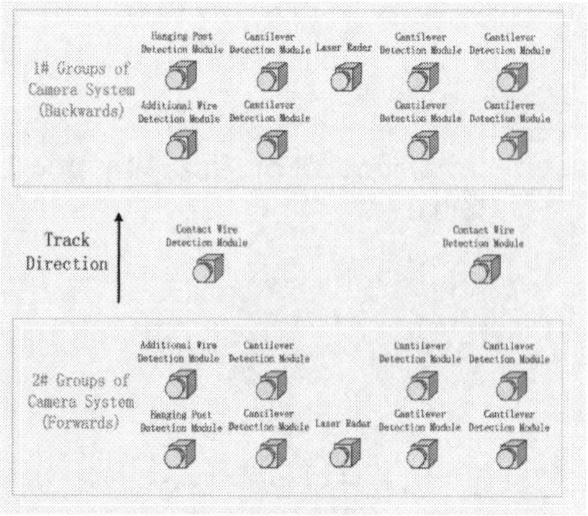

Figure 3. Camera system arrangement.

Most catenary components concentrate at catenary posts, and we have to capture images at a proper distance to each post. Two laser radars are utilized here to give an accurate trigger signal for the camera system. The laser radars continuously measure the distance from the camera system to next catenary post, and a trigger signal will be generated and distributed to the camera system, when the distance reaches a pre-define value.

The final design of the camera system is showed in Fig. 4.

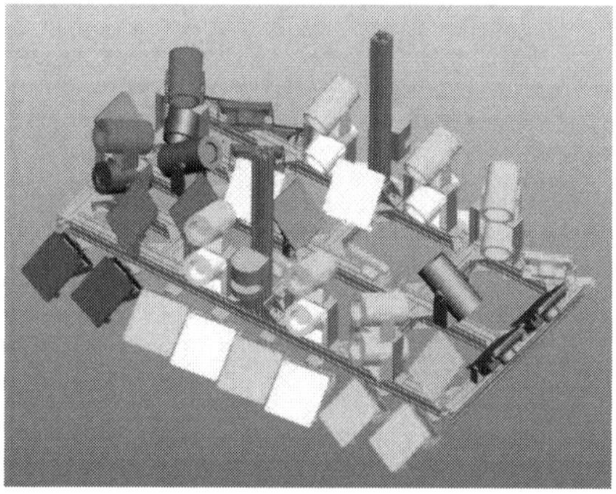

Figure 4. Camera system design.

Table 1 gives the specifications of each detection module in the camera system.

Table 1. Specifications of camera system.

Detection Modules	Camera Resolution (megapixel)	Number of Cameras	View Directions
Cantilever	29	12	Forwards & backwards
Contact Wire	25	2	Left & right

 DOI:10.1051/matecconf/201712401006

Additional Wire	16	2	Forwards & backwards
Hanging Post	16	2	Forwards & backwards
Laser Radar	/	2	Forwards & backwards

The arrangement of the machine cabinet is illustrated in Fig. 5. There are camera control unit and light control unit, which control the power supply and trigger signal of cameras and lighting device respectively. Trigger and control unit is in charge of the signal processing of laser radars, also the generation and distribution of trigger signals. The final image data acquisition and storing is managed by a data acquisition server. There is also an UPS here in order to guarantee the stable power supply of the system.

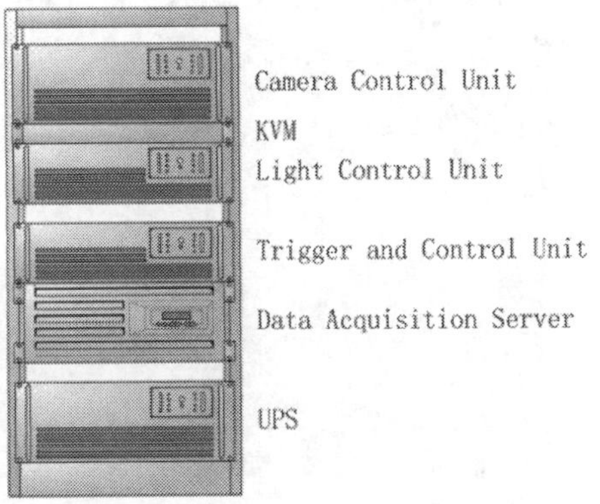

Figure 5. Machine cabinet.

The graphic user interface of the acquisition system is showed in Fig. 6.

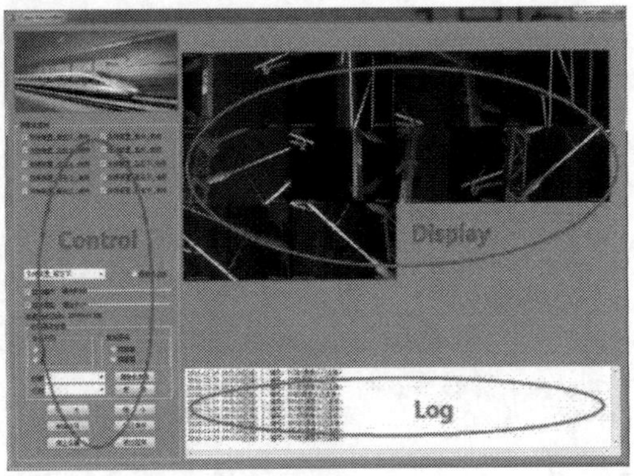

Figure 6. Graphical user interface.

3 Defects pre-made and data acquisition

The data acquisition was performed on some selected sections of a typical high-speed railway which was under construction. Straight lines, curves, stations, tunnels were all involved, and the acquisition was performed both in day and at night, in order to ensure the representativeness and completeness of the database.

Hundreds of different types of defects had been manually made on different kinds of catenary components beforehand. Super-high-resolution image data were captured both with and without the defects, also both in day and at night. The pre-made defects were restored after the acquisition, so as to guarantee the safety and quality of catenary.

Fig. 7 shows the process of manually-making defects, and the data acquisition is showed in Fig. 8.

Figure 7. Defects pre-made on catenary.

Figure 8. Data acquisition.

4 Visual inspection database

The final visual inspection database is summarized in Table 2. It covers different kinds of catenary components, and plenty of different types of defects data, in different inspection conditions.

Table 2. Visual inspection database.

Types	Number of posts	Mileage (km)	Number of images
Normal data	1665	83.2	19980
Defects data	87	4.4	1044
Total	1752	87.6	21024

The data acquisition was performed both in day and at night, which is summarized in Table 3.

Table 3. Visual inspection database.

Inspection conditions	Number of posts	Mileage (km)	Number of images
In day	364	20.0	4368
At night	1388	67.6	16656
Total	1752	87.6	21024

Some enlarged normal data are showed in Fig. 9~11. Some defects data are showed in Fig. 12~17, with the position of defects labelled using red ellipses.

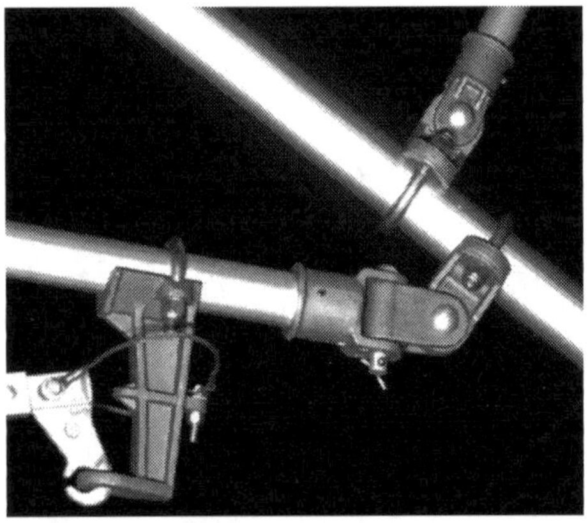

Figure 9. Normal data (enlarged).

Figure 10. Normal data (enlarged).

Figure 11. Normal data (enlarged).

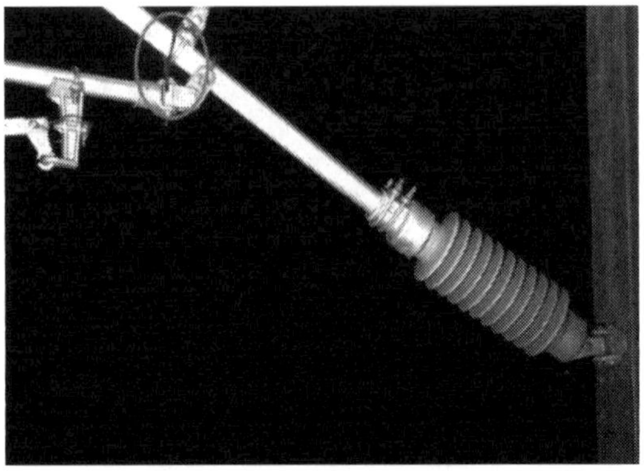

Figure 12. Data with pre-made defects.

Figure 13. Data with pre-made defects.

Figure 14. Data with pre-made defects.

Figure 15. Data with pre-made defects.

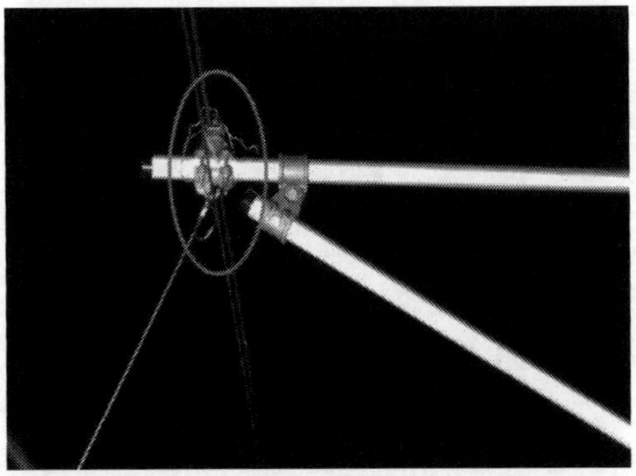

Figure 16. Data with pre-made defects.

Figure 17. Data with pre-made defects.

5 Summary

Technologies of computer vision and visual inspection have been applied into the inspection of catenary on high-speed railways. However, the automatic analysis of visual inspection data relies on a high quality visual inspection database, which is very difficult to be collected in the conventional manner. In this paper, an on-site data acquisition system was proposed, and plenty of image data are captured with different types of defects pre-made on different kinds of catenary components. The final visual inspection database will be of great use in the future development and test of recognition algorithms for catenary.

Acknowledgement

This research was partly supported by the National Key Basic Research Program of China (973 Program) (Grant No. 2013CB329406), China Railway Corporation Program on Scientific Research and Development Project (Grant No. 2015T003-A), and Beijing IMAP Technology Co., Ltd Program on Scientific Research and Development Project (Grant No. 2016JJXM11).

References

1. F. Kiessling, R. Puschmann, A. Schmieder, et al. "Contact lines for electric railways (second edition)." Publicis Kommunikations Agentur GmbH, Germany (2009).
2. D. A. Forsyth and J. Ponce. "Computer vision: a modern approach." Prentice Hall, America (2003).
3. E. R. Davies. "Machine vision: theory, algorithms, practicalities (third edition)." Elsevier, Singapore (2005).
4. Yan-guo Wang, Zhongguo Sun, Shiping Gu, et al. "Real-time measurement for the angle of steady arm on high-speed railways." Proc. of the 10th World Congress of Railway Research (2013).
5. Wei Zhou, Zhongguo Sun, Shengwei Ren, et al. "Measurement method for geometric parameters of overhead contact line based on multi-view stereovision." China Railway Science, 36(5): (In Chinese). (2015) 104-109.
6. U. Richter and R. Schneider. "Automatic optical inspection for catenary." Railway Engineer (Chinese Edition), 7(1), (2001) 24-28.
7. U. Richter. "Contact and non-contact comprehensive catenary inspection system." Railway Technology Review, (2009) 8-10.
8. Yan-guo Wang, Wei Zhou, Dapeng Xie, et al. "Automatic Visual Inspection for Catenary on High-speed Railways." Proc. of the First International Conference on Rail Transportation (2017).

Estimating Bluetooth mac scanner based pedestrian flow characteristic by taking the through pedestrian flow as a case study

Qing Lan[1], Bowen Gao[2], Zhigui Chen[2], and Sicong Zhu[3]

[1]*Communication and Transportation Department of CangzhouJieFangXi Road No.77, Cangzhou, Hebei Province, China.*
[2]*Beijing Jiaotong University, China Haidian District, Beijing 100044, P.R. China*
[3]*MOE Key Laboratory for Urban Transportation Complex Systems Theory and Technology, Beijing Jiaotong University, China Haidian District, Beijing 100044, P.R. China*

Abstract. The Bluetooth scanner has been widely used as vehicle traffic detector in recent years. A new solution based on the Bluetooth scanner has been deliver to detect pedestrian traffic by estimating the pedestrian flow characteristic. The equipment is tuned to scan Bluetooth devices more frequently. Consequentially, the unique Bluetooth MAC can be detected multiple times by the scanner. The numeric simulation also suggests median detection is the chosen one to calculate traveling time. In this study, Bluetooth MAC scanning, as a non-intrusive technology, is tested in a pedestrian transfer tunnel, including antenna compatibility working within buildings. The penetration rate also increases slightly. Finally, the proposed solution offers a mature technology for pedestrian study, integrating scanner configuration, antenna selection and the traveling time calculation method.

1 Introduction

Rapid deployment of smartphones as all-purpose mobile communication systems has led to the fast adoption of wireless communication systems such as Wi-Fi and Bluetooth. The media access control (MAC) of Bluetooth devices can be viewed as a unique identification. The Bluetooth scanner utilized as a traffic detection technology has emerged in the past decade [1]. At the current time, networks of Bluetooth MAC scanners, which detect and match the MAC addresses of Bluetooth devices within wireless networks, have been widely utilized to collect vehicle traffic data [2] [3].

On the other hand, the organization of pedestrian flows in large public buildings presents a major challenge. In traffic terminals, such as railway stations and subway stations, systems with information about current crowd flows are able to support the control and management of pedestrian flows and can reduce traveling time and management cost [4]. Meanwhile, shopping malls or other commercial real estate expect a higher pedestrian density and a longer dwelling time. According to Kardi Teknomo [5], pedestrian detection technology can be classified into two categories, namely, manual counting and video image processing. Recently, the Bluetooth scanner has been deployed in different locations to measure pedestrian volume and traveling time [6].

This paper aims to deliver a solution for monitoring pedestrian movements, especially in large traffic terminals, by means of stationary Bluetooth scanners. The proposed solution consists of theoretical numeric simulation, hardware configuration (including antenna selection), and Bluetooth scanner parameter tuning. In the remainder of the paper, an overview of existing efforts is presented, followed by a discussion of the methodology and a numeric simulation. Results from the study site are described and discussed, followed by concluding remarks.

2 Background

According to IEEE 802.15.1 protocol [7], the connection between a Bluetooth scanner and a Bluetooth device is established by the process described in Figure 1.

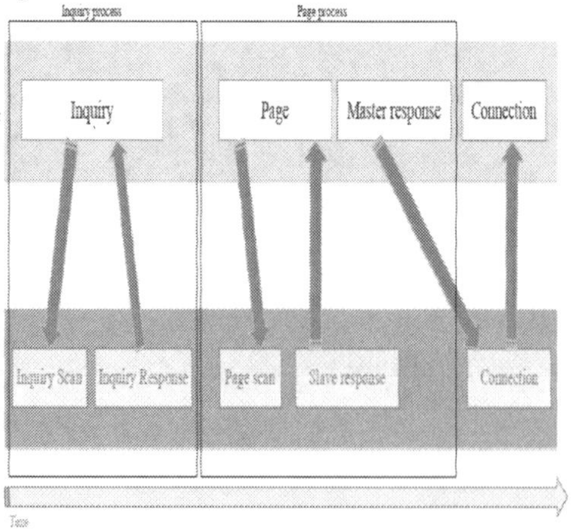

Figure 1. Bluetooth connection process [8]

The Bluetooth scanner can collect the encrypted MAC address of device and corresponding detection time. The traveling time can be calculated by using logarithmic differentiation of detection time collected by two proximate Bluetooth scanners.

$$T(m,1,2) = t(m,2) - t(m,1) \quad (1)$$

where

T(m,1,2): the mth MAC-ID traveling time from zone 1 to zone2;

t(m,k): the mth MAC-ID detection time at k^{th} zone ;

The Bluetooth scanner detects active Bluetooth devices within its scanning coverage. Due to the frequency hopping process mentioned below, the acquired traveling time is a zone-to-zone traveling time, rather than a point-to-point traveling time such that obtained by conventional automatic vehicle location (AVL) technology.

In this study, the Bluetooth data has the following three fields:

- ID: a unique encrypted MAC address of the Bluetooth device detected
- Time stamp: time when the scanner records the Bluetooth device
- Frequency of detection: frequency of being detected by scanner to pass through the scanner coverage.

In addition to the traveling time, this paper focuses on the frequency of detection on a single MAC basis, which is highly correlated with the scanner's signal quality. This is presented in detail below in the third section of the paper.

3 Methodology

3.1 Descriptions of Bluetooth Scanner Working Process

The Bluetooth scanner "sniffs" data packages sent by proximate Bluetooth devices. To simulate the scanner working process, a probabilistic model is established by making some assumptions regarding the operations of the scanner and Bluetooth device. Initially, the scanner is assumed to remain in its inquiry state, and it hops on each frequency and scans for 0.625 millisecond. The second assumption is that the Bluetooth device passing by the scanner enters its inquiry scan state every 1.28 s and listens on channels for 11.25 milliseconds, as described by Welsh et al. [9]. By making these assumptions, when the Bluetooth device is within the range of the scanner, 18 frequencies of the total 32 frequencies that belong to the Bluetooth spread spectrum will be scanned during a single inquiry period, of which the probability is equal to 56.25% [10].

3.1.1 Bluetooth Inquire and Radio Scanning Cycle

In contrast to existing studies [11], the scanning cycle is set to 1.28 s which leads to a higher detection frequency. From the large number of detections for a specific Bluetooth MAC, only one is selected and used to determine traveling time. The varied methods regarding detection timing selection for differential operation in Equation 1, namely, early detection, median detection, and final detection, are compared and examined by a numeric simulation proposed in this section. For the other detections, two pieces of information are available, namely, detection time and approximate location of the detected Bluetooth device when within scanning coverage. Hence, this paper defines the redundancy of detections as ambiguity. The utility of ambiguity is presented in the following numeric simulation and analysis of the experiment's dataset.

3.1.2 Signal Propagation and Antenna Selection

The performance of the Bluetooth scanner, as a non-intrusive detection technology, is substantially correlated with signal quality[12]. Antenna selection and signal propagation are two critical factors. The signal propagation is highly subject to the environment, especially the indoor environment. In the antenna selection criteria, two main factors are considered: *polarization* and *antenna gain*. Polarization refers to the orientation of the electric field vector in the radiated wave. Antenna gain is defined as the power output in a particular direction, compared to that produced in any direction by a perfect omnidirectional antenna (i.e., an isotropic antenna). In this research, the selected antenna needs to be polarized to enable an effective range that is as long as possible to create ideal signal coverage, as demonstrated in Figure 1. The selected antenna also needs an appropriate level of antenna gain with the scanner able to avoid capturing MACs for Bluetooth devices outside the testing fields. The study presented in this paper tested and compared two categories of antennae when working within buildings. The frequencies of cluster size are examined to see how these values may be of use in describing a particular location.

3.2 Numeric Simulation

A numeric simulation is established based on the following assumptions.

- Ideal signal propagation and device detection
- High frequency scanning : Scanning cycle is set to 1.28s
- Uniform pedestrian speed: 5km/h (1.40 m/s)
- One Bluetooth device for each pedestrian

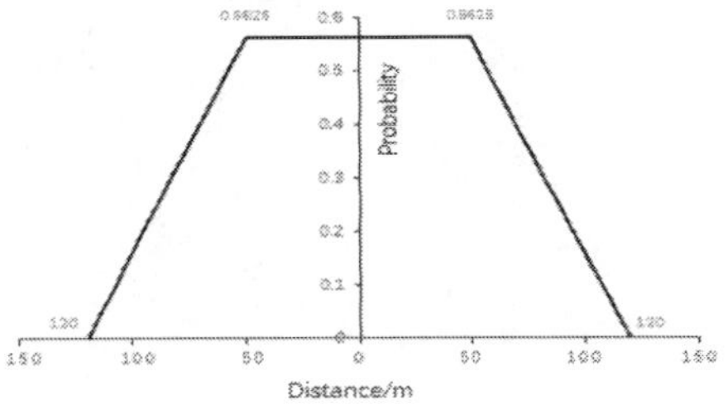

Figure 2. Location based detection probability

The Figure 2 illustrates the probability that a Bluetooth device is detected by a Bluetooth scanner in

 DOI:10.1051/matecconf/201712401007

relation to the distance of the device from the scanner, based on assumptions of the ideal signal propagation proposed by Moghaddam and Hellinga [13].

The Bluetooth scanning interval is 1.28 s. Based on the third assumption, the location and time of the detection are recorded when a pedestrian is traveling through the Bluetooth scanner's antenna coverage area. Through the use of the location-based probability of detection (as defined in Figure 2), it is determined whether each pedestrian had been detected. The simulation runs 100,000 times [14] [15] A simulation can be formulated by the following Pseudo-code:

```
For each single pedestrian
  For each step of the scanning cycle
      Generate a random number
        If the random number < probability of being detected, the device is detected
            Record time step and location
        Else the random number >Probability of being detection
            Otherwise
        End if
   End
End
```

Due to the assumption 3, the location and time of the detection is recorded when pedestrian traveling through the Bluetooth scanners antenna coverage area. The high volume of MAC detections are classified into clusters. A *cluster* is defined as a collection of data records with the same matched MACs sorted sequentially by time, shown in Figure 3. Clusters of MACs are defined for individual Bluetooth scanner, and they represent the number of times a single unique MAC is captured as a pedestrian passes through the Bluetooth scanner's coverage area. Three typical detection timings may be chosen to calculate pedestrian traveling time, namely, initial detection, median detection, final detection. As the detection probability function is symmetric in Figure 2, two out of the three scenarios, the initial one and the median one, are compared and evaluated in terms of uncertainty. Hereby, for each scenario, two detection locations of same timing, generated by two independent simulations are selected for traveling time differentiation, shown in Figure 4. In theory, the independent detection location should be static ideally to calculate the point to point traveling time. Hence, the measurement error is the gap between the two detection locations generated by the same simulation scenario.

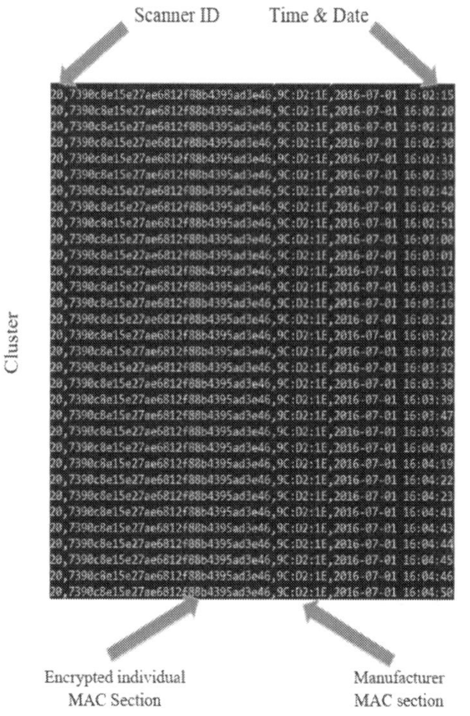

Figure 3. Cluster of Bluetooth device MAC

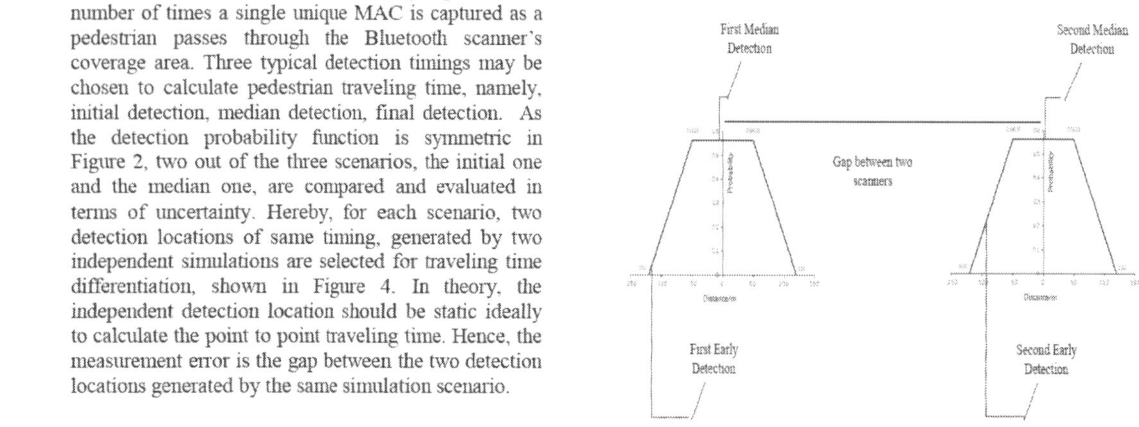

Figure 4. Numeric simulation: Measurement uncertainty

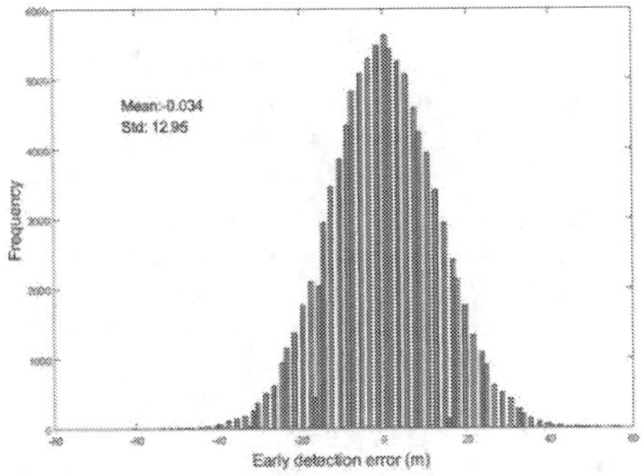

(a)

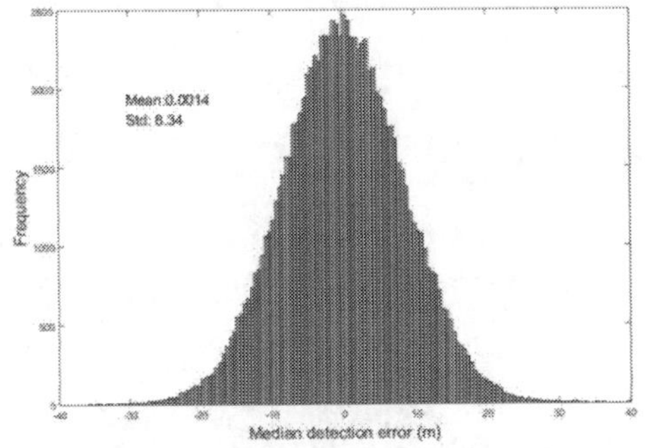

(b)

Figure 5. Simulation results (a): Early detection traveling time error (b): Median detection traveling time error

3.2.1 Simulation Findings

Due to the signal unreliability at the marginal areas of scanner coverage, the location of early detection (final detection) is highly unstable. The Figure 5 (a) illustrates that the uncertainty of early detection scenario is higher that its counterpart. According to the simulation result, the median detection location is approximate to scanner's location where the signal is the most ideal. Hence, it is promising to select the median detection out of cluster to calculate the travelling time, as shown in Figure 5 (b).

4 Field experiments

A series of field tests have conducted to examine the solution. The parameters are tuned up to realize the numeric simulations in section 3.

4.1 Equipment

A MAC scanner device was used to collect MAC addresses of Bluetooth devices as they passed by the scanner. Regular Bluetooth devices are normally restricted to a range of about 10 m, but an increasing number of devices are able to read MAC addresses to a range up to 300 m (*4*). Such a device and antennas, shown in Figure 6, have been developed by the Smart Transportation Applications Research laboratory at the Queensland University of Technology and was used in this study. The device is used in conjunction with a 5-dBi (decibel isotropic) omnidirectional antenna to obtain a range of roughly 50 m. The scanning cycle is 1.28 s to be considered continuously present. Directional antenna diverts the signal in a particular direction to farther distances. While directional antennas can be of great value for certain indoor applications, Directional antennas used for the indoors typically have a lower gain. This results in less ability to reject or reduce the interference signals received from directions outside the primary lobe area [16].

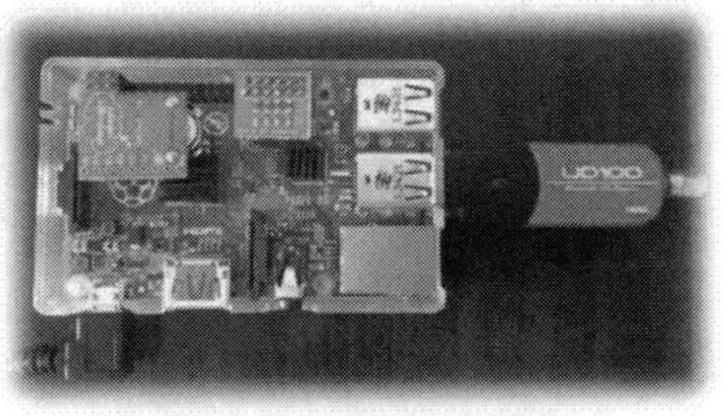

(a)

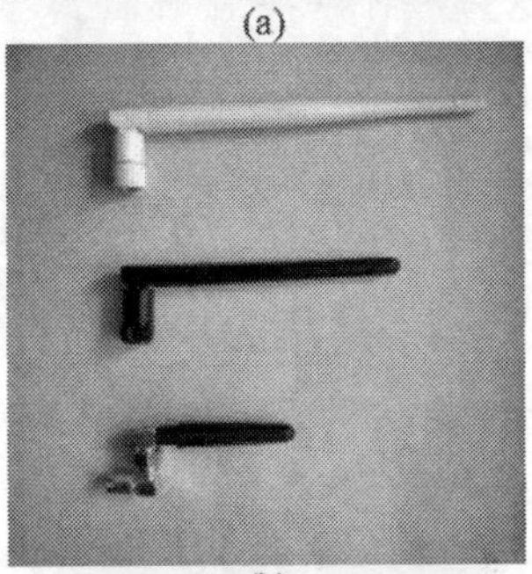

(b)

(c)

Figure 6. Bluetooth scanner and accessories (a) Bluetooth scanner (b) Omni-directional antennae (from top to bottom 10-dBi gain , 5-dBigain, 3-dBi) (c) Directional antenna

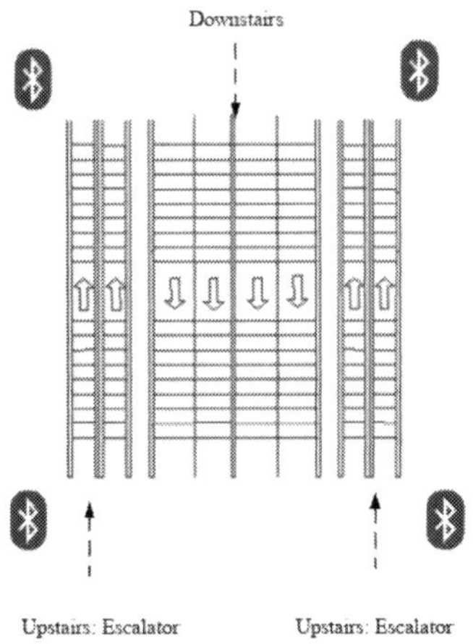

Figure 7. Experiment site: transfer tunnel layout

4.2 Pre-experiment

In the Xizhimen subway interchange station Beijing, the four locations chosen for detector mounting were at both ends of a pedestrian transfer tunnel linking Route 13 with Route 2 and Route 4, shown in Figure 7.

The *penetration rate* was the main antenna performance measure and was computed as follows:

$$P = N / M \quad (2)$$

where

P - the penetration rate

N - the number of MACs matched at both ends

M - the number of through pedestrians collected manually

One-hour manual counts were done within the range of the installed scanners to determine the approximate penetration rates at each location at peak hour during which the Xizhimen subway station has extremely high pedestrian flows of approximately 20,100 pedestrians per hour in both directions. Several similar experiments were conducted at the same spot. The dataset shows that the scanners were able to capture roughly 2% of the population travelling through the tunnel.

4.3 Field Test Data and Analysis

Four scanners were mounted at four different sites at both ends of the tunnel. Two of them were equipped with directional antennae, while the others were equipped with omnidirectional one. During peak hour from 17:40 to 18:40, test results were collected by the two pairs of scanners equipped with directional antennae and their omnidirectional counterparts, are shown and compared in Table 1.The datasets were extracted and stratified into clusters defined in section 3.The study focused on the pedestrian flow through the tunnel. Thus, only the MACs that were matched at both scanners were taken into account.

The Figure 8 demonstrates the distribution of traveling time. The distribution also demonstrates that a few MACs have been detected at extremely high frequency in Figure 8 (a) (c). Therefore, the traveling time is calculated by the differentiation of the two initial detection times collected by two scanners with omnidirectional antennae, while the median detection times are used to calculate traveling time for the dataset collected by the directional scanning antennae. As a longer tail of cluster size can be found in the dataset collected by the omnidirectional antennae rather than the dataset obtained by the directional antennae. It indicates that MACs with a high dwelling time are not likely to roam in the designated area. The reasonable explanation is that the omnidirectional one may capture some MACs continuously outside the test location, due to its open coverage. The comparison between two categories of antennas suggests the directional antenna can create a wider effective range for MAC detection. These phenomena demonstrate a methodological improvement. Both the higher penetration rate and the better shaped distribution suggest the current solution realizes the ideal condition to some extent, based on the assumptions listed in section 3. As a result, the traveling time calculated by the omnidirectional dataset seems to be biased, due to the unstable signal at marginal areas of the signal coverage. The omnidirectional data also shows that the travelling time is more dispersed than that of its directional counterpart, as shown in Table 1, Figure 8, and Figure 9.

Table 1. Field Test Results

Antenna	No of matched IDs	Direction	Ave. Traveling Time (s)	Std (s)
Omni-directional	474	U*	68.7	24.7
		D*	61.3	31
Directional	516	U*	79.9	24.1
		D*	78	18.2

U* stands for upstairs D* stands for downstairs

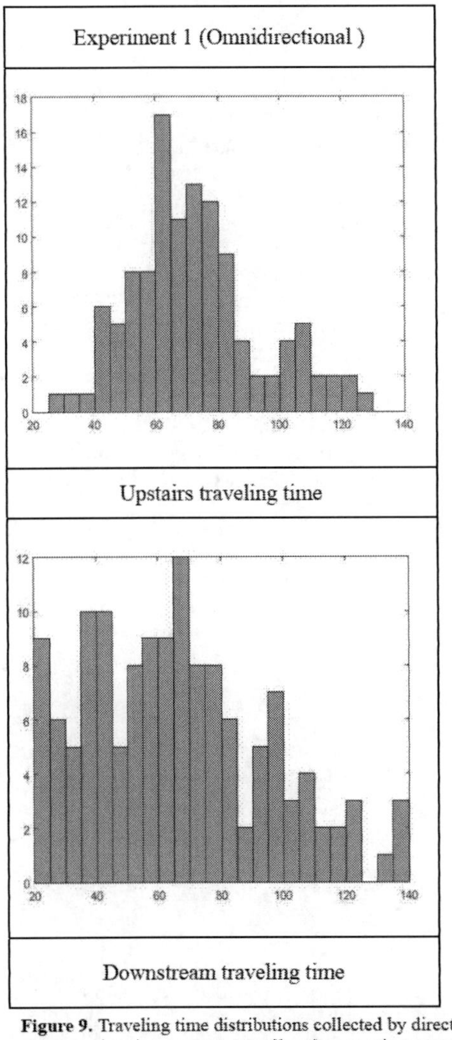

Figure 9. Traveling time distributions collected by directional antennae. (x axis represents traveling time, y axis represents frequency

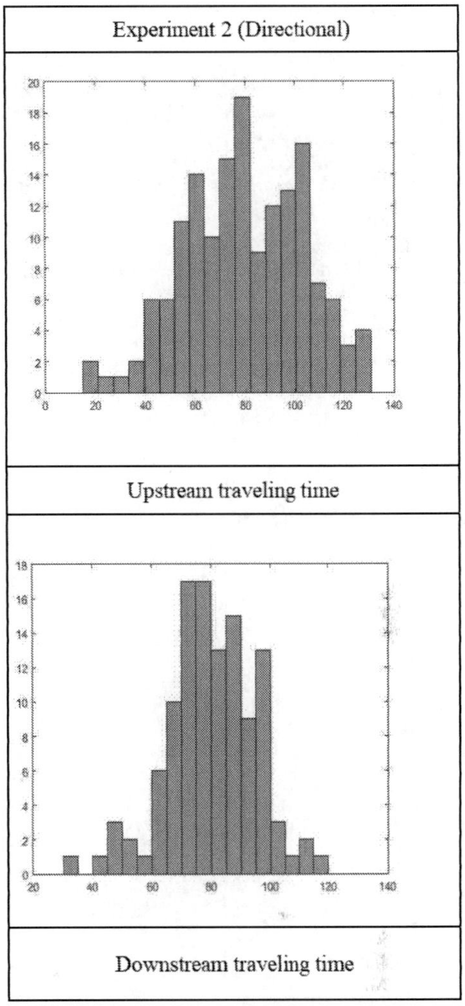

Figure 8. Traveling time distributions collected by Omni-directional antennae. (x axis represents traveling time, y axis represents frequency

5.Conclusions

By detecting and matching Bluetooth MAC addresses, the study presented in this paper develops and delivers a solution to measure pedestrian volume and speed. The study makes several contributions in the area of pedestrian detection, putting forward the measurement of theoretical development. It proposes a new high frequency scanning method to be used within complex structures. Meanwhile, median detection is chosen from each MAC cluster to calculate traveling time. In addition, the study improves pedestrian detection with specific antennae in an indoor environment. Therefore, the solution delivers a more accurate and reliable solution based on the use of a Bluetooth scanner to measure pedestrian flow characteristics. In the future, the penetration rate can be improved by introducing a Wi-Fi scanning module. This scanning technology can be applied extensively in different areas of pedestrian studies. Further testing in the field is also necessary to validate the data with data obtained by indoor positioning technology.

Acknowledgements

The authors acknowledge technical support from the Smart Transport Research Centre, Queensland University of Technology, Australia. The authors acknowledge financial support from Beijing Jiaotong Talent Fund No: T15RC00050

References

1. H. Sawant, J. Tan, Q. Yang and Q. Wang, Intelligent Transportation Systems, Proceedings. The 7th International IEEE Conference on, 767-772, IEEE (2004).
2. A. Bhaskar, M. Qu and E. Chung, IEEE Transactions on Intelligent Transportation Systems,99. 1-10 (2014).
3. A. Bhaskar, M. Qu, A. Nantes, M. Miska and E. Chung, Inter. J. Intel.Trans. Sys. Re. 13, 119-130 (2014).
4. L. Schauer, M. Werner and P. Marcus, Proceedings of the 11th International Conference on Mobile and Ubiquitous Systems: Computing, Networking and Services, 171-177, ICST.
5. K. Teknomo, Diss. Tohoku Univ (2002).
6. J. van den Heuvel, A. Voskamp, W. Daamen and S. P. Hoogendoorn, Traffic and Granular Flow'13, p. 73-82, Springer, (2015).
7. I. C. Society 802.15.1-2005 - IEEE Standard for Information technology-- Local and metropolitan area networks-- Specific requirements, 3 Park Avenue, New York, NY 10016-5997, USA (2005)
8. A. Bhaskar and E. Chung, Transportation Research Part C, 37, 42-72 (2013).
9. E. Welsh, P. Murphy and P. Frantz, International Conferece on Fundamentals of Electronics Communications and Computer Sciences (ICFS).
10. C. Bakula, W. H. Schneider IV and J. Roth, J. Trans. Eng. 138, 43-49 (2011).
11. Y. Malinovskiy, N. Saunier and Y. Wang, Transportation Research Record, 137-149 (2012).
12. N. Abedi, A. Bhaskar and E. Chung *36th* ATRF (2013).
13. S. Moghaddam and B. Hellinga, Transportation Research Record, 2395 111-122 (2013).
14. L. K. Muthén and B. O. Muthén, Structural Equation Modeling 9, 599-620 (2002).
15. S. Zhu and L. Ferreira, Transportation Research Part D 21, 19-25 (2013).
16. CISCO Omni Antenna vs. Directional Antenna http://www.cisco.com/c/en/us/support/docs/wireless-mobility/wireless-lan-wlan/82068-omni-vs-direct.html (2007)

Irregular GIS Curve Fitting based High Speed Railway Earthquake Influence Range Calculation Model

Zhaobing Hu[1], Zhiming Liu[1] and Lianbao Yang[2]

[1]*China Academy of Railway Sciences, Signal & Communication Research Institute, 100081 Beijing, China*
[2]*China Academy of Railway Sciences, Railway Big Data Research and Application Innovation Center, 100081 Beijing, China*

Abstract. In this paper, to guarantee that the train can take measures to reduce the damage caused by the earthquake, it propose an irregular GIS curve fitting based high-speed railway earthquake influence range calculation model. Firstly, this model eliminates the abnormal points, calculates feature points and finds demarcation points of the high-speed railway GIS curve to get the processed point collection in Mercator coordinate. Secondly, though using the processed point collection, this model applies least square polynomial segmentation fitting method to implement complex high-speed GIS curve fitting. Thirdly, calculate the earthquake influence rang on high-seed railway line, according to the scope of the earthquake equation and the high-speed railway GIS curve fitted equation. Finally, the paper selects the Beijing South to Dezhou East high-speed railway section which is part of Beijing-Shanghai line as a case study, which proves that the model can calculate the earthquake influence scope on the railway line offering decision support for train operation to ensure safety.

1 Introduction

By the end of 2016, China's total mileage of high-speed railway was more than 20,000 kilometers [1, 2], accounted for over 60% of the world's total mileage of high speed railway, which ranked first in the world. China is located in the earthquake area which is vulnerable to earthquake disasters [3]. When the earthquake occurs, within its influence scope, high-speed railway line will vibrate and distort as the ground, leading to the trains jumping up and down, which will cause a devastating disaster [4, 5].

At present, the discretization of high-speed railway lines' GIS curve mainly through the direct method and the indirect method for the data acquisition [6]. Both ways use a finite number of coordinate points to characterize the railway line but without a mathematical equation to represent. The main process of curve discretization of high-speed railway lines is easy to produce abnormal points and lose feature points, which are caused by the system error or random selection of high-speed railway GIS curve points, leading to the coordinate points deviate from the correct position gravely. How to eliminate abnormal points, calculate feature points and on this basis to achieve the high-speed GIS curve fitting is an important problem urgent to tackle [7, 8].

The high-speed railway lines are mainly composed of line segment and circular arc. In view of commonly used methods of irregular curve fitting [9, 10] and combined with the characteristics of the high-speed railway lines, this paper puts forward an irregular GIS curve fitting based high-speed railway earthquake influence range calculation model. The model mainly includes two parts. The first part is to use the least square polynomial fitting method [11] to get the mathematical model of GIS curve including eliminate abnormal points, calculate feature points, find demarcation points. The second part is based on the mathematical model in combination with influence scope of earthquake equation to get the earthquake influence range of the high-speed railway line. As shown in the Fig 1.

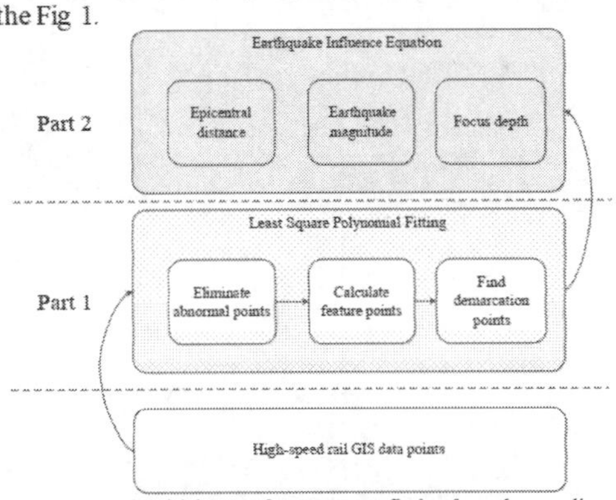

Figure 1. Irregular high-speed GIS curve fitting based on radius of high speed railway earthquake calculation mode

Finally, this paper uses Beijing South to Dezhou East high speed railway section which is part of the Beijing-Shanghai high-speed railway line as a case to implement the proposed model. The R-square and Adjusted R-square

are all above 0.99, shows the excellent goodness of fit by the least square polynomial fitting method proposed above. Moreover though simulate the earthquake happens and the curve fitted, this paper validates the correctness of the high-speed railway earthquake influence range calculation model.

2 Irregular high-speed lines GIS curve fitting

Irregular GIS curve fitting of high-speed railway line is to select the appropriate type of fitting method to the coordinate points. Curve fitting model consists of linear and nonlinear fitting and the polynomial fitting belongs to the linear fitting. Exponential function and trigonometric function belong to nonlinear fitting. In base of weather the fitting method can generate parameter equation or not, curve fitting, can also be divided into the parameter fitting and nonparametric fitting. The parameter fitting is mainly based on the least square fitting. Nonparametric fitting mainly use interpolation method such as cubic spline interpolation which can pass all data points and smoothness is better, but it is difficult to get the parameter equation. High-speed railway line is mainly made up of line segments and the circular arc with the large radius. Meanwhile the polynomial fitting can adapt to the characteristics of high-speed railway line. Therefore, in order to get parameters equation model, and guarantee the good fitting effect, this article adopts the method of least square polynomial fitting.

For the discretization of high-speed railway lines GIS curve, the discretization coordinate points may have abnormal points and redundancy points, which serious influence the goodness of curve fitting. Thus the first step of curve fitting is to eliminate abnormal points and reduce the redundancy points. To the complex high-speed lines curve also need to extract the boundary point. This chapter will introduce each part of the process in detail.

2.1 Eliminate abnormal points

Abnormal points refer to the points of GIS curve deviates from the position gravely. The abnormal point elimination is according to the distance ratio as shown in Fig 2.

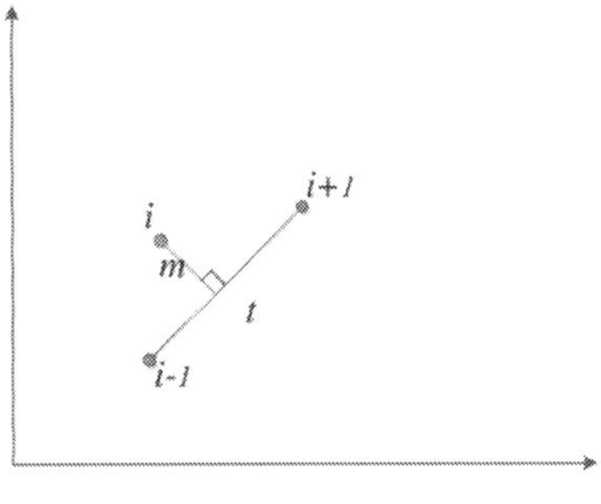

Figure 2. The distance ratio to elimate abnormal points

Among them, m represents the point i to point $i-1$ and point $i+1$ straight line distance, t represents the distance between point $i-1$ and point $i+1$. The method to calculate the distance is shown in (1)-(3).

$$m=\sqrt{\frac{\left((x_i-x_{i-1})(y_{i+1}-y_{i-1})-(y_i-y_{i-1})(x_{i+1}-x_{i-1})\right)^2}{(x_{i+1}-x_{i-1})^2+(y_{i+1}-y_{i-1})^2}} \quad (1)$$

$$t=\sqrt{(x_{i+1}-x_{i-1})^2+(y_{i+1}-y_{i-1})^2}. \quad (2)$$

$$l=\frac{m}{t}. \quad (3)$$

Generally l value of GIS curve is small, but the l value of abnormal point will increase significantly.

2.2 Calculate feature points

Feature points are generally recognized as the major points of the curve, which can describe the curve effectively. The method of selecting feature points mainly consists of angle test method and a polygon approximation method. This paper adopts an improved feature extraction method of vector curves, which is more simple and accurate than Douglas-Peucker algorithm [12].

Suppose the points of the high-speed railway GIS line are $A_0(x_0,y_0), A_1(x_1,y_1),\cdots,A_m(x_m,y_m)$. The slope of $A_{i-1}A_i$ is as follows.

$$k_{i-1,i}=(y_i-y_{i-1})/(x_i-x_{i-1}) \quad (4)$$

The slope of A_iA_{i+1} is as follows.

$$k_{i,i+1}=(y_{i+1}-y_i)/(x_{i+1}-x_i) \quad (5)$$

Set the threshold ε be a small value, if the

$$\left|k_{i,i+1}-k_{i-1,i}\right|\le\varepsilon \quad (6)$$

Which means that the point i and point $i+1$ slope is very similar, the point A_i is abandoned then. Otherwise the point A_i is selected as the feature point.

The selection of feature points is mainly evaluated by the data compression and fidelity, which is measured by compression ratio and integral square error commonly. The compression ratio is calculated as (7).

$$CR=\frac{n_1}{n_2} \quad (7)$$

n_1 is the number of original data points and n_2 is the resulting feature points. ISE is the integral error caused by the approximation curve, calculated by (8).

$$ISE = \sum_{i=2}^{n_2-1} m^2 \qquad (8)$$

2.3 Find demarcation points

Demarcation points represent the key position points of the curve, such as turning point. The demarcation points are mainly to select the largest slope point among the feature points.

2.4 The least square polynomial fitting

The least square polynomial fitting is mainly according to the given set of data points in the space of polynomial:

$$\phi = span\{1, x, x^2, ..., x^n\} \qquad (9)$$

To confirm $c_0, c_1, c_2, ..., c_n$, get the math model:

$$\varphi(x) = \sum_{k=0}^{n} c_k x^k, \forall c_k \in R \qquad (10)$$

Guarantee it the minimum square error:

$$\sum_{i=0}^{m} (y_i - \varphi(x_i))^2 = \min \qquad (11)$$

Let

$$Y = (y_0, y_1, ..., y_m) \qquad (12)$$

$$C = (c_0, c_1, ..., c_{n+1}) \qquad (13)$$

And

$$A = \begin{bmatrix} 1 & x_0 & \cdots & x_0^n \\ 1 & x_1 & \cdots & x_1^n \\ \vdots & \vdots & \vdots & \vdots \\ 1 & x_m^n & \cdots & x_m^n \end{bmatrix}_{(m+1)\times(n+1)} \qquad (14)$$

So (11) can transfer to:

$$\|Y - AC\|_2^2 = \min \qquad (15)$$

After get the (15) partial derivatives and let the gradient extreme value to be 0 then it changes to:

$$A^T AC = A^T Y \qquad (16)$$

Then by matrix operation:

$$C = (A^T A)^{-1} A^T Y \qquad (17)$$

Above all, the equation of high-speed railway line under the least square polynomial method is obtained.

3 High speed railway earthquake influence range calculation model

At the time of the earthquake happens, it produces P wave and the S wave, which is also known as the longitudinal wave and transverse wave. The P wave's vibration direction is parallel to the direction of earthquake and its transmission speed is very fast. However the S wave's vibration direction is vertical to the direction of earthquake and its transmission speed is very slow. P wave is more destructive than the S wave. High-speed railway earthquake monitoring and early warning system is based on the P wave and S wave velocity difference to realize earthquake early warning. The principle of high-speed railway earthquake monitoring and early warning is shown in Fig 3.

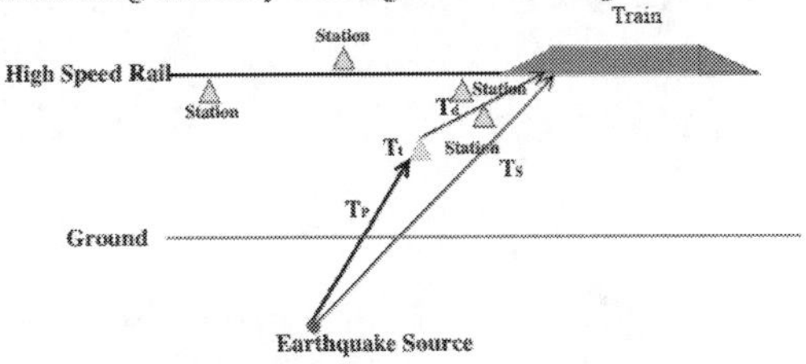

Figure 3. Principle of high-speed railway earthquake monitoring and early warning

T_s, T_p, T_d represents S wave, P wave and electromagnetic wave transmission time respectively, T_t refers to the stations identify seismic waves and start to send electromagnetic waves time. In order to make the earthquake early warning be efficient, the data processing time and the sum of T_s, T_p, T_t must be as small as possible. When T_s is greater than the sum of T_s, T_p, T_t the early warning and effective and the greater the better.

When the earthquake happens, it can be represented an equation of circle, with epicenter as the center and the impact scope as radius. In combine with the fitted curve function, the influence scope of earthquake on high-speed railway line can be transformed into the problem of solving the equation as shown in (18).

$$\begin{cases} \varphi(x) = \sum_{k=0}^{n} c_k x^k \\ E(x) = (x - x_e)^2 + (y - y_e)^2 = R^2 \\ x_{\min} \le x \le x_{\max} \\ y_{\min} \le y \le y_{\max} \end{cases} \qquad (18)$$

$\varphi(x)$ is the fitted high-speed railway GIS, line equation $E(x)$ is the earthquake influence range equation. R is the epicentral distance, which can be calculated according to the parameters such as magnitude and focal depth etc. $(x_{\min}, y_{\min})$ is the minimum coordinate value of high-speed railway GIS line and $(x_{\max}, y_{\max})$ is the maximum. When calculate the intersection points, then the earthquake specific influence range on the line should be accurately calculated.

4 Case study

To validate the proposed model, this paper selects Beijing South to Dezhou East high-speed railway section which is part of the Beijing-Shanghai high-speed railway line as a test case. At the same time, the railway line is through the earthquake zone as shown in Fig 4.

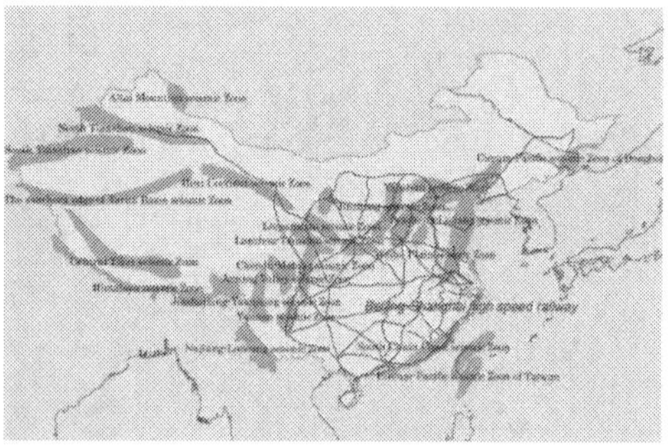

Figure 4. Principle of high-speed railway earthquake monitoring and early warning

After converting it to the ArcGIS Mercator plane coordinate, there has 5788 sampling points and the gis curve line in is shown as the Fig 5.

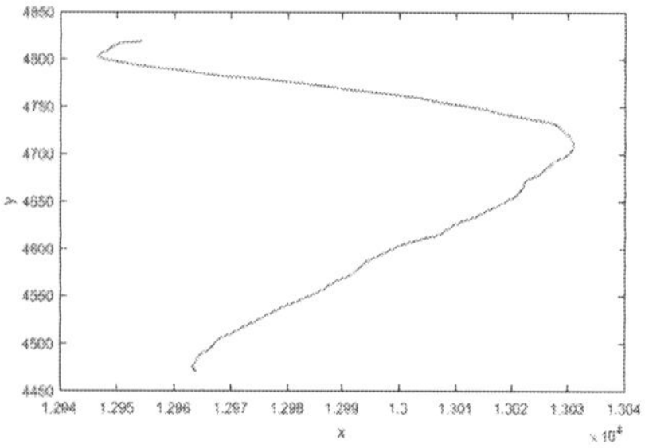

Figure 5. Beijing South-Dezhou East high-speed railway gis curve in mercator coordinate

4.1 Eliminate abnormal points

According to the 2.1's method, calculate the l value of each point and get the histogram figure as shown in Fig 6.

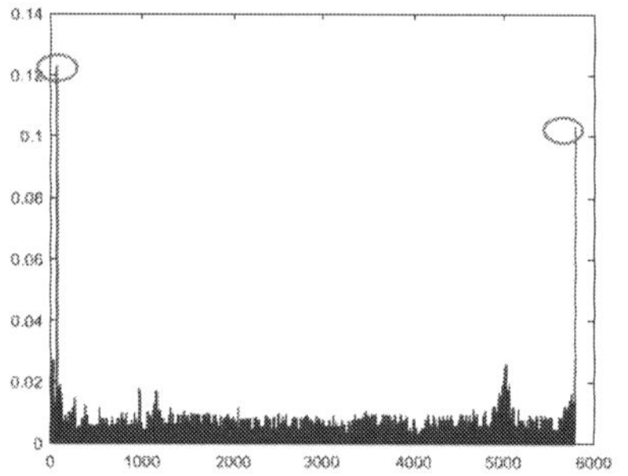

Figure 6. The l value of Beijing South-Dezhou East high-speed railway gis curve

There are two points' l value are bigger than others in Fig 6, which are abnormal points that need to eliminate. The Mercator coordinate of the abnormal points are (12952.26988,12952.26988) and (12963.85045, 12963.85045).

4.2 Calculate feature points

According to the 2.2's method, calculate the absolute slope value of each point as shown in the Fig 7.

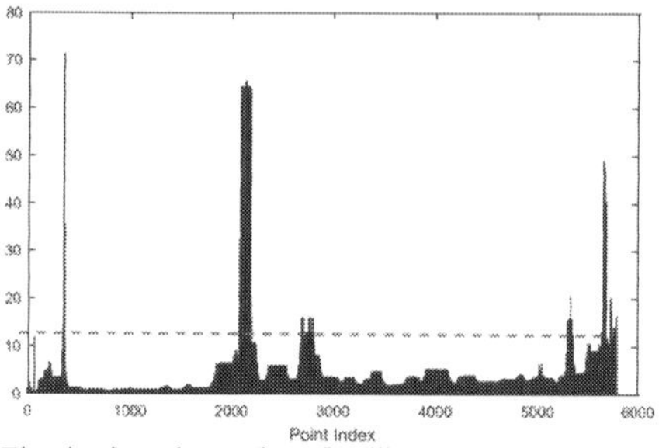

Figure 7. The absolute slope value of Beijing South-Dezhou East high-speed railway gis curve

The absolute slope values of the coordinate points are mainly lower than 10, so choose the threshold l > 10 and extract a total of 391 feature points. Moreover, the feature points can represent the original 5676 points with less distortion as shown in Fig 8.

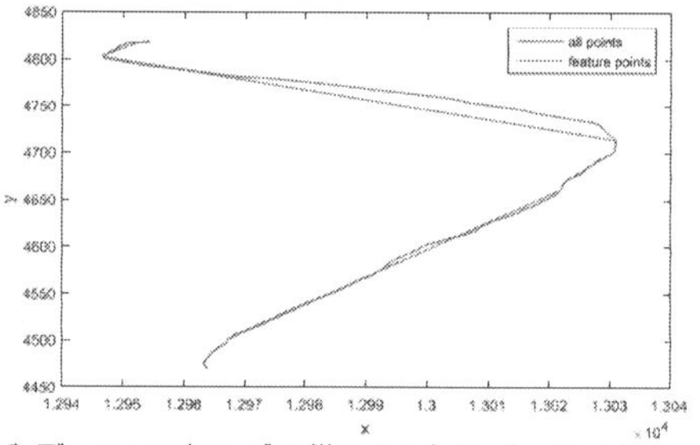

Figure 8. The comparison of Beijing South-Dezhou East high-speed railway gis curve by feature points and all points

4.3 Find demarcation points

According to the 2.3's method, choose the threshold l > 60 and extract the demarcation points as shown in Fig 9.

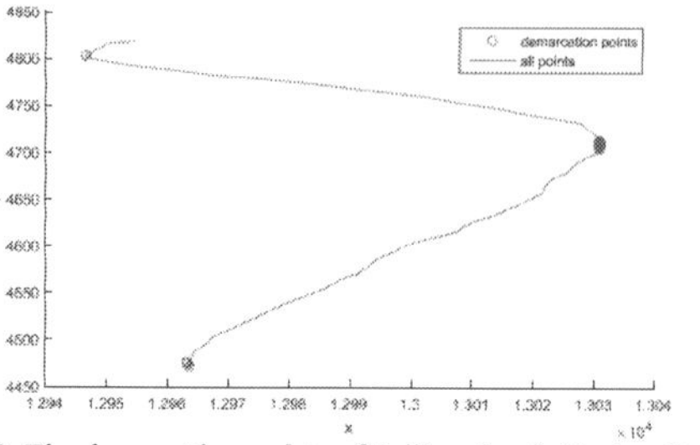

Figure 9. The demarcation points of Beijing South-Dezhou East high-speed railway gis curve

DOI:10.1051/matecconf/201712402001

4.4 The least square polynomial fitting

For the processed Mercator coordinate points set of Beijing South-Dezhou East high-speed railway GIS curve, adopt the least square polynomial fitting, which cannot obtain good fitting effect. The fitted figure is shown in Fig 10 (a) and the fitted polynomial equation is (19).

$$\varphi(x) = 44.59x^7 + 4.409x^6 - 195.5x^5 + 42.72x^4 + 201.2x^3 - 67.7x^2 + 18.92x + 4638 \quad (19)$$

By analyzing the parameters of fitting effect, such as the SSE, R square , Adjusted R square and RMSE, which are shown in Table 1. The R square and Adjusted R square both are about 0.46 ,which are far away from 1.So it's necessary to use the demarcation points calculated above to do piecewise fitting. The fitted figure result is shown in Fig 10 (b-d) and the fitted polynomial equation is (20).The R-square is nealy 1and SSE is very sall,which are shown in Table 1.

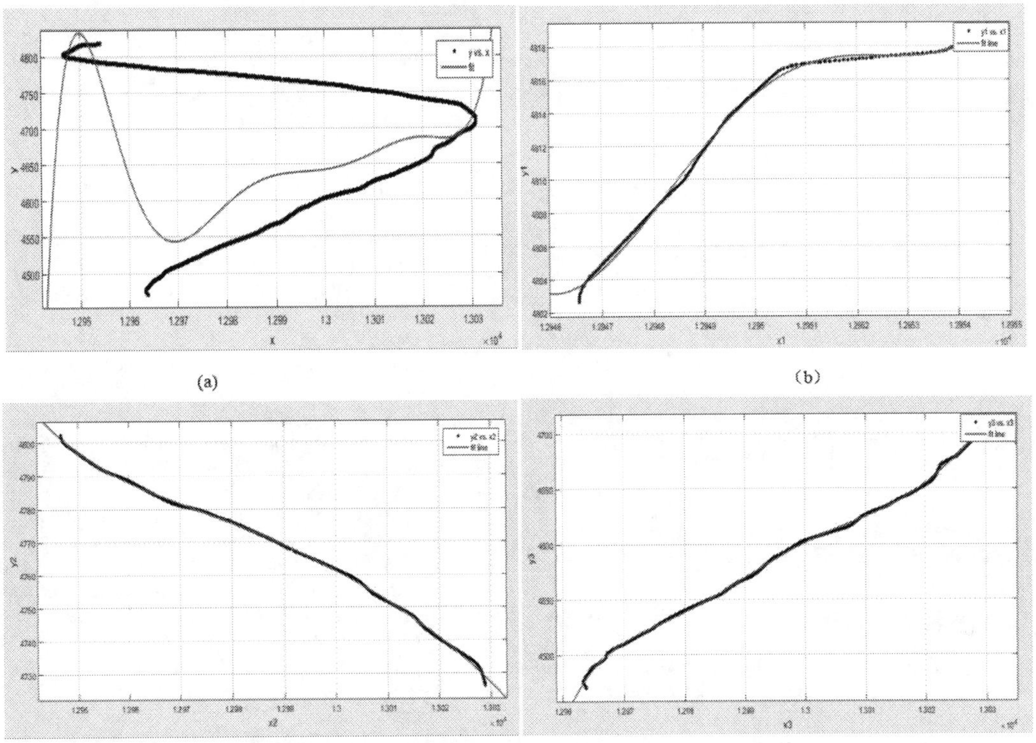

(a) (b)

(c) (d)

Figure 10. Beijing South-Dezhou East high-speed railway gis curve fitting results

Table 1. Goodness of fit by direct fitting and piecewise fitting

Fit Method	SSE	R-square	Adjusted R-square	RMSE
Direct fitting	34510000	0.4634	0.4627	77.28
Piecewise fitting	21.38	0.9973	0.9973	0.2486
	406	0.9994	0.9994	0.513
	10380	0.9994	0.9994	1.7

$$\varphi(x) = \begin{cases} 0.579x^4 - 0.6665x^3 - 3.087x^2 + 6.151x + 4814, 12946.5614196 \le x \le 12954.3304068 \\ -0.3694x^5 + 0.4256x^4 - 0.6263x^3 - 4.045x^2 - 18.09x + 4769, 12946.5614196 \le x \le 13028.842108 \\ -0.853x^8 - 1.867x^7 - 2.29x^6 + 15.41x^5 + 15.15x^4 - 22.67x^3 - 16.14x^2 + 70.55x + 4586, 12963.8816191999 \le x \le 13028.842108 \end{cases} \quad (20)$$

4.5 Earthquake influence range calculation

This paper simulates a earthquake that happens with the magnitude of 5.9 and the epicentral distance of 63 KM. By using the proposed model, the influence range of the earthquake on the high-speed railway line was obtained and as shown in Fig 11.

5 Conclusion

The model proposed by this paper still need to be improved to reduce the earthquake damage to the high-speed railway. The future work of this paper is to simplify the equation of the high-speed railway line and improve the efficiency of the calculation speed to quickly get the influence scope of earthquake on the high-speed railway line.

Also, trying to use the big data and cloud computing technology to improve the efficiency of earthquake early warning computing is also urgent to study.

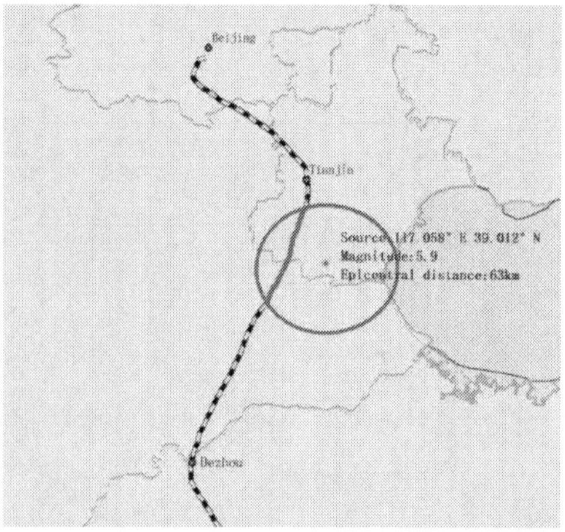

Figure 11 Influence range of the earthquake on the Beijing-Shanghai East high-speed railway gis curve

Acknowledgement

The authors would like to thank the support of National Natural Science Foundation of China under Grant no. U1534202.

References

1. K. Takagi, "Development of high-speed railways in China". Japan Railway & Transport Review, (2011)
2. A. Ryder, "High speed rail", Journal of Transport Geography, **22**, pp. 303-305,(2012)
3. L. Huang, W.L. Bao, K. Chen, "The comparison analysis of Chinese public perception of earthquakes on different time scales", Nat. Hazards, 73 (2),pp. 613–62,(2014)
4. Z. Jin, S. Pei, X. Li, H. Liu, S. Qiang, "Effect of vertical ground motion on earthquake-induced derailment of railway vehicles over simply-supported bridges", Journal of Sound and Vibration, vol. **383**, pp. 277-294,(2016)
5. J. Koseki, M. Koda, S. Matsuo, H. Takasaki, T. Fujiwara, "Damage to railway earth structures and foundations caused by the 2011 off the Pacific Coast of Tohoku Earthquake", Soils and Foundations, vol. **52**, pp. 872-889 ,(2012)
6. P. Bo, G. Luo, K. Wang, "A graph-based method for fitting planar B-spline curves with intersections", Journal of Computational Design and Engineering, vol. **3**(1), pp. 14-23,(2016)
7. J. Wang, L. Feng, "Curve-fitting models for fossil fuel production forecasting: Key influence factors, Journal of Natural Gas Science and Engineering", vol. **32**, pp. 138-149,(2016)
8. Diogo E.V. Andrade, Rubens R. Fernandes, Tainan G.M. Santos, Emanuel Vitor Ceccon, Ana C.B. da Cruz, Admilson T. Franco, Cezar O.R. Negrão, "Curve-fitting equation for prediction of the start-up stress overshoot of an oil-based drilling fluid", Journal of Petroleum Science and Engineering, vol. **146**, pp. 902-908, (2016)
9. S. Biswas, D. Ghoshal, R. Hazra, "A new algorithm of image segmentation using curve fitting based higher order polynomial smoothing", Optik - International Journal for Light and Electron Optics, vol. **127**, pp. 8916-8925, (2016)
10. M. Marji, P. Siy, "A new algorithm for dominant points detection and polygonization of digital curves", Pattern Recognition, vol. **36**, pp. 2239-2251, (2003)
11. A. Björck, "Numerical Methods for Least Squares Problems", SIAM, Philadelphia, (1996)
12. K. Ebisch, "A correction to the Douglas–Peucker line generalization algorithm", Computers & Geosciences, vol. **28**, pp. 995-997, (2002)

DOI:10.1051/matecconf/201712402002

Experimental Research on Shear Ductility of Regional Confined Concrete Beam

Qinggui Wu[1] Xinming Cao[1] Guyue Guo[1]

[1]*College of Civil Engineering, Guizhou University ,Guiyang 550025, China*

Abstract. In efforts to study the shear ductility of regional confined concrete beam, 5 reinforced concrete beams were tested to examine its shear performance. These beams has the same shear span ratio, concrete strength, different ratios of tension reinforcement and shapes of stirrup. The purpose of the test is studying the effects of stirrup shape and tension reinforcement ratio on failure mode and shear ductility. The test shows that the regional confined part can be used as an independent part and the rest of the beam is good to work together so that the ductility of the beam is more one time higher than that of the normal confined concrete beam. The related laws of the effect of tension reinforcement ratio and stirrup shapes on beam's shear ductility were founded

1. Introduction

Reginal Confined Concrete (RCC) is a kind of confined concrete conception and was put forward by Cao xinming (2013). RCC has been developed into relative mature theory. It has been employed into many engineering practice and brought about great economical merits.

A measure of the ductility of a structure is the displacement ductility coefficient. It defined as the lateral deflection at the end of the post-elastic range divided by the lateral deflection when yield is first reached (park 1975). Ahmad Sh (1994) has defined the shear ductility coefficient by imitating the analysis method of normal section ductility .Yi Weijian (2009) has done the research on regularity of shear ductility of normal confined concrete（NCC）beam. In his work, the experimental data shows that the shear ductility of NCC beams reduces along with the increase of tension reinforcement ratio.

NCC beam's bearing capacity is often improved by increasing the tension reinforcement ratio while its section size and concrete strength cannot be increased. Unfortunately, the high ratio of tension reinforcement leads to the low ductility of NCC beam. The concrete of NCC beam in compression zone is broken before the tension reinforcement reaching its yield. This failure mode of beam is forbidden appearing in engineering.

However, as is shown in Fig.1, the reinforcement is putted in compression zone of RCC beams and the ratio of tension reinforcement is increased. In this way, the reinforcement concrete beam's bearing capacity is improved. But there no data to show whether the shear ductility of RCC beams decreases along with the growth of longitudinal reinforcement ratio. In effort to study the shear ductility of RCC beam the monotonic static load test of 4 RCC beams and 1 NCC beam was carried out.

2. Experimental Instruction

2.1 Specimens

Experimental beams' cross section size is 150mm×300mm and are 1500mm long. Concrete strength is 40MPa in the test for concentric compression. The yielding stresses and ultimate stresses of longitudinal reinforcement, which's diameter is 10mm, are 530MPa and 660 MPa respectively. The yielding stresses and ultimate stresses of longitudinal reinforcement, which's diameter is 22mm, are 465MPa and 632MPa respectively. The yielding stresses and ultimate stresses of longitudinal reinforcement, which's diameter is 25mm, are 442MPaand 597 respectively. And the yielding stresses and ultimate stresses of stirrup, which's diameter is 8mm, are 490MPa and 614 respectively. Details of specimen are listed in the table 1 and figure 1.

2.2 Test procedure

The test was taken at structural laboratory of Fourth Engineering Company of China. Beams were loaded on Counter force test bench using 1000KN oil jack. The loading equipment is shown in figure 2. And the loading procedure is shown in figure 3.

2.3 Arrangement of measuring points

In effort to study the working performance of beams, experimental beams' deflection are measured by 5 displacement sensors. They are putted on two beam bearings, loading points and the middle of beam. Strain measuring points is arranged in the region where the inclined crack is predicted to expand. Mutual vertical strain gauges on the surface of beams is used to measure the concrete strain. Experimental beam displacement and strain measuring points are shown in figure 4.

Table 1. Specimen parameter

Specimen	Concrete strength	Ratio of tension reinforcement ρ s/%	Ratio of stirrups ρ v/%	Shear span ratio
NCC-1	54.42	5.39	0.021	1.23
RCC-1a	55.42	5.39	0.021	1.23
RCC-2a	56.42	4.78	0.021	1.23
RCC-1b	57.42	5.39	0.021	1.23
RCC-2b	58.42	4.78	0.021	1.23

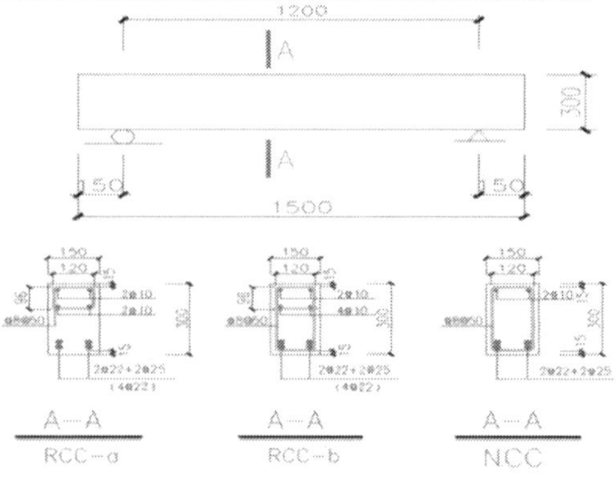

Figure 1. Dimension of specimen and configuration of reinforcement

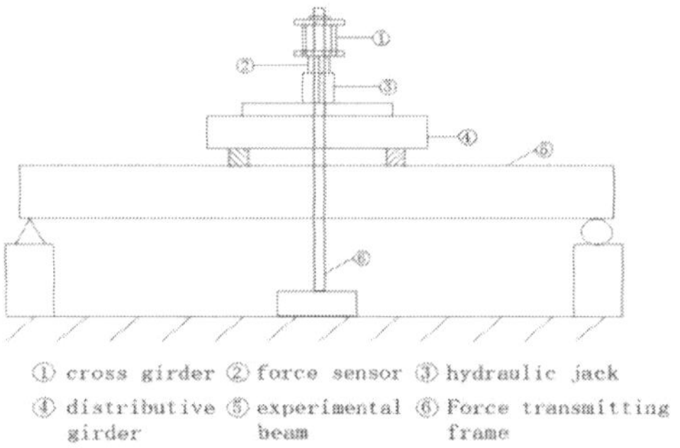

Figure 2. Test setup

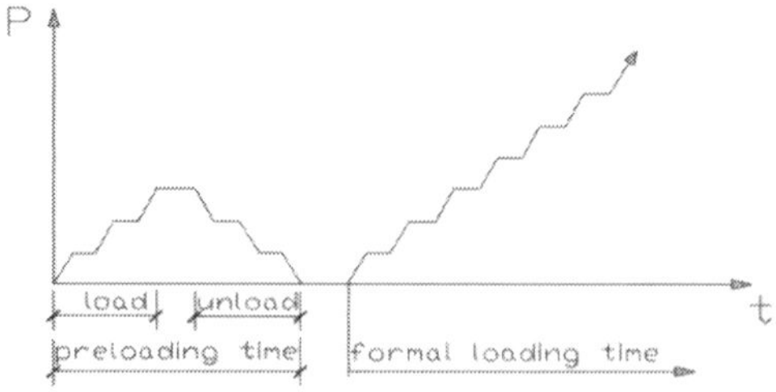

Figure 3. Loading history

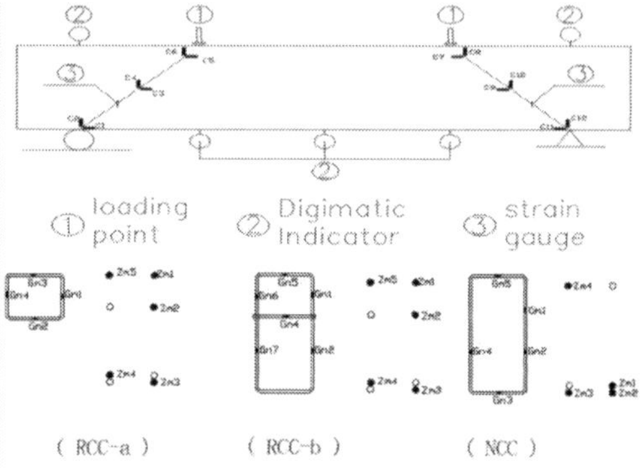

Figure 4. Arrangement of Strain gauges and displacement sensors

3. Experimental phenomena and analysis of load-deflection curve

The load-deflection curve of RCC-1a and RCC-1b beams can also be divided into three stages: elastic stage, elastoplastic stage and failure stage. But the load-deflection curve of RCC beams was different from NCC beam. In the elastic plastic stage, there were three different slope of the curve. The three different slopes of the curve can be seen in the load-deflection of RCC-1b, and RCC-2b beam (Fig.6). The failure stage of NCC beams showed good ductility, as shown in Fig.6.

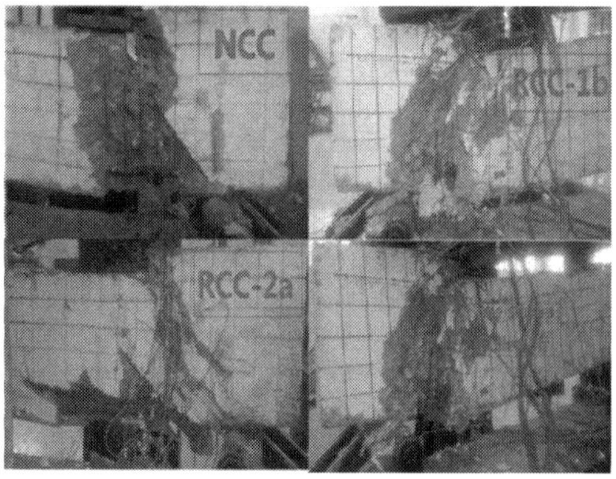

Figure 5. Failure modes for beams

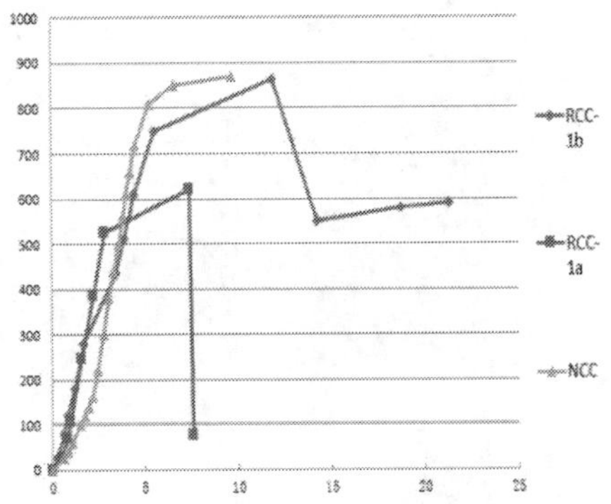

Figure 6. Load-deflection of beams with different stirrup shapes

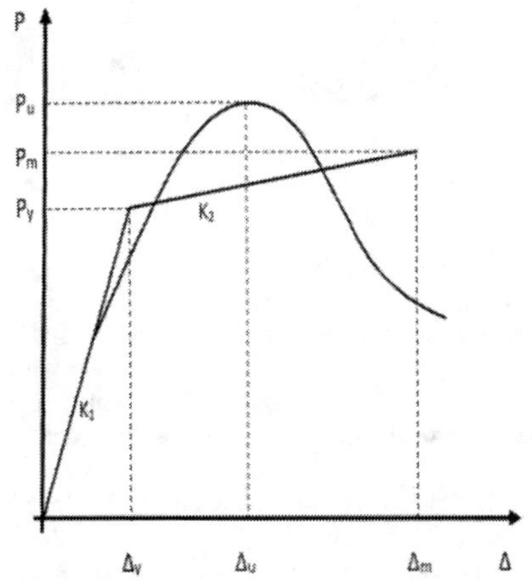

Figure 7. Equivalent method of equivalent shear ductilitycoefficient

4. Analysis of Shearing Ductility Coefficient

T According to the analysis method of the ductility of flexural test beam (Feng 2005), the ductility of shearing test is calculated. Based on the energy equivalent principle, the load-deflection curve is simplified into two linear load-deflection curve. The simplified load-deflection curve has been showed in figure 7. In equation form(weijian 2009), it is:

$\mu=\Delta m/\Delta y$ (1) $E = (P_y^2/2K_1) + 1/2(P_y + P_m)(\Delta m- Py/K_1)$ (2)

$P_y=K_1\Delta_y$ (3)

$P_m=K_1 (\Delta m-\Delta_y)$ (4)

$K2=0.05K_1, Py=0.75P_u$

E—the area of the load deflection and the transverse coordinate

Δ_y—Simplified yield displacement

Δ_m—Simplified limit displacement

P_y—Simplified yield load

P_m—Simplified ultimate load

P_u—Ultimate load

K_1—Simplified initial stiffness

K_2—Simplified degenerated stiffness

By the Eqns 1, 2 , 3 and 4, the shear ductility coefficient (μ) can be obtained. Shearing ductility coefficient (μ) should be standardized to μ, which's load-deflection curve has significant level line and be listed in Table 2.

Table 2. Equivalent shear ductility coefficient of specimen

Specimen	Ratio of tension reinforcement ρs/%	Equivalent ductility coefficient μ	Calibration ductility coefficient μ'
RCC-1a	5.4	4.75	2.63
RCC-2a	4.8	7.05	3.27
RCC-1b	5.4	7.43	3.81
RCC-2b	4.8	4.11	1.96
NCC	5.4	3.66	1.83

5. Effect of tension reinforcement ratio on shear ductility coefficient

The shear ductility coefficient of the experimental beam changes with the variation of longitudinal reinforcement ratio (Table 2). It can be obtained that the shear ductility coefficient of RCC-a beam decreased with the increase of tension reinforcement ratio. The tension reinforcement ratio increased from 4.8% to 5.4%, the equivalent ductility coefficient (μ) decreased from 7.05 to 4.75. It Has reduced by 32.6%.The calibration ductility coefficient (μ') reduced From 3.27% to 2.63% and reduced by 19.6% (Fig.8). The reason is that with the seam shape of the stirrup, the greater the tension reinforcement ratio, the stiffness of the RCC-a beams increases. That leads to the reduction of rotation capacity of the experimental beam. So the RCC-a beam has poor ductility. For the influence of the tension reinforcement ratio on the shear ductility, RCC-a beam has the seam regular with NCC beam.

From Figure 8, the shear ductility coefficient of RCC-b increases with the increase of the tension reinforcement ratio. The tension reinforcement ratio increases from 4.8% to 5.4%, and the equivalent ductility coefficient (μ) increased from 4.11 to 7.43. It increased by 32.6%. Calibration ductility coefficient (μ') increased from 1.96 to 3.81. It increased by 94.4%. And the ductility coefficient increased by nearly one time.

The effect of tension reinforcement ratio on shear ductility coefficient of RCC-b beams is Contrary to the phenomenon and results in the test of normal confined concrete beams (weijian 2009). The reason is that there is a regional confined area in RCC-b beam compared with the normal confined concrete beam. Due to the upper

 DOI: 10.1051/matecconf/201712402002

concrete of RCC beam strengthened by stirrup and longitudinal reinforcement, the upper concrete of RCC beam belongs to the well confined concrete. The redistribution of internal force is caused along with the emergence of plastic hinge in lower concrete. And the regional confined part has a collaborative work with the tension reinforcement. In that way, the shear ductility of reinforced concrete beam is improved. Therefore, it showed the effect of tension reinforcement ratio on the shear ductility coefficient is different with normal confined concrete beam. However, the RCC-a without stirrups throughout the beam's section, the tension reinforcement is not connected. The regional confined part and the lower longitudinal reinforcement cannot form a loaded body when the weak confined concrete destroyed, which leads to their collaborative work performance has not been fully played. Therefore, the law of influence of tension reinforcement ratio on shear ductility coefficient of RCC-b beam and RCC-a beam is different.

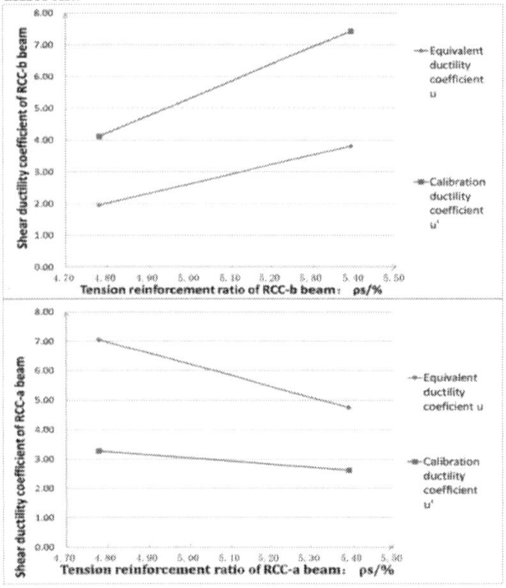

Figure 8. Influence of ρs on shear ducltility coefficient of RCC beam

6. Effect of stirrup shapes on shear ductility coefficient

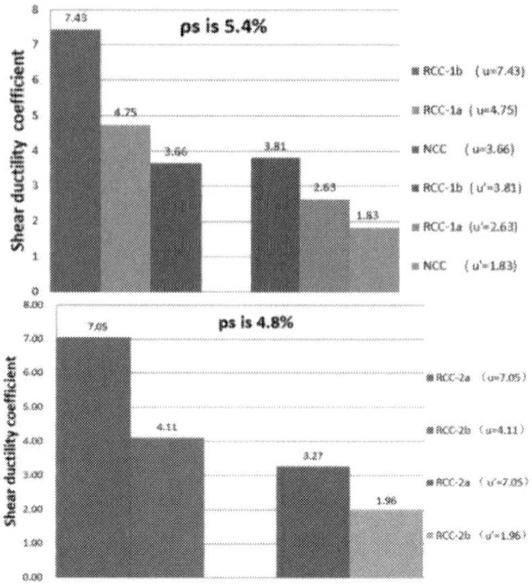

Figure 9. Effect of different stirrups on shear ductility coefficient of specimens

The tension reinforcement ratio in normal confined concrete beam is limited to 2.5% in the code (2004) because the high tension reinforcement ratio would lead to the low ductility of normal confined concrete beam. Fortunately, the RCC beams with high tension reinforcement ratio has desirable ductility (Fig.9). Therefore, the beam's size can be decreased to save building space with the application of RCC beam.

From figure 9, it can be seen with the tension reinforcement ratio 5.4%, equivalent shear ductility coefficient μ and calibration of the shear ductility coefficient μ' from NCC, RCC-a to RCC-b beam gradually increases and the increase rate is gradually increased. The shear ductility coefficient of RCC-1a beam is 30% higher than NCC beam's. What's more, the shear ductility coefficient of RCC-1b beam is 103% higher than NCC beams. The beam calibration shear ductility coefficient (1.96) does not meet the requirements of seismic ductility. However, the equivalent shear ductility coefficient and the calibrated ductility coefficient of two kinds of regional confined beams meet the seismic demand and have a certain surplus. The shear ductility of RCC-a beam is relative poor when the tension reinforcement ratio is 4.8%. The range of the optimal tension reinforcement ratio of the RCC beam needs to be further studied. The test data shows that deformation and energy dissipation capacity of RCC beam is better than that of the NCC beam. What's more, with the increase of tension reinforcement ratio, the deformation and energy dissipation capacity will be more prominent when compared with the NCC beam. Therefore, it can be inferred that NCC beam can be made certain contribution in seismic engineering.

DOI:10.1051/matecconf/201712402002

7. Conclusion

As a new type of reinforced concrete beam, Regional Confined Concrete beam is different from normal confined concrete beam in the mechanical properties and failure mode. Because of the change of confined mechanism, its mechanical properties can be improved.

(1) The shear ductility coefficient of RCC-a and RCC-b is higher than that of NCC beam. It proves that the seismic performance of regional confined concrete beam is more excellent than that of normal confined concrete beam

(2) The shear ductility coefficient of RCC-b beam increases along with the increase of tension reinforcement ratio. And it contradicts with that of NCC and RCC-a beam.

(3) Because of the experimental beams is not enough, the experimental data is not enough to draw the formula of the ductility and the optimum tension reinforcement ratio, which will be solved in the following research.

(4) With the application of RCC beam, the limit of tension reinforcement ratio in code can be broaden. And the size of reinforcement concrete beam may be decreased to save the costing of architecture safely.

(5) Experimental data proves that the regional confined part improves the reinforced concrete beam's brittle shear failure, the seismic performance and the energy dissipation capacity. The regional confined concrete will has broad application prospect in the construction of anti-seismic structure.

Acknowledgements

This work was carried out at structural laboratory of Fourth Engineering Company of China. And it was supported by Compilation of Code for Design of Regional Confined Concrete Structure of The Guizhou Local Standard (Qian Jian Ke Tong No. [2013]367). Besides, we would like to thank Weixue, Yan Kunhong, Zhou Jianying, Zou Qinwei and Zaolin for their help.

References

1. Buildings Department of the HKSAR Government. *Code of Practice for Structural Use of Concrete 2004*, Buildings Department, the HKSAR Government. Hong Kong(2004)
2. Cao, X. M., Huang, X. W., Mo, Z. G., & Tian, H. Y.. "Introduction of regional confined concrete", *Advanced Materials Research*, 671-674, 697-703(2013)
3. Feng, P., Lie-Ping, Y. E., & Huang, Y. L. "Deformability and new performance indices of flexural members", *Engineering Mechanics*, 22(6), 28-36 (2005)
4. Agency, F. E. M.. *Nehrp guidelines for the seismic rehabilitation of buildings (fema publication 273)*, Wbdg Org (1997)
5. Park, R., Paulay, T. *Reinforced Concrete Structures*, John Wiley &Sons Inc, p203-208 (1975).
6. Weijian, Y. I, & Yanmei, L. "Experimental study on shear behavior of high-strength concrete beams with high-strength stirrups", *Journal of Building Structures*, 30(4), 94-101 (2009).
7. Xie, Y., Ahmad, S. H., Yu, T., Hino, S., & W. Chung. "Shear ductility of reinforced concrete beams of normal and high-strength concrete" *Aci Structural Journal*, 91(2), 140-149 (1994)

Congestion Tolling for Mixed Urban Freight and Passenger Traffic

Chaoda Xie[1] and Xifu Wang[1]

[1] School of Traffic and Transportation, Beijing Jiaotong University, 100044 Beijing, China

Abstract. This paper investigates the welfare effects of optimal tolling on urban traffic congestion, in a bottleneck model, with mixed freight and passenger users. The users' marginal utility of time is considered to be varying with time. Under both no-toll equilibrium and socially optimal tolling, the users are found to sort their arrival time according to the increasing rates of marginal utility at the destination. The optimal toll that maximizes social welfare does not change each user's indirect utility relative to the no-toll equilibrium, but completely removes the queue, which also removes the barrier of freight carriers to accept congestion pricing by relating their marginal utilities directly to the toll. When the toll is equally rebated, the proposed social optimal tolling is a Pareto improvement relative to the no-toll equilibrium. Those more productive users also suffer more in both no-toll equilibrium and optimal tolling, which indicates that a differentiated redistribution of toll revenues could be an incentive to improve productivity.

1 Introduction

This paper focuses on the dynamics of traffic congestion induced by two distinct transportation demands, i.e. passengers and freights. Freight vehicles are evidently found to be competing with passenger cars for the capacity of urban roads. Empirical evidence [1] shows that the that freight traffic is a significant proportion of the total traffic volume at the major interstate crossings of the Port Authority of New York and New Jersey (PANYNJ), and the flow of freight and passenger traffic are overlapped. GPS data of Tokyo Metropolis Area [2] indicates that trucks are concentrating in the city center during rush hours. In a survey of Italian cities [3], the peak of freight traffic often overlaps with the peak of passenger traffic, during 8AM-11AM and 3PM-5PM, due to regulatory concerns. A report of Beijing, China, indicates that over 80% of freight traffic volume in terms of the number of trips occurs during daytime in the central area of the city, including 11.7% occurs during peak hours of passenger traffic [4].

Despite the existence of mixed freight and passenger traffic, the study of the dynamics of urban traffic congestion begins by treating the users as commuters only. As the pioneer work, [5] first depicts the scheduling preference of commuters as their travel cost when passing through a single bottleneck in the city center. The model assumes that commuters valued the travel time per se at a rate of α. They are supposed to arrive at an exogenous preferred arrival time (PAT), and are subjected to a scheduling cost at a rate of β per time unit for early arrival or γ per time unit for late arrival, which is well-knows as the $\alpha-\beta-\gamma$ specification. In light of this simple and tractable scheduling preference specification, the congestion dynamics with respect to the heterogeneity of passenger traffic has been extensively discussed. Limited number of user classes and continuously distributed user classes are considered by [6] and [7] respectively, but the ratio of β/γ is fixed for the purpose of analytic convenience. [8] discusses equilibrium properties by assuming a more general scheduling delay cost function, but the PAT remains exogenous. The specification of PAT with an indifference band has also been considered in a number of studies to allow for flexibility of arrival time (e.g. [9], [10]).

However, it is not generally true that the preferred arrival times can be exogenously treated and are unrelated with scheduling cost, since the PAT itself is ambiguous without the mechanism underlying the preference decision. The seminal work of [11] first connects the scheduling preference with the utilities generated by the activities conducted before and after the travel. This specification relates the PAT with the marginal utility of being at the base and the destination, which facilitates the analyze of PAT decision behavior as well as more general time-varying marginal utilities. The time-varying marginal utility has been empirically identified among commuters, as evidenced by [12–14]. By elaborating the model of [11], [15] and [16] notes that if the travel duration is fixed and is irrelevant to time, then the traveler will optimally choose to travel between the times when his marginal utility of being at the base equals his marginal utility of being at the destination. [17] provides a useful review on the connection between the time-varying marginal utility specification and the common α-β-γ preference. [18] further associates the agglomeration of co-workers with their marginal utilities, as well as commuters' scheduling preferences. However, none of these literatures deal with the heterogeneity of marginal utilities, and the type of user are also limited to commuter.

Freight carriers differs from the private commuters in at least two ways. First, they tend to use vehicles with lager capacity, collecting several goods for multiple receivers each trip to lower the marginal cost of delivery. Second, the fact that commuters travel time is usually not paid by their employers. Therefore, higher value of travel time per vehicle of freight users than that of private commuters has been empirically found [19]. On the other hand, a freight delivery trip to the city center often ends at an area where a number of receivers spatially close to each other. Due to the limited capacity of urban facilities and the heterogeneity of receivers' preferences, it is usually not possible for the receivers who locate at the city center arrive at their workplace simultaneously. As a result, the numbers of available receivers distribute unevenly over a specific period of time, which will cause the carrier's productivity to vary over time at the destination. This is exactly the same mechanism as the time-varying productivity of commuters with the number of working dependents [20]. It is therefore possible to treat the commuting and freight users by their heterogeneous marginal utilities.

Given these findings, it is now clear that the dynamics of urban road congestion involves both freight and commuting user who are heterogeneous users in terms of their marginal utilities, and that their marginal utilities is likely to vary over time. This naturally raise two questions: First, how the users with heterogeneous time-varying marginal utilities behave when queuing at a bottleneck, compared with homogeneous users? Second, if their behavior affects welfare distribution of the first-best pricing strategy?

To deal with these two questions, this paper introduces the user heterogeneity to the time-varying marginal utility model [11]. The scheduling preference of freight and commuting users is defined by a marginal utility function with a contentiously distributed increasing rate. This methodology is built in light of [16] that shows a way of analyzing the properties of congestion when the users' travel distances to the urban bottleneck are continuously distributed. By applying the model framework of [11], they find that users sort their arrival time at the bottleneck according to the travel distance to the bottleneck, and that the distant users tend to gain more in optimal tolling scheme. Our paper differs from [16] in two major respects. First, it concerns user heterogeneity in marginal utility of time, instead of travel distance. Second, the achieved utility function of user is parametrized, which allows for a continuously distributed rate of marginal utility. We thereby find users sort their arrival time at the bottleneck by the increasing rate of marginal utilities at the destination under both no-toll equilibrium and socially optimal tolling. The social optimal toll removes the queue completely and does not change social welfare when no toll revenue is rebated. The finding implies that equal redistribution to all users can be a Pareto improvement.

In the context of mixed freight and passenger traffic, extending the model of [16] by introducing the heterogeneity of marginal utility also relates with the attitude of freight carrier towards congestion charging. Since the carrier calculates the cost of delivery per trip instead of the number of receivers they serve, the congestion pricing policy with a time-dependent cordon toll, which is the most conventional way in practice, cannot enter the marginal utility of freight carrier. The toll, therefore, becomes a fixed cost and does not vary with the number of receivers per trip. When the charge rate of delivery has to be set to the marginal cost under a complete competitive market, the receivers cannot get any price signal from the cordon toll to choose less congested delivery times [21]. To cope with this fact, [22] proposes a time-distance pricing scheme to show an incentive of off-peak delivery to the carriers, which, however, not applicable when receivers are densely concentrated in the city center. In the setting of the present paper, the above problems are solved by setting the optimal toll to be varying with the amount of available receivers. Indeed, this paper does not intent to discuss the market equilibrium between the freight carriers and the receivers. Throughout the discussion, the charge rate to the receivers is assumed to be fixed. We also limit the freight traffic to light freight vehicles physical externalities (e.g. size and acceleration performance) are similar to private passenger vehicles.

In Section 2, the scheduling model of freight and passenger users is described in a unified formulation. Section 3 shows the conditions when the no-toll equilibrium uniquely exist. Then the socially optimal tolling that exactly removes the equilibrium queue and the corresponding welfare effects are discussed in Section 4. A numerical example is simulated in Section 5, where the distribution of productivity increasing rate is bimodal. The setting of distribution is to reflect the largely differentiated empirical value of travel time and schedule delay time estimated in freight and passenger users. Section 6 concludes the key findings and identifies the possible extension to be introduced. The proofs of propositions are all deferred to the Appendix B.

2 A unified model

Consider that N road users consisting of freight carriers and private commuters travel along a single urban corridor from the suburb to the city center. The freight carrier loads a batch of packages to be delivered to multiple receivers. Commuter each drives a car and travels alone. Before arriving at the destination, they travel without congestion first and then passing through a bottleneck with a constant capacity of φ vehicles per time unit. All users follow the FIFO(first-in-first-out) discipline. The freight carriers and commuters achieve their utility by spending their time with relating parties at their respective bases and the destination. Both freight carriers and commuters incur no further penalties other than the loss utility of time, when their arrivals at the destination delays. The agglomeration of their respective relating parties varies with time of day, therefore, at either ends of the travel, users' marginal utilities of time (MUT) are time-varying. The exogenous agglomeration magnitudes are assumed to be decreasing at their respective base and increasing at the destination. The receivers locate very closely in the city center, then the travel time between different receivers is ignored. The commuters also do not need additional travel time when interacting with their co-workers.

Both the carriers and commuters desire to make maximum use of their time at both the base and destination within an exogenously given time window, denoted by D, which can be interpreted as the traveling time and the two periods of time adjacent to it. The achieved utility within the time window is denoted by $u(d,t)$, where d and t are respectively the time when the travel starts and ends. Suppose that the users have completely no gain during the time spending on the move and could only achieve utility gain before d and after t. We consider that the road is allowed to be tolled by τ measured in utility unit, and that the toll is not returned to the user. Then an individual user can attain utility of $u(d,t)-\tau$ during D. Since the travel demand is assumed to be fixed, we then define a social welfare function as the total utilities of all users. All other utility losses are assumed to be constant and thus able to be ignored without affecting the result qualitatively.

The marginal utility of time at the base is assumed to be $h(s)=\beta_0+\beta_1 s$, and that at the destination to be $w(s)=\gamma_0+\gamma_1 s$. Then u can be described as

$$u(d,t) = \int_{T_1}^{d}(\beta_0+\beta_1 s)ds+\int_{t}^{T_2}(\gamma_0+\gamma_1 s)ds$$
$$= H(d)+W(t)$$

where $T_1, T_2 \in D$ are constants, $H(d)$ and $W(t)$ are the cumulative utility at the base and the destination respectively. We assume that $H'(d)=\beta_0+\beta_1 d>0, H''(d)<0$, and that $W'(t)=-\gamma_0-\gamma_1 t<0, W''(t)<0$. The achieved utility $u(d,t)$ of a single user is therefore strictly concave.

The users have scheduling preferences that are only heterogeneous in terms of marginal utility of time at the destination. The increasing rate of marginal utility at the destination γ_1 follows a distribution with density f and cumulative distribution F, both of which have support on $\Gamma=[\gamma_L,\gamma_H]$. With the assumptions on $H'(d)$ and $W'(t)$, there must be an intersection of the two marginal utility functions. We define the time when $H'(d)$ and $W'(t)$ intersect as the ideal arrival time(IAT), which is the optimal arrival time when the travel time being reduced to zero. Given we only consider γ_1 varies among users, their IATs are identical.

In order to examine the impacts of tolling on users with heterogeneous MUT, we assume that all the users have to travel the same T time units before arriving at the bottleneck. The arrival time at the bottleneck is denoted by a, and the arrival rate at the bottleneck by $\rho(a)$. When $\rho(a)$ is larger than φ, a vertical queue is generated at the entrance of bottleneck. Let a_0 and a_1 be the time when

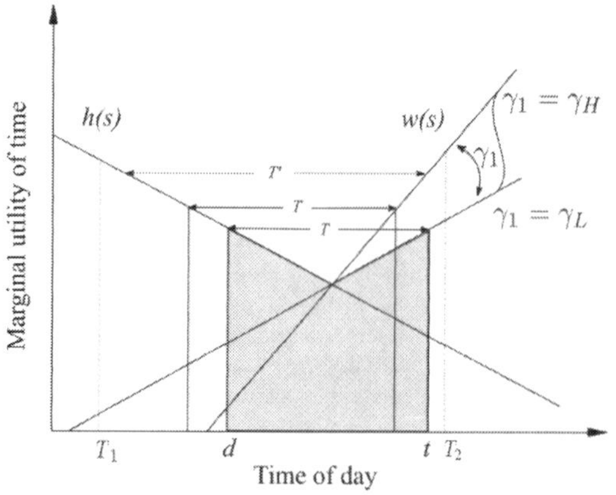

Figure 1. Graphical solution to the utility maximization problem of users with different marginal utility increasing rates

the first and the last user join the queue respectively, then the cumulative arrival $R(a)=\int_{a_0}^{a}\rho(s)da$, and the total travel time of the user who arrives at the bottleneck at time a will be $a_0+\frac{R(a)}{\varphi}-(a-T)$.

A single user with fixed γ_1 travels through the corridor and the bottleneck without congestion can achieve utility $u(a-T,a)$, where a is the time he arrives at the bottleneck as well as the destination. To maximize the utility, he will choose the optimal arrival time $a_*(\gamma_1)$ which solves the utility maximization problem arg $\max_a$ $u(a-T,a)$. A graphical solution to the maximization problem is illustrated in Figure 1. The shaded area gives the minimum utility loss when the user γ_L schedule his arrival time t at the optimal arrival time $a_*(\gamma_1)$. The optimal arrival time $a_*(\gamma_1)$ is found to have the following property.

Proposition 1 *If* $H'(d)=\beta_0+\beta_1 d>0, W'(t)=-\gamma_0-\gamma_1 t<0$ *and* $H''(d)<0, W''(t)<0$ *hold for all* $t,d\in D, d\le t$, *and if, but not necessarily if,* $\gamma_0-\beta_0\le 0$ *holds, then the optimal arrival time* $a_*(\gamma_1)$ *uniquely exists and* $a'_*(\gamma_1)<0$.

This proposition identifies the feature that travelers sort their optimal arrival times according to the increasing rates of their marginal utilities. The negative $a'_*(\gamma_1)$ indicates that the traveler whose MUT at the destination increases faster will prefer to arrive earlier when traveling without congestion.

In particular, a freight carrier is more likely to gain a higher increasing rate of marginal utility at the destination than a single commuter, owing to the dense agglomeration of payloads in freight vehicles. In the meanwhile, the receivers of the payload are not likely to arrive at the city center simultaneously. Therefore, the agglomeration process of receivers leads to an increase to the marginal utility of shippers. The proposition shows us that the higher increasing rate of marginal utility at the destination is one of the reason that drives carrier to arrive earlier at the bottleneck than commuters. [1] evidences that the peak hour for freight vehicles is typically one hour earlier than for all traffic in the morning with similar travel distance from suburbs to the urban area.

3 No-toll equilibrium

Consider each user schedules his travel by taking the decisions of all other users as given. The conditions when Nash equilibrium exists and the basic properties can be described by the following propositions.

Proposition 2 *Assume that* $a_*(\gamma_1) < \frac{N-F(\gamma_1)}{\varphi} + a_*(\gamma_H)$ *holds for all* $\gamma_1 \in \Gamma, \gamma_1 < \gamma_H$. *When users taking Nash equilibrium strategy, the queue begins and ends during an interval* $[a_0, a_1]$, *where* $a_1 = a_0 + N/\varphi$, *such that* $a_0 \le a_*(\gamma_H)$ *and* $a_1 \ge a_*(\gamma_L)$.

The condition ensures that if the user with the highest increasing rate of marginal utilities arrive the bottleneck at his optimal arrival time, all the other users with lower increasing rate of marginal utilities arrive later than their optimal arrival time. The queue which starts at a_0 and ends at a_1 always exists during $[a_0, a_1]$ where the first and last user experience exactly no queue. The first user arrives at the bottleneck earlier and the last arrives later than his optimal arrival time. This phenomenon can be visualized in Figure 1, when travel time increased from T to T'. We have so far shown the basic queueing property when the heterogeneity of marginal utility exists. The following proposition shows the properties of arrival time, and the analytical solution of scheduling utility.

Proposition 3 *Assume that* $a_*(\gamma_1) < \frac{N-F(\gamma_1)}{\varphi} + a_*(\gamma_H)$ *holds for all* $\gamma_1 \in \Gamma, \gamma_1 < \gamma_H$. *The no-toll equilibrium exists and is unique. Users who derive increasing marginal utility at rate* γ_1 *arrive at the bottleneck at time* $a(\gamma_1)$ *and its derivative satisfies*

$$a'(\gamma_1) = -\frac{f(\gamma_1)}{\varphi}\frac{\gamma_0 + \gamma_1\left(a_0 + \frac{N - F(\gamma_1)}{\varphi}\right)}{\beta_0 + \beta_1(a - T)} < 0. \quad (1)$$

The scheduling utility under equilibrium is

$$u\left(a - T, a_0 + \frac{N - F(\gamma_1)}{\varphi}\right)$$

The proposition shows that the arrival time a at the destination is a function of the increasing rate γ_1 of marginal utility at the destination under no-toll equilibrium. The derivative of a is strictly negative which indicates that those who gain higher increasing rate of marginal utility at the destination arrive earlier. Given the assumption of FIFO(first-in-first-out) discipline, the bottleneck preserves the arriving order at the entrance. The time of their arrival at the destination is also in the order of their γ_1 at the destination.

According to the arrival schedule, the first user who arrives at bottleneck as well as the destination at time $a(\gamma_H) = a_0$ has the highest increasing rate γ_H. The user derives an increasing rate γ_1 of his marginal utility at the destination arrives at the destination after $N - F(\gamma_1)$ users at the bottleneck at time $a_0 + (N - F(\gamma_1))/\varphi$, while those who gain lower increasing rate of the marginal utility arrive later. Lastly, the user with the lowest increasing rate γ_L of marginal utility joins the queue, and arrives at the destination at time $a_1 = a_0 + N/\varphi$.

4 Optimal tolling

In this section, we first find the socially optimal toll $\tau(a)$, then discuss the welfare effects by the optimal toll on users with different increasing rates of MUT.

Proposition 4 *A socially optimal toll exists if, but not necessarily if,* $a'_*(\gamma_1) < -\frac{f(\gamma_1)}{\varphi}$. *The arrival at the bottleneck begins at* $a_{\tau 0}$ *which solves*

$$0 = \int_{a_{\tau 0}}^{a_{\tau 0}+N/\varphi} [\beta_0 + \beta_1(a - T) - \gamma_0 - \gamma_1 a]\, da.$$

The optimal toll completely removes the queue, which satisfies

$$\tau'(a_\tau(\gamma_1)) = \beta_1(a_\tau(\gamma_1) - T) + \beta_0 - \gamma_1 a_\tau(\gamma_1) - \gamma_0,$$

and

$$\tau(a_{\tau 0}) = \tau(a_{\tau 1}).$$

$\tau(a_{\tau 0}) = 0$ *gives one of the optimal tolls. The users arrive the bottleneck and the destination simultaneously at*

$$a_\tau(\gamma_1) = \frac{N - F(\gamma_1)}{\varphi} + a_{\tau 0}.$$

The order of the arrival at the destination remain unchanged relative to no-toll equilibrium.

Infinitely number of optimal tolls can exist when the demand elasticity is ignored, where $\tau(a_{\tau_0}) = 0$ is one of them. The proof shows that the optimal toll keeps the sorting property of arrival under no-toll equilibrium unchanged. The formulation of arrival time at the destination is also identical to the one under no-toll equilibrium.

Along with the commuters, the carriers are also facing the optimal toll that varies with the increasing rate of MUT at the destination γ_1. The toll now enters the marginal cost when delivering an additional order to a receiver. It is then of great interest to discuss if anyone is getting better or worse off by tolling when this important barrier for freight users to accept the toll is removed. The next proposition also answers the question if the arrival interval is able to be shifted by the optimal toll.

Proposition 5 *With the conclusion of , consider that an socially optimal toll can be achieved by* $\tau(a_{\tau 0}) = \tau(a_{\tau 1}) = 0$. *The arrivals in social optimum are the same as that under no-toll equilibrium:* $a_\tau(\gamma_1) = a(\gamma_1)$. *Relative to the no-toll equilibrium, the socially optimal toll does not change each user's indirect utility, when the toll revenues are not redistributed to the users.*

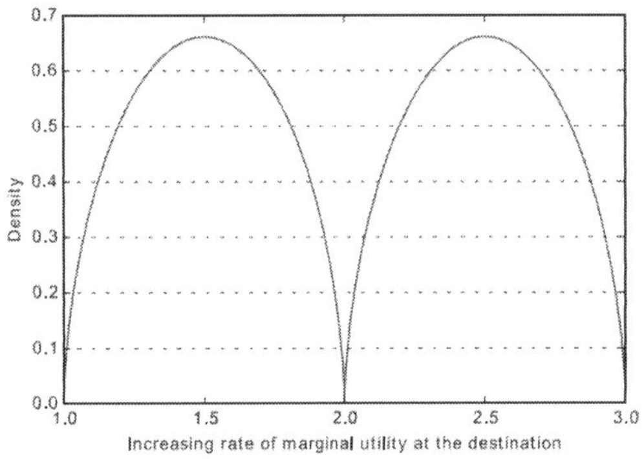

Figure 2. The density of increasing rates of marginal utility at the destination.

Under the no-toll equilibrium, the user with highest MUT increasing rate at the destination departs and arrives first to avoid the queue, and the one with lowest MUT increasing rate arrives last. If the queue were removed, the first user could arrive later, and the last one earlier. However, a social optimal toll that removes the queue requires the travelers pay exactly the same monetary cost as the scheduling utility gain. This explains the reason why the arrival interval under socially optimal tolling remains unchanged. This observation is very similar to the conventional bottleneck model with homogeneous user [23] and the one with heterogeneous users whose cost of arriving late differs [6].

It is also worthy to relate the current finding with that of [16]. The heterogeneity of distance to the bottleneck in the model of [16] causes unequal gain and loss under the first-best tolling scheme. The first-best pricing scheme under their setting does not equal the time saving benefit. User who travels longer distance to the bottleneck whose MUT at the destination is higher than those short distance traveler, therefore they benefit more from the elimination of the queue.

5 Numerical example

A numerical simulation is presented here to visualize the theoretical model by freight and passenger users with continuously distributed increasing rates of MUT at the destination. The users are assumed to be a continuum with mass 1. The schedule preference is specified by the following achieved utility:

$$u(d,t) = \int_{T_1}^{d} (\beta_0 + \beta_1 s)ds + \int_{t}^{T_2} (\gamma_0 + \gamma_1 s)ds,$$

where β_0 is assumed to be equal to γ_0 without losing generality. T_1 and T_2 are arbitrarily set to 0 without affecting the result in qualitative. The capacity of bottleneck φ is 0.02 users per minute, which means all users can pass the bottleneck in 50 minutes.Free flow travel time T before arriving at the bottleneck is arbitrarily set to 1 minute. The increasing rates γ_1 of MUT at the destination follow a bimodal distribution, composed of two beta distributions, each with mass 0.5. One of the two has support on [1,2] and the other on [2,3]. The density of γ_1 is shown by Figure 2.

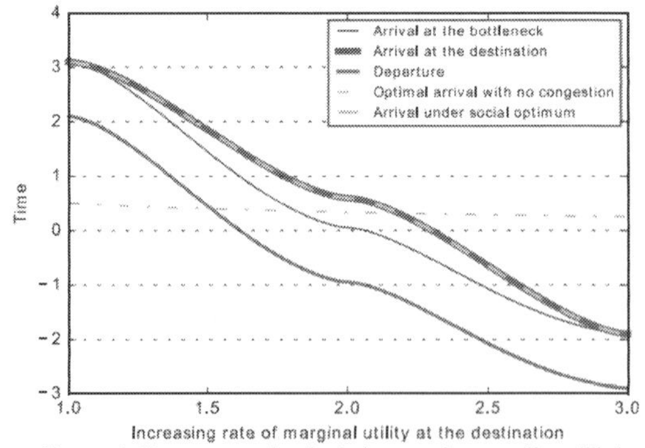

Figure 3. Departure and arrival times under no-toll equilibrium and social optimum

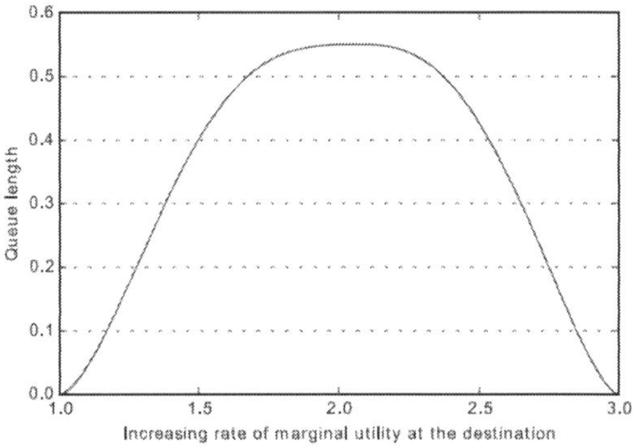

Figure 4. Queue length

The simulation calculates for each γ_1, then performs a search of a_0 by solving $a_1 = a_0 + N/\varphi$. Under social optimum, $a_{\tau 0}$ is found numerically to maximize mean scheduling utility as well as social welfare. Simulation result of arrival times are illustrated in Figure 3.

The dot-dash line shows the situation when all users travel without congestion where only free flow travel time T is needed. As states, the optimal arrival times are sorted by the descending order γ_1. Since β_0 is assumed to be equal to γ_0, then $a_*(\gamma_1) = \frac{\beta_1 T}{\beta_1 - \gamma_1}$. The first user with the highest γ_1 schedules his individual optimal arrival time at $a_*(\gamma_H) = 0.25$.

Under no-toll equilibrium, as shown by the upper thick line, the user with the highest increasing rate of MUT at the destination shifts his arrival time to a_0 to avoid the queue, which is earlier than his optimal arrival time $a_*(\gamma_H)$. He arrives at the destination as soon as he gets to the bottleneck, as there is no queue for him. At the other end of the queue, the user gains the lowest increasing rate γ_L of MUT arrives last at time $a_1 > a_*(\gamma_L)$. He also experiences exactly no queue either. The thin line is the arrival time at the bottleneck, while the lower thick line gives departure time from their bases. Since all user are supposed to travel the same time before arriving at the bottleneck, the departure time function has the same derivative as the arrival time at the bottleneck, and the traveler with lower γ_1 always departs later.

 DOI:10.1051/matecconf/201712402003

Figure 4 gives a closer look at the queue length as a function of γ_1. The density of γ_1 is lower than the bottleneck capacity near $\gamma_1 = 2$. But the two curves $a_0 + \frac{N-F(\gamma_1)}{\varphi}$ and $a_*(\gamma_1)$ still intersect only once, which means the density is not sparse enough to evolve another peak.

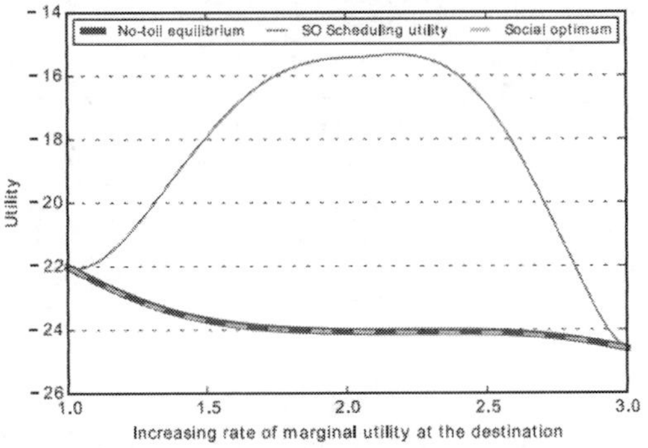

Figure 5. Utilities of users with different increasing rate of marginal utilities at the destination

Next consider the situation under social optimum. Since the functional form $\frac{N-F(\gamma_1)}{\varphi} + a_{\tau 0}$ remain unchanged, the order of arrival at the destination is identical to that in no-toll equilibrium. Moreover, the constant $a_{\tau 0}$ does not shift the arrival interval either earlier or later, compared with that of no-toll equilibrium. Therefore, the arrival time under social optimum shown by the upper dash line in Figure 3 coincides with the arrival time under no-toll equilibrium. Consequently, the departure time in social optimum does not move.

Figure 5 presents the achieved utilities of users with different increasing rates of MUT at the destination. The indirect utility under no-toll equilibrium shown by the lower thick curve consists only scheduling utility. The scheduling utility in social optimum is shown by the upper thin line. The lower dash line gives the indirect utility in social optimum, which exactly coincides with the indirect utility under no-toll equilibrium. The difference between them is thus the socially optimal toll. As stated in Proposition 5, the welfare gain in terms of scheduling utility is completely offset, and no user can benefit from the tolling, when the toll revenues are not returned to the user.

It is noteworthy that the indirect utilities under both no-toll equilibrium and social optimum are a monotonically decreasing function of γ_1. The users with higher productivity increasing rate achieve less utility instead of more, as they incur higher scheduling cost. It implies that a higher rebate to those more productive users can motivate them to improve marginal utility.

6 Conclusion

This paper introduces the heterogeneity of preference to the time-varying marginal utility scheduling model. The model is then applied to treat the dynamics of urban traffic congestion incurred by both freight and passenger traffic. By directly relating the productivity of freight carriers to the amount of toll, the proposed optimal tolling removes the fundamental barrier for freight carriers to accept congestion pricing. The first-best pricing strategy without rebate is found inducing no welfare gain or loss to any of the users, in spite of the heterogeneity of marginal utilities. This finding gives an important policy implication that an equal revenue rebate of a social optimal toll can generate a Pareto improvement, relative to the no-toll equilibrium.

Our interesting result is that the user who gains higher productivity increasing rate failed to achieve higher utility, which indicates that a differentiated redistribution of toll revenues can be an incentive to improve productivity. This shows the possibility of two interesting extensions to our current model. The first would be to examine the second-best pricing policy to freight and passenger traffic, for example in the form of freight dedicated lane [24] or pricing a portion of lanes [25]. The second is to make the productivity increasing rate in our model endogenous. Compared with commuters, it is more possible for freight carriers to decide their marginal utilities at the destination, as they could choose an appropriate load per trip. By capturing this responsive behavior, the dynamics of congestion and the effect of tolling policy can be more comprehensively examined.

Appendix A. Main notations

Symbol	Description
PAT	Exogenous preferred arrival time
N	Total number of road users
φ	Capacity of the bottleneck in the city center
MUT	Marginal utility of time
D	A specific period of time within a day
T	Travel time from the base to the bottleneck
u	Achieved utility
τ	Time-varying toll at the bottleneck
h	Marginal utility of time at the base
w	Marginal utility of time at the destination
β_0	Constant of the marginal utility of time at the base
β_1	Increasing rate(slope) of MUT at the base
γ_0	Constant of the marginal utility of time at the destination
γ_1	Increasing rate(slope) of MUT at the destination
H	Cumulative utility at the base
W	Cumulative utility at the destination

IAT	Ideal arrival time when travel time is zero
a	Arrival time at the bottleneck without tolling
d	Departure time from the base
t	Arrival time at the destination
ρ	Arrival rate at the bottleneck
R	Cumulative arrival at the bottleneck
a_*	Optimal arrival time without queueing
$f(\gamma_1)$	Density of γ_1
$F(\gamma_1)$	Cumulative distribution of γ_1
γ_H	Upper bound of γ_1
γ_L	Lower bound of γ_1
a_0	Time when the queue begins
a_1	Time when the queue ends
a_τ	Arrival time at the bottleneck under socially optimal tolling

Appendix B. Lemmas and proof of propositions

Proof of Proposition 1 The strictly convexity assumption on u assures that the optimal arrival time $a^*(\gamma_1)$ uniquely exists. Here the the first order condition for the utility maximization is that

$$0 = \beta_0 + \beta_1(a - T) - \gamma_0 - \gamma_1 a. \quad (2)$$

Differentiate the first order condition with respect to γ_1 to find that

$$a'_*(\gamma_1) = \frac{\beta_1 T + \gamma_0 - \beta_0}{(\beta_1 - \gamma_1)^2}.$$

Note that $H''(d) = \beta_1 < 0$, and $\gamma_0 - \beta_0 \le 0$, then $a'_*(\gamma_1)$ must be negative.

Lemma 1 *If $a_0 \le a_*(\gamma_H), a_0 + N/\varphi = a_1 \ge a_*(\gamma_L)$, and if there is no queue at time a_1, then taking the behavior of all other users as given, any user will choose to arrive at the bottleneck in the interval $[a_0, a_1]$.*

Consider an arbitrary queue starts at a_0 and ends at a_1, and a user with an increasing marginal utility at the destination at rate $\gamma_1 \in \Gamma$, where $a_0 \le a_*(\gamma_H) \le a_*(\gamma_1) \le a_*(\gamma_L) \le a_1$. Then the user prefers to arrive at the bottleneck at a_0 to any other earlier time, since earlier arrival time will not reduce the queueing time anymore but decrease the utility. He also prefers a_1 to any other time later, since the queue ends at time a_1. Therefore there will not be any one choosing to arrive outside the interval $[a_0, a_1]$.

Lemma 2 When $a_*(\gamma_1) < \frac{N-F(\gamma_1)}{\varphi} + a_*(\gamma_H)$ holds for all $\gamma_1 \in \Gamma, \gamma_1 < \gamma_H$, the equilibrium arrive time as a function of γ_1 is given by (1) and

$$-\int_{\gamma_L}^{\gamma_H} a'(\gamma_1) d\gamma_1 = \frac{N}{\varphi}. \quad (3)$$

Then $a(\gamma_1)$ satisfies $a(\gamma_H) \le a_*(\gamma_H)$ and $a(\gamma_L) \ge a_*(\gamma_L)$.

Given the conclusion of, $a_*(\gamma_1) < 0$. Then if any user arrives later than his optimal arrival time, his marginal utility at the destination when he arrives must be larger than that at the base when he departs, which implies $a'(\gamma_1) > -\frac{f(\gamma_1)}{\varphi}$. If $a(\gamma_1) > a_*(\gamma_1)$ for all γ_1, then $a_1 - a_0 > N/\varphi$, which contradicts (3).

Suppose $a(\gamma_H) > a_*(\gamma_H)$. There exists $\xi < \gamma_H$ such that $a_(\xi) = a_*(\xi)$. Then for all $\gamma_1 > \xi$, $a(\gamma_1) > a_*(\gamma_1)$ and $a'(\gamma_1) > -\frac{f(\gamma_1)}{\varphi}$ hold. Therefore by $a_*(\gamma_1) < \frac{N-F(\gamma_1)}{\varphi} + a_*(\gamma_H)$, we have $a(\xi) > a_0 + \frac{N-F(\gamma_1)}{\varphi} > a_*(\gamma_H) + \frac{N-F(\gamma_1)}{\varphi} > a_*(\xi)$, which is a contradiction to $a_(\xi) = a_*(\xi)$.

Then assume that $a_1 < a_*(\gamma_L)$. If $a_0 + N - F(\gamma_1)/\varphi < a_*(\gamma_1)$ holds for all γ_1, then $a'(\gamma_1) < -\frac{\gamma_1}{\varphi}$ for all γ_1, which contradicts (3). So there exists ζ such that $a_0 + \frac{N-F(\zeta)}{\varphi} = a_*(\zeta)$. Then by $a_1 - a_1 = N/\varphi$ and $a_*(\gamma_1) < \frac{N-F(\gamma_1)}{\varphi} + a_*(\gamma_H)$, we have $\frac{N}{\varphi} - \frac{N-F(\zeta)}{\varphi} = a_1 - a_*(\zeta) < a_*(\gamma_L) - a_*(\zeta) < \frac{N}{\varphi} - \frac{N-F(\zeta)}{\varphi}$, which is a contradiction.

Therefore $a(\gamma_H) \le a_*(\gamma_H)$ and $a(\gamma_L) \ge a_*(\gamma_L)$ must hold.

Proof of Proposition 2 Let $[a_0, a_1]$ be the minimum arrival interval required by all N users in no-toll equilibrium. By Proposition 1, $a_*(\gamma_1) < 0$ for all γ_1, then $a_0 \le a_*(\gamma_1)$ for all γ_1, otherwise there could be some users who are supposed to arrive within interval $[a_0, a_1]$ can shift their arrive to an earlier time at their optimal arrival time without queueing and then increase utility. At the end of the queue, one could similarly have $a_1 \ge a_*(\gamma_1)$ for all γ_1. Therefore, $a_0 \le a_*(\gamma_L)$ and $a_1 \ge a_*(\gamma_H)$.

Next we show that the length of the interval when no-toll equilibrium must be N/φ. If $a_1 - a_0 > N/\varphi$, there must be a period of time when the bottleneck is not fully utilized. Given $a_*(\gamma_1) < \frac{N-F(\gamma_1)}{\varphi} + a_*(\gamma_H)$, someone who is supposed to arrive later in the period can shift to an earlier arrival time when the bottleneck was not fully utilized, and this will lead to strict increase of utility and contradict the equilibrium. If $a_1 - a_0 < N/\varphi$, there will be a residual queue at a_1 (see [26] for further discussion on the condition when the property of no residual queue holds). The user in the residual queue can arrive later at the bottleneck to reduce queueing time and arrive the destination at the same time as they wait in the residual queue.

Proof of Proposition 3 To prove $a'(\gamma)$ has a unique solution, we first assume the equilibrium exists for the ease of discussion. As shown by Proposition 2, the queue starts at a_0 and ends at a_1. Given there will always be queue during $[a_0, a_1]$, for a user with increasing rate of marginal utility γ_1, the achieved utility can be represented by

$$u\left(a-T,\frac{R(a)}{\varphi}+a_0\right) = \int_0^{a-T}(\beta_0+\beta_1 s)ds+\int_{\frac{R(a)}{\varphi}+a_0}^{0}(\gamma_0+\gamma_1 s)ds$$

Then the first order condition of maximizing the utility of the individual is

$$0=\beta_1(a-T)+\beta_0-\left[\gamma_0+\gamma_1\left(\frac{R(a)}{\varphi}+a_0\right)\right]\frac{\rho(a)}{\varphi} \quad (4)$$

and the second order condition is

$$\beta_1-\gamma_0\frac{\rho'(a)}{\varphi}-\gamma_1\frac{R(a)\rho(a)}{\varphi^2}-\gamma_1\left(\frac{\rho(a)}{\varphi}\right)^2-\gamma_1\frac{\rho'(a)}{\varphi}a_0 < 0$$

Then the equilibrium utility $u\left(a(\gamma_1)-T,\frac{R(a(\gamma_1))}{\varphi}+a_0\right)$ can be achieved when the arrival time at the bottleneck $a(\gamma_1)$ solves the utility maximization problem.

To see how equilibrium utility varies with the increasing rate of marginal utility, we could differentiate u with respect to γ_1, which yields

$$\frac{\partial}{\partial\gamma_1}u\left(a(\gamma_1)-T,\frac{R(a(\gamma_1))}{\varphi}+a_0\right) = a'(\gamma_1)\beta_1(a(\gamma_1)-T)-\gamma_1\left(a_0+\frac{R(a)}{\varphi}\right)\frac{\rho(a)}{\varphi} -\frac{1}{2}\left(a_0+\frac{R(a)}{\varphi}\right)^2-\gamma_0\frac{\rho(a)}{\varphi}. \quad (5)$$

By the concavity assumptions on u, (5) can vary from negative to positive, which relates with cumulative arrival $R(a)$.

To examine the property of $a'(\gamma_1)$, we differentiate the first order condition (4) with respect to γ_1, which yields

$$0=a'\cdot u''(a)-\frac{\rho(a)}{\varphi}\left(\frac{R(a)}{\varphi}+a_0\right).$$

Since u is assumed to be strictly concave, we have $u''(a)<0$, and by $\rho(a)>0$, therefore $a'(\gamma_1)<0$. Note that $a(\gamma_1)$ has an inverse $\gamma_1(a)$, which gives $a(\gamma_1(a))=a$, we then have $\gamma_1'(a)=1/a'(\gamma_1)<0$. Therefore $R(a)=N-F(\gamma_1)$. Differentiate $R(a)$ with respect to a shows that

$$\rho(a)=-\frac{f(\gamma_1)}{a'(\gamma_1)}. \quad (6)$$

Substitute $\rho(a)$ into the first order condition (4), we find that

$$a'(\gamma_1)=-\frac{f(\gamma_1)}{\rho(a)}=-\frac{\gamma_0+\gamma_1\left(a_0+\frac{N-F(\gamma_1)}{\varphi}\right)}{\beta_0+\beta_1(a-T)}\frac{f(\gamma_1)}{\varphi}.$$

Two conditions must hold for the existence of equilibrium: (a) No one can improve his utility by unilaterally changing his arrival time at the bottleneck, and (b) all users must arrive. Since the user's individual scheduling utility u is assumed to be concave, a user has a unique maximum scheduling utility taking all others behavior as given. In the meanwhile, the arrival rate at the bottleneck given by (1) and $\int_{\gamma_H}^{\gamma_L}a'(\gamma_1)d\gamma_1=N/\varphi$ has the unique solution, which does not depend on the existence of equilibrium. If someone is able to increase individual utility by changing his arrival time at the bottleneck, then $a'(\gamma_1)$ cannot have an unique solution, which contradicts to (1). Therefore, condition (a) must hold. Condition (b) requires that $a(\gamma_L)=a_0+N/\varphi$. If a_0 exists and is not unique, $a'(\gamma_1)$ will either change non-monotonically or remain unchanged when a_0 changes. Then by differentiating $a'(\gamma_1)$ with respect to a_0, we have

$$\frac{\partial a'(\gamma_1)}{\partial a_0} = \left(-\frac{f(\gamma_1)}{\varphi}\right)\left[\frac{\gamma_1[\beta_1(a-T)+\beta_0] -\beta_1\frac{\partial a(\gamma_1)}{\partial a_0}\left[\gamma_1\left(a_0+\frac{N-F(\gamma_1)}{\varphi}\right)+\gamma_0\right]}{[\beta_1(a-T)+\beta_0]^2}\right]. \quad (7)$$

By the assumptions on u, $\frac{\partial a'(\gamma_1)}{\partial a_0}$ is strictly negative if $\frac{\partial a(\gamma_1)}{\partial a_0}\geq 0$. Note that $\frac{\partial a(\gamma_1)}{\partial a_0}=\frac{\partial(a(\gamma_1)-a_0)}{\partial a_0}+1$, and that

$$\frac{\partial(a(\gamma_1)-a_0)}{\partial a_0}=\int_{\gamma_H}^{\gamma_1}\frac{\partial a'(\gamma_1)}{\partial a_0}d\gamma_1.$$

If $\frac{\partial(a(\xi)-a_0)}{\partial a_0}\geq 0$ holds for all $\xi>\gamma_1$ can be proved, then $\frac{\partial a'(\xi)}{\partial a_0}<0$ would hold for all $\xi>\gamma_1$ by the above strictly negative condition of (7). Note that $a'(\gamma_1)$ is continuous, since $[a_0,a_1]$ is the smallest interval that contains all the arrivals at the bottleneck, then $\frac{\partial a'(\xi)}{\partial a_0}$ must be continuous as well. At point $\xi=\gamma_1$, $\frac{\partial a'(\gamma_1)}{\partial a_0}\leq 0$ must hold. Therefore we could have $\frac{\partial(a(\gamma_1)-a_0)}{\partial a_0}=\int_{\gamma_H}^{\xi}\frac{\partial a'(\xi)}{\partial a_0}d\xi+\int_{\xi}^{\gamma_1}\frac{\partial a'(\gamma_1)}{\partial a_0}d\gamma_1>0$.

Substitute $\frac{\partial a(\gamma_H)-a_0}{\partial a_0}=0$ into , we have $\frac{\partial a'(\gamma_H)}{\partial a_0}<0$. Then there exist a small neighborhood $V(\gamma_H)$ around γ_H such that $\frac{\partial(a(\gamma_1)-a_0)}{\partial a_0}\geq 0$ for all $\gamma\in V(\gamma_H)$. By the above discussion, therefore, $\frac{\partial(a(\gamma_1)-a_0)}{\partial a_0}>0$ holds for all $\gamma_1<\gamma_H$, where $\frac{\partial(a(\gamma_L)-a_0)}{\partial a_0}>0$. So there is only one a_0 that satisfies $a(\gamma_L)-a_0=N/\varphi$, which ensures that only one equilibrium can exist.

Next we show there indeed exists a_0 such that $a(\gamma_L)=a_0+N/\varphi$, then condition b holds. By Proposition 2, $a_0\leq a_*(\gamma_H)$, $a_1\geq a_*(\gamma_L)$, and $a_1-a_0=N/\varphi$ hold under no-toll equilibrium. Then $a(\gamma_1)\leq\frac{N-F(\gamma_1)}{\varphi}$ must holds, since there is always a queue during $[a_0,a_1]$. Here we then discuss the situation when a_0 varies outside the above range.

Consider first $a_0+N/\varphi<a_*(\gamma_H)$. By the above discussion, $\frac{\partial(a(\gamma_1)-a_0)}{\partial a_0}=1+\frac{\partial(a(\gamma_1-a_0))}{\partial a_0}>1$ holds for all $\gamma_1<\gamma_H$, and $\frac{\partial a(\gamma_H)}{\partial a_0}=1+\frac{\partial(a(\gamma_H)-a_0)}{\partial a_0}=1$. Therefore when $a_0<a_*(\gamma_H)-N/\varphi$, we have

$$a'(\gamma_1) = -\frac{\gamma_0 + \gamma_1\left(a_0 + \frac{N - F(\gamma_1)}{\varphi}\right)}{\beta_0 + \beta_1(a - T)}\frac{f(\gamma_1)}{\varphi} < -\frac{f(\gamma_1)}{\varphi}$$

and

$$\frac{\gamma_0 + \gamma_1\left(a_0 + \frac{N - F(\gamma_1)}{\varphi}\right)}{\beta_0 + \beta_1(a - T)} > 1,$$

then

$$\int_{\gamma_H}^{\gamma_L} a'(\gamma_1)d\gamma_1 > \int_{\gamma_H}^{\gamma_L} -\frac{f(\gamma_1)}{\varphi}d\gamma_1 = \frac{N}{\varphi}.$$

If $a_0 > a_*(\gamma_L)$, we could otherwise obtain $\int_{\gamma_H}^{\gamma_L} a'(\gamma_1)d\gamma_1 < N/\varphi$. Then $a(\gamma_L) - a(\gamma_H)$ can achieve both larger and smaller values than N/φ, which establishes the existence of equilibrium.

By Lemma 2, when a_0 attains $a_1 - a_0 = N/\varphi$, $a_0 < a_*(\gamma_H)$ and $a_1 > a_*(\gamma_L)$ hold. Then by Lemma 1, no one will choose the arrival time outside the interval $[a_0, a_1]$. Together with Lemma 1 and the previous discussion that (7) is strictly negative, it is now sufficient to conclude that (1) defines a unique equilibrium that no one can improve his utility by unilaterally changing arrival time at the bottleneck within the interval $[a_0, a_1]$.

Proof of pproposition 4 The socially optimal toll exactly removes the queue. If there is a queue at some point of time, the user will delay the arrival and social welfare is then improved. Therefore we have $R_\tau(a) = \varphi(a - a_{\tau 0})$ and $\rho_\tau = \varphi$. Under the optimal tolling, a user with the increasing rate γ_1 of marginal utility arrives at the bottleneck at time $a_\tau(\gamma_1)$. The first order condition for utility maximization is

$$\tau'(a_\tau(\gamma_1)) = \beta_1(a_\tau - T) + \beta_0 - \gamma_1 a_\tau - \gamma_0, \quad (8)$$

and the corresponding second order condition is

$$\tau''(a_\tau(\gamma_1)) > \beta_1 - \gamma_1.$$

Differentiating the first order condition with respect to γ_1 yields

$$0 = \tau'' - \beta_1 a'_\tau + \gamma_1 a'_\tau + a_\tau.$$

By the second order condition and the assumptions of $\gamma_1 > 0 > \beta$, $a_\tau'(\gamma_1)$ must be negative, which means the order of arrival will remain unchanged compared with the one under no-toll equilibrium. Then we have

$$R_\tau(a_\tau(\gamma_1)) = N - F(\gamma_1) = \varphi(a_\tau(\gamma_1) - a_{\tau 0}),$$

where $a_\tau(\gamma_1) = \frac{N-F(\gamma_1)}{\varphi} + a_{\tau 0}$. To find the solution to an optimal toll, note first that the inverse of a_τ is $\gamma_{1\tau}(a) = F^{-1}(N - \varphi(a - a_{\tau 0}))$. Substituting $\gamma_{1\tau}$ into the first order condition (8) yields the derivative of an optimal toll $\tau'(a)$ which satisfies

$$\tau'(a) = \beta_1(a - T) + \beta_0 - F^{-1}(N - \varphi(a - a_{\tau 0}))a - \gamma_0,$$

such that

$$\tau(a) - \tau(a_{\tau 0}) = \int_{a_{\tau 0}}^{a} \tau'(s)ds = \int_{a_{\tau 0}}^{a} \beta_1(s - T) - F^{-1}(N - \varphi(s - a_{\tau 0}))sds.$$

Any toll satisfies (9) will exactly remove the queue. Users under such queue arrives at the bottleneck as well as at the destination within the interval $[a_{\tau 0}, a_{\tau 1}]$. Then we have the mean scheduling utility under optimal tolling

$$Eu = \int_{\gamma_L}^{\gamma_H} u\left(\frac{N - F(\gamma_1)}{\varphi} + a_{\tau 0} - T, \frac{N - F(\gamma_1)}{\varphi} + a_{\tau 0}\right) f(\gamma_1)d\gamma_1$$

Differentiate the mean scheduling utility with respect to a_0 yields

$$\frac{\partial Eu}{\partial a_0} = \int_{\gamma_L}^{\gamma_H}\left[\begin{array}{c}\beta_0 + \beta_1\left(\frac{N - F(\gamma_1)}{\varphi} + a_\tau 0 - T\right)\\ -\gamma_0 - \gamma_1\left(\frac{N - F(\gamma_1)}{\varphi} + a_\tau 0\right)\end{array}\right] f(\gamma_1)d\gamma_1$$

$$= \varphi\int_{a_{\tau 0}}^{a_{\tau 1}} (\beta_0 + \beta_1(a - T) - \gamma_0 - \gamma_1 a)\, da, \quad (10)$$

and the second order derivative is

$$\frac{\partial^2 Eu}{\partial (a_0)^2} = \int_{\gamma_L}^{\gamma_H} (\beta_1 - \gamma_1) f(\gamma_1)d\gamma_1.$$

Since $H' > 0 > W'$, there must exist γ_1 such that (10) equals zero, and (11) is strictly negative. Therefore Eu is strictly concave as a function of a_0. When $\frac{\partial Eu}{\partial a_0} = 0$, Eu reaches a global maximum, where the start time of the arrival under an optimal toll $a_{\tau 0}$ is the unique solution. Then we rewrite $\frac{\partial Eu}{\partial a_0} = 0$ as

$$0 = \int_{a_{\tau 0}}^{a_{\tau 0} + \frac{N}{\varphi}} (\beta_0 + \beta_1(a - T) - \gamma_0 - F^{-1}(N - \varphi(a - a_{\tau 0}))a)\, da.$$

Substitute (9) in to the above equation, and denote $a_{\tau 0} + N/\varphi$ by $a_{\tau 1}$, then we have $\tau(a_{\tau 0}) = \tau(a_{\tau 1})$.

If the toll satisfies $\tau(a_{\tau 0}) = \tau(a_{\tau 1}) = 0$, then by Lemma 1 in the appendix, we could eliminate the possibility that users may choose an arrival time outside the tolling interval. To see the reason, note first that $\beta_1 - \gamma_1 < 0$ by the assumption of the model. Then by (2) and (8), it is clear that $a(\gamma_1) \gtreqless a_*(\gamma_1)$ is a sufficient and necessary condition of $\tau'(a) \lesseqgtr 0$. If $'_*(\gamma) < -\frac{f(\gamma)}{\varphi}$, there must exist only one $\zeta \in \Gamma$ such that $a(\zeta) = a_*(\zeta)$.Therefore $a_{\tau 0} \le a_*(\gamma_H), a_*(\gamma_L) \le a_{\tau 1}$.

Proof of Proposition 5 The utility difference between the case with and without optimal tolling of a user with MUT increasing rate at γ_1 can be written as

$$\Delta u(\gamma_1) = u(a_\tau(\gamma_1) - T, a_\tau(\gamma_1)) - \tau(a_\tau(\gamma_1)) - u\left(a(\gamma_1) - T, \frac{N - F(\gamma_1)}{\varphi} + a_0\right).$$

By differentiating $\Delta u(\gamma_1)$ with respect to γ_1, and substiute the first order condition under no-toll equilibrium and under socially optimal tolling, we have

$$\frac{\partial \Delta u(\gamma_1)}{\partial \gamma_1} = -\frac{1}{2}\left(\frac{N - F(\gamma_1)}{\varphi} + a_{\tau 0}\right)^2 + \frac{1}{2}\left(\frac{N - F(\gamma_1)}{\varphi} + a_0\right)^2.$$

Suppose that $a_{\tau 0} < a_0$. Then if $\frac{N-F(\gamma_1)}{\varphi} + a_{\tau 0} < \frac{N-F(\gamma_1)}{\varphi} + a_0 < 0$ or $\frac{N-F(\gamma_1)}{\varphi} + a_{\tau 0} < -\left(\frac{N-F(\gamma_1)}{\varphi} + a_0\right) < 0$, for all γ_1, $\frac{\partial \Delta u(\gamma_1)}{\partial \gamma_1} < 0$ must hold. Then the two curve $\frac{N-F(\gamma_1)}{\varphi} + a_{\tau 0}$ and $\frac{N-F(\gamma_1)}{\varphi} + a_0$ must intersect at some

$\xi \in \Gamma$. Therefore for any $\gamma_1 > \xi$, $\frac{N-F(\gamma_1)}{\varphi} + a_{\tau 0} > \frac{N-F(\gamma_1)}{\varphi} + a_0$ must hold, which is a contradiction. If $0 < \frac{N-F(\gamma_1)}{\varphi} + a_{\tau 0} < \frac{N-F(\gamma_1)}{\varphi} + a_0$, or $\frac{N-F(\gamma_1)}{\varphi} + a_{\tau 0} < 0, \frac{N-F(\gamma_1)}{\varphi} + a_0 > 0$ and $\frac{N-F(\gamma_1)}{\varphi} + a_{\tau 0} > -\left(\frac{N-F(\gamma_1)}{\varphi} + a_0\right) < 0$, then $\frac{\partial \Delta u(\gamma_1)}{\partial \gamma_1} < 0$ must hold for all γ_1. Therefore $a_{\tau 1} < a_1$, and both $\Delta u(\gamma_L)$ must be negative, which contradicts Proposition 2.

The contradictions can be similarly found if $a_{\tau 0} > a_0$. This then ensures $a_{\tau 0} = a_0$ must hold.

References

1. P. Vilain, P. Wolfrom, Transp. Res. Rec. J. Transp. Res. Board, **1707**, 64–72 (2000)
2. T. Kenmochi, H. Oka, R. Tani, D. Fukumoto, Y. Hagino, T. Hyodo, *IBS Annual Report* (2016).
3. MDS Transmodal Limited, Chester, UK., DG MOVE European Commission: Study on urban freight transport, Final report (2012) [Accessed: 17-May-2016]
4. Beijing Municipal Commission of Transport, China, Institute of Transportation Science, Ministry of Transport, China, and Beijing Municipal Bureau of Statistics, China, Report on urban freight demand in central Beijing (in Chinese) (2011) [Accessed: 17-May-2016]
5. W. S. Vickrey, *Am. Econ. Rev.*, **59**, 2, 251–260 (1969)
6. R. Arnott, A. de Palma, and R. Lindsey, *J. Transp. Econ. Policy*, **28**, 2, 139–161 (1994)
7. V. A. C. van den Berg, E. T. Verhoef, *Transp. Res. Part B Methodol.*, **45**,1, 60-78 (2011)
8. R. Lindsey, *Transp. Sci.*, **38**, 3, 293–314, (2004)
9. K. A. Small, *Am. Econ. Rev.*, **72**, 3, 467–479, (1982)
10. S. Mun, M. Yonekawa, *J. Transp. Econ. Policy*, **40**, 3, 329–358 (2006)
11. W. Vickrey, *Highw. Res. Rec.*, 476, 36–48 (1973)
12. Y.-Y. Tseng, E. T. Verhoef, *Transp. Res. Part B Methodol.*, vol. 42, no. 7–8, pp. 607–618, Aug. 2008.
13. M. Börjesson, J. Eliasson, J. P. Franklin, *Transp. Res. Part B Methodol.*, **46**, 7, 855–873 (2012)
14. K. Hjorth, M. Börjesson, L. Engelson, M. Fosgerau, *Transp. Res. Part B Methodol.*, **81**, 1, 230–251 (2015)
15. M. Fosgerau, L. Engelson, *Transp. Res. Part B Methodol.*, **45**, 1, 1–8 (2011)
16. M. Fosgerau, A. de Palma, *J. Urban Econ.*, **71**, 3, 269–277 (2012)
17. K. A. Small, *Econ. Transp.*, **1**, 1–2, 2–14 (2012)
18. M. Fosgerau, K. Small, University of California-Irvine, Department of Economics, Working Paper 131403 (2014)
19. Department for Transport, UK., WebTAG: TAG data book, December 2015 - Publications - GOV.UK, (2015) [Accessed: 09-Feb-2016]
20. J. V. Henderson, *J. Urban Econ.*, **9**, 3, 349–364, (1981)
21. J. Holguín-Veras, *Transp. Res. Part Policy Pract.*, **42**, 2, 392–413 (2008)
22. J. Holguín-Veras, *Transp. Res. Part Policy Pract.*, **45**, 8, 802–824 (2011)
23. R. Arnott, A. de Palma, R. Lindsey, *J. Urban Econ.*, **27**, 1, 111–130 (1990)
24. A. de Palma, M. Kilani, R. Lindsey, *J. Urban Econ.*, **64**, 2, 340–361 (2008)
25. J. D. Hall, Pareto improvements from Lexus lanes: the effects of pricing a portion of the lanes on congested highways, Working paper (2013) [Accessed: 18-May-2016]
26. A. de Palma, M. Fosgerau, *Eur. J. Oper. Res.*, **230**, 2, 313–320 (2013)

 DOI:10.1051/matecconf/201712402004

Research on Short-term Prediction Model of Freeway Operation Situation

Xiaodan Zhang[1]

[1] *Highway Traffic Safety Engineering Research Center, Research Institute of Highway Ministry of Transport, Beijing, China*

Abstract. Based on the traffic flow data and accident data of Beijing-Tianjin-Tanggu freeway, the security situation short-term prediction model was established in the paper. Firstly, we established the risk prediction database, and developed the pre-analysis software of basic data; secondly, the traffic flow data between 10 to 15 minutes prior to the time of accident were aggregated at 5-minute level, and the volume, speed, occupancy as well as their statistical parameters were selected; finally, based on the correlation analysis results of parameters, the multi-parameters Logistic regression model was established. The results indicate, the change of traffic flow parameters and their statistics can effectively predict the possibility of accident, in which the average value of speed of small car, the standard deviation of volume of large car and the average value of volume difference between large car and small car at 5-minute level have a significant impact on the risk of accident.

1 Introduction

Intelligent transportation system (ITS) is widely used in the global scope, which makes the traffic managers have large real-time traffic situation data. Many researchers and practitioners have been fully aware that all the advantages of ITS will not be recognized without realizing the ability of the traffic flow short-term prediction[1] (Brian L. Smith, 2002). The traffic flow prediction model can provide such a kind of ability, and can provide forward-looking traffic management as well as comprehensive travel information service.

At present, the object of traffic forecast is mainly urban road, and it focuses on the unblocked reliability and efficiency[2-7], including the assessment of traffic operation situation, short-term traffic forecast, road traffic condition judgment and so on, but the researches on the traffic flow operation security situation prediction are rare.

In China, due to the lack of detailed accident data and microscopic traffic flow data as support, resulting in a serious shortage in the real-time traffic flow security analysis theory, leading to the current freeway security management lag behind the real-time traffic situation prediction in our country. The United States and Canada began to carry out researches on the traffic accident detection algorithms and the traffic flow harbinger characteristics before the traffic accident from 1990s[8-15]. Among them, Chris adopted the speed differences between upstream and downstream, and the variances of the cross-section speed as the characterization factors of the traffic flow real-time risk discrimination[9], the results of which was referenced by Kansas state highway agency of USA in 2006. However, the main shortcomings were unknown of the risk reason and subjectivity of the risk rank assessment, and its scientificity needs to be further examined.

So, the prediction method of freeway operation situation based on short-term traffic flow multi-parameters regression is researched in this paper, thus achieving the short-term prediction of the traffic flow operation situation, and the results are helpful to reduce the risk of accident, decrease traffic accidents, and improve the operation security of freeway.

2 Data processing

Beijing-Tianjin-Tanggu freeway is about 142.69 kilometers, has 28 microwave detectors, and the average spacing of detectors is 5 kilometers, in which, the coils of Beijing section is more intensive, but some of the data is missing. All the traffic flow data as well as the traffic accidents data in Beijing-Tianjin-Tanggu freeway are extracted in this paper, in which, the traffic accidents include time, location, type, reason, and et. al., and the traffic flow data are speed, volume and occupancy of divided-lane at 1-minute level.

First, we select the nearest detector data within 2 kilometers prior to or back to the location of accident, in order to screen data for first round, in which, the choice of accidents should avoid these caused by external factors such as weather, linear conditions, drivers and vehicles as much as possible, only in this way we can accurately excavate the rule of the traffic flow fluctuation affecting on the accidents, and in consideration that the relationship between the causes and effects of single-vehicle accidents and the traffic flow may not be strong under high service level, so the select of accident samples

DOI:10.1051/matecconf/201712402004

is more emphasis on the multi-vehicle accidents in large traffic volume (under C-classical service level); and then, we screen data for second round: check the quality of the traffic flow data, including deleting and processing the singular values such as speed with 0 km/h, diagnose the outliers by the data spatio-temporal graph method and the statistical method, and correct the abnormal data values by the simple difference method and the filtering method, in order to improve the accuracy and reliability of the analytical results, thus selecting the speed, volume and occupancy of detectors with better data quality to pair with accidents, and extract the traffic flow data of the control groups by 1:4 ratio.

The control group data meet the following requirements: the date is different with the corresponding accident; the time, week, and location are the same as the corresponding accident; the control group has no accidents at this location at the same day. Then, we select the control groups with better data quality according to the same method mentioned above, so as to establish the database required in this paper.

Because the location of traffic accident recorded by the police department is a cross-sectional stake number, therefore, it is necessary to aggregate the divided-lane traffic flow data into a cross-sectional traffic flow data, thus taking the cross-sectional data as the foundation of the research, so, the divided-lane data are transformed into the cross-sectional data by weighting method: $q_s=\sum_{i=1}^{n} q_i$, $k_s=\frac{\sum_{i=1}^{n} q_i\cdot k_i}{\sum_{i=1}^{n} q_i}$, $v_s=\frac{\sum_{i=1}^{n} q_i\cdot v_i}{\sum_{i=1}^{n} q_i}$, in which, q_i , k_i , v_i are respectively volume, occupancy, speed of divided-lane, n is the number of lanes.

3 Mathematical model

3.1 Logistic regression model

Binary Logistic regression model is commonly used to quantitatively analyze the impact of explanatory variables on binary dependent variables, also can be used to estimate the occurrence probability of a category of the dependent variables, and from the traffic flow operation results, the dependent variables just can be divided into two categories: accidents and non-accidents. The probability of the accident corresponding to one sample data is:

$$P(x_i)=\frac{1}{1+e^{-x_i'\beta}},\quad i=1,2,\cdots,n \tag{1}$$

The linear expression after logit transform is:

$$\ln\frac{P(x_i)}{1-P(x_i)}=x_i'\beta,\quad i=1,2,\cdots,n \tag{2}$$

Where, $P(x_i)$ represents the probability of traffic accident; $x_i'\beta$ represents the linear combination of explanatory variables: $x_i'\beta=\beta_0+\beta_1 x_{1i}+\cdots+\beta_k x_{ki}$, x_{ki} is the value of variable k in i -th sample; β_0 is the regression intercept; $\beta_1,\beta_2,\cdots,\beta_k$ are the regression coefficients of explanatory variable x_{ki} ; $\beta_0,\beta_1,\beta_2,\cdots,\beta_k$ can be calculated by the maximum likelihood estimation method:

$$\ln L(\beta,x_i)=\sum_{i=1}^{n}[\beta_0+\beta_1 x_{1i}+\cdots+\beta_k x_{ki}-\ln(1+e^{\beta_0+\beta_1 x_{1i}+\cdots+\beta_k x_{ki}})] .$$

3.2 Logistic model testing

In Logistic regression, likelihood ratio test, Akaike information criterion (Akaike Information Criterion, AIC) and Schwarz criterion can be used to reflect the goodness of the model fitting.

We adopts AIC to reflect the fitting effect of the final model in this paper, that is $AIC=-2LL+2(K+S)$, in which, K is the number of independent variables of the model, S is the total number of response variable categories minus 1, the range of $-2LL$ is 0 to infinity, which is the smaller the better. Under the same conditions, the smaller value of AIC indicates the better model fitting.

The prediction accuracy for classification is usually used to reflect the prediction accuracy of the model. Using Logistic model to predict the classifications needs specifying a threshold of probability, that is, when the probability calculated by Logistic model is greater than a specified threshold, it is discriminated as the traffic accident, and when the probability is less than a specified threshold, it is discriminated as the security state with non-accident. The threshold value decides the forecast accuracy of each category and the total samples, the current researches commonly use the proportion of a category in the whole samples as the threshold value of this category prediction. Because of researching the prediction method of freeway traffic operation situation short-term in this paper, the proportion of accidents in the whole samples is adopted as the threshold value.

4 Research results and analysis

4.1 Data preparation Logistic regression model

In order to predict the traffic accident in advance, the calibrated traffic flow data within 10~15 minutes prior to the time of the accident is extracted in this paper, meanwhile, the traffic flow data of the control group in the corresponding period is extracted for each accident. Through screening of data quality, we ultimately retain 33 groups of accident samples and 132 groups of non-accident samples as the control groups for modeling research, and divide them into 2 categories: accident and non-accident, that is the value of dependent variable is 1 indicating accident and 0 meaning non-accident.

Firstly, the original traffic flow data of divided-lane are aggregated into a cross-sectional traffic flow data, secondly, in order to avoid the data noise due to the short acquisition interval, the data are converged with 5-minute level in order to get averages and standard deviations.

4.2 Model Parameters Selection

We extract 14 statistical parameters as the chosen model parameters, as shown in Tab. 1:

Table 1. Model parameters

Variable Symbol	Symbolic Interpretation
lvavg	the average value of speed of large car at 5-minute level
svavg	the average value of speed of small car at 5-minute level
lvsd	the standard deviation of speed of large car at 5-minute level
svsd	the standard deviation of speed of small car at 5-minute level
gavg	the average value of occupancy at 5-minute level
gsd	the standard deviation of occupancy at 5-minute level
lqavg	the average value of volume of large car at 5-minute level
sqavg	the average value of volume of small car at 5-minute level
lqsd	the standard deviation of volume of large car at 5-minute level
sqsd	the standard deviation of volume of small car at 5-minute level
vdavg	the average value of speed difference between large car and small car at 5-minute level
vdsd	the standard deviation of speed difference between large car and small car at 5-minute level
qdavg	the average value of volume difference between large car and small car at 5-minute level
qdsd	the standard deviation of volume difference between large car and small car at 5-minute level

4.3 Modelling steps

The binary Logistic model of the freeway traffic security operation situation deduction based on short-term traffic flow multi-parameters is established by using statistical analysis software with R programming language in this paper, the steps are as follows:

(1) We use the correlation analysis method to examine the correlation between variables, which makes the highly correlated variables be not into the Logistic model;

(2) We select the reasonable explanatory variables for modeling according to the backward selection method of the Logistic regression, the basic steps are as follows:

a) First, all the variables are contained in the model;

b) Then, calculate the z test values of all variables, and get the corresponding P values;

c) Last, find the largest P value, if the P value is greater than the significance level α_{out}, this variable is eliminated;

d) Go back to step b) for the next round of elimination.

Among them, the significance level of the reserved values is set as: including variables is $P \le 0.05$, excluding variables is $P > 0.1$.

4.4 Logistic model regression and analysis

According to the above modeling steps (1), the results are shown in Tab. 2.

Generally, if the correlation coefficient between two parameters is > 0.6 or < -0.6, they are strongly related to each other, and they cannot enter the model at the same time. It can be seen that:

lvsd has a strong correlation with *sqsd*, *vdsd*, *vdavg*, and *lqavg*;

gavg has a strong correlation with *gsd*, *sqavg*, and *qdavg*;

gsd has a strong correlation with *gavg*;

lqavg has a strong correlation with *lvsd*;

sqavg has a strong correlation with *qdavg*, *qdsd*, *lqsd*, and *gavg*;

lqsd has a strong correlation with *sqavg*, and *qdsd*;

sqsd has a strong correlation with *vdsd*, *lvsd*, and *vdavg*;

vdavg has a strong correlation with *sqsd*, *vdsd*, and *lvsd*;

vdsd has a strong correlation with *sqsd*, *lvsd*, and *vdavg*;

qdavg has a strong correlation with *sqavg*, *qdsd*, and *gavg*;

qdsd has a strong correlation with *sqavg*, *qdavg*, and *lqsd*.

By comparing, we select *lvavg*, *svavg*, *lvsd*, *svsd*, *gsd*, *lqsd*, and *qdavg* as the variables of the model, and according to the above modeling steps (2), the results are as follows in Tab. III:

It can be seen from the z value, the significance levels of parameters are all $P < 0.05$, indicating that *svavg*, *lqsd* and *qdavg* have significant effect on the traffic accident risk of the freeway.

Odds ratio of the traffic flow parameters can be used to quantify the impacts of different traffic flow parameters on the risk of accident. The odds of an event is defined as the ratio of the probability with occurrence and the probability without occurrence, therefore, for the Logistic regression model:

$$\log[\frac{\Pr ob(event)}{\Pr ob(nonevent)}] = \beta_0 + \beta_1 X_1 + \cdots + \beta_k X_k \quad (3)$$

$$Odds = \frac{\Pr ob(event)}{\Pr ob(nonevent)} = e^{\beta_0 + \beta_1 X_1 + \cdots + \beta_k X_k} \quad (4)$$

It can be seen that when the i-th independent variable changes a unit, the change of odds is $Exp(\beta_i)$, that is:

$$Odds\ ratio = \frac{Odds(X_i + 1)}{Odds(X_i)} = Exp(\beta_i) \quad (5)$$

If the coefficient of an independent variable is positive, it means that odds will increase, and this value will be greater than 1; if the coefficient of an independent variable is negative, it means that odds will decrease, and this value will be less than 1; when the coefficient of an independent variable is 0, this value is equal to 1. The change percentage of odds ratio is $Exp(\beta_i) - 1$.

From Tab. 3, the odds ratios of *svavg*, *lqsd* and *qdavg* are respectively 0.96659, 1.49283 and 0.84473, indicating that when *lqsd* changes a unit, the risk of traffic accident will increase 49.3%, however when *svavg* and *qdavg* change a unit, the risk of traffic accident will decrease 3.3% and 15.5% respectively.

According to the analysis, when *lqsd* increases, it indicates that the difference between the volume of large car and the average value of volume of large car increases, that is the distribution of volume of large car is more discrete, which easily causes the traffic flow be not stable: congestion state a while, or free flow state a while,

Table 2. Correlation coefficients of variables

	lvavg	*svavg*	*lvsd*	*svsd*	*gavg*	*gsd*	*lqavg*	*sqavg*	*lqsd*	*sqsd*	*vdavg*	*vdsd*	*qdavg*	*qdsd*
lvavg	1	0.4937	-0.5472	-0.0728	-0.4874	-0.4464	0.3994	-0.1670	-0.0490	-0.4065	-0.3287	-0.4065	-0.1666	-0.2354
svavg	0.4937	1	-0.0094	-0.4246	-0.5947	-0.4970	0.0672	-0.1815	-0.0063	0.0651	0.1899	0.0651	-0.2569	-0.1977
lvsd	-0.5472	-0.0094	1	0.1581	-0.1039	-0.0828	-0.6033	-0.0393	-0.0303	0.8476	0.6059	0.8476	-0.0575	0.0688
svsd	-0.0728	-0.4246	0.1581	1	-0.2486	-0.1071	-0.2588	-0.4052	-0.4437	0.3720	0.4672	0.3720	-0.2678	-0.3330
gavg	-0.4874	-0.5947	-0.1039	-0.2486	1	0.6510	0.0584	0.6334	0.4389	-0.2841	-0.4409	-0.2841	0.6128	0.5728
gsd	-0.4464	-0.4970	-0.0828	-0.1071	0.6510	1	-0.0237	0.1312	0.0928	-0.1915	-0.3110	-0.1915	0.1591	0.1178
lqavg	0.3994	0.0672	-0.6033	-0.2588	0.0584	-0.0237	1	0.0833	0.1247	-0.5648	-0.5534	-0.5648	0.0930	-0.0139
sqavg	-0.1670	-0.1815	-0.0393	-0.40515	0.6334	0.1312	0.0833	1	0.6650	-0.2170	-0.3912	-0.2170	0.9164	0.8637
lqsd	-0.0490	-0.0063	-0.0303	-0.4437	0.4389	0.0928	0.1247	0.6650	1	-0.1653	-0.3398	-0.1653	0.5984	0.6192
sqsd	-0.4065	0.0651	0.8476	0.3720	-0.2841	-0.1915	-0.5648	-0.2170	-0.1653	1	0.7182	1	-0.2022	-0.0787
vdavg	-0.3287	0.1899	0.6059	0.4672	-0.4409	-0.3110	-0.5534	-0.3912	-0.3398	0.7182	1	0.7182	-0.3562	-0.2534
vdsd	-0.4065	0.0651	0.8476	0.3720	-0.2841	-0.1915	-0.5648	-0.2170	-0.1653	1	0.7182	1	-0.2022	-0.0787
qdavg	-0.1666	-0.2569	-0.0575	-0.2678	0.6128	0.1591	0.0930	0.9164	0.5984	-0.2022	-0.3562	-0.2022	1	0.6598
qdsd	-0.2354	-0.1977	0.0688	-0.3330	0.5728	0.1178	-0.0139	0.8637	0.6192	-0.0787	-0.2534	-0.0787	0.6598	1

Table 3. Regression analysis results

Variable	Coefficient (β_i)	Standard error	*z* value	Significance level ($\Pr(>\|z\|)$)	$Exp(\beta_i)$
Constant	0.48457	0.78750	0.615	0.04383	1.62348
svavg	-0.03398	0.01080	-3.145	0.00166	0.96659
lqsd	0.40067	0.15989	2.506	0.01221	1.49283
qdavg	-0.16874	0.08121	-2.078	0.03772	0.84473
The fitting parameters of the model: sample size=165、 $-2LL$=150.37 、 AIC=158.37					

Table 4. Regression model prediction accuracy

Actual Classification	Forecast Classification		Total (rate)
	Accident (rate)	Non-accident (rate)	
Non-accident	45 (34.09%)	87 (65.91%)	132 (100%)
Accident	20 (60.61%)	13 (39.39%)	33 (100%)

thereby increasing the risk of traffic accident; the vehicle speed at the location of the accident decreases quickly, and it propagates from downstream to upstream in the form of shock wave, therefore the speed of upstream coil will appear mutation, so *svavg* decreases, indicating that the risk of accident increases; when *qdavg* increases, that is the difference between the volume of large car and the volume of small car increases, easily causing the instability of traffic flow, so the risk of accident will increase too.

So, the change of *svavg*, *lqsd* and *qdavg* between 10 to 15 minutes prior to the time of the accident are the most effective to predict the probability of accident, which are used as the risk characterization factors of real-time safety assessment for the traffic flow operation.

The final model of the security situation deduction is as follows:

$$P(x_i)=\frac{1}{1+e^{-x_i\beta}}=\frac{1}{1+e^{-(0.48457-0.03398svavg+0.40067lqsd-0.16874qdavg)}} \quad (6)$$

Where, *svavg*, *lqsd* and *qdavg* are the average value of speed of small car, the standard deviation of volume of large car and the average value of volume difference between large car and small car between 10 to 15 minutes prior to the time of the accident respectively.

After specifying a reasonable threshold value, the calibrated model can predict the risk of freeway traffic accidents in real-time. Because in the total samples, the proportion of accidents is 20%, so the threshold is set to 0.2 in this paper, namely when the probability of the model output is greater than 0.2, it is discriminated as the traffic accident; and when the probability of the model output is less than 0.2, it is discriminated as the safety state with non-accident. The model prediction accuracy is shown in Tab. 4.

As shown in Tab. 4, the Logistic model based on the traffic flow data of Beijing-Tianjin-Tanggu freeway can predict 60.61% of accidents and 65.91% of non-accidents, and the total prediction accuracy is 64.85%. Therefore, the accident risk prediction model established in this section can use the real-time traffic flow data to predict the risk of traffic accident of freeway well. In order to reduce the false-positive rate in actual applications, it can improve the prediction threshold value of the model according to the specific human and material resources allocation situation, such as increasing to 0.4 or 0.5.

5 Conclusions

The traffic flow data of the nearest detectors prior to or back to the location of the typical accidents in Beijing-Tianjin-Tanggu freeway are extracted in this paper, on the basis, we select the binary Logistic regression method to establish the security risk prediction model and realize the real-time prediction of the probability of accident by using the average value of speed of small car, the standard deviation of volume of large car and the average value of volume difference between large car and small car between 10 to 15 minutes prior to the time of the accident, which providing the more scientific support for the traffic control and the traffic emergency management decisions of freeway.

Acknowledgment

Fund: Applied Basic Research Project of Ministry of Transport (2013319223170).

References

1. Beijing Traffic Research Center. *TMC Beijing Live-presentation Meeting Information* **12**, 15-18 (Beijing, 2005).
2. G. Wang, X. Guo, Y. Jiang, J. Hou. Journal of Highway and Transportation Research and Development **27(4)**, 155-158 (2010).
3. Y. Liang, Y. Chen, F. Ren. Journal of Highway and Transportation Research and Development **22(12)**, 105-108 (2005).
4. K. Chen, W. Zou, X. Li. Highways & Automotive Applications **3**, 59-61 (2012).
5. J. Chen, W. Zhou. Journal of Chang'an University (Natural Science Edition) **22(4)**, 52-54 (2002).
6. H. Guo, H. Shi. Journal of Highway and Transportation Research and Development **22(8)**, 102-105 (2005).
7. Q. Xu, R. Yang. Journal of Highway and Transportation Research and Development **22(12)**, 131-134 (2005).
8. Weil R, Wootton J, Garcia-ortiz A. Mathematics Computer Modeling **27(9)**, 257 (1998).
9. Lee Chris Choongho. *School of Civil Engineering, The University of Waterloo*, (Ontario, 2004).
10. Abdelaty M, Uddin N, Abdalla F, et al. Transportation Research Record **1897**, 88-95 (2004).
11. Abdelaty M, Pande A. Journal of Safety Research **36(1)**, 97-108 (2005).
12. Abdelaty M, Uddin N, Pande A. Transportation Research Record **1908**, 51-58 (2005).
13. Lee C, Hellinga B, Saccomanno F. Transportation Research Record **2749**, 67-77 (2003).
14. Lee C, Saccomanno F, Hellinga B. Transportation Research Record **1784**, 1-8 (2002).
15. Hossain M, Muromachi Y. *The 89th Annual Meeting of the Transportation Research Board*, 1-15 (Washington, D. C., 2010)

Application of the Analytic Hierarchy Process to Identify the Most Suitable Lessor of Freight Car Finance Leasing

Lei Lei[1], Yin Zhou[2] and Yongqing Xiao[2]

[1]*Graduate School, China Academy of Railway Sciences, Beijing 100081, China;*
[2]*Transportation and Economic Institute, China Academy of Railway Sciences, Beijing 100081, China*

Abstract. Finance leasing (also "equipment leasing") saves the cost, improves the efficiency and benefit, larger the manufacture supply channels, which is an optimal solution for equipment supply with uncertain freight demand. The article collects the definitions of Finance Leasing based on the four pillars theory of finance leasing, also divides the lessors in Freight Car finance leasing into three categories according to their major business: manufacturers, banks as the representative financial institutions, firms that specialized in finance leasing. To identify the most suitable lessor for each railway department, an indicator system is built and operated by Yaahp (a software) based on Analytic Hierarchy Process.

1 Introduction

In recent years, a number of national policies have been made to relax the social capital investment railway options, intended to promote the reform of railway investment and financing system. According to the World Bank 2014 China Transportation Special Top 10 (Private Capital for Railway Development), to attract more private sector investment to the railway sector, the most fundamental step is to create railway markets (e.g., equipment leasing) and entities (e.g., leasing companies, railway infrastructure companies, etc.) that are profitable and financially sustainable.

Under downward pressure on China's economy, demand of coal and steel has been declined in recent years and so has the railway freight volume. With the guidance of the management principle, less cost and more efficient, the China Railway Corporation (CRC) had stored up some locomotives and freight cars before October in 2016. That month, the demand of railway traffic sharply increased mainly driven by the policy that forbidding cars overloaded running on the road, the coals supply quantities and price were shooting up. Facing this challenge, CRC has an urgent problem, the shortage of freight car. In order to improve the railway traffic volume, railway administrations carrying the major goods of coal decided to relieve some stored up freight cars and even purchased some additional to support the transport capacity. On the other hand, the decline demand for rail freight in recent years has reduced the amount of new purchasing and old repairing freight cars, which leads the general overcapacity of vehicle manufacturing and maintenance enterprises.

Financial leasing can utilizes the railway equipment manufacturing enterprises idle capacity and helps the railway operators to overcome the shortage of equipment at the same time [1]. Because mobile equipment is recognized as a financial leasing assets [2-5], the objective of this paper is to explore the suitable lessor of freight car finance leasing using analytic hierarchy process.

The rest of the paper proceeds as follows: Section 2 reviews the literature about finance leasing in China. Section 3 establishes lessors selection analysis framework, describes our methodology and applying it to the problem. We present the strategy analysis results of lessor selection in Section 4 and conclude in Section 5.

2 Literature review of railway finance leasing

Finance lease also known as modern leasing or equipment leasing. The definition is not unified yet [6], there is no obvious solution as to where to draw the line on a consistent basis for all countries [7]. Based on the universally recognized theory of 'four pillars of financial leasing, concerning law, accounting standards, supervision and tax policies, introduced by famous international leasing experts Sudhir P. Amembal. The definition of financing lease in China for four pillars (Table 1) focus on different parts. The Ministry of Finance pays more attention to the risk, compensation, ownership and other closely related aspects for accounting, while *the Contract Law* focuses on the description of the rights and obligations of both parties. The administrative organs cares similar to *the Contract Law*, but a bit more rigorous, and tax departments emphasize ownership period and others closely related to tax treatment methods. In addition, the definition of finance leasing from accounting standards is various between countries. Fortunately, Chinese *Accounting Standard for Business Enterprises No. 21 - leasing*, to a large extent, is reference to the International Leasing Accounting Standards IAS 17 and the lease accounting standards of other countries, such as the Unite State. [8] The finance leasing in this paper is equipment leasing, that the lessor and the lessee have a financial contract of equipment that allow the separation of ownership and use of an asset.

Table 1. Definitions about the finance leasing

Institution	Policy / Document	Related Definition
The Ministry of Finance	*Accounting Standard for Business Enterprises No. 21 – leasing*	Finance lease is a lease that essentially transfers all the risks and rewards associated with the ownership of the asset. Its ownership may be or not be eventually transferred.
Standing Committee of the NPC	*Contract Law of the People's Republic of China* (1999)	The finance leasing contract is a contract that the lessors purchase the lease items from sellers according to lessees' target sellers and lease items, provide the use right to lessees and get rents paid by lessees.
Ministry of Commerce	*Provisional Measures on Administration of Domestic Leasing Investments of Qualified Foreign Institutional Investors* (2005)	The term "finance leasing business" as mentioned in the present Measures refers to the business of the lessors leasing the use right to the lessee on the basis of the lessees' choice of the seller and the leased items, and collecting rent from lessees.
China Banking Regulatory Commission	*Measures for the Administration of Finance Leasing Companies* (2014)	The term "financial leasing" as mentioned in the present Measures means that the lessors should lease items obtaining from the supplier to lessees according to lessees' selection or requirement and receive rents, all these should be in accordance with the contract.
The State Administrat-ion of Taxation	*Notification About Turnover Tax of Finance Leasing* (2000)	Finance leasing refers to the equipment leasing business has the nature of financing and the transfer of ownership: Lessors purchase equipment and lease to the lessee according to the lessees' required specifications, models, performance and other conditions. Within the contract period, the ownership of the equipment belongs to the lessor, the lessee has only the right of use. On the expiry of a contract, the lessee has the right to purchase the equipment by the residual value to own the ownership of the equipment.

Source: From the policy documents in this table.

Original in World War II, the finance lease has been studied extensively in the extant literature, but there are many puzzles that have not been solved so far. For example, the debt and lease are complementary or substitutional. On the other side, the academic communities basically reach consensus on that the finance leasing with industry bias exists, especially in the air transport and retail trade, services and public utilities. [9]

Although carried out late comparatively, Chinese finance lease is developing fast since 2010 and China has become second largest leasing market in the world. During 2009 to 2012, rate of annual growth is up to 61%, and business volume reached $88.9 billion in 2014. [7] Due to the date qualification, theoretical research lags behind the actual business in China. More quantitative academic research compared with qualitative research has been carried out there.

The financial leasing research of railway focuses primarily upon the necessity. Du Meijie (2010) compares the difficulties and costs between finance lease and bank loan. It is suggests that the finance leasing is not only a supplement of current railway financing, but also works well in optimizing the financing structure.[4] Wang Zhaocheng (2003) analyzes the current situation of railway technology and equipment leasing in China and foreign countries, has a conclusion that the modern leasing plays an important role in Chines railway development.[10] Chen Bo Shi, Ding Huiping (2006) points out there is a capital transmission relationship between the railway equipment manufacturing enterprises and rail transport enterprises. Because in current capital market environment in China, both sides are facing the same challenges, that is the single financing channels cannot meet the financing needs, which providing an opportunity for the development of finance lease. [11] Only a few qualitative research are conducted about the necessity (or called advantage). To analyze the economic feasibility of the finance lease model of railway transportation equipment, Gan Kaiyuan (2006) compares the net cash value under some assumptions. [3]

The economic feasibility of the railway finance leasing is discuss as well. The Chinese finance lease was still in its childbearing period, going through a tortuous suffering in 2008. Che Tanyan (2008) shows great concern for the finance lease whether it can catch up with the takeoff of equipment manufacturing industry, considering the difficulties of funding for leasing enterprises and the risk of outstanding payments among lessees. [12]

Involved with the specific strategies, scholars study from both the lessor perspective. Ma Yuanbing (2012) explores the regulation and risk management ability improvement for the rail leasing company.[13] He Lin (2008) analyses risk factors of the railway finance leasing and has a conclusion that the credit risk of the finance lease is higher than the bank loan for the lessee. [14]

3 Freight car finance leasing selection process

The main division of the lessors based on the major business, usually divided into three categories: manufacturers, banks as the representative financial institutions and firms that specialized in finance leasing.

(1) Manufacturers

The international leasing of the railway industry started late. At present, there's few finance leasing enterprises with the background of manufacturers, and the only one named CRCC Investment & Leasing Co., Ltd. is a wholly owned subsidiary of CRRC Corporation Limited (CRRC). Although there are railcar producing and leasing companies are multinational, such as General Electric Company (GE) and TTX Company, multinational manufacturers and lessors are less likely to enter the domestic railway finance leasing market in view of the policy "Made in China".

(2) Financial institutions represented by banks

The amount of finance leasing companies controlled effectively by the financial institutions, especially with banking background, grow up so fast. Companies including ICBC Leasing, China Development Bank Leasing, Minsheng Financial Leasing, Bank of Communications Financial Leasing provide leasing services to transportation services, such as aviation, ship and marine equipment, but there are basically no lessors engaged in freight car leasing.

These companies mainly owned by the national commercial banks, with a natural advantage in the risk prevention, are far superior to other two categories (manufacturers and specialized firms) in the lessees' risk rating, contracts management, the application of laws and regulations. Furthermore, they have more abundant human resources and dedicated departments to manage risk generally.

(3) Firms specialized in finance leasing

In fact, only a small number of specialized firms in railway equipment finance leasing exist in China.[3] They are a Sino-foreign joint venture named CHINA RAILWAY LEASING, a foreign invested enterprise named IMC-GATX FINANCIAL LEASING (SHANGHAI). Due to the incomplete rail finance leasing market, IMC-GATX China focuses on shipping leasing now, only CHINA RAILWAY LEASING is still keeping the locomotive rental business, which might benefit from its investor's major market share in railway industry, a state-owned enterprise with a long history named China Railway Materials Company. A private company named Guangzhou Zhongche Railway Vehicles Equipment had been shifted the business to the vehicle air-conditioning equipment manufacturing and was purchased by state-owned enterprise.

As a traditional and worldwide financial leasing object, railway car leasing needs different lessor around countries and companies. In North American countries, the main lessors are manufacturers, while in British they are companies controlled by the banks. [4] So it's absolutely necessary to identify the most suitable lessor for Chinese railway and find the practicable solution of cooperation between lessors and lessees.

3.1 Lessors selection analysis framework

From the Section 2 of paper, former literature did not so much work according to the detailed classification of railway lessors. On the one hand, it's because the finance leasing market in our country is not perfect or complete, the enterprises with high market share involved in finance leasing have not yet opened up the railway leasing market. On the other hand, the railway is a government regulation industry [15] and has the lowest degree of marketization in the field of transportation [16]. The development of finance leasing of rail is much slower than that of aviation, shipping, medical, engineering equipment and others.

In this paper, the competitiveness of railway freight car lessors stems from financing cost (FC), supporting services (SS), management capacity (MC) and four pillars theory of finance leasing (FP).(Shen Yanli, 2008 [17] , Zhen Lin, 2015 [18]). Analysis framework of lessor selection is established (Fig. 1). The four pillars theory has been discussed in Section 2. The most influencing factors of financing cost include interest rate, lease term, margin and rent payment method [19]. The most influencing factors are service quality, cooperation experience, product type. And management ability are reflected as risk management and efficiency of contracts implementation.

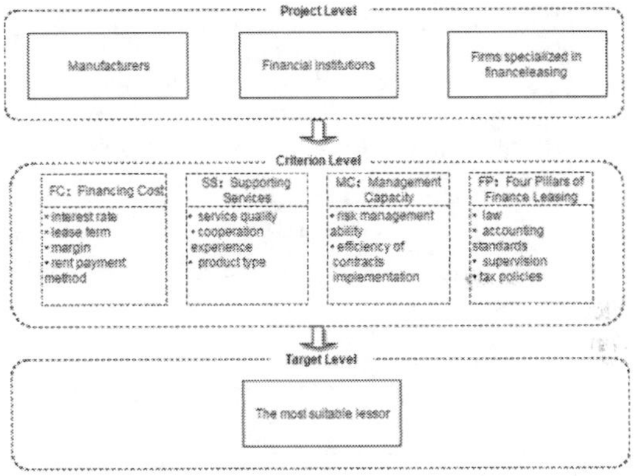

Figure 1. Analysis framework of lessors selection

3.2 Date and method

Delphi method, Analytic hierarchy process (AHP), DEA and Fuzzy Comprehensive Evaluation are alternative evaluation methods. Refer to Dhir S, Marinov M V, Worsley D. (2015) [20], this paper selects AHP designed by Thomas L.Saaty in the 1970s as the evaluation method. The reason for choosing AHP for application in this study was due to its primary purpose being for decision making of strategic level.

According to the general steps of AHP: First, defining the problem and desired solution. Secondly, structuring of the hierarchy. Third, construction of the pairwise comparison matrices. Fourth, complete all the pairwise judgments using the fundamental scale. Fifth, transferral of the judgments into matrices. Sixth, perform steps 3, 4 and 5 for all pairwise comparisons. Seventh, use the synthesis method to produce the eigenvectors and priorities for each matrix and hence calculate the final priorities. [21]

We invited twenty experts in railway management departments to do the questionnaire survey and fortunately twenty are effective questionnaires recovered. We analyze this program by Yaahp, a business software specialized for AHP method.

3.3 Application of the analytic hierarchy process

3.3.1 constructing the analytic hierarchy

The hierarchical analysis structure is the same as Fig. 1, the first layer corresponds to the project level, the second layer corresponds to the criterion level, the third layer corresponds to the target level.

3.3.2 Weight ascertainment

First, construct the judgment matrix. According to the commonly used 1 ~ 9 scales of comparisons (Saaty (1999) [21], Du Dong (2008) [22]), experts compare and numerically value each comparison of criteria (see Criterion Level in Fig. 1) using questionnaire. We construct the pairwise comparison matrix by collecting the experts questionnaire. And the resulting matrix could be seen as Table 2-6.

Secondly, test consistency for each matrix through characteristic root test. Calculate characteristic root λ_{max} of equation ($AW = \lambda_{max}*W$, A and W are matrix), then get priorities of criteria after the normalization. More detail calculation steps can be seen in Du Dong's book (Du Dong, 2008) [22].

Table 2. Criteria 1- Pairwise Comparison Matrix

The most Suitable Lessor	FC	SS	MC	FP	W_i
FC	1	1/3	1/3	1/7	0.0752
SS	3	1	1/3	1/5	0.1512
MC	3	3	1	1/5	0.5083
FP	7	5	5	1	0.2653

Note: FC is short for financing cost, SS is short for supporting services, MC is short for management capacity and FP is short for four pillars theory of finance leasing.

Yaahp computing result is λ_{max}=4.1981, C.R. (Consistency Ratio) = 0.0742 < 0.10, pass the consistence test.

According to the importance order (Table 2), judgements that MC is the strongest factor to identify the most suitable lessor, followed by FP, SS and FC.

Table 3. Criteria 2- Priorities of Finance Cost (FC)

Financing Cost (FC)	Margin	Rent Payment Method	Lease Term	Interest Rate	W_i
Margin	1	3	3	1/3	0.2653
Rent Payment Method	1/3	1	1/3	1/5	0.0752
Lease Term	1/3	3	1	1/3	0.1512
Interest Rate	3	5	1/3	1	0.5083

Yaahp computing result is λ_{max} =4.1981, C.R. = 0.0742 < 0.10, pass the consistence test.

Table 4. Criteria 3-Priorities of Supporting Services (SS)

Supporting Services (SS)	Service Quality	Cooperation Experience	Product Type	W_i
Service Quality	1	1/3	3	0.2583
Cooperation Experience	3	1	5	0.637
Product Type	1/3	1/5	1	0.1047

Yaahp computing result is λ_{max}=3.0385, C.R. = 0.0370 < 0.10, pass the consistence test.

Table 5. Criteria 4-Management Capacity (MC)

Management Capacity (MC)	Efficiency of Contracts Implementation	Risk Management Ability	W_i
Efficiency of Contracts Implementation	1	7	0.25
Risk Management Ability	1/7	1	0.75

Yaahp computing result is λ_{max} =2.0000, C.R. =0.0000 < 0.10, pass the consistence test.

Table 6. Criteria 5- Four Pillars of Finance Leasing (FP)

Four Pillars of Finance Leasing (FP)	Law	Tax Policies	Supervision	Accounting Standards	W_i
Law	1	7	5	7	0.6471
Tax policies	1/7	1	1/3	3	0.1032
Supervision	1/5	3	1	3	0.1908
Accounting Standards	1/7	1/3	1/3	1	0.059

Yaahp computing result is λ_{max} =4.2281, C.R. =0.0854 < 0.10, pass the consistence test.

The resulting matrixes complete consistency check and each criteria's priorities (W_i) are calculated (see Table 2-6).

3.3.3 Analysis and evaluation of results

Fig. 2 depicts the final calculated priorities for the lessors produced by Yaahp. The most suitable lessor being manufacturer, followed by the financial institutions and then the specialized leasing companies, based on the AHP analysis conducted. During the selection, the lessee is most concerned about the management capacity, followed by the four pillars, supporting services and financing costs, which supports the idea that using only accounting factors to decide the lessor is imperfect

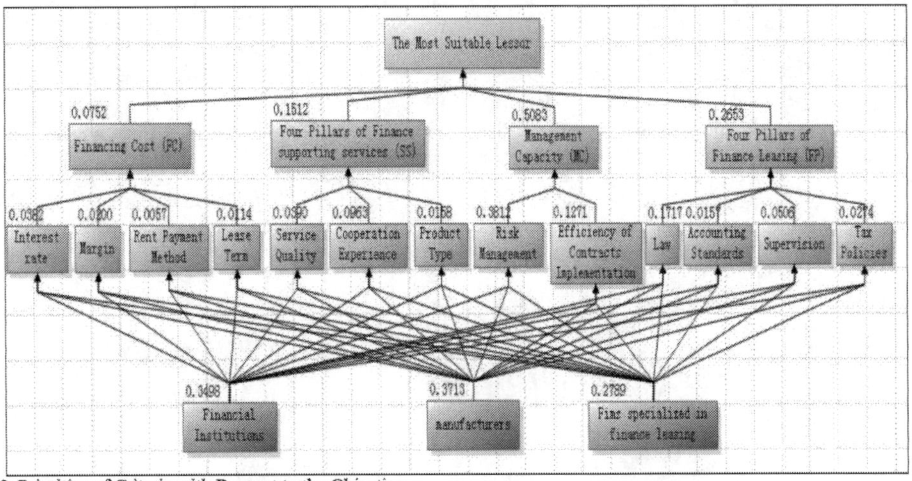

Figure 2. Priorities of Criteria with Respect to the Objective

4 Strategy analysis of most suitable lessors

For CRC, locomotives, freight cars and high-speed trains mainly purchased through competitive bidding procurement organized by the firm's headquarters, and rarely small part of rail equipment could be purchased independently by the railway administrations in CRC and other private companies. Different lessees with distinct rights should pick up which kind of lessors are discussed in the following section.

(1) At this stage, CRC carrying out cooperation with CRCC in finance lease is the best choice. According to the results of AHP analysis in Section 3, to carry out cooperation with the manufacturer contributes to lower transaction costs, which is the same finance lessor selection as North America and Europe. However, since the monopoly of both manufacturing and using rail cars in China, it would be easy to form a bilateral monopoly market if CRC were to lease freight cars and others from CRCC, and then the rails would have to pay the excess rent. In 2006, all three British railway leasing companies were accused of over-pricing behavior by Department for Transport.

(2) For the long-term stage, the financial institutions combined with manufacturers could cooperate with CRC to develop leasing business. Manufacturers lack the experience of preventing financial risks, while finance leasing companies can prevent risk effectively, have enough funds and lower financing costs, but lack management experiences and professional persons about the leasing objects. These two sides united to form alliance is the inevitable and preferable choice for the further development of finance leasing. For more information about the benefits of this type alliance, see Chen Boshi and Ding Huiping (2006) [11].

In addition, railway enterprises with sufficient capital could establish internal finance leasing companies. A precedent as China Railway Group Limited set up an internal financial leasing company named China Railway Capital Financing Leasing in 2015 to do equipment management, equipment manufacturing transformation and upgrading, innovation of investment and finance. Besides, the financial institutions in CRC could become a shareholder of joint-stock company doing finance leasing to make full use of the resources.

(3) Under special conditions, the railway administrations or the jointly owned railway companies would cooperate with firms specialized in finance leasing as a supplement. For most of the jointly owned railways and regional railways, it is hard to issue bonds or get concessional loans. A realistic financing channel that might solve the funding problems for equipment purchase is finance lease. As specialized financial leasing agencies are highly efficient and close to the local railway companies geographically, they could response to emergency quickly which would meet local transport demand, so it's also a good choice for the railway administrations, the jointly owned railway companies and regional railways.

5 Conclusions

Nowadays Chinese finance leasing business develops rapidly and maturely in the fields of aviation, shipping and rail transportation, but the railway finance leasing is lagging behind other transportation mode, both in the development speed and scale. Although only a few rail car lessors are active in market currently, they are diverse in form, which are Sino-foreign joint ventures, state-owned enterprises and multinational companies. And banks and other financial institutions are potential lessors as well. It seems that Chinese rail car finance leasing market is a considerable potential market.

At this stage, the best cooperation is between CRC and manufacturers named CRCC. And the financial institutions combined with the manufactures will be beneficial supplements References in the long term. Considering the risk of uncertain freight demands, the cooperation of some railway administrations and the specialized firms would be another choice. If the finance leasing market improved and completed constantly, the cooperation between rails and suitable lessors could further stimulate the market power to provide excellent advisory services for rails, to improve the rails capital structure and the management level.

Only China Development Bank Leasing is a listed company, so data acquisition is difficult. It suggests that the evaluation progress could add financial indicators and the lessor's finance leasing mode could be carried out in the case of sufficient data. Furthermore, the bargaining game process of the monopoly manufacturer lessor and the lessee should be worth looking into.

Acknowledgement

Thanks to Wang Huaixiang, Song xiaoman, Zhu Jiafa, Zhang Mengdi for their invaluable guidance during the review process. We also gratefully acknowledge the financial support of Technology Research and Development Program of China Railway (Grant No. 2016F018 and No.J2016Z009). All errors are the responsibility of the authors.

References

1. B. Chen, Q. Li, H. Ding. Chinese Railways **9**, 40 (2003)
2. M. Lawrence, G.O. *Private Capital for Railway Development in China Transport Topics No.10.* World Bank (2014)
3. K. Gan. *The Innovation Research for Financing China's Railway by Financial Lease Model*, 28 (2007)
4. Du Meijie. China Economic & Trade Herald **20**, 42 (2010)
5. Y. Zhang. Railway Transport and Economy **3**, 5 (2008)
6. H. Zhang. *Financing Leasing and its Application in the Railway*, 3 (2007)
7. White Clarke Group. *Global Leasing Report*, 5-7 (2015)
8. L. Zhu. Commerial Accounting **1**, 18 (2014)
9. Morais A I. Academia Revista Latinoamericana de Administración **26(3)**, 432 (2013)
10. Z. Wang. Chinese Railways **10**, 11-14 (2003)
11. B. Chen, H. Ding. Review of Investment Studies **12**, 25-29 (2006)
12. T. Che, W. Hu, X. Wang. Journal of Dalian Maritime University: Social Science Edition **2**, 36-39 (2008)
13. Y. Ma. *Risk Management Study of Freight Car Lease Business for ZB Company*, 27-41 (2012)
14. L. He, X. Xiao, H. Ding. Journal of Beijing Jiaotong University Social Sciences Edition **2**, 78 (2008)
15. B. Chen. *Research on Innovating and Realizing of Fiannce Lease in China Rail Transportation Equipment*, 78 (2008)
16. X. Kang. Journal of Railway Engineering Society **8**, 123 (2013)
17. Y. Shen. *A Study on the Competitiveness of Chinese Financing Lease*, 26-32 (2006)
18. Z. Lin, *Factors Influencing Automobile Financial Leasing and Risk Control– An Empirical Study on China Automobile Leasing Marke*, 41 (2015)
19. D. Han. Low Carbon World **1**, 144-145 (2015)
20. Dhir S, Marinov M V, Worsley D. Case Studies on Transport Policy **4**, 431-448 (2015)
21. Saaty T L. IEEE Transactions on Systems, Man, and Cybernetics SMC **15(3)**, 450-452 (1985)
22. D. Du, Q. Pang, Y. Wu. *Modern Model of Comprehensive Evaluation and case study*, (2008)

Design of a Multi-layer Lane-Level Map for Vehicle Route Planning

Chaoran Liu , Kun Jiang, Diange Yang and Zhongyang Xiao

State Key Laboratory of Automotive Safety and Energy, Collaborative Innovation Center of Intelligent New Energy Vehicle, Department of Automotive Engineering, Tsinghua University, Beijing 100084, China

Abstract. With the development of intelligent transportation system, there occurs further demand for high precision localization and route planning, and simultaneously the traditional road-level map fails to meet with this requirement, by which this paper is motivated. In this paper, the three-layer lane-level map architecture for vehicle path guidance is established, and the mathematical models of road-level layer, intermediate layer and lane-level layer are designed considering efficiency and precision. The geometric model of the lane-level layer of the map is characterized by Cubic Hermite Spline for continuity. A method of generating the lane geometry with fixed and variable control points is proposed, which can effectively ensure the accuracy with limited number of control points. In experimental part, a multi-layer map of an intersection is built to validate the map model, and an example of a local map was generated with the lane-level geometry.

1 Introduction

Digital maps play an important role in current vehicle applications, e.g. vehicle localization and route planning. With more attention on intelligent transportation system, advanced driving assistance system and even driverless vehicle, the improvements in accuracy and richness are required in digital map technology [1]. When it is available to determine the vehicles' position precisely in lane level, there are a lot of probable benefits we may obtain, i.e. the transportation officials and researchers may determine distinctions in traffic conditions for different lanes on a freeway via probe vehicles [2].

Most of existing digital maps are based on road level data, which may illustrate basic information and provide useful applications for users but ignore some precise details that are essential for vehicles' high precision localization and path guidance. The improvements of accurate localization sensors such as RTK GPS and other on-board sensors such as IMU and LiDAR made the enhanced maps become possible. On the purpose of increasing utility of digital maps in advanced vehicle applications, creating the lane-level map is becoming a prevailing interest among researchers worldwide [3]. Compared with the traditional road level map, the lane level map is enhanced with massive data, which are capable of improving the accuracy from about 10 meter to decimetre level or even centimetre level and illustrating more precise geometry in line with the real situations.

The map model is supposed to be created for routing planning before geometrical representation because vehicle navigation is established on a network consisting of vertices (e.g., intersections) and edges (e.g., roads, lanes). Thus, the map model is supposed to be an abstract representation of the map database [4]. In terms of map model, there are a number of map standards containing specific explanations on the map model which we can refer to. In traditional road level map, GDF, KiWi, Navteq are mainstream digital map standards which have been employed widespread for decades [5], however, these road-level model are not directly applicable to lane level. RNDF is the road network definition file designed for DARPA urban challenge, in which the basic structure segment-lane-waypoint are included to provide basic information to driverless vehicles [6]. Nowadays, OpenDRIVE and NDS are well-known map standard providers, but their complicated map model and inaccessibility to public (NDS) cause difficulties for practical uses. Besides the standards above, Qing Zhu et al. [7] establish a 3-D road network which is composed of three parts: roadway centerline, carriageway, and lane. Tao Zhang et al. [4] design a lane-level map model including details on intersections and lanes. Jiang Liu et al. [8] generate enhanced intersection model and use circular to illustrate the virtual lane inside the intersection. Thus, few studies are able to provide an applicable map model with completeness and simplicity for route planning.

Another significant issue about lane-level maps is the geometrical representation of the lanes. Expanding from traditional maps, some lane illustrations in enhanced maps continue using the polyline [9]. Du Jie et al. [2] employ the piecewise polyline to approximate the centreline in a lane. Betaille et al. [3] in their lane-level enhanced map apply the clothoids for the description of the road lanes in a digital map. Circular arc spline is also applied in the map geometrical model [10]. Compared to there methods mentioned, Cubic Hermite Spline (CHS) may possess some advantages, because it can generate various lines, such as, straight lines, arcs, and even clothoids [11], therefore the whole map would share one type of line to accurately generate the geometrical shapes of different lanes. And there have been several studies employing the CHS to describe the shape of lanes, however, the method mentioned in [4] applies CHS to the whole lanes and even virtual lanes in the intersections without distinguishing the control points with different functions.

This paper is organized as below. Firstly, in section 2, the multi-layer lane-level map model is proposed, which includes three layer and other significant details. Then a method of geometrical representation is established by applying the CHS. Finally, the experimental validations are performed in section 4.

2 Multi-layer lane-level map model

In terms of convenience of geographical illustration and efficiency of further applications like route planning, we propose a multi-layer lane-level map model, which contains three layers: road level layer, intermediate layer, and lane level layer.

Road-level layer in this map model reserves most of traditional mathematical expressions for current existing map model, containing roads and intersections, which is intended for making use of ripe routing algorithms based on road level maps. The intermediate layer acts as a bridge between the upper layer and the lower one, in which relationship between some sets is stored for routing planning and other applications. And the third layer is designed to express the lane-level details, not only the lane sets on a road and more detailed intersections, but also geometrical elements such as high-precision points on the centre line of a lane, lane lines and so on. An example is Figure 1

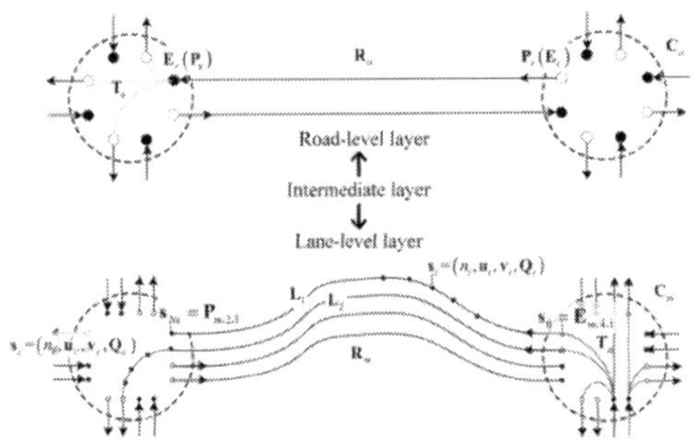

Figure 1. Multi-layer map model

2.1 Road-level layer

According to intuition that a road network is supposed to include roads and intersections, thus, a road-level layer is expressed as follow,

$$\mathbf{W}_a = (\mathbf{C}_a, \mathbf{R}_a) \quad (1)$$

where, $\mathbf{W}_a$ is the whole road-level network, $\mathbf{C}_a$ is the set of road-level intersections, and $\mathbf{R}_a$ is the set of roads.

The road-level intersection is defined as,

$$\mathbf{c}_a = (\mathbf{P}_c, \mathbf{E}_c, \mathbf{T}_a) \quad (2)$$

where, $\mathbf{P}_c$ is the set of road-level nodes entering this intersection, and $\mathbf{E}_c$ is the set of road-level nodes leaving this intersection. $\mathbf{T}_a$, road-level traffic matrix, indicating if there is a topological connection between two nodes, defined as,

$$\mathbf{T}_a = \begin{array}{c} \\ E_{c,1} \\ \vdots \\ E_{c,j} \end{array} \begin{array}{c} P_{c,1} \ \cdots \ P_{c,i} \\ \begin{bmatrix} \mathbf{t}_{a,1,1} & \cdots & \mathbf{t}_{a,i,1} \\ \vdots & \ddots & \vdots \\ \mathbf{t}_{a,1,j} & \cdots & \mathbf{t}_{a,i,j} \end{bmatrix} \end{array} \quad (3)$$

where, the element $\mathbf{t}_{a,i,j}$ represents whether a vehicle could drive from the entering node $P_{c,i}$ to leaving node $E_{c,j}$, and owns mathematical expression $\mathbf{t}_{a,i,j} = (f_a, m_a)$, as shown in Figure 2, in which f_a denotes whether (=1) or not (=0) a vehicle may drive from $P_{c,i}$ to $E_{c,j}$, and m_a is the way of passing, e.g. turning left, turning right, going straight, U-turn.

The road-level road is unidirectional, which means in this model, there generally are two roads between two adjacent intersections. A road is expressed as,

$$\mathbf{r}_a = (\mathbf{P}_r, \mathbf{E}_r, \mathbf{Q}_a) \quad (4)$$

where, $\mathbf{P}_r$ is the set of road-level nodes entering this road, and $\mathbf{E}_r$ is the set of road-level nodes leaving this road. $\mathbf{Q}_a$ includes the road class k_r and road length l_r.

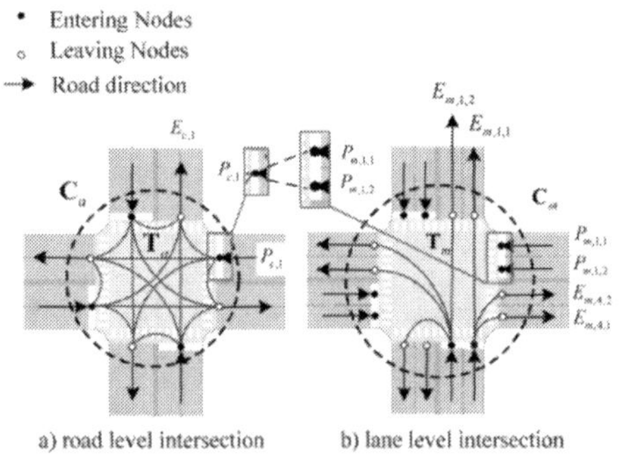

Figure 2. Traffic matrix and intermediate layer

2.2 Intermediate layer

In this layer, a logic connection between upper and lower layer is designed. On the purpose of providing higher precision data and conserving the advantages of traditional maps, the intermediate layer serves as a library of corresponding relationship from road level to lane level.

In an intersection $\mathbf{c}$, $\mathbf{P}_c$ is the set of its road-level incoming nodes. And without losing generality, let us assume this intersection has four road-level incoming nodes from different incoming directions, which could expressed as $\mathbf{P}_c = (\mathbf{P}_{c,1}, \cdots, \mathbf{P}_{c,4})$.

And generally there are at least one lanes for each road at an intersection. As Figure 2 shows, the road-level node $\mathbf{P}_{c,1}$ represents two lane-level nodes $\mathbf{P}_{m,1,1}, \mathbf{P}_{m,1,2}$.

Thus, intermediate layer is expressed as follow,

$$\mathbf{P}_{c,i} = (\mathbf{P}_{m,i,1}, \cdots, \mathbf{P}_{m,i,k_i}) \quad (5)$$

2.3 Micro layer in lane level

Lane-level layer should not only possess more data with higher accuracy which could supplement the deficiency of road-level layer, but also provide information of complete road network for navigation.

To achieve this goal, the lane-level layer is defined as,

$$\mathbf{W}_m = (\mathbf{C}_m, \mathbf{R}_m) \quad (6)$$

where, $\mathbf{C}_m$ is the set of lane-level intersections, and $\mathbf{R}_m$ is the set of roads.

The road-level intersection is defined as,

$$\mathbf{c}_m = (\mathbf{P}_m, \mathbf{E}_m, \mathbf{T}_m) \quad (7)$$

where, $\mathbf{P}_m$ is the set of lane-level nodes entering this intersection, and $\mathbf{E}_m$ is the set of lane-level nodes leaving this intersection. $\mathbf{T}_m$, lane-level traffic matrix, indicating whether there is a topological connection between two lane-level nodes, which is defined as,

$$\mathbf{T}_m = \begin{array}{c} \\ E_{m,1} \\ \vdots \\ E_{m,j_m} \end{array} \begin{array}{c} \begin{array}{ccc} P_{m,1} & \cdots & P_{m,i_m} \end{array} \\ \begin{bmatrix} \mathbf{t}_{m,1,1} & \cdots & \mathbf{t}_{m,i_m,1} \\ \vdots & \ddots & \vdots \\ \mathbf{t}_{m,1,j_m} & \cdots & \mathbf{t}_{m,i_m,j_m} \end{bmatrix} \end{array} \quad (8)$$

where, the element $\mathbf{t}_{m,i_m,j_m}$ implies whether and how a vehicle could drive from the entering node P_{m,i_m} to leaving node E_{m,j_m} ,and its mathematical expression is $\mathbf{t}_{m,i_m,j_m} = (f_m, m_m, \mathbf{S}_c)$, in which f_m denotes whether (=1) or not (=0) a vehicle may drive from P_{m,i_m} to E_{m,j_m}, and m_m is the way of passing which shares the same definition with that in road-level. The $\mathbf{S}_c$ represents the set of control points on the centre line of the virtual lane between two lane-level nodes in an intersection, as Figure 2.

The lane-level road $\mathbf{r}_m$ contains all lanes on the road, thus we define,

$$\mathbf{r}_m = (\mathbf{L}, c_{in}, c_{out}, \mathbf{Q}_r) \quad (9)$$

where, $\mathbf{L}$ is the set of lanes, and c_{in} and c_{out} are the indices of entrance and exit intersections according to traffic directions on the road respectively, and $\mathbf{Q}_r$ is the attributes of the road.

Each lane in the road, i.e. element in the set of lanes, is defined as,

$$\mathbf{l} = (\mathbf{S}_l, \mathbf{Q}_l) \quad (10)$$

where, $\mathbf{S}_l$ is the set of control points on the centre line of the lane, and $\mathbf{Q}_l$ is the set of attributes, including lane width l_w, lane length l_e and speed limit V.

2.4 Node and control point

There are nodes in road-level and lane-level (Notice that nodes in intermediate layer come from the other two layers.), which only exist at the connection knots between intersections and roads(lanes) and are defined as,

$$\mathbf{N} = \{\mathbf{u}_N, \mathbf{v}_N\} \quad (11)$$

where, $\mathbf{N}$ represents the entering or leaving nodes in this map model, including $\mathbf{P}_c$, $\mathbf{E}_c$ in the road-level intersections, $\mathbf{P}_r, \mathbf{E}_r$ in the road-level roads, and $\mathbf{P}_m, \mathbf{E}_m$ in the lane-level intersections. $\mathbf{u}_N$ is the location of the node, and $\mathbf{v}_N$ is the traffic direction of the node.

Besides nodes, control points also play an important role in layered map, especially lane-level layer. Control points in intersections and lanes both contain precise positions, tangent vectors, indices, and other attributes, and are defined as,

$$\mathbf{s} = \{n_s, \mathbf{u}_s, \mathbf{v}_s, \mathbf{Q}_s\} \quad (12)$$

where, n_s is the serial number, and $\mathbf{u}_s$ is the position which is two or three-dimension vector, and $\mathbf{v}_s$ is the tangent vector on the node, and $\mathbf{Q}_s$ involves attributes, such as lateral sequence number of the lane, line types of left lane line and right lane line for control points on a lane.

Particularly, in a common scenario, the nodes at the start/end of a lane share the same positions and tangent direction with the first and the last control points.

3 Geometrical representation

The detailed geometry in road level is not necessary because the detailed information and precise geometry is presented in lane level and it is easy to obtain the lane-level geometrical information via intermediate layer from road level and on the other hand, the road level nodes are easy to locate, simple lines such as straight line could illustrate the connection and topology in road level completely. Thus, according to our multi-layer model, the lane-level geometry is discussed as below.

3.1 Lane-level geometry

In order to achieve the global conformity and continuity, as we mentioned before, CHS curve is employed in this part. CHS is a spline in which each segments between two control points is a three order polyline represented by Hermite form.

Assuming that two given control points are $\mathbf{s}_i$, $\mathbf{s}_{i+1}$, where $\mathbf{s}_i = (n_i, \mathbf{u}_i, \mathbf{v}_i)$, and definitions of symbols are the same with Equation (12), then the CHS generated by these two points is presented as,

$$Q(n) = f(x) = \begin{bmatrix} 2x^3 - 3x^2 + 1 \\ x^3 - 2x^2 + x \\ -2x^3 + 3x^2 \\ x^3 - x^2 \end{bmatrix}^T \begin{bmatrix} \mathbf{u}_i \\ \mathbf{v}_i \\ \mathbf{u}_{i+1} \\ \mathbf{v}_{i+1} \end{bmatrix} \quad (13)$$

where, $x = (n - n_i)/(n_{i+1} - n_i)$.

DOI:10.1051/matecconf/201712403001

It is easy to get the $Q(n_i)=\mathbf{u}_i$, $Q(n_{i+1})=\mathbf{u}_{i+1}$, $Q'(n_i)=\mathbf{v}_i$, $Q'(n_{i+1})=\mathbf{v}_{i+1}$, thus at each control point the first order continuity is realized, which may offer convenience for vehicle path planning due to global tangential continuity.

3.2 Control Points Choosing

The control points serve as the shape points in the lane level geometry, and simultaneously some of them are supposed to be feature points, of which on two sides there are some different attributes, at particular positions, e.g., the control points at the start/end of a lane. Though this kind of control points own the same form with the common control points, because these do not change with other variable control points, we call them the fixed control points.

3.2.1 Fixed Control Points

The fixed control points $\mathbf{S}_f$ are determined by the lane attributes. Between two adjacent fixed control points the lane attributes, except the shape of the lane, keep the same, and the lane attributes include speed limit, lateral sequence number of the lane, line types of lane lines, and etc. Some details are displayed as below.

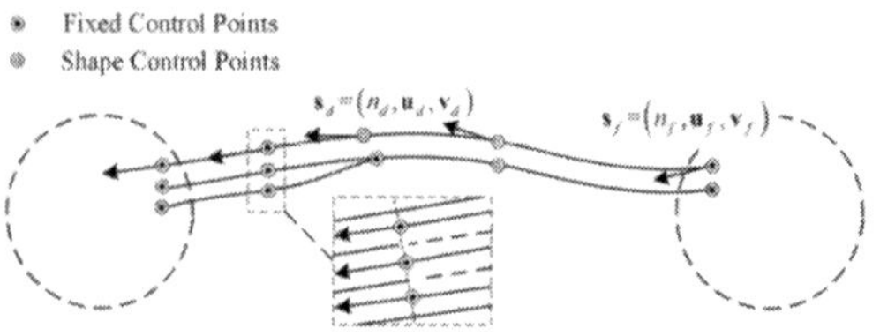

Figure 3. Control Points on lanes (Red dashed line indicates the position where the type of lane lines changes.)

3.2.2 Variable Control Points

The determination of shape control points occurs in two adjacent fixed control points in a lane. Assume that a series of sampled points representing the centreline of a lane are known, expressed as $\mathbf{S}_t=(\mathbf{s}_{t,1},\cdots,\mathbf{s}_{t,j},\cdots,\mathbf{s}_{t,N_t})$, and the first and last points are fixed points, i.e. $\mathbf{s}_{t,1},\mathbf{s}_{t,N_t}\in\mathbf{S}_f$, a stepwise approach on determining the series of shape control points $\mathbf{S}_d=(\mathbf{s}_{d,1},\cdots,\mathbf{s}_{d,i},\cdots,\mathbf{s}_{d,N_d})$ is proposed.

In this stepwise method, the distances between two adjacent shape control points are variable, which is determined by,

$$T_h \ge e = \max_{j=1,\cdots,N}\min_{x\in[0,1]}\left\|\mathbf{u}_t(j)-f_{\mathbf{s}_{d,i},\mathbf{s}_{d,i+1}}(x)\right\|_2 \tag{13}$$

Where, T_h is the threshold designed, which can specify the accuracy of the geometrical representation, and e is the measurement of the gap between CHS curve generated and corresponding sampled point, and $\mathbf{u}_t(j)$ is the position extracted from $\mathbf{s}_{t,j}$ which are located among two adjacent shape control points, as shown in Figure 4.

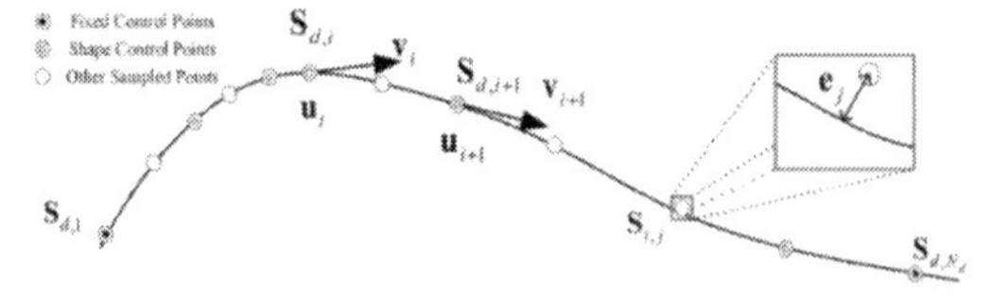

Figure 4. Shape control points choosing

The stepwise approach is started at $\mathbf{s}_{t,1}=\mathbf{s}_{d,1}$, then from each chosen shape control point, successively choose the next shape control point, the details are listed in Table 1.

Table 1 Pseudocode: stepwise method for shape control points

set $\mathbf{S}_d=(\mathbf{s}_{t,1})$, $\mathbf{S}_t=(\mathbf{s}_{t,1},\cdots,\mathbf{s}_{t,j},\cdots,\mathbf{s}_{t,N_t})$;
while 1
if $\mathbf{S}_t=\varnothing$ return $\mathbf{S}_d$;
set $\mathbf{S}_{d,c}=\mathbf{S}_d(end)$, $k=2$;
while 1
$\mathbf{S}_{d,n}=\mathbf{S}_t(k)$;
$\mathbf{s}_{d,c},\mathbf{s}_{d,n}\rightarrow f_{\mathbf{s}_{d,c},\mathbf{s}_{d,n}}(x)$;
$e=\max_{j=1,\cdots,k}\min_{x\in[0,1]}\left\|\mathbf{u}_t(j)-f_{\mathbf{s}_{d,c},\mathbf{s}_{d,n}}(x)\right\|_2,\mathbf{u}_t(j)\in\mathbf{S}_t(j)$;
if $e\ge T_h$,break;
else $k=k+1$;
if $\mathbf{S}_t(k)=\varnothing$,break;
end if;
end if;
end while;
$\mathbf{s}_{d,n}=\mathbf{S}_t(k-1)$;
$\mathbf{S}_d=\mathbf{S}_d\cap\mathbf{s}_{d,n}$,$\mathbf{S}_t=\mathbf{S}_t/(\mathbf{s}_{d,c},\cdots,\mathbf{s}_{d,n})$;
end while;

Thus, we get the complete series of control points, including fixed control points and other variable shape control points, which are combined to illustrate the lane-level geometry.

4 Experimental Validation

Two experiments are conducted to validate our theoretical model and geometrical representation respectively.

In order to demonstrate the validity of multi-layer map model, we combine the aerial photography and ground calibration to establish a local map including a complete intersection which is displayed as Figure 5. The left graph is the road-level layer with the background of an aerial photo, in which the points symbolized by star are road-level nodes. The right one is the lane-level map, and the details on lanes corresponding to the road they belonging to are shown. Besides, the traffic matrix and other information are stored in TAB format which can be viewed in MapInfo and employed easily by other applications.

a) road-level layer in a map b) lane-level layer in a map

Figure 5. An example of map based on multi-layer map model

In the other experimental test, our test vehicle is equipped with RTK GPS and IMU system. The RTK GPS is BD982 from Trimble, and IMU is produced in Oxts, the model is RT2502 which can measure the position and orientation accurately. In order to measure the centreline of lanes on a road, the vehicle in our test is required to drive along the centreline as close as possible. Then the vehicle state space is established for applying Kalman Filter to obtain the precise results of locations regarded as the true value of points on centrelines. The nodes and fixed control points are determined by static vehicle with equipment running. When all the fixed lane-level control points are measured, the stepwise method of variable shape control points is executed to pick up the other control points for lane-level geometrical representation.

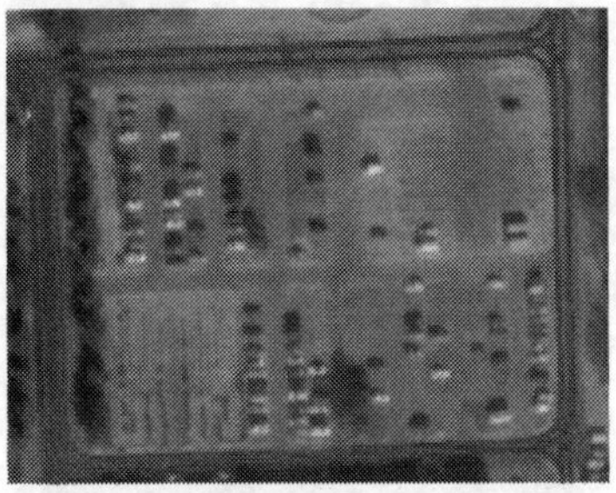

Figure 6 A lane level local map with geometry

As Figure 6 shows, the geometry representation can illustrate the shape of lanes and even intersections, increasing utility compared to the traditional geometry.

4 Conclusions and future researches

This paper proposes a multi-layer map model for route planning, in which we integrate the traditional road-level map with enhanced lane-level map, we also design an intermediate layer to connect the other two layers. According to advantages of CHS curve in presenting lane-level geometry, a stepwise approach to generating lanes is discussed and a geometrical system including fixed control points and shape control points is performed. A test to validate the feasibility of the map model and geometrical representation is conducted.

There are some ongoing researches based on this multi-layer map model. We have finished the route planning algorithm based on this layered map model [12], and due to lane-level layer in this model contains all detailed information for routing, a route planning method directly employing lane-level map is also designed [13]. Then, there is a large-scale map we are building in a South-eastern province, China, which is completely based on this our layered model.

Acknowledgement

This work is supported by International Science & Technology Cooperation Program of China under contract No.2016YFE0102200.

References

1. K. Wevers, S. Dreher. "Digital Maps for Lane Level Positioning". *15th World Congress on Intelligent Transport Systems and ITS America's 2008 Annual Meeting*, (2008).
2. J. Du, M.J. Barth. "Next-generation automated vehicle location systems: positioning at the lane level". IEEE Trans. Intell. Transp. Syst. **9**(1): 48–57, (2008).
3. D. Betaille, R. Toledo-Moreo. "Creating enhanced maps for lane-level vehicle navigation". IEEE Trans. Intell. Transp. Syst. **11**(4):786-798, (2011).
4. T. Zhang, S. Arrigoni, M. Garozzo, et al. "A lane-level road network model with global continuity". Transp. Res. Part C, 71:32-50, (2016).
5. T. Zhang, D. Yang, T. Li, et al. "An improved virtual intersection model for vehicle navigation at intersections," Transp. Res. Part C, **19**(3):413–423, (2011).
6. J. Bohren et al., "Little Ben : The Ben Franklin Racing Team's Entry in the 2007 DARPA Urban Challenge," J. F. Robot., **25**(9):598–614, (2008).
7. Q. L, Y. Zhu, "Hierarchical lane - oriented 3D road - network model," Int. J. Geogr. Inf. Sci., **22**(5):479–505, (2008).
8. J. Liu, B. Cai, Y. Wang, et al. "Generating Enhanced Intersection Maps for Lane Level Vehicle Positioning based Applications," *13th COTA Int. Conf. Transp. Prof.*, **96**(Cictp):2395–2403, (2013).
9. EDMap Consortium, "Enhanced Digital Mapping Project Final Report," (2004).
10. A. Schindler, G. Maier, F. Janda. "Generation of high precision digital maps using circular arc splines," *IEEE Intell. Veh. Symp*, 246–251. (2012).
11. A. Chen, A. Ramanandan, J.A. Farrell. "High-precision lane-level road map building for vehicle navigation." *IEEE Position Location and Navigation Symposium.* 1035-1042, (2010).
12. C. Liu, K, Jiang, Z. Xiao, Z. Cao, D. Yang, "Lane-Level Route Planning Based on a Multi-Layer Map Model. " to be published.
13. C. Liu, T. Zhang, S. Arrigoni, K. Jiang, F. Chelia, D. Yang, "Addressing the route planning problem on the lane-level road network," to be published.

Review on Driverless Traffic from Management Perspective

Tingting Chen[1], Fei Chen[1] and Cheng Chen[1]

[1]School of Transportation, Southeast University, 210096 No.2 Sipailou, Nanjing, P.R. China

Abstract. The move towards automated driving is gaining ground. This paper reviews the development process of self-driving technology and discusses the safety and efficiency advantages of autonomous vehicles. The discussion shows that the existing traffic management system, including transport infrastructures and regulations, should be changed accordingly to maximize the advantages of autonomous driving. Thus, this paper subsequently gives an insight of the traffic management from three aspects: fully self-driving traffic infrastructures, mixed traffic infrastructures and regulations. First, it is summarized in detail what should be adjusted in intersections, parking lots, pedestrian crossings, ramps, signs and markings. With the transformation of traffic infrastructures, the advantages of driverless car will be more pronounced on account of increased capacity, reduced delay and land use. Also, this paper indicates that the implementations of strict product liability for self-driving car manufacturers and no-fault tort liability for users are applicable to automated vehicle accidents.

1 Introduction

1.1 Development of automated driving

Autonomous car is also referred as self-driving car or wheeled robot, which can sense surroundings via sensor without human input, and be driven by the intelligent autopilot based on computer system. The concept was firstly raised by American industrial designer Norman Bel Geddes in 1939. Then, the United States, Britain and other countries carried on the authentication. In 1977, the first autonomous car was made by S.Tsugawa and his colleagues in Mechanical Engineering Laboratory, Tsukuba, Japan. Of which the speed can get to 30km/h.

With the deepening of research, the definition of automation is being consummated continuously. In 2013, NHTSA defined it into five levels: no automation, function-specific automation, combined function automation, limited self-driving automation, full self-driving automation [1]. Now, the development of general self-driving car is in the third stage, while the "Autopilot" of Tesla is in the fourth stage and Google self-driving car the fifth. It is estimated that self-driving car will be partially used in 2020, and will fully come into society up to 2040 [2].

1.2 Impacts

Automated driving is usually thought as "disruptive technologies"[3]. It not only brings great convenience to traffic, but also influences existing laws、industrial structure、layout of power energy, resulting in big social effects. The emphasis of this paper is to discuss the impacts in management.

2 Safety and efficiency of automated driving

2.1 Traffic safety

According to NHTSA's investigation: nearly 90% accidents resulted from driver's mistakes [4] including distraction, speeding, drunk driving and situation misjudging. Compared with manned driving, unmanned driving has obvious advantages:

• Accurate operations

Self-driving car can strictly implement orders, which determines that it can avoid human errors and reduce the possibility of accidents. According to IIHS's estimation [5], accident rate is possible to be reduced by 1/3 if all cars are equipped with lane departure warning, over-speed warning, distance warning and adaptive headlamps.

• No physical defects

Self-driving car is sensitive and fatigue will not appear, which means great reduction of accidents.

Of course, it should be recognized that potential security risks also exist if autonomous car is applied, such as sensor errors or software mistakes. However, the probability of machine failures is smaller than that of human mistakes, and it is easier to avoid machine failures. In summary, man-made traffic accidents will be greatly eliminated and traffic safety will be improved.

2.2 Traffic efficiency

In addition to excellent safety performance, traffic efficiency is also of great advantages [6].

2.2.1 Fully self-driving traffic

The advantages of increasing traffic capacity and shortening commuting time are obvious when road is universally occupied by self-driving cars. Firstly, the sensitivity and accuracy of computer system are far beyond that of human. Secondly, utilizing connected car, information can be transferred among vehicles.

2.2.2 Mixed traffic

This will presumably be the normal state, in which conventional and automated vehicles interact with each other, for quite some time even. It is, therefore, of great practical significance to gain a precise understanding of traffic efficiency under this situation.

Wagner. P [7] constructed a mixed traffic flow model, and analyzed the impacts of the introduction of driverless cars on traffic by simulation experiments. The study showed that under the existing traffic conditions, the greater the proportion of autonomous cars, the more significant increase in capacity and the higher passing rate appear. But compared to the pure self-driving environment, mixed traffic still has the following problems:

- Self-driving cars cannot effectively reduce the startup delay at a crossing. This is because the traditional car delay breaks the continuity of autonomous vehicles.
- In the mixed traffic, traditional cars' limitation in car-following determines there will be no high-speed high-density traffic queue (which can be also called platoon) which can only be realized in a pure autonomous environment.

Therefore, in the mixed traffic, its greater advantage is excellent safety, and the followed is efficiency.

3 Traffic management

While self-driving vehicles have shown great advantages in terms of safety and efficiency, the realization of these advantages requires not only advanced technology, but also corresponding traffic management approaches. However, the safety and efficiency of self-driving vehicles are limited by the existing traffic facilities, laws and regulations nowadays. In the future, along with the promotion of autonomous vehicles, the forms, the using rules of transport infrastructures and the accident liability must be changed accordingly.

3.1 Infrastructure management in fully autonomous traffic

3.1.1 Intersection management

As a major bottleneck for traffic efficiency and safety, intersection attracts many researchers' attentions all the time. In most existing intersections, traffic lights are installed to coordinate conflicting traffic flows. However, the efficiency and safety of such system is doubted: 44% of collisions in the U.S. are within the intersection area and delays induced by traffic lights can be high [8].

In 2008, Dresner and Stone [9] of Texas University at Austin proposed the concept of intelligent intersection. Automated vehicles are coordinated to cross intersections without traffic lights, fully utilizing the advanced sensing, communication and manoeuvre capacities of automated vehicles. This concept is widely accepted by researchers for its efficiency improvements (throughput, average delay, etc.)and many improved approaches are proposed [10-17], which can be categorized into planning-based approaches [9-12] and hybrid approaches [13-17].

1) Planning-based approaches

In planning-based approaches, an intersection controller finds collision-free trajectories for all vehicles. Then vehicles should follow the trajectories to cross the intersection. Any deviation from planned trajectories may cause collisions. The difficulty of planning method is to generate trajectories for all cars, which is a great challenge to the computing capability of intelligent intersection. J. Lee et al. [12] argued trajectory generation is a complex large-scale nonlinear constrained optimization problem that can be performed using optimization methods, such as Active-set Method [18], Interior Point Method [18] and Genetic Algorithm [19]. Dresner and Stone [9] modeled a "reservation based" approach to intersection management enabled by driverless technology. The vehicles could reserve trajectories from the intersection controller and the controller will centrally decide to accept or deny the reservations. These centralized, planning-based approaches are quite efficient, while autonomous cars' behavior is not subject to their own control and autonomy cannot be ensured. However, the unique needs of each vehicle are difficult to integrate into the planning method as this will necessarily introduce a large number of parameters, adding more complexity to already complex system. So the vehicle can only be fully subject to the intersection controller. The egoistic needs of each individual vehicle might be ignored.

2) Hybrid approaches

• Priority-based approaches

Hybrid approaches plan some high level priority relations defining the relative orders of vehicles. Obeying these relations, vehicles are able to cross intersections safely. L. Makarem et al.[13] proposed an approach based on navigation function to define priority. When the two cars are both in the crossroad area, the vehicle's navigation function will send their own route information to the intersection controller, and the probability of collision between the two cars will be calculated according to the vehicles' locations, driving directions and speed. If there is a possibility of collision, the vehicle nearer the collision point will be assigned greater priority and be informed to accelerate while the other one will slow down to avoid crash. In figure 1. there is a collision point between vehicle one and vehicle two. Consequently, vehicle two should be accelerated and vehicle one should slow down. If priority relations are properly defined in such approaches, the balance can be achieved between cooperation and egoistic need.

The method may avoid collision under normal situation, except for some extreme cases. For example: vehicles collide as unexpectedly braking due to the presence of a pedestrian. To solve such problems, H. Kowshik et al.[14-16] adopted a more control theoretical approach to formulate safety constraints. A bad set that comprises all collision configurations can be defined. Vehicles are then controlled in a way that they will always avoid the bad set.

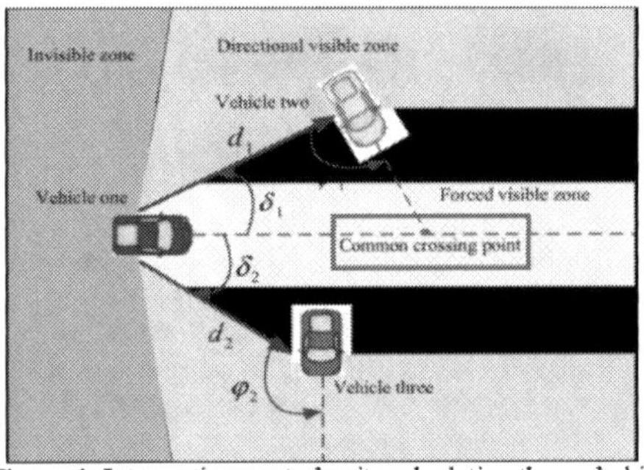

Figure 1. Intersection control units calculating the probability of collision [13]

• Autonomous intersection models

Arnaud et al.[8] established a autonomous intersection model shown in Figure 2, which includes:

- A roadside unit as the intersection controller is responsible for generating the permissions for the vehicles entering the intersection;
- Inner dotted box represents the intersection area where vehicles without permission will not be admitted;
- Outer area refers to the cooperative area, where vehicles start to communicate and cooperate with the intersection controller.

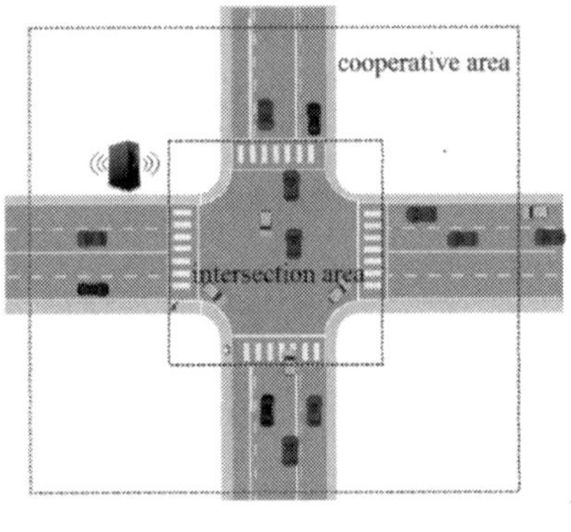

Figure 2. Schematic diagram of the autonomous intersection

Vehicles that enter the cooperative area notify the intersection controller with their presence and the controller adds it to the waiting sequence. At the same time the controller maintains a permissions list. The controller works in discrete time. At each time step, the intersection controller will add several vehicles to the permissions list from the waiting sequence, then the vehicles are permitted to pass through the intersection. Once passing through the intersection, they will be released from the permission sequence. In addition, by adjusting the speed, the vehicles can cross the intersection area without waiting, improving passing efficiency.

Three organizations, including MIT, have jointly created Slot-Based Systems [20], which are similar to those commonly used in aerial traffic: (1) Based on well-designed scheduling algorithm, the systems are able to allocate safe arrival time slots at cross areas to vehicles;(2) Vehicles control their speed to reach the intersection at the beginning of the assigned time slot, avoiding traffic conflicts and without stopping to wait. Both Slot-Based Systems and the above autonomous intersection model proposed by Arnaud focus on the interaction between vehicle and intersection controller. Though the two models share similarity in control methods, there are still differences:

- In both models, vehicles experience the same two states: the first is waiting and controlling the speed, the second is through the intersection. In the autonomous intersection model established by Arnaud, the above two states are clearly distinguished by the intersection controller for the clear division of the cooperative area and intersection area, while the boundary between these two states is not obvious in Slot-Based Systems.
- The order of permission assignment between these two models is quite different. The autonomous intersection model is controlled by the intersection controller. Thus the vehicle first controls the speed to wait for permission. After admitted, the vehicle can finally cross the intersection; In Slot-Based Systems, the vehicle first obtain the access, and then adjust speed to cross the intersection.

The above analysis shows that traffic flow in Slot-Based Systems is more smooth. Forming platoons of vehicles and serving all vehicles in the platoon before giving way to a conflicting flow is more efficient from a capacity point of view. Based on the generalized queue theory, the researchers found that the Slot-Based Systems capacity can be doubled and the delay can be significantly reduced compared to the traffic-light controls [20]. But the premise of Slot-Based Systems is that the roads are basically occupied by autonomous cars, thus there may be decades to achieve this system.

3.1.2 Parking management

The emergence of driverless vehicles will make urban parking management model adjust accordingly. The ideal management model is: when owner arrives at the destination, the autonomous car can find their own safe and reliable parking space without human operation.

Volkswagen has already taken some projects testing parking in garages with autonomous vehicles [21], [22]. The driver hands the vehicle over at the garage entrance. The parking function is activated with a smartphone app. The car receives the route data to the nearest available space from a central garage computer via WLAN and drives to it autonomously.

The possibility of saving space is named as an essential argument for the use of automated parking systems. These should be made more space-efficient, primarily by replacing ramps and aisles with lift shafts, and lowering storey heights, but also by increasing parking density [23], [24]. Kowalewski.S [25] reckoned on up to 60 % more parking spaces on the same area by using automated parking. It will relieve parking pressure in some functional centers (urban commercial centers, scenic areas, etc.) and greatly reduce the cost of urban construction and parking facilities [25-27].

3.1.3 Pedestrian crossing management

Autonomous driving leads to expectations that capacity on transport routes will be freed up with platoon. Fernandez [28] assumed vehicle density increase up to 500 %. Brownell [29] estimated over 250 % for highways and some 180 % for inner-city streets.

High density means that the area required for flowing traffic could be reduced, by reducing the number of lanes [30]. Lane width could also be cut in comparison to the present size, due to the accurate lane-keeping technology of autonomous vehicles. These could have an impact on transport users such as pedestrians. Although the distance of crossing street decreases, densely flowing traffic may make it tougher for pedestrians. In order to guarantee the conveniences of pedestrians, the installation of intersection-free crossings such as under and overpasses would be a necessary consequence.

3.1.4 Other transportation facilities management

Automated vehicles propose specific demands for road quality, traffic conditions, traffic signs, safety facilities, etc. [31]:

- Lane markings need to be clearly identifiable by automated driving systems and their vision sensors. For lane assistance systems, the quality of lane markings is crucial. Lane markings should be detected by the employed sensors irrespective of weather conditions (e.g. rain,snow, etc.) or road conditions (e.g. presence of cracks, potholes).
- The color, luminance and shape of the various marks must be readable for vehicle sensors
- Old lane marking remnants should be well removed, so they do not create any confusion.
- Defined and visible road profile and continuous road edges or kerbs are needed
- Electronic speed limitation is indispensable.
- Roadside V2I/I2V communication infrastructure (via wireless G5 or cellular networks) should be employed.

These requirements are found in the self-driving vehicle tests. The solutions to these problems can also accelerate the pace of putting automated vehicles into operation.

3.2 Infrastructure management in mixed traffic

At present, most of the researches on self-driving traffic management are concentrated in fully self-driving traffic, and there are few studies in mixed traffic. The existing strategy is to separate unmanned vehicles from manned vehicles, thus effectively taking the advantages of autonomous vehicles. One way to achieve this aim is to create an autonomous vehicle dedicated lane.

As the efficiency of the traffic depends largely upon the on-ramps and off-ramps of freeway and the signalized intersections on the urban roads, the focus of the mixed traffic management research is on the two issues. Bernhard [32] has conducted some studies on them.

3.2.1 Motorways

The on-ramps and off-ramps are the most important elements that affect the traffic efficiency of motorways, and the following management measures can be carried out [32]:

- One measure is looking forward to further developing already-emerging technical solutions with assistance functions, such as the merging assistant, particularly in regard to the possibilities of machine cooperation.
- Solutions for adjustments of facility structure and regulatory mode are still to be developed. For example, one interesting scenario is where autonomous traffic can be directed onto separate lanes between motorway intersections. This separation is then removed in intersection areas where autonomous and human-controlled vehicles drive in all lanes and each may perform all maneuvers (autonomous, highly assisted or human-driven) at a predetermined low speed. It will be ensured that autonomous vehicles can form platoon, without affecting the vehicle to change route.

3.2.2 Urban roads

Urban traffic flow is complex, so the rules of right of way should still be adhered to under mixed traffic. Signal control system could set a separate phase for autonomous vehicles in which self-driving vehicles from all directions can cross the intersection simultaneously [32]—the maneuvers of the conflicting flows in the intersection area would be negotiated independently by the autonomous vehicles. All other road users would be controlled by the existing signal phases. By such differentiated management strategy, it is expected to maximize the automated driving advantage under mixed traffic.

3.3 Laws and regulations

A crash involving Tesla Model S occurred on May 7, 2016, Florida, USA, due to the autopilot's failure of identifying the front vehicle, which resulted in Tesla owner's death. NHTSA is still conducting technical verification on the same type of Tesla Model S , in order to affirm who is responsible for the accident. As the first fatal traffic accident involving automated car in the U.S, not only reveals that the driverless technology is not completed, this incident also indicates that before the application of self-driving car, the principle of driverless traffic accident liability division and other related regulations are necessary to be formulated in advance.

3.3.1 Accident liability division

Since both car manufacturers and car users are involved in the driverless traffic accident, the division of responsibility should be respectively discussed .

1) Responsibility of car manufacturers

Under the conventional traffic, once an accident happened, it is usually the driver not the car manufacturer who will be held accountable because car is completely controlled by the driver. On the contrary, after the application of driverless cars, they can control themselves and the drivers no longer have the access. The accident involving self-driving car is usually caused by the defects of the car itself, so it is unreasonable that the driver should be completely responsible for the accident. At that time, the car manufacturer will take more product liability for the accident compared to the previous situation [4].

The product liability of car manufacturers can be judged by two principles, including strict product liability and negligence product liability.

- Strict product liability: As long as the existence of defects leads damage to men and property, The car manufacturer will assume the reparation liability wherever the fault is subjective or not, except for the statutory exemption mistakes. Under the principle, the consumer will get compensation as long as he proves the relationship between the product and the manufacturer and points out the defects of the product, which is beneficial for the consumer to cite evidence.
- Negligence product liability: This principle, bringing great difficulties to the consumers, requires the car users not only to point out the flaws, but also to prove that the producers do make a subjective mistake, which is demanding for the consumers to cite evidence.

If autonomous car is applied, the principle of strict product liability can meet the psychological needs of consumers [33]. In the manufacturers' point of view, the cost of taking liability, brought by this principle, may weaken the enthusiasm of driverless car manufacturers. In fact, manufacturers, including Volvo, Google and Mercedes, are committed to compensate for the loss of driverless car accidents. Taking the responsibility initiatively will eliminate consumers' doubts about the new technology and help to promote the automated vehicles. In addition, the increased cost can be transferred to the price of the car.

2) Responsibility of car users

The tort liability of car users can be also judged by two principles: no fault liability and fault liability.

If the principle of fault liability is insisted on, the injured party will have to prove that the driverless car user had made a subjective fault. As the user was not at fault, the victim can do nothing but turn to sue the manufacturer for product liability. Under that circumstance, they have to confront expensive litigations, lengthy trial period and adduction difficulty. In contrast, non-fault criterion liability is easier to implement as victims can receive compensation directly from insurance companies and the litigation expense is eliminated. In conclusion, the principle of no-fault liability is more applicable to the tort liability authentication of driverless vehicle users.

3.3.2 Traffic laws

In fact, there have been some tentative laws for autonomous vehicles. In 2011, Nevada took the lead in passing legislation, allowing self-driving vehicle tests to be carried out. Subsequently, Florida, California, Washington, the District of Columbia etc, have also allowed the tests in their states. Nevada has also proposed rules for driverless vehicle sellers and users to ensure the reliability of vehicles [2]. Sellers must obtain certificates of compliance from car manufacturers or related technology accreditation bodies and users need to obtain specific licenses. However, these regulations do not illustrate the principle of liability division for the accident clearly.

On December 9, 2016, the first comprehensive statute in the United States, covering the test, use and sale of driverless vehicle, was promulgated in Michigan. The law stipulates that the liability in an autonomous car accident will be deduced based on the principle of no-fault in the Michigan Insurance Act, and the loss will be paid by the insurance companies. At the same time, the manufacturers will be held accountable under the product-liability law. The establishment of the law, setting a benchmark for the legislation of administration over autonomous vehicles, stimulates the initiation of lawmaking in other regions and benefits the further development of driverless technology.

4 Conclusions and prospect

This paper reviewed the development of self-driving technology and discussed the advantages of driverless vehicles in safety and efficiency. On the basis of the above discussion, it is recognized that appropriate traffic management measures are still needed to guarantee the superiorities. Therefore, this article, mainly from traffic facilities and laws, summed up the following key points on the autonomous traffic management.

1) Driverless vehicles can not only effectively improve traffic safety, but also significantly enhance traffic efficiency. However, it is worth noting that, in the mixed traffic flow combining conventional, connected, and automated vehicles, the advantage of traffic efficiency is not so distinct that setting dedicated lanes are necessary to maximize the preponderance of self-driving vehicles.

2) In the case of coexistence of autonomous vehicles and manned vehicles, the researches focused on the on-ramps and off-ramps of freeways and the signalized intersections on urban roads. For ramps of expressways, merging assistant as well as adjustments of facility structure and regulatory mode will take effect. For urban signalized intersections, exclusive signal phase for driverless vehicles can take full advantage of them and enhance traffic efficiency adequately.

3) If self-driving vehicle is applied universally, the management pattern of intersection will be changed completely, from traffic signal into intelligent intersection. The control of intelligent intersection can be divided into planning-based approaches and hybrid approaches.
- Planning-based approaches: The approaches manage intersections by planning collision-free trajectory.
- Hybrid approaches: This kind of approaches includes priority-based approaches and autonomous intersection models. The priority-based approaches define relative orders and bad-sets to control vehicles, while the autonomous intersection models assign permissions to those vehicles through intersections.

Automated intersections adapted to autonomous technology not only strengthen safety, but also reduce delays and the pressure of urban traffic capacity. Among above approaches, autonomous intersection models may be optimal for the reason that autonomous intersection models, different from planing-based approcahes, take the egoistic needs of each individual vehicle into consideration and deceleration is not required in the model compared with priority-based approaches. Moreover, in the autonomous intersection models, the vehicle will running more smoothly in Slot-Based Systems without obvious boundary.

4) With the development of self-driving technology, other transportation facilities, including parking lots, pedestrian crossing facilities, signs, markings, and communication facilities, also need to adapt themselves to the automation demands. For example, intelligent parking lots are required to achieve automatic parking management.

5) In terms of laws and regulations, car manufacturers will assume more product liability than they do now. What's more, principle of no-fault liability is more suitable for both manufacturers and users.

Researches on driverless traffic management remain to be further studied, although plenty of corresponding traffic management approaches have been proposed. Moreover, there are many other obstacles to the promotion and implementation of autonomous vehicles, including the immaturity of technology, the resistance of the traditional automotive industry, and ethical difficulties. Once these problems are resolved, it will not be far for self-driving vehicles to enter the market.

References

1. D. C. National Highway Traffic Safety Administration, U. S. Department of Transportation (2013).
2. Adeel Lari, Frank Douma & Ify Onyiah.16 Minn. J. L. Sci. & Tech. 735 (2015).
3. Fung . Technical Report (2013).
4. NHTSA.Preliminary Statement of Policy Concerning Automated Vehicles (2013)
5. M. James Anderson, Nidhi Kalra, D. Karlyn Stanley, Paul Sorensen, Constantine Samaras, Oluwatobi A. Oluwatola. Autonomous Vehicle Technology. A Guide for Policymakers (2014).
6. Todd Litman,TRB, 15-3326 (2016).
7. Wagner, P. Traffic control and traffic management in a transportation system with autonomous vehicles, In Autonomous Driving (2016).
8. Arnaud De La Fortelle, Xiangjun Qian. ITS World Congress (2015).
9. K. M. Dresner, P. Stone, J. Artif. Intell. Res.(JAIR), vol.31, pp. 591–656 (2008).
10. K. Dresner, P. Stone, Springer, pp. 129–138 (2006).
11. I. H. Zohdy, R. K. Kamalanathsharma, H. Rakha, ITSC, pp. 1109–1114 (2012).
12. J. Lee, B. Park, K. Malakorn, J. So, Transp. Res. Part C Emerg. Technol, vol. 32, pp. 193–206 (2013).
13. L. Makarem, M. H. Pham, A. G. Dumont, D. Gillet, TRB (2012).
14. H. Kowshik, D. Caveney, P. R. Kumar, Veh. Technol. IEEE Trans, vol. 60, pp. 804–818 (2011).
15. K. D. Kim, P. R. Kumar, IEEE Trans. Automat. Contr., vol. 59, pp. 3341– 3356 (2014).
16. M. R. Hafner, D. Cunningham, L. Caminiti, D. Del Vecchio, IEEE Trans. Intell. Transp. Syst., vol. 14, pp. 1162–1175 (2013).
17. M. A. S. Kamal, J. Imura, A. Ohata, T. Hayakawa, K. Aihara, ITSC, pp. 922–927 (2013).
18. J. Nocedal, S. Wright. Numerical Optimization. Springer (2000).
19. K.Ozbay, H. Yang, B. Bartin, S. Mudigonda. Transportation Research Record, pp.105–113 (2008).
20. Tachet R, Santi P, Sobolevsky S, Reyes-Castro LI, Frazzoli E, Helbing D,et al. PLoS ONE (2016)
21. Continental AG: Der Fahrer entscheidet, das Auto übernimmt (2014).
22. Audi: Automatisch ins Parkhaus (2012).
23. Fagnant, D. J. Kockelmann, Eno Foundation (2013).
24. Irmscher,I. (n.D.): Benutzerfreundliche automatische Parksysteme – Besondere Anforderungen –Planung – Einsatz.
25. Kowalewski, S. Überlassen Sie das Parken Ray. Deutschlandradio Kultur (2014).
26. W. J. Mitchell, E. Boronni-Bird, L. D. Burns, The MIT Press (2010).
27. LEG Stadtentwicklung GMBH: Mobilitätshandbuch Zukunftsstandort Phoenix West.Dortmund (2008).
28. C. K. Brownell: Shared Autonomous Taxi Networks: An Analysis of Transportation Demand in NJ and a 21st Century Solution for Congestion (2013).
29. C. K. Brownell & A. Kornhauser, TRB (2013).
30. P. Fernandes, IEEE Transactions on Intelligent Transportation Systems, vol.13 (2012)
31. Isabela Mocanu, Philippe Nitsche, Peter Saleh. 25th PIARC World Congress (2015).
32. B. Friedrich, "The Effect of Autonomous Vehicles on Traffic" In Autonomous Driving, Chapter 16, pp. 317-334(Springer Berlin Heidelberg, 2016).
33. S. Stephen. "Product Liability Issues in the U.S. and Associated Risk Management" In Autonomous Driving, pp.523-570 (2016).

Island in an island – The suggestions for transportation improvement plan for Haidian Island, Haikou, Hainan

Rosalind Juo Ling Sia[1], Xiaodan Zhu[1]
[1] *Tourism College, Hainan University, Haikou Hainan*

Abstract: Haidian Island, which situated at the Northern part of Haikou City of Hainan Province, is an island within a city. Haidian Island is unique in term of it's development which centered around an university, the Hainan University, besides some others important landmarks, such as Haikou city hospital, Baishamen municipal park, Golf Driving Range etc. All commercials, residential, recreational activities etc are planned to serve Hainan University in particular. The study, taking 'Haidian Island Area Development Control Plan' as case study, would like to look into the importance of transportation and traffic planning. The study used observation, site investigation and traffic study methods to gather data needed. Firstly the study analyzed the current state of transportation system for Haidian Island in accordance to the Island Development Control plan and Haikou master plan and identified the problems. Then, the study made some recommendations for these problems. The study highlighted the important of non-motorized, cycling and walking as the main transportation system for an education-based island and as supportive to domestic tourism activities found. The transportation planning suggested by the study took 'green and low-carbon' approaches considered the role of University as the core activity in the island.

1 Introduction

As a city with more than 200 million populations, even though is not a mega city, Haikou still stuck with transportation problems; there are some traffic circumstances such as congestion at peak time, poor road design and lack of integration between vehicle-human-landscape etc[1]. Haidian island, which forms a part of Haikou city, is the education hub for Hainan province, shares the similar transport problems.

Today, under the pressure of the fast-growing population and increasing number of tourists in Haidian island, transportation improvement has become essential. Transportation system is not merely a commuting channels but the vein of a city[2]. Transportation system encourages human interaction[3] and enlightens city liveliness[4]. Transportation plan then, plays an important role in ensuring these veins to function efficiently and serve it people well. This paper aimed to analysis the existing transportation problems in the ***Haidian Island Area Development Control Plan*** and then, put forward some suggestions to improve the plan, that meets the current development of Haidian island.

2 Research area

Hainan is an island which is located at the southern tip of china, is a tropical island which is famous for its summer sunshine and coconut tree. After the declaration of ***international touristic island*** [5], tourism has become the major industry that brings benefit to the economy of this island.

Haikou is the capital of Hainan province which is the center of administrative, economy and culture of Hainan. The research area of this paper, Haidian island is a small island situated at the northern part of Haikou (Figure 1). Haidian island is a typical delta island located in the Nandu river estuary. Haidian island is the biggest island in Haikou. There are Hainan University and Haikou hospital on this small island. Haidian island and Haikou city are separated by the Nandu river but they are linked by three major bridges, namely Century Bridge, Renmin bridge and Heping(Peace) bridge[6]. Haidian island, with the total land area of approximate 14 kilometer square[7], is hammer-head in shape, has flat terrain with load of river system, lakes, irrigation canals and ditches, that makes the island look really picturesque.

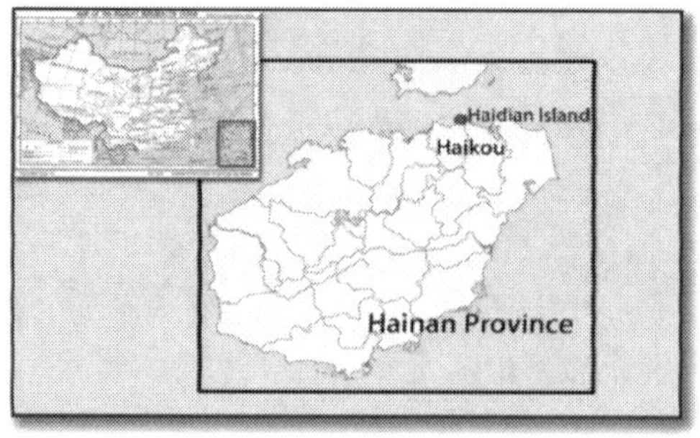

Figure 1. Location of Haidian Island at Haikou city of Hainan Province. (Source: Author illustration)

MATEC Web of Conferences **124**, 03003 (2017) DOI:10.1051/matecconf/201712403003
ICTTE 2017

3 Research methods

The study applied two general methods; observation and site investigation to obtain data needed. Firstly, the researchers look into all kinds of literature and master plans that related to Haidian island and decide on research topic. Then, to gather specific data on transportation, the author employed site investigation-cum-observation survey method to gather information on current situation, followed by traffic survey on vehicles and pedestrians on road during evening peak time around 17:30 o'clock to 18:30 o'clock.

4 Spatial planning and current transportation problems

4.1 The Haidian development control plan

The latest '**Haikou city master plan, 2005-2020**' planned Haidian island as the education hub for Haikou city[8] in specific. Currently, the highest education institution of Hainan province, the Hainan University (HaiNu) is situated on this small island. The University is surrounded by mostly low-rise residential and community commercial shops, as well as a local park and others auxiliary facilities that cater for the resident needs.

According to the '**Haidian Island Area Development Control Plan**' (after this refers as **'The Plan'**), the development area covered a total of 1148.78 hectares with estimated population of 25.6 million, which including permanent residents of 15.5 million and not including of approximate 3 million students of Hainan University per annum[9]. The major landuse of Haidian island are designated for residential, however, there are some landmark building in between, including the Haikou people's Hospital, Baishamen Park besides Hainan University.

As stated in The Plan, the spatial distribution for Haidian island area are divided by 'one centre, one belt, two axes, five core areas' concept[9], as showed on Figure 2. **One centre** refers to Baishamen park as green leisure centre for the islanders and tourists; **One belt** refers to the round island, outer ring road which form the sightseeing and recreation route; **Two axes** refers to two major trunk roads, Haidian Wuxi road and Renmin Avenue; **Five core areas** refers to four residential zone and one educational institution, i.e. Hainan university zone, the west residential zone, the northeast residential zone, the northeast residential zone and the southeast residential zone.

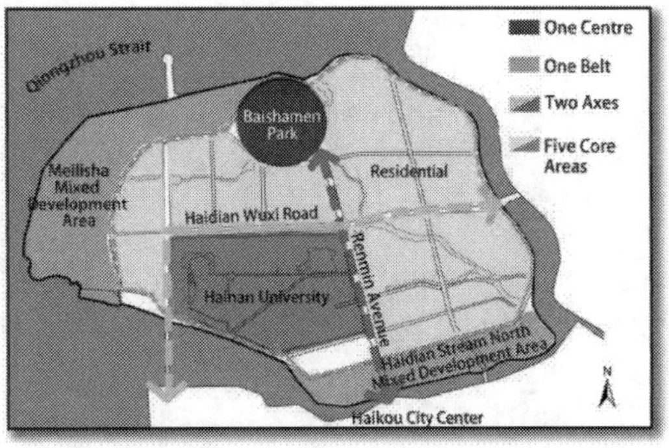

Figure 2. Spatial division of Haidian Island Area as suggested in Haidian Island Area Development Control Plan (Source: [9])

4.2 The transportation plan

The landuse area for roads and open space reserves are planned to be 218.59 hectares, about 19.03% of total developed area (1148.78 hectares) of Haidian island[9]. There are 9 areas of transportation planning stated in '**The Plan**'. In this section, the paper first presented the contents stated in the 'Haidian Island Area Development Control Plan' (**The Plan**) for each areas; then, discussed the differences occured during the implementation of the plan which formed the **current situation** of each areas.

Area 1: Road network plan

The Plan: There are four types of road network in Haidian island; trunk distribution road, arterial distributions road, local distribution road and community access road. As showed in Figure 3, there are two types of trunk distribution road, which Haidian Wuxi road and Century Avenue are the score trunk roads while Heping (Peace) Avenue and Renmin Avenue are thr general trunk roads. These four roads also act as the external traffic that connect Haidian island to other areas within Haikou city. Arterial roads are those roads that connected to the trunk roads, which consist of Haidian Road 2, Road 3, Haidian middle Road 4, Seaview Road, Huandao Road etc. Table 1 showed the designated road widths and speed limits for the above mentioned road system.

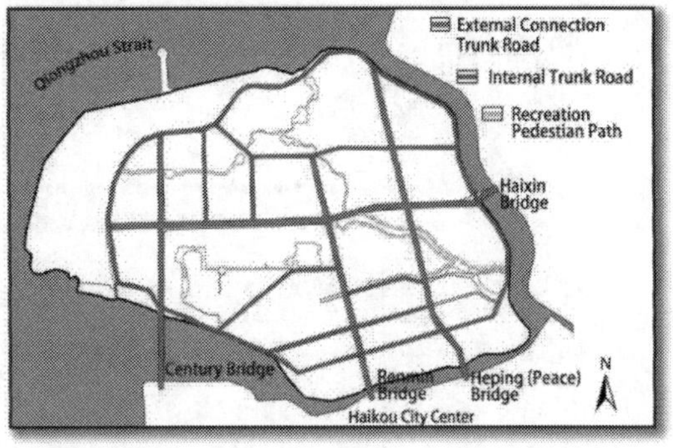

Figure 3. The road system of Haidian island(Source: [9])

Current situation: All designated roads had been constructed to serve the islanders daily usage. The existing roading networks of 4 different hierarchies roads are built according to the plan and municipal engineering requirement. The road widths of these roads are appropriated and no serious traffic jam observed even during peak hour.

Table 1. Road types, speed limits and widths of Road system in Haidian island

		Designated Speed (km/h)	Road Width(m)
1	Core Trunk distribution Road	60-80	50-70
2	General Trunk distribution Road	50-60	40-60
3	Arterial Distribution Road	30-40	30-40
4	Local distribution road/ community access road	-	12-25

(Source: [9])

Area 2: Road cross section design

The Plan and Current situation: all cross sections of the proposed roads are designed and built according to the requirement.

Area 3: Static traffic system

The Plan: There will be six(6) public parking areas and two(2) bus stations proposed for Haidian island. The location of public parking areas are not specified in the plan while the bus stations had been determined to be located at south of Baishamen wastewater treatment plant and east of Haidian Road 2.

Current situation: Bus station at south of Baishamen wastewater treatment plant had been built and in-use but the other one at East of Haidian Road 2 is not constructed but replaced to a new location underneath the Century Bridge as showed in Figure 4 and 5.

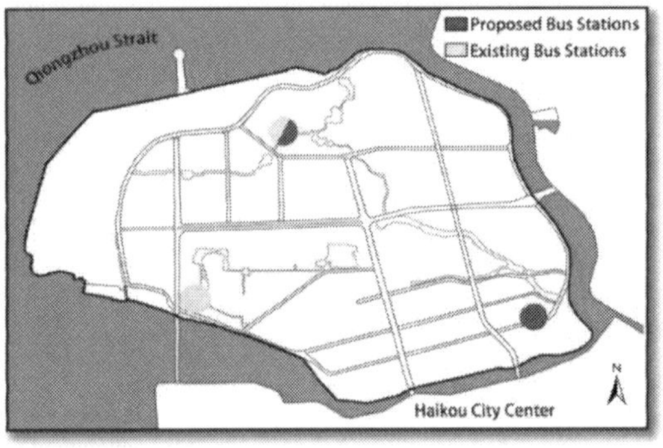

Figure 4. Bus station location (Source: [9], author added-on)

Area 4: Slow transit system

The Plan: walking, cycling and skating are three major slow transportation systems that highlighted in the Plan. In line with the '*green and low carbon*' development concept, the government intended to build the basic infrastructures that accomplished with the above mentioned slow systems for healthy lifestyle of the islanders. Table 2 showed the proposed development areas for these slow transportation systems.

Figure 5. Bus station underneath Century Bridge. (Source: author photograph)

Table 2. Proposed walking, cycling and skating system for Haidian island

	Transit system	Area covered
1	Walking system	Bihai avenue, Haida Road, Yixin Road, Baisha River, Moon River
2	Cycling system	Classic Route: Baishamen Park, Hainan University, Heping Avenue and Haidian Road 1 etc; Ring Road: Bihai Avenue, Haidian Road 1 etc
3	Skating system	Northern Seaside of Baishamen Park

(Source: [9])

Current situation: Till now, cycling and pedestrian paths are constructed gradually according to the plan. However, some of the paths had been cut off, especially those at the southern part of Hainan University and along Haidian stream caused by rapid housing development that affects the width of road reserve to be reduced.

Area 5: Water and cruise transport

The Plan: A number of cruise jetties are proposed to be located along the coastline surrounding Haidian island. No fix location is specified on plan but according to '**Haikou Yacht and Cruise development plan (2011 – 2020)**', two marinas were proposed to be built at the western part of Haidian island, one of it is a private marina while the other one is an integrated terminal (Figure 6) [10] .

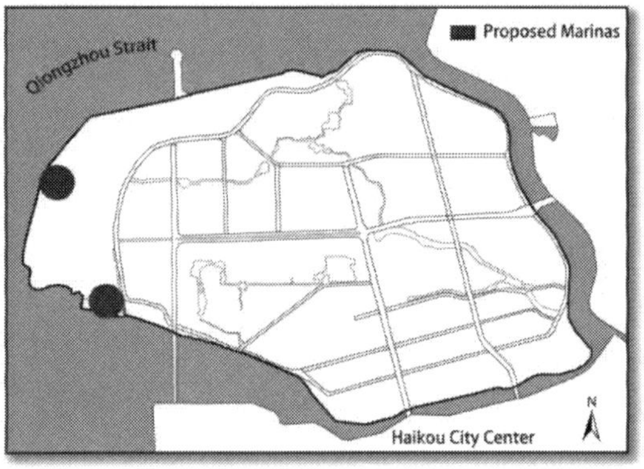

Figure 6: Proposed marinas in Haidian island (Source : [10])

DOI: 10.1051/matecconf/201712403003

Current situation: Up-to-date, there is one marina under construction at the south-western part of Haidian island, which is on the location for the proposed 'Danna marina'. Also, along the Haidian stream at southern part of Haidian island, there is a harbor for local fishing boats/ships (Figure 7).

Figure 7. (Above) Danna marina under construction and (Below) existing habor for fishing boats
(Source: author photograph)

Area 6: Traffic control planning

The Plan: The use of truncation area at junction for development is prohibited. Opening for ingress and egress are proposed to be every 200m for trunk road and 100m for arterial road.

Current situation: No fouls are found.

Area 7: Vertical traffic planning

The Plan: In order to reserve sufficient space for future drainage system, flood protection, traffic safety and municipal engineering work etc, the plan determined the control standard for road elevation and slope should be larger than 0.5% or at least 0.3% if situation is difficult. For new developed area, elevation and slope could be higher 0.5 meter than the normal level.

Current situation: This rule had been followed accordingly.

Area 8: Road intersection

The Plan: There are four intersections found in Haidian island with the design of footbridge on top for pedestrian and T-junction on ground for vehicles. In addition, all intersections of trunk-trunk road, trunk-arterial road would use traffic light as control elements while for local distribution road and community access road, right turning road are designed to deliberate traffic.

Current situation: Table 3 showed the existing intersections and it's traffic flow during site investigation. The observation result showed that Haidian Road 2 and 3 has heavy human and non-motorized vehicles flow while Haidian Road 5 has more motorized vehicles flow during peak hours.

Area 9: Footbridge proposal

The Plan: The plan suggested for two overhead foot bridges to be built at; i) intersection of Haidian Road 2 – Renmin Avenue and ii) intersection of Haidian Road 5 – Renmin Avenue.

Current situation: Up-to-date, the Haidian Road 2 – Renmin Avenue overhead bridge had been completely constructed (Figure 8) but the other overhead footbridge at intersection of Haidian Road 5-Renmin Avenue had not been constructed.

Figure 8. overhead footbridge at intersection of Haidian Road 2 –Renmin Avenue (Source: Author photograph)

4.3 The problems

The study found several problems from the implementation of the 'Haidian Island Area Development Control Plan' as discussed above which could be summed up as below.

Table 3. Human, Motorized and Non-motorized vehicles flow at Intersections during Peak hour

Evening Peak hour (17: 30-18: 30)	**Human Flow (p/h)**	**Motor vehicle Flow (pcu/h)**	**Non-motorized vehicle Flow (Bicycle and electric bike) (pcu/h)**
Intersection of Haidian Wuxi Road and Century Avenue	300	3720	1680
Intersection of Haidian Wuxi Road and Renmin Avenue	300	3540	5880
Intersection of Haidian Road 3 west and Renmin Avenue	2880	1560	4140
Intersection of Haidian Road 2 West and Renmin Avenue	1560	2820	5940

(Source: Author observation result)

4.3.1 Long one-way routes

Two major trunk roads, Haidian Wuxi road and Renmin Avenue, are both long and one-way traffic road for motorized vehicles with six carriageways (three lanes each for one side). Vehicles that wish to cross over to the opposite side of the road need to travel to the intersection at both end of these two roads to make the turning (Figure 9), that caused congestion at the intersection with through and u-turn traffic[11]. In addition, there is one u-turn opening at Renmin Avenue are designed at traffic light junction also caused some serious issues. i. Collison of traffic between on-coming traffic and u-turn traffic; ii. U-turn radius not enough for even normal car to make a turn within one turning (Figure 10).

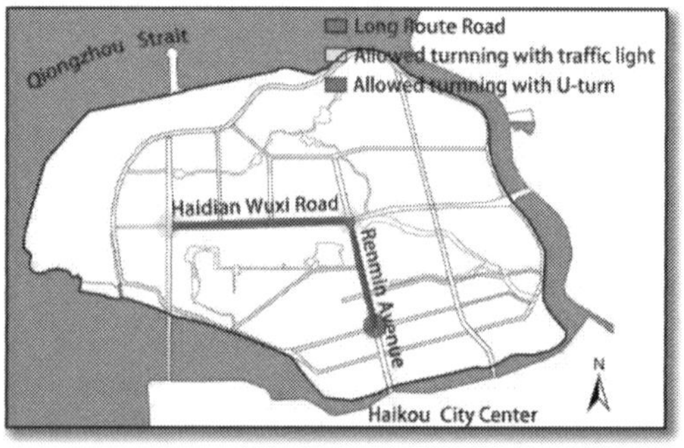

Figure 9. Long route without turning at Haidian Wuxi Road and Renmin Avenue (Source: author survey)

Figure 10. Dangerous turning at Renmin Avenue (Source, author photograph)

4.3.2 Uncontrolled right turning of vehicles

As stated above, right turn traffic flow is not been controlled by traffic light at any roads, is a practice in China[12]. This practice caused motorized vehicles to collide with other road users especially pedestrians[13]. Pedestrians cannot cross over the street even with zebra crossing (Figure 11).

Figure 11. pedestrians at zebra crossing cannot cross the street caused by non-stop right turn vehicles. (Source: author photograph)

4.3.3 Congestion at intersections

Even though the site investigation found that there were no heavy traffic flow of motorized vehicles even during peak hour, but there is still, congestion at intersection, particularly with non-motorized transport (especially the electric bike) and pedestrians (Figure 12)[14]. This congestion become worst with uncontrolled right turning vehicles and usually caused minor accidents here and there at the intersections.

Figure 12. Congestion at intersection of Haidian Road 2-Renmin Avenue during evening peak hour. (Source: author photograph)

4.3.4 Incomplete pedestrian and cycling paths

From the site investigation, there are some significant issues with pedestrian and cycling paths within the island, which are:

- Break off of pedestrian and cycling paths due to expansion of road or reduction of road widths.
- Cut off of pedestrian and cycling paths due to motorized vehicles parked on the pedestrian paths and road side car parking that occupied the cycling paths.
- Collision between pedestrian and bicycle on pedestrian paths.

4.3.5 Lack of touristic direction signboards

There are insufficient of signboards within the island which provide proper and exact direction to road users, especially the tourists. Baishamen park and Hainan University are two major attractions in Haikou city which attracted local and international tourists to visit especially by foot or bus. No direct public bus services from city centre to these attractions caused the tourists to ask around for directions when come nearby to these locations. Therefore, touristic signboards are important to guide them not to get lose and get organized[15].

5 Improvement plan

To enhance the 'Haidian Island Area Development Control Plan', the paper put forward some suggestions as below.

5.1 Awareness in driving and pedestrian safety for the people

DOI: 10.1051/matecconf/201712403003

Drivers (of motorized and non-motorized transport) and pedestrians are the major users of the road. Both parties should work together to ensure safety of each others. Awareness campaign should be carried out time to time in public or the university, to remain people of maintains good habits on road[16], [17]. Road safety is not just slogan to look at but to practice on. Chinese road users need to improve their on-road behavior to guarantee smooth traffic, particularly the pedestrians. Obviously, pedestrians on Haidian island would cross the road without following traffic instructions which messes up the road condition and eventually caused accidents to happen. Governmental authorities, private enterprises and public should work hand-in-hand to boost the awareness of general public.

5.2 Prohibition of young drivers

Nowadays, the riding of electric bike in china does not have any age restriction. Neither to obtain a driving license. As the usage of electric bikes are very common in China, some restrictions should be implied to ensure safeness. The rider for electric bike should followed the provision for 'application and usage of motor vehicle driver's license' which stated the minimum age to applied for license is 18 years old and should passed the all relevant examinations [18]. As we can see today, young electric bike riders, especially primary and middle school students which below the minimum age should not be allowed on road because they does not have enough knowledge, self-control and consciousness in making real time reaction to immediate traffic situation.

5.3 Completion of round island cycling and pedestrian paths

'Green city, low corban' concept encourages more walking and cycling, less driving. In conjunction to this, round island cycling and pedestrian paths need to be completed, for at less two purposes; i. islander daily usage and ii. Tourist, such as suggested by Cheng for Xiamen[19]. Figure 13 showed the proposed cycling and pedestrian paths for Haidian island, basically round island. Walk paths should be extended in all directions within the island while the cycling paths can be specified into two categories, daily route and tourist route.

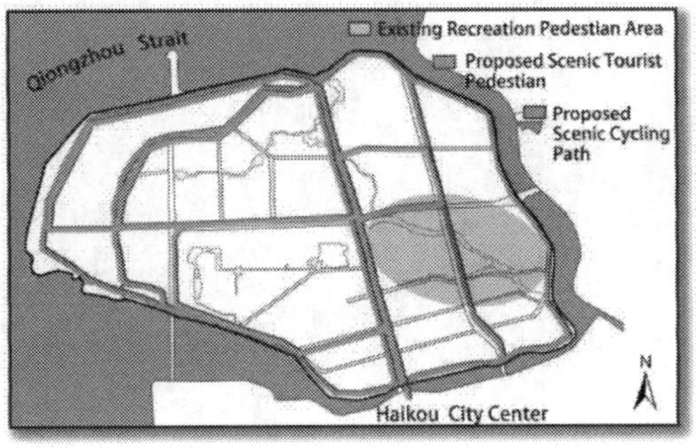

Figure 13. Proposed round island cycling and pedestrian paths (Source: author illustration)

5.4 Introduction to touristic routes and signages

There are several touristic attractions in Haidian island such as Baishamen park, Hainan University while Haikou Qilou Arcade street and Haikou old city centre is located at the adjacent location, which formed the 'Golden Triangle for Haikou Day-trip' destination. Therefore, guide-orientated signages should be located along the way, from the old city centre to Baishamen park, through Hainan University. In order to exalt tourism activities on Haidian island, touristic routes should be created to help tourist enjoys scenery view of Haidian island.

5.5 Promote the use of non-motorize and public transit

Again, green and low carbon are the concept for tomorrow city, Haikou too has the same vision. Some improvement actions are recommended to enhance this vision, including:

- Increase the regularity of bus service duration and familiarize the bus drivers with knowledge of all bus routes and ability to give advices to passenger with smiling face at all time.
- Improve basic infrastructures for bus stops and facilities along the way to bus stops, rain shed, sitting bench, lighting, trash bin, route schedule, directional and guide tile etc that provide cozy environment and encourages people to use public transport (Figure 14).
- Bike sharing project is now undergoing in China which is a big step forward in promoting low carbon concept, however, awareness for people to use and park in right manner is important to maintain the bike sharing to be sustainable (Figure 15).
- Electric bike users should be advised to follow motorcycle rules and regulations on driving and parking as both of these vehicles share many common similarities.

Figure 14. Public bus shelter integrated with pedestrian path (Source: [20])

Figure 15. Bike sharing at Haidian island
(Source: author photograph)

6 Conclusion

Mass transit and Non-motorize transport are the trend of future that responses to sustainable urban transportation planning[21-23]. People are advised to adapt with the shift from small and private vehicles that pollute the environment to an energy saving and healthy lifestyle of cycling and walking. The city is changing, bike sharing project shows a good start for people to move around, work and leisure. More effects on 'green travel, low carbon' are going to implement by the authorities but more important of all, is the awareness, adaptation and change of attitudes of the people to follow the path of the government, in the initiative of sustainable urban transportation planning for better future. The Haidian island improvement plan as suggested by the researchers take into consideration landuse, people and authourities, integrating non-motorized vehicles with cycling and pedestrian, is able to bring better living environment for the islander from now and to the future.

References

1. H. Pan, The characteristics of Traffic Problems in Development Middle cities. Urban Planning Review, 1998, (2):24 -29, (1998).
2. B. Wang, Application of New Technology in Urban Transportation Planning. Transport World, (14):62-63, (2015).
3. Y. Yan, Brief discussion on road reconstruction and construction technology. Science and Wealth, (4):304, (2012).
4. H. Hua, New urban district traffic planning strategy under low carbon traffic. Knowledge of Architecture, (B04):67&69, (2013).
5. Government of China, Several opinions of the State Council on promoting the construction and development of Hainan International tourism island. Retrieved on 2017-04-26, http://www.gov.cn/zwgk/2010-01/04/content_1502531.htm, (2010).
6. Hainan History Network, Hainan Profile. Retrieved on 2017-04-27, http://www.hnszw.org.cn/data/news/2014/12/75985/, (2008).
7. S. Liu, Look at the world, Haidian island (In Mandarin). Hainan Daily,2007-08-16 (15), (2007).
8. Haikou Municipal Government (2005) Haikou City Master planning, 2005 -2020. Haikou: Haikou Municipal Government. June, (2005).
9. Haikou Urban Planning Bureau.Development control planning for Haidian Island Area, Haikou. Retrieved on 2017-04-22, http://www.hkup.gov.cn/ghcgshow.asp?smallclassName=%BF%D8%B9%E6&id=7010&Page=2, (2012).
10. Haikou Tourism Development Committee, Haikou Yacht and Cruise Industry Development Plan (2011 – 2020), Haikou: Haikou Tourism Development Committee, (December, 2011).
11. R. Liu, Queuing Molde for U0turn mouth of Bi-directional streets. Journal of Chongqing Jiaotong University (Natural Science), 28(5):921 -925, (2009).
12. Y. Bao, Z. Zhou, & Y. Xu, A Conflict Model involving right-turning vehicles and cross-walking pedestrians at signalized intersections. Journal of Transport Information and Safety, (3):41-49, (2016).
13. Y. Ni & K. Li, Dealing with conflict between pedestrian and right turning vehicle in signalized intersection. Computer and Communication, (1):22-26, (2007).
14. Y. Liu, Studies of Improvement for traffic safety at Urban Signalized Intersections based on cyclist have priorities. Xi'an: Chang'an University, (2012).
15. X. Wang, Knowledge of Signage. Forest & Humankind, (12): 88-95.
16. S. Chai, Popularize traffic safety knowledge among universities students. Charming China, (15):34, (2011).
17. J. Li, Traffic safety into university campus. Road traffic Management, (1):74, (2016).
18. The ministry of public security of the People's Republic of China, The decision to the amendment of application and usage of motor vehicles license. Retrieved on 2017-04-25, http://www.mps.gov.cn/n2254314/n2254409/n4904353/c5171192/content.html, (2016).
19. J. Cheng, Landscape creation on seaside road in cities: with landscape design on Huangdao Road in Amoy as an example. Planner, (5):71-74, (2004).
20. Santa Clara Valley Transport Authority. Bus stop and passengers facilities standards. Retrieved on 2017-04-25. http://vtaorgcontent.s3-us-west-1.amazonaws.com/Site_Content/VTA%20Bus%20Stop%20Passenger%20Fac%20Stds%202010.pdf, (2010).
21. R. Sia, B. Wu, & J. Li, Mass transit as urban tourist transport. Applied Mechanics and Materials, 97-98:1131-1134, (2011).
22. H. Guo, The city needs friendly slow traffic. Neijiang Technology, (2):121-122, (2016).
23. L. Qian, The fundamental to solve urban transport problems is to develop public transport in scale. Beijing City Planning & Construction Review, (6):25-27, (1998)

DOI:10.1051/matecconf/201712403004

Study on Comprehensive Decision of Urban Rail Transit Network Planning

Liu Sheng[1]
[1]*Lanzhou jiao tong University, Lanzhou, China*

Abstract. Based on the research and analysis of urban rail transit network planning, Determining Evaluation Index System of Urban Mass Transit Network.On this basis, the evaluation index system of network planning is established. this paper analyzes and compares the planning scheme of urban rail transit network in a city, and combines the analytic hierarchy process and the gray correlation method so that the alternatives can be discharged in good or bad order.The ranking results can objectively reflect the actual situation of urban rail transit network planning, in order to solve the problem of optimal selection of urban rail transit network planning, a set of objective scheme is proposed to avoid the decision deviation caused by the defects of single decision method.

1 Introduction

Urban rail transit network plan decision is the key step in the urban rail transit network planning, which is the basis of project optimization. The planning of urban rail transit network planning is mainly based on qualitative analysis [1], [2].This paper is based on the analysis of urban rail transit network planning, using analytic hierarchy process (AHP) combined with the GRAY correlation method, and to establish a scientific and rational evaluation index system, the comprehensive analysis of each scheme for qualitative and quantitative solution should selected from the different options for the best program.

Urban rail transit network evaluation is

different from the general urban road network and regional highway network evaluation.To evaluate the urban rail transit network, we must consider various characteristic factors and establish a complete, scientific and comprehensive evaluation system with the actual situation of the urban rail network. In addition, the scale of the rational control index system is also an important aspect [3].According to the above principle,using AHP basic principle, to establish a set of evaluation index system[4], [5] In the specific planning, according to the need for the following table to the appropriate increase or decrease.

2 The urban rail transit network based on Ahp - Gray comprehensive decision method

The factors that affect the urban rail transit network are expressed as hierarchical structure, and the indicators have qualitative and quantitative indicators.In the following,we combine the AHP method and the GRAY method to establish the decision model of urban rail transit network.

2.1 Index weight assignment method-AHP

Analytic hierarchy process (AHP) was proposed in the 1970s by the United States operations researcher T.L.satty for the first time. The method is a combination of quantitative and qualitative methods.it is often used in combination with other evaluation methods to determine the weight of evaluation index [6,7].

AHP method of the basic steps:(1)Create a hierarchical hierarchy model;(2) To construct judgment matrix;(3) Determine the hierarchical weight value;(4) Consistency test of the judgment matrix;

The matrix deviates from the consistency indicator C.I., as in equation (1).

$$C.I. = \frac{\lambda_{max} - n}{n - 1} \quad (1)$$

Random Consistency Ratio C.R., as in equation (2).

$$C.R. = \frac{C.I.}{R.I.} < 0.1 \quad (2)$$

(5) Calculate the weight of the elements of each layer.

2.2 The optimal decision-making program-GRAY

Gray correlation method(GRAY) is a multi-factor integrated decision-making method.It can make full use of the existing information to reduce the error by using the Gray correlation degree as the measure, and provide a simple and applicable method for the optimization decision making[8], [9] In the above analysis, this paper establishes the optimization model of urban rail transit network planning and evaluation decision-making system, and puts forward a decision method based on Gray relational decision-making method.

2.2.1 Determine the decision matrix

Assume that A is a set of alternatives of decision-making $A = \{a_1, a_2, ..., a_m\}$, V is a set of decision matrices $V = \{v_1, v_2, ..., v_n\}$, the attribute of the scheme a_i is $X_{ij}(i = 1,2, ..., m ; j = 1, 2, ..., n)$, then the matrix $X = (x_{ij})_{m \times n}$ is the decision matrix of the set A of the scheme.

2.2.2 Evaluation index matrix standardized treatment

In the standardization of qualitative indicators and quantitative indicators to be used when the different treatment methods.

Quantitative function of the benefit index V_j:

$$x'_{ij} = \frac{x_{ij}}{\max\limits_{1 \le i \le m} x_{ij}} \quad (3)$$

Quantitative function of cost index V_j:

$$x'_{ij} = \frac{\max\limits_{1 \le i \le m} x_{ij} + \min\limits_{1 \le i \le m} x_{ij} - x_{ij}}{\max\limits_{1 \le i \le m} x_{ij}} \quad (4)$$

Quantitative function of qualitative index V_j:

$$x'_{ij} = 1 - \frac{|q_{ij} - x_{ij}|}{\max |q_{ij} - \min\limits_{1 \le i \le m} x_{ij}, \max\limits_{1 \le i \le m} x_{ij} - q_{ij}|} \quad (5)$$

2.2.3 Gray correlation coefficient of the program

The gray correlation coefficient of the scheme is based on the relevance of the evaluation scheme index vector and the relative optimal index vector as the criterion of the decision-making scheme, which is derived from the gray relational decision theory[8].The gray correlation coefficient is:

$$\xi_{ij} = \frac{\min\limits_{i}\min\limits_{j} |x_{0j} - x_{ij}| + \rho \max\limits_{i}\max\limits_{j} |x_{0j} - x_{ij}|}{|x_{0j} - x_{ij}| + \rho \max\limits_{i}\max\limits_{j} |x_{0j} - x_{ij}|} \quad (6)$$

In the formula, ρ is the resolution coefficient, generally take 0.5.

According to the above analysis, we can see that the gray relational coefficient matrix is

$$R = \begin{vmatrix} \xi_{11} & \xi_{12} & \cdots & \xi_{1n} \\ \xi_{21} & \xi_{22} & \cdots & \xi_{2n} \\ \cdots & \cdots & \cdots & \cdots \\ \xi_{m1} & \xi_{m2} & \cdots & \xi_{mn} \end{vmatrix} \quad (7)$$

2.3 Comprehensive decision-making model

Evaluation index system of n relative to the overall goal of weight vector $W=(w_1, w_2, ..., w_n)$,Then the weighting degree of the relative optimal scheme is related to each other R':

$$R'=RW=(r_1, r_2 ..., r_n) \quad (8)$$

In the formula, $r_i = \sum\limits_{j}^{n} r_{ij} w_j, i = 1,2,..., m$.

Therefore,when $\max(r_1, r_2 ..., r_n)$,Scheme for the planning of the optimal solution.

3 Examples of comprehensive decision-makin method validation

3.1 Construction of line network evaluation system

A Case Study of Urban Rail Transit Network Planning. The network planning have three programs, specific data and parties as listed in Table 2.

Table 2. Data of Urban Rail Transit Planning

Index	Program 1	Program 2	Program 3
Line density	0.73	0.71	0.75
Line Passenger Intensity	2.8	2.7	2.6
Convenient links to core cities	8	7	7
Transfer coefficient	1.41	1.39	1.34
Line length	107.93	107.08	117.14
Staging construction rationality	8	7	7
Engineering difficulties or ease of use	7	7	8
Feasibility of project implementation	8	7	7
the adaptability of urban landscape	8	7	7
In line with the overall urban development plan	8	8	7
the adaptation of urban land use planning	8	8	7
Promote social development	8	7	7
Improve the people's living standards	8	7	7

From the technical evaluation, economic evaluation, environmental evaluation, social evaluation, combined with the actual situation of a city network construction evaluation index system, see Figure 1.

DOI:10.1051/matecconf/201712403004

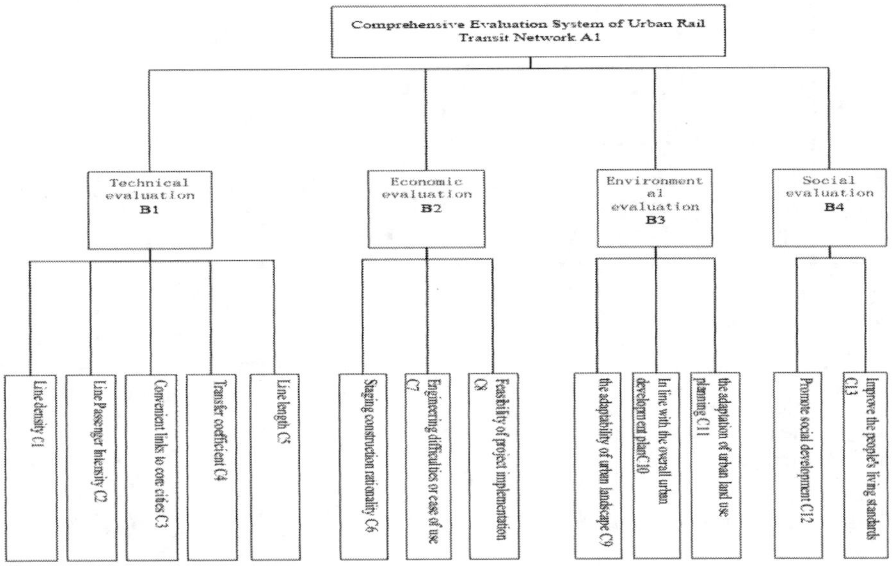

Figure 1. Evaluation Index System of Urban Rail Transit Network

3.2 Determination of the weight of indicators

The AHP method is used to quantify the importance of each evaluation index by using the 1-9 scale method through expert evaluation [10]. The Square root method is used to calculate the weight vector.

Two of each factor to establish judgment matrix and the matrix consistency index as shown in Table3, Table 4 shows the weight of the index system of the evaluation index system.

Table 3. Pairwise Judgment Matrix and Matrix Consistency Index

Calculate the content	A(B1,B2,B3,B4)	B1(C1,C2,C3,C4,C5)	B2(C6,C7,C8)	B3(C9,C10,C11)	B4(C12,C13)
Judgment matrix	1 4 3 5 1/4 1 1/3 1 1/3 3 1 2 1/5 1 1/2 1	1 1/2 4 3 3 2 1 7 5 5 1/4 1/7 1 1/2 1/3 1/3 1/5 2 1 1 1/3 1/5 3 1 1	1 1/5 1/3 5 1 3 3 1/3 1	1 1/3 1/3 3 1 1 3 1 1	1 1/3 3 1
The geometric average	0.549 0.106 0.236 0.109	0.263 0.475 0.055 0.090 0.110	0.105 0.637 0.258	0.143 0.429 0.429	0.250 0.750
Maximum eigenvalue	4.058	5.073	3.039	3	2
Relative consistency index	0.019	0.018	0.0191	0	0
Consistency indicator	0.021<0.1	0.016<0.1	0.033<0.1	0<0.1	0<0.1

Table 4. Urban index system single layer relative weight

Index system stratification		Single layer weight		
Comprehensive Evaluation System of Urban Rail Transit Network A	Technical evaluation B1	Line density C1	0.549	0.263
		Line Passenger Intensity C2		0.475
		Convenient links to core cities C3		0.055
		Transfer coefficient C4		0.090
		Line length C5		0.110
	Economic evaluation B2	Staging construction rationality C6	0.106	0.105
		Engineering difficulties or ease of use C7		0.637
		Feasibility of project implementation C8	0.236	0.258
	Environmental assessment B3	the adaptability of urban landscape C9	0.236	0.143

		the adaptability of urban landscape C10		0.429
		In line with the overall urban development plan C11		0.429
	Social evaluation B4	Promote social development C12	0.109	0.250
		Improve the people's living standards C13		0.750

The combined weights of the layers are calculated, i.E. W=(0.144,0.261,0.030,0.050,0.060,0.011,0.068,0.0273,0.034,0.101,0.101,0.027,0.0818).

3.3 According to the gray method to give each program indicators and indicators of optimal correlation matrix

3.3.1 Determine the decision matrix

$$X=\begin{vmatrix} 0.73 & 2.8 & 8 & 1.41 & 107.93 & 8 & 7 & 8 & 8 & 8 & 8 & 8 & 8 \\ 0.71 & 2.7 & 7 & 1.39 & 107.08 & 8 & 7 & 7 & 7 & 8 & 8 & 7 & 7 \\ 0.75 & 2.6 & 7 & 1.34 & 117.14 & 7 & 8 & 7 & 7 & 7 & 7 & 7 & 7 \end{vmatrix}$$

3.3.2 After the normalized processing, the comprehensive decision matrix is obtained

$$X'=\begin{vmatrix} 0.97 & 1 & 1 & 0.95 & 0.921 & 0.88 & 1 & 1 & 1 & 1 & 1 & 1 & 1 \\ 0.95 & 0.96 & 0.88 & 0.96 & 0.914 & 0.88 & 1 & 0.88 & 0.88 & 1 & 1 & 0.94 & 0.94 \\ 1 & 0.93 & 0.88 & 1 & 1 & 1 & 0.88 & 0.88 & 0.88 & 0.88 & 0.88 & 0.88 & 0.88 \end{vmatrix}$$

3.3.3 Gray relational judgment matrix R, by the formula (6),

$$R=\begin{vmatrix} 0.67 & 1 & 1 & 0.55 & 0.432 & 0.33 & 1 & 1 & 1 & 1 & 1 & 1 & 1 \\ 0.55 & 0.06 & 0.33 & 0.06 & 0.411 & 0.33 & 1 & 0.33 & 0.33 & 1 & 1 & 0.5 & 0.5 \\ 1 & 0.462 & 0.33 & 1 & 1 & 1 & 0.33 & 0.33 & 0.33 & 0.33 & 0.33 & 0.33 & 0.33 \end{vmatrix}$$

W=(0.144,0.261,0.030,0.050,0.060,0.011,0.068,0.0273,0.034,0.101,0.101,0.027,0.0818).

3.3.4 Integrated decision-making

$R'=W^TR=$ (0.855,0.481,0.540)

$maxR'=max(r_1, r_2, r_3)=max(0.885,0.481,0.540)=0.885$.

According to the above operation, we can see that the comprehensive decision value of scheme 1 is obviously larger than that of scheme 2 and scheme 3, so the scheme is optimal;Option 2 is the worst, scenario 3 is second, but there is no obvious superiority compared to scenario 1. Therefore, the city rail transit network planning theory should choose the first program.

4 Conclusion

In this paper, the combination decision-making model of AHP-GRAY,using the Analytic Hierarchy Process to determine the weight of evaluation indicators, taking into account the main and objective factors, the scientific description of the actual problem,Based on the establishment of gray relational decision model, combined with examples, the urban rail transit network planning in the three pre-selection program decision-making, and gives the optimal program.This method provides a simple and convenient method for decision-making of line network scheme.

References

1. H. Ning, P. Xiangyang.Decision - making Analysis of Urban Rapid Rail Transit Planning[J].China Journal of Highway and Transport.12(3):73-81, (1999).
2. G. Xiucheng, *et al*.Study on Evaluation Method of Urban Rail Line Network Planning[J].Research on Urban Rail Transit,(4),pp,24-27, (2000).
3. C. Xumei, L. Fengjun, The study on the criterion system for comprehensive system for evaluation of The urban[J].City Planning Review,25(10),pp.61-64,(2001).
4. J. Jialun, Research on Comprehensive Evaluation Index System of Urban Rail Transit Network Planning Scheme [J]. Journal of Systems Engineering and Practice.(3),pp.129-133, (2004).
5. C. Guotai, C. Tingting, Z. Kun. Construction of Evaluation Index System for Human 's Comprehensive Development Based on Correlation Principal Component Analysis[J].Systems engineering theory and practice, 32(1),pp. 111-119, (2012).
6. W. Lianfen,et al.An Introduction to Analytic Hierarchy Process[M].Beijing:China Ren min University Press, (1995).
7. T.L. Satty, The Analytic Hierarchy Process: Planning,Priority Setting ,Resource Allocation[M]. New York:MC Graw-Hill, Inc., (1980).
8. L. Sifeng, D. Yaoguo, F. Zhigeng.Gray System Theory and Its Application[M]. Beijing:Science Press, (2004).
9. M. Chaoqun, W. Yuping, C. Kuanmin.etal. Comprehensive evaluation for urban rail transit network based on gray weight relation[J].Journal of Chang an University: Natural Science Edition,(3),pp.84,87,110, (2007)
10. M. Shuhong.Study on the Adaptability of Regional Highway Traffic and Economic Development[D].Chang an University, (2002).

Analytical study of the lane-by-lane variation for lane-utility and traffic safety performance assessment using NGSIM data

Zhandong Xu

School of Transportation and Logistics, Southwest Jiaotong University, 610031 Chengdu, China

Abstract. Highlights the shortcoming of using aggregate data as a whole with the assumption of similar lane-use-pattern over lanes, traffic variations are attentively investigated to explore traffic dynamics distribution both in lane-utility in efficiency and safety performance lane-by-lane. In this paper, above two dimensions are examined for each of five mainline lanes on Highway 101 in Angeles using Next Generation Simulation (NGSIM) data. The lane-specific data is statistically processed to reflect lane utility in lane flow parameters along the location and identify Lane Risk Coefficient by verifying degree of consistence vehicles operation with 'Assured Clear Distance Ahead' criterion. It appears completely new that the most-middle lanes are similar utilized under congested conditions compared to two sides of the lanes, as well as the quite close values of LRC to reveals an uniformly safety level between Lane 2 and Lane 3. The findings of this study are intended to be relevant with the calibration of microscopic simulation models and suggest implementing a proper variable speed toward under-utilized shoulder Lane 5 but with worst safety performance.

1 Introduction

It is commonly recognized that macroscopic traffic dynamics acquiesce in strong assumption that all lanes behave with uniformity of traffic flow variables, such as flow, density and speed that represent the average across all available lanes. However, critics have also argued that a complete traffic flow theory must take into account lane specific behaviour because it affects capacity and transitions between uncongested and congested traffic flows. Highlights the shortcoming of using aggregate and average data, non-equilibrium over lanes can occurs and fails to identify or forecast different traffic states without a distinction over lanes [1]. In addition, the Highway Capacity Manual (HCM, 1994) acknowledges the inherent error for neglecting the variation of each lanes and aggregating them in the 2010 edition, which explicitly indicating that the traffic distribution of a multi-lane roadway depends on a number of factors such as traffic regulation, traffic composition, driving behaviour and the origin-destination patterns [2].

Other studies have been extensively focus on the lane flow distribution (LFD) to statistically explore the relative impacts of lane-by-lane patterns with location and daily variability and examine the lane variability across lanes [3]. Their studies against the lane-specific variability typically confirmed the relationship between traffic flow parameters across lanes using curve fitting techniques. However, most researches focus on an equilibrium traffic flow without the influence of merging traffic and safety consideration. This raises the problem how lane-utility changes along the location of roadway with merging flow.

Based on which, quite a few of researches investigated whether a vehicle is in consistent with Assured Clear Distance Ahead (**ACDA**) criterion for safety performance assessment [4]. It is confirmed that unbalanced lane-using patterns will result in variability of safety performance of vehicles operating on different lanes. However, little research has been done to investigate safety performance among several available lanes at a stretch of freeways. Thus, this paper intends to assess the safety coefficient across different lanes considering the lane variation on freeway. In order to do so, there are two aspects of lane use pattern have to be analysed. The first one is to statistically verify significant variation across lanes using detected data from the Next Generation Simulation (**NGSIM**) program. The second one is to assess the vehicle operation and safety performance on each individual lane and makes an evaluation lane-by-lane.

This paper is organized as follows. Section 2 presents the studied freeway area and a detail description of data items. Section 3 present the lane traffic distribution along the driveway, this is followed by analysis of lane-by-lane safety performance assessment in Section 4. This paper concludes by Section 5 with the implication of the results.

2 Site and data

This study focused on the program of the Next Generation Simulation (NGSIM) southbound direction of

U.S. freeway 101 (Hollywood Freeway) as illustrated in Figure 1. The freeway 101 is located in Angeles, California and mounted with video cameras for recording vehicle trajectory dataset. The site is approximately 640m in length with five mainline lanes, one-way throughout the section including the on-ramp at Ventura Boulevard and the off-ramp at Cahuenga Boulevard. We named the five mainline lanes from leftmost to the right as Lane 1 to Lane 5.

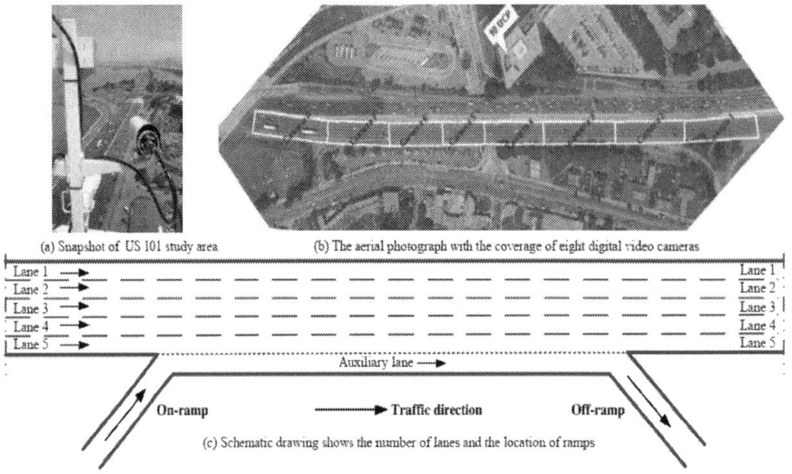

Figure 1. Traffic data collection via cameras on a section of about 640m on US Highway 101.

The program has collected high-quality and lane-specific thousands of vehicle paths in a total of 45 minutes between 7:50 a.m. and 8:35 a.m. on June 15, 2005. Further, the selection of this time frame determines the study focused on experienced heavy congestion (LOS E). Video measurements of individual vehicle dynamics including vehicle position (coordination x-y), vehicle length, occupied lane and etc., were recorded at every 0.1 second, ensuring the purpose of this study using lane-specific data.

3 Lane traffic analysis

Traditionally, the capacity and service quality of freeway is evaluated by using the entire freeway traffic data, neglecting the capacity distribution of more precisely traffic variations between different lanes. Not many speculations from lane-level view are raised to analyse the lane-using patterns and distributions across lanes. In this paper, the three microscopic traffic flow parameters, including traffic flow, speed and occupancy, are investigated to explore lane-level capacity distribution.

3.1 Traffic flows

The single direction of studied freeway consists of five mainline lanes, and a total of 7496 vehicles in 45 minutes are statistically added up from the classification of three types of traffic composition (i.e. small car and truck) and traffic demand. According to the whereabouts of the traffic flow without considering zero ramp-to-ramp traffic, it can be divided into three categories: freeway-to-freeway traffic, freeway-to-ramp traffic and ramp-to-freeway traffic. The traffic flow rate per lane can be expressed as the number of vehicles per hour occupied that lane through the section. Table 1 summarizes statistical calculation results of traffic flow across lanes.

Table 1. Vehicle types and traffic demand distribution among individual lanes.

	Type (veh)		Traffic Demand (veh)		
	Car	Truck	Freeway to freeway	Freeway to ramp	Ramp to freeway
Lane1	1219	7	1218	1	7
Lane2	1466	4	1437	2	31
Lane3	1492	13	1426	2	77
Lane4	1541	47	1388	11	189
Lane5	1638	69	1085	223	399
Total	7356	140	6554	239	703

•Outer lane carries more flows This is because of origin-to-destination purposes, which is inevitable for those vehicles to use. Meanwhile, heavier vehicles (e. g. truck or bus) using the rightmost lanes not only for route patterns, also corresponding to their less-powerful to be overtaken and mass-loaded characters so as to suggest their driving preference at outer lane.

•Freeway-to-freeway vehicles tend to use the leftmost lanes to avoid those turning vehicles, in contrast to most of merging vehicles driving on Lane 4 and Lane.

In general, it can be clearly observed that the traffic flows on median Lane 1 and shoulder Lane 5 are significant different with adjacent lanes, with 19.88% and 7.5% of growth respectively. The traffic flow per lane of lane-using variation reflects the traffic flow parameters do not have uniform volumes across available lanes. Specifically speaking, different characteristics of roadway, such as traffic composition, traffic regulation and rules, driving behavior and so forth will be the causative factors for such traffic flow distribution.

3.2 Traffic speed

As Kerner (2004) argues that not all roadway capacity is used as long as the traffic speeds are not equal across lanes, which implies non-equilibrium of speed over lanes is harmful to total flows [6]. The sensors have collected instantaneous velocity of vehicles at every 0.1s in traffic dataset. Time mean speed (TMS) for vehicle k is calculated considering n midpoints where the vehicle travelling on it, we will have the following:

$$\overline{u_k} = \frac{1}{n}\sum_{t}^{n} v(t_k) \quad (1)$$

The TMS corresponding to each lane is segmented to obtain basic statistic such as mean, standard deviation, median and skewness, as summarized in Table 2.

Table 2. Basic statistics of TMS for different lanes.

	Mean (km/h)	Standard Deviation (km/h)	Median (km/h)	Skew-ness
Lane 1	33.49	10.93	31.76	1.42
Lane 2	36.67	11.75	35.22	0.74
Lane 3	37.01	12.13	35.45	0.69
Lane 4	37.74	12.03	36.29	0.54
Lane 5	39.56	10.92	39.30	0.14

From the table it can be seen that the mean and median value of TMS show an increasing trend from leftmost lane to another. And the standard deviation is roughly 11.5 km/h. The decrease in positive values of skewness with outer lane position indicates that the distribution of TMS is more skewed toward the right at the leftmost Lane 1. To statistically confirm the difference of TMS between two adjacent lanes, hypothesis testing using Kolmogorov-Smirnov Test is performed to prove speed adjustment variation of drivers across lane. The null and alternate hypothesis is defined as follows:

H_0: The distribution of TMS is no different between two adjacent lanes.

H_1: The distribution of TMS is different between two adjacent lanes.

The significant level of hypothesis test is set to be 0.95. The results including the KS-Statistic and P-Value are given in Table 3.

Table 3. Hypothesis test for TMS between adjacent two lanes.

	KS-Statistic	P-Value	Hypothesis
Lane 1 & Lane 2	0.182	$\rightarrow 0^+$	Reject H_0
Lane 2 & Lane 3	0.038	0.230	Accept H_0
Lane 3 & Lane 4	0.072	0.011	Reject H_0
Lane 4 & Lane 5	0.140	$\rightarrow 0^+$	Reject H_0

Note: $\rightarrow 0^+$: Infinitely close to 0 in positive.

The results of the hypothesis test indicate that the distribution of Lane2 and Lane3 is indeed small variations than other adjacent lanes. We conduct a range of KS-Tests for each lane to examine likelihood statistic hypothesis, and the results demonstrate that TMS accords best fitting with lognormal distributions of Lane 1-4 and normal distribution of Lane 5, which is shown in Figure 2.

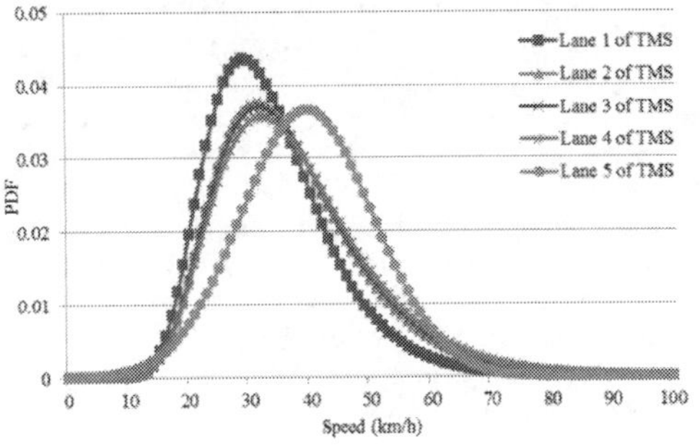

Figure 3. Best fitted distributions for TMS of five lanes.

From Figure 3, it can be seen that as lane location closing to the right the highest expected outcome decreased and gradually stabilized to 3.6%. It can also be observed that the distribution are increasingly skewed to the right with lane position coming towards the left, which confirms that an increasing proportion of drivers have a tendency to slow down to be more prudent. The figure also demonstrates distributions of three middlemost lanes have tiny variation reciprocally, and relatively great variety arises at two bilateral lanes. However, the higher averaged speed on Lane 5 indicate drivers to behave more aggressive with merging traffics, which is harmful to cause hidden dangers.

3.3 Occupancy

Time occupancy ρ is defined as the percent of time vehicles occupy the section to the total detected time interval, which can directly reflects the lane-utility along the location. To formally describe the considered traffic density of section along each lane, we denote $\rho_{i,x}$ to represent time occupancy at the location of x in i^{th} lane. If the vehicle k has passed through x of i and l_k is the length of subjective vehicle k, the elapsed time will be added up to calculate final observing $\rho_{i,x}$ with equation (2), and illustrated in Figure 4.

$$\rho_{i,x} = \sum_{k}(t_{k,x+l_k} - t_{k,x}) / T_{total}, \ where \ \ Lane(t_k) = i \quad (2)$$

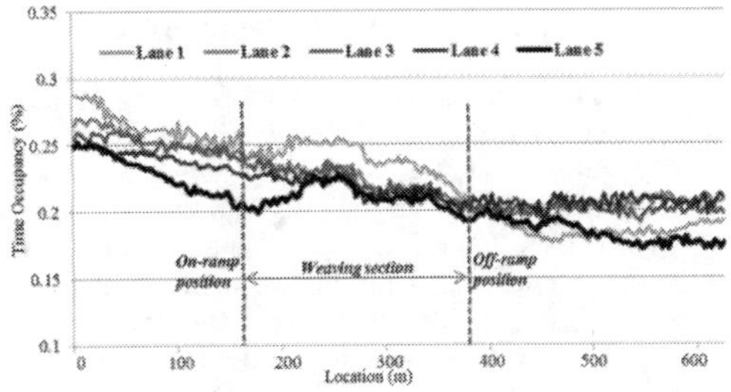

Figure 4. Time occupancy along the carriageway over lanes.

As we can clearly observe from Figure 4, the values of time occupancy range between 15% and 30% for five lanes considered. The fluctuation within a narrow range is corresponding with vehicles lane-changing behavior and possibly inaccurate positioning errors.

The figure also confirms that an unbalanced lane-utility exists across lanes. As for **median Lane 1**: the occupancy has been at a relatively high level at upper reaches with slower speed, on which may result in semi-congestion. Thereby, more motorists will aspire to higher speed travelling at lower density lane, which can account for the sharp decline trend. The **middle of three lanes** (i. e. Lane 2, Lane 3 and Lane 4) has little variations reciprocally: the curves show a closing to linear decreasing trend (a total of 6%-7%) with approaching downstream to off-ramp position, and after through that the time occupancy keeps maintaining at nearly 20%. The occupancy on **shoulder Lane 5**: decreases (nearly 5%) from origin to on-ramp position, which is because drivers tend to change to the left lanes making room near the on-ramp. The succedent increasing with 3% indicates a considerable proportion of vehicles merging onto the lane.

3.4 Findings

The statistical results demonstrate that traffic condition at the same period of time vary across available lanes. In conference to these past studies at heavy traffic flow, we have similar conclusions to some extent that the shoulder Lane 5 is under-utilized compared to inner lanes with higher occupancy. And the median Lane 1 carries a large part of occupancy at congestion flow. Once a certain degree of congestion has set in, the prohibition of overtaking at left-side no longer stands and the traffic in three middle lanes become almost equal regardless of whether in flow, speed or density.

By analysing traffic flow lane-by-lane, we have found that the median Lane 1 is utilized most and shoulder Lane 5 is just the opposite in congested condition. The processes demonstrate no difference between Lane 2 and Lane 3 and imply equilibrium of lane use of the traffic using Kolmogorov-Smirnov analysis. In addition, comparing to inner lanes without being influenced by merging flow, the extremely unstable but aggressive vehicles at Lane 5 are likely to introduce hidden danger along the driveway. The assessment of such lane-by-lane safety performance is conducted in next section.

4 Safety performance assessment

As is stated above, traffic flow distributions are destined to make a difference on driving behaviors in real traffic environment, further causing the non-equilibrium of safety performance across lanes. However, few previous studied have been able to draw on structured research into the opinions and attitudes of safety performance evaluation at lane-level view, not to mention exploring impacts of lane flow on safety performance variability. In this section, we will extend speed-based rule to consider safety performance of motorists for individual lane, generally known as the Assured Clear Distance Ahead Rule (ACDA).

4.1 Methodology

The ACDA rule requests drivers operate their vehicle at a limited speed and sufficient spacing from the vehicle ahead by the objective purpose to ensure collision avoidance. In other words, the safety operation of an ego vehicle is capable to be identified whether drivers follow the ACDA criterion along the driveway, beware of future hazard behavior of the preceding vehicle may suddenly slow down or even stop.

Considering two vehicles, the following vehicle veh_f and the preceding vehicle veh_p, are travelling on the same lane. Without loss of generality, we are focusing on extreme safe operation of vehicles in the driveway so that we simply assume the maximum permitted deceleration a_{max} for all vehicles. Assuming that veh_p unexpectedly brakes at the constant maximum negative acceleration from the beginning of braking process and the following vehicle continues driving with constant velocity v_f before fully brakes after a reaction delay time ε. We denote Δx_0 as the initial spacing gap of two vehicles and l_p as the length of preceding vehicle. The stopping distance x_p and x_f that vehicles travel during braking from initial velocity to zero are obtained by general kinematic solution,

$$\begin{cases} x_p = \dfrac{1}{2}\dfrac{v_p^{\ 2}}{a_{\max}^{\ -}} \\ x_f = v_f * \varepsilon + \dfrac{1}{2}\dfrac{v_f^{\ 2}}{a_{\max}^{\ -}} \end{cases} \quad (3)$$

We introduce the risk factor φ to investigate whether the ego vehicle (i. e. veh_f) maintains a safe spacing distance and how hazard condition it would be when the front vehicle stops suddenly, which can be written as,

$$\varphi(t, veh_f, veh_p) = \frac{x_f + l_p - x_p}{\Delta x_0} \quad (4)$$

Consequently, we will have $\varphi \leq 1$ for the requirement not to strike the rear bumper of *vehp*, and vice versa. To evaluate the overall hazard state of a specific lane, we believe it would be better not only considered the total offensive time temporally, also the severity of offense spatially and its nonlinear increasing hazard features. For example, a vehicle with the risk factor equaling to four at a particular time is double higher than the risk of two. Thus, we define an exponential function to take into account instantaneous criterion-violated performance of the vehicle operation along the driveway,

$$f(x) = e^{\alpha(x-1)} - 1 \quad (5)$$

In equation (5), the parameter α is determined by the range of risk factor to limit the hazardous weight interval at (0, 1). We further introduce ϕ_i to represent the hazard level of ith lane, and formula to calculate the Lane Risk Coefficient (**LRC**) is

$$\phi_i = \sum_k \sum_{t_a} \varphi(t_a, veh_f, veh_p) f(\varphi), \quad \varphi > 1 \quad (6)$$

4.2 Evaluation

We further have to choose a proper reaction time and maximum deceleration for the braking process of vehicles. Argument against human reaction time have been investigated and summarized in [7], and it is shown in their studies that reaction time to full braking are all around 2s obtained from regardless of simulators or a test public road. The International Standards Organization [8] reports that maximum permitted rate of deceleration is specified as $5m/s^2$, and longitudinal human-driving braking is substantially less than the maximum deceleration rate of $8.5m/s^2$, even in the situation for crash avoidance or mitigation [9].

The histogram for risk factor φ in deceleration of $8.5m/s^2$ to illustrate risk distribution across lanes is shown in Figure 5.

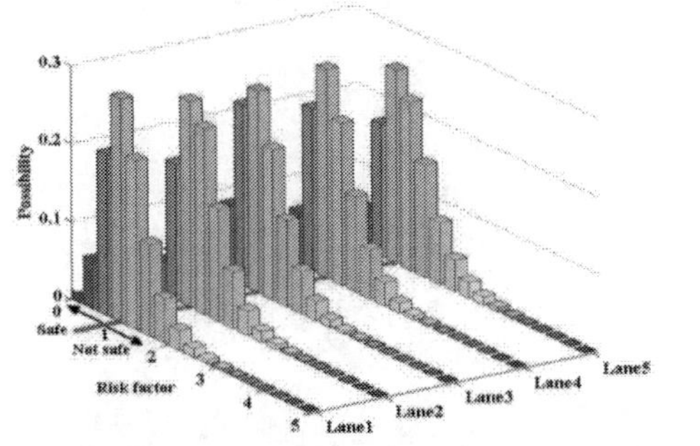

Figure 5. Histograms of risk factor φ across lanes.

The demand safe distance of vehicles to recorded distance within the range of zero to five. As seen from Figure 5, the distribution of risk factor over lanes do not have significant differences in shape, which means the safety performance is mainly determined by the amount of ACDA-violation track points of vehicles.

Since it cannot be directly inferred that how hard all drivers will pressed the brakes to handle with such emergency situation. Thus, we conduct sensitive analysis by changing maximum deceleration vary from $5m/s^2$ to $10m/s^2$. The snapshot of road surface shows a good dry condition, and without loss of generality, it is assumed that all vehicles are the same maximum deceleration rate and reaction delay time. The LRC curve and unsafe proportion of vehicle trace points is shown in Figure 6.

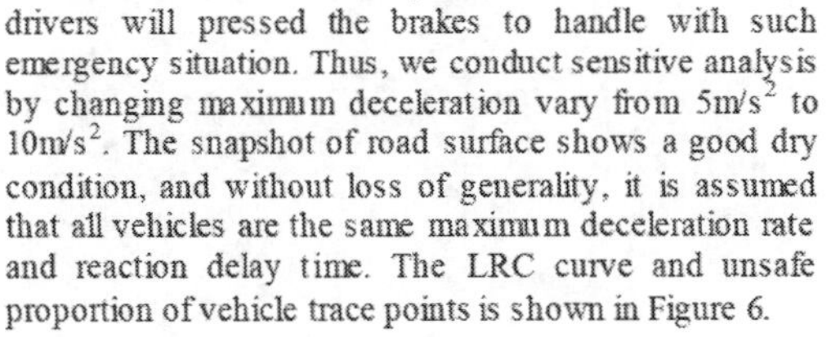

Figure 6. a) Unsafe proportion of trace points violation with ACDA criterion; b) LRC curve with maximum-permitted values of deceleration.

As introduced in equation (4), once drivers have a greater reduction in speed, they are more capable of avoiding collision in critical situation with smaller value of risk factor. It can be clearly observed that the percentage of tracepoints of vehicles driving in unsafe manner increases slightly from around 51% at Lane 1 to around 59% at Lane 5.

We can found that LRC decreases with the maximum-permitted deceleration value increasing, which demonstrates that vehicles operating behave more safely. Right lanes carry higher rate of conflicting flows with respect to the left. This indicates that the higher averaged speed of the outside lane, not only causing a more difficult situation for merging process and leading to congestion of a capacity drop, also reduce the safety status because of merging with smaller gaps in the traffic. Interestingly, the values of LRC are quite close in three middle lanes (especially between in Lane 2 and Lane 3), so that we have reason to deduce a similar safely-lane-use pattern between the three lanes.

5 Conclusions

It appears completely new that the most-middle lanes are similar utilized under congested conditions compared to two sides of the lanes, as well as the quite close values of LRC to reveals an uniformly safety level between Lane 2 and Lane 3. This may have close relation with relatively unrestrained lane choice in middle lanes. More importantly, since the shoulder Lane 5 is under-utilized but with most hazardous lane use characteristic, toward which to suggest implementing a proper variable speed limit will be beneficial to overall safety performance.

In this paper, many of statistical results are in some extent consistent to a few previous literatures. However, limitation in the available reference data did not allow for an extensive quantitative comparison, and as such, the analysis was limited to a qualitative examination of the results and can only explain site-specific traffic in congestion flow. A more complete theory of lane use behavior might also consider such factors as lane and ramp configurations, average vehicle spacing, and lane-changing behavior. Further studies on lane specific behavior and ramp configurations will be summarized in our next study to demonstrate. We also have following insight of this study,

- The study has empirical evidence that demonstrate the use of lane-specific information is a critical factor that should be incorporated within microscopic traffic simulation models. It is felt that the findings are intended to be relevant to the calibration of microscopic models.
- The lane-by-lane safety assessment reveals that non-uniform flow parameters with significant difference may adversely affect lane-using performance in real traffic. It is hoped that the consideration may lead to a better simulation of traffic conditions and thus aid in the successful operation of freeway traffic management.
- Critical for lane level routing analysis once we can fully understand the traffic dynamics on lane level, thus the BPR function or friction factor could be produced based on lane level analysis.

References

1. N. Chiabaut, L. Leclercq, C. Buisson, "From heterogeneous drivers to macroscopic patterns in congestion [J]". Transportation Research Part B: Methodological, **44**, pp. 299–308 (2010)
2. Highway capacity manual (HCM), Transportation Research Board, Washington D.C (2010)
3. A. Duret, A. Soyoung, B. Christine, "Lane flow distribution on a three-lane freeway: General features and the effects of traffic controls [J]". Transportation Research Part C: Emerging Technologies, **24**, pp. 464-464 (2012)
4. A. Pompigna, F. Rupi, "Capacity and Lane Distributional Effects on a Three-Lane Carriageway Section: a Case Study on the Italian Freeway Network", *Sidt Scientific Seminar* (2015)
5. P. Maravelias, "Tactics for safety and optimization", Advanced highway driving, (2015)
6. B S. Kerner, "Empirical Complex Pattern Formation Caused by Peculiarities of Freeway Infrastructure [M]", *The Physics of Traffic*, Springer Berlin Heidelberg, **58**, pp. 365-372 (2004)
7. D V. Mcgehee, E N. Mazzae, G H S. Baldwin, "Driver Reaction Time in Crash Avoidance Research: Validation of a Driving Simulator Study on a Test Track [J]". Marine Pollution Bulletin, **44**, pp. 1157-1165 (2000)
8. International Standard Organization (ISO), "Full speed range adaptive cruise control system, Standard #22179, Geneva (2009)
9. DB. Fambro, K. Fitzpatrick and RJ. Koppa, "Determination of stopping sight distance [M]", National cooperative highway research program (NCHRP), *National Academy Press* (1997)

MATEC Web of Conferences **124**, 04002 (2017)
ICTTE 2017
DOI:10.1051/matecconf/201712404002

Safety Performance Evaluations for the Vehicle Based Movable Barriers Using Full Scale Crash Tests

Minsoo Jin[1], Sukki Lee[2], Wonil Park[2] and Insong Park[3]

[1] *Korea Institute of Civil Engineering and Building Technology, and Department of Transportation Engineering, University of Seoul, Korea*
[2] *Korea Institute of Civil Engineering and Building Technology, Korea*
[3] *Smart Air Chamber Co., Ltd., Korea*

Abstract. The present study aims to develop a prototype of large-size movable barriers to protect roadside workers from incoming vehicles to the road work area with the following functions: maximization of work space in the right and left directions, convenient mobility, and minimization of impact without modification of the inside of movable barriers into traffic lanes and perform safety performance assessment on passengers through full scale crash tests. The large movable barrier was divided into folder type and telescope type and the development stage was now at the prototype phase. A full scale crash test was conducted prior to certification test at a level of 90%. The full scale crash test result showed that both types of folder type movable barrier and telescope type movable barrier satisfied the standard of the passenger safety performance evaluation at a level of 90%.

1 Introduction

The Highway Safety Manual(HSM, 2010) in the USA suggested that traffic accidents occur due to three factors: human factors(93%), vehicle factors(34%) and roadway/environment factors(13%) [1].

Due to the education, enforcement of law, and technological development to reduce traffic accidents, the number of traffic accidents has been decreased overall in recent 10 years but a sole-car accident type and accidents during road works have been increased rather, which needs to establish a management and prevention measure

In particular, the number of casualties during road works is 27 out of 583 workers in the regional office in the Ministry of Land, Transport and Maritime who work for maintenance of general national roads as of 2012. This number means approximately 5% of the road workers were dead or injured, which was much higher proportion of casualties than that of fire-fighting officers (0.97%) classified as one of the high risk groups. The number of traffic safety-related accidents of road workers was 114 cases and 152 workers were damaged (10 fatal and 142 injured workers), which were produced due to careless driving (e.g., mobile phones or digital multimedia broadcasting (DMB) manipulation, forward-looking neglect, and drowsy driving etc.) while driving on road work sections in general national roads from 2003 to 2012. This result means that one out of four workers (25%) has experienced accidents during their average work period (10 years) [2, 3].

For national expressways, more workers were dead or injured from 2008 to 2013 due to worse traffic environment against road workers than that of general national roads, which resulted in 56 fatal casualties and 15 injuries of general drivers and 47 deaths and 15 injuries of road workers (103 deaths and 30 injuries in total). That is, about 22 casualties in road works occur in a year.

For general national roads, new employment of road workers is limitedly added so the number of road workers has been decreased gradually from 773 in 2004, 625 in 2008, and 583 in 2013 whereas the road maintenance and repair jobs were not decreased thereby inputting road workers for traffic control to road works due to the insufficient number of road workers [3].

In recent years, much attention has been paid to road maintenance due to the increases in demand on road services and traffic volume and the construction records of road works in Seoul during 2002 to 2004 showed that 320 road works a day have been conducted on average and this trend has been increased. The degradation in inherent functions of road due to road works is closely related to not only traffic flow but also safety problems of road users and has become a significant threat to the safety of road workers particularly [4].

The present study aims to develop a prototype of large-size movable barriers to protect roadside workers from incoming vehicles to the road work area with the following functions: maximization of work space in the right and left directions, convenient mobility, and minimization of impact without modification of the inside of movable barriers into traffic lanes and perform safety performance assessment on passengers through full scale crash tests.

DOI:10.1051/matecconf/201712404002

2 Review and concept for development

2.1 Review

Caltrans was developed a Balsi Beam trailer system consists of a semi-truck tractor and a trailer in 2003. The trailer has two telescoping expandable steel beams along each side of the trailer that rotate to provide a stackable two beam protection barrier on either side of the trailer. The beams can be rotated to either side depending on the location of the work that beams.

FHWA(Federal Highway Administration) recognized the necessity of road maintenance worker's injury prevention. So FHWA and Mobilebarriers LLC had conducted study of a movable barrier to protect workers in a side collision accidents and developed and operating MBT-1.

The Smart Air Chamber Co., Ltd. developed H-beam type movable protective car based on Article 4 of Automobile Management Act to protect road workers in freeway workzone.

2.2 Concept for development

Based on the literatures and website based review, we established the basic concept to develop the movable safety barriers. The first consideration we made is the idea that the type of barriers needs to be varied according to the work zone characteristics. The second is that the barrier should meet the required safety performance designated by Government of Korea.

With the concept, we designed large-sized movable barriers presuming that barriers can be differently applied according to the characteristic of work zone. Large-sized movable barriers are again classified into two types such as Folder Type and Telescope Type according to the kinematic mechanism, and the former type has foldable feature and the latter type has expansion and contraction operative feature.

Figure 1 shows the folder type of movable barrier, which has the series of the block units. While moving, the block unit is folded to meet the vehicle code designated by Government such as the maximum length of the vehicle. When the vehicle begins to service, the folded block is stretched to the certain length to protect the work zone effectively.

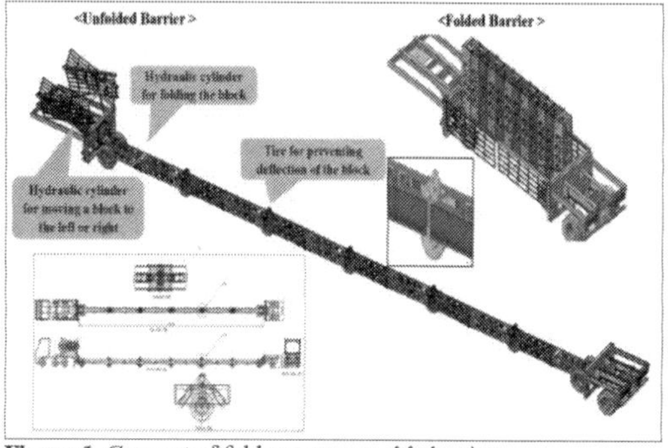

Figure 1. Concept of folder type movable barrier

Figure 2 shows the telescope type of movable barrier, which has the same feature while moving but the barrier beam is expanded from the rear side of the trailer is moving backward [5].

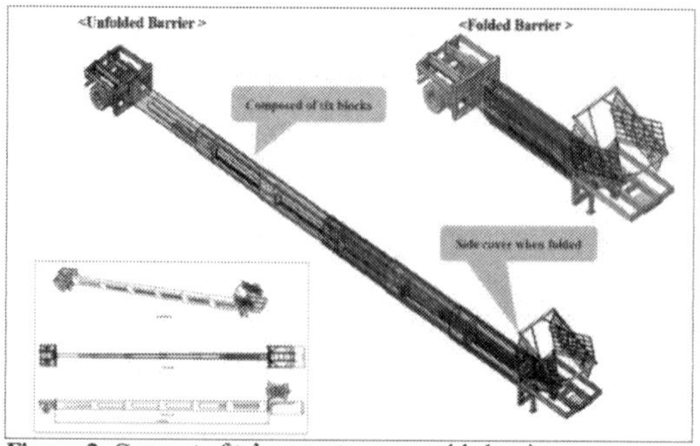

Figure 2. Concept of telescope type movable barrier

3 Full scale crash tests

The full-scale crash test of barriers should be conducted in accordance with the "Guideline for Safety Barriers (Safety facility of barriers for vehicles, 2014)" and performance test should be performed by the performance evaluation institution of road safety facility in order to certify the result of performance evaluation. Table 1 presents the test conditions to evaluate the safety of occupants.

Table 1. Test conditions for occupant protections

Level	Impact Velocity (km/h)	Impact Wight(kg)	Impact Angle(deg.)
SB1	60	900	20
SB2, SB4	80		
SB3	100		
SB5, SB6, SB7	100		
SB3-B, SB5-B	120		

3.1 Criteria of occupant protections

The criteria of occupant protection performance evaluation are presented in Table 2.

Theoretical Head Impact Velocity(THIV) refers to an instantaneous relative velocity between the "theoretical occupant head" and the vehicle when the vehicle is decelerated after crash with facility and the head is free flight thereby crashing to the virtually set surface in the inner space in the vehicle assuming that the occupant head and vehicle are free flight at the impact speed.

DOI:10.1051/matecconf/201712404002

Post-impact Head Deceleration(PHD) refers to the calculated maximum of 10 m/s average acceleration of vehicle after THIV is calculated assuming that an occupant maintains a contact with a virtually-set surface inside the vehicle and receives the same acceleration of the vehicle.

Acceleration Severity Index(ASI) refers to a value that represents a sum of each ratio of 50ms average acceleration in the x, y, and z directions, respectively. It is expressed as a dimensionless scalar. It has a positive value always and if the value is larger than one, a risk of occupant becomes larger.

Table 2. Vehicle occupant impact severity criteria

Criteria	Unit	Index severity
THIV	m/s (km/h)	9 (33)
PHD	g	20
ASI	-	1.9

3.2 Field setting of crash tests

A full-scale crash test field was constructed by Smart Air Chamber Co., Ltd. that conducted the present study with their own investment in Ansung Gyeongi-do. The test field has been utilized as a full-scale test of front, side, and rear of vehicles.

Figure 3. Full scale crash test bed(Smart Air Chamber co., Ltd.)

A length of road test drive road, the maximum driving speed, and the maximum weight of tow truck were 150m, 120km/h, and eight ton, respectively. During the crash test, the program is automatically interlinked thereby collecting THIV, PHD, and ASI values of speed, acceleration, and angular velocity. In addition to the construction, certification of full-scale crash test field that is internationally certified or equivalent level has been planned.

3.3 Field setting of crash tests

No crash analysis standards on movable barriers in Korea have been reported. Thus, this study applied and analyzed based on the longitudinal barrier of "Guideline for Safety Barriers".

A goal level of large-sized movable barrier through the present study was set to SB2 grade. The crash test was a prototype stage yet so a full-scale crash test was conducted at a level of 90% prior to certification test. The impact speed and angle of the movable barrier were set to 70 km/h and 20 degree and an impact vehicle was set to a passenger car(900kg).

Table 3 and Figures 4 to 9 present THIV, PHD, and ASI values as results of the full-scale crash test.

The full scale crash test result showed that both types of folder type and telescope type movable barrier satisfied the standard of the passenger safety performance evaluation at a level of 90%. Compared to folder type and telescope type, folder type had lower THIV, PHD, ASI than telescope type.

Table 3. Result of occupant protection

Category	Folder type	Telescope type
THIV(km/h)	15.58	22.36
PHD(g)	2.20	13.72
ASI	0.52	1.11

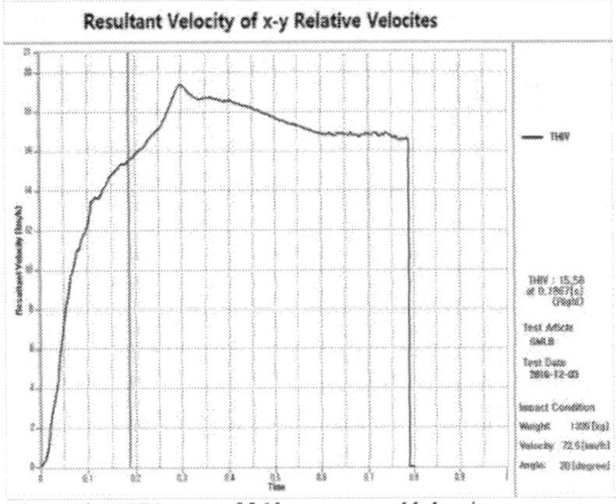

Figure 4. THIV curve of folder type movable barrier

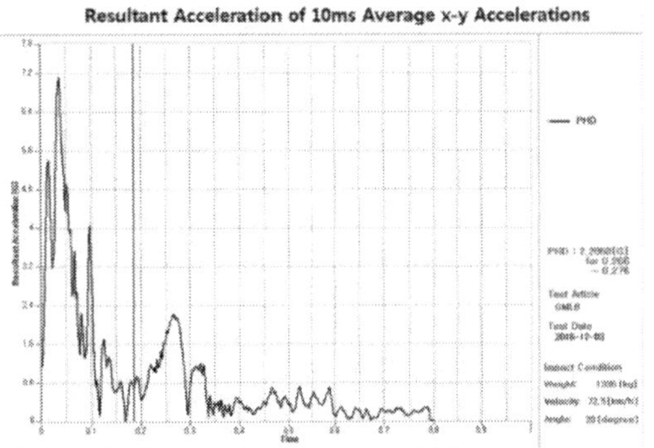

Figure 5. PHD curve of folder type movable barrier

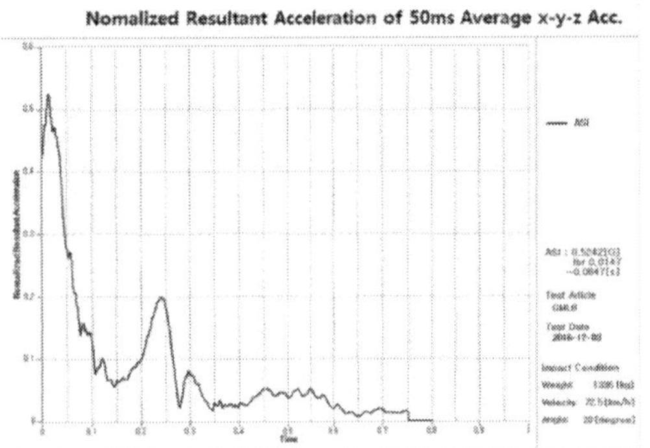

Figure 6. ASI curve of folder type movable barrier

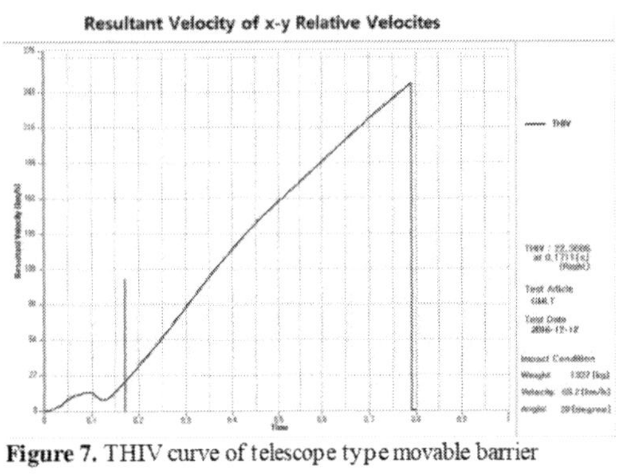

Figure 7. THIV curve of telescope type movable barrier

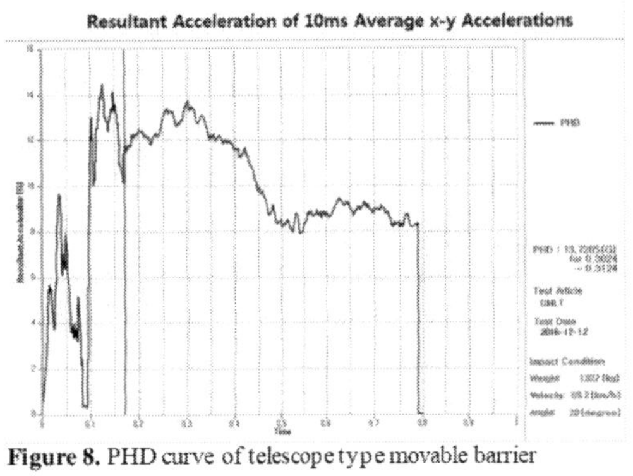

Figure 8. PHD curve of telescope type movable barrier

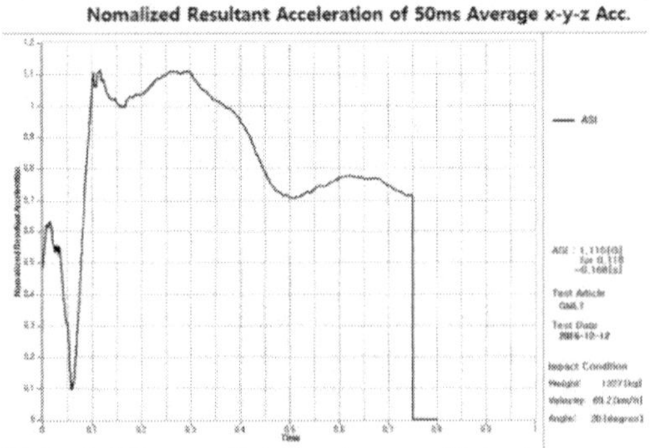

Figure 9. ASI curve of telescope type movable barrier

4 Conclusion

This study developed a prototype of large-sized movable barrier and conducted a performance evaluation of occupant protection.

It is significant in that it is first study to carry out the performance evaluation and analysis on the movable barrier which is under development for the safety of road workers in Korea.

The performance of the movable barrier was tested by 900kg passenger car, 70km/h impact velocity, 20 degree impact angle. The test procedure was carried out according to the test procedure of "Guideline for Safety Barriers". The conclusions of the present study are as follows:

^ The full scale crash test result showed that both types of folder type movable barrier and telescope type movable barrier satisfied the standard of the passenger safety performance evaluation at a level of 90%.

^ The THIV, PHD, ASI value of folder type movable barrier showed that 15.58, 2.20, 0.52.

^ The THIV, PHD, ASI value of telescope type movable barrier showed that 22.36, 13.72, 1.11.

No crash analysis standards on movable barriers in Korea have been reported. Thus, the present study has a limitation that the test results were analyzed based on the longitudinal barrier. And in order to enter the market, it is necessary to consider not only the safety of collision but also the cost of maintenance and manufacturing costs.

In addition, analysis on exit box that analyzes behaviors after crash between a vehicle and large-sized movable barrier should also be conducted.

For the future study, safety performance should be verified through certified authority test after product development to increase a safety level for road workers.

Acknowledgment

This research was supported by a grant from "Development of Accident Risk Mitigating Technology for Vertical Structure Collision on Road and Road Workers(grant number 17TLRP-C096228-03)" funded by the Ministry of Land, Infrastructure and Transport, Republic of Korea.

References

1. AASHTO, *Highway Safety Manual* (2010)
2. J.G. Kim, C.H. Yang, and D.G. Yoon, "Prioritization of Potential Technology for Establishing a Safe Work Zone Environment", *International Journal of Highway Engineering(IJHE)*, **17(6)**, 117-126 (2015)
3. Korea Institute of Civil Engineering and Building Technology, *Development of Accident Risk Mitigating Technology for Vertical Structure Collision on Road and Road Workers*, Internal Document (2015)
4. J.Y. Go, "The Study on Permission and Management System Improvement of Traffic Control System on Road works", *Master's Thesis, University of Seoul* (2005)
5. M.S. Jin., S.K. Lee, J.P. Moon, Y.S. Kim, and I.S. Park, "Vehicle Based Movable Barriers for Work Zone Safety", *Journal of Emerging Trends in Computing and Information sciences(CIS)*, **7(3)**, 168-173 (2016)
6. Ministry of Land & Transportation(MOLIT), *Guideline for Safety Barriers* (2014)
7. AASHTO, *Manual for Assessing Safety Hardware* (2009)
8. TRB, *NCHRP Report 350:Recommended Procedures for the Safety Performance Evaluation of Highway Features* (1993)
9. http://www.dot.ca.gov
10. http:/www.mobilebarriers.com

A Novel Approach for Identification and Ranking of Road Traffic Accident Hotspots

El-Said M. M. Zahran , Soon Jiann Tan, Yok Hoe Yap, Ena K. A. Rahman and Nurulhikmah H. Husaini

Centre for Transport Research, Universiti Teknologi Brunei, Brunei Darussalam

Abstract. Road Traffic Accidents (RTA) are known to be one of the main causes of fatalities worldwide. One useful approach to improve road safety is through the identification of RTA hotspots along a road, so they can be prioritised and treated. This paper introduces an approach based on Geographical Information System (GIS) to identify and prioritise RTA hotspots along a road network using historical RTA data. One particular urban road in Brunei with a historically high rate of RTAs, Jalan Gadong, was selected as a case study. Five years of historical RTA data were acquired from the relevant authorities and input into a GIS database. GIS analysis was then used to identify the spatial extension of the RTA hotspots. The RTA hotspots were ranked according to three different schemes: frequency, severity and socio-economic impact of RTAs. A composite ranking scheme was also developed to combine these schemes; this enabled the prioritisation and development of intervention and maintenance programmes of the identified RTA hotspots. A visualisation method of the RTA spatial distribution within each identified RTA hotspot was also developed to determine the most risky road stretches within each hotspot, which is important for treatment prioritisation when limited resources are available.

1 Introduction

In 2013, the number of human injuries and fatalities resulting from Road Traffic Accidents (RTAs) alone was estimated to be 1.25 million worldwide [1, 2]. These RTAs have adverse impacts, directly and indirectly, on both national economics as well as health consequences and are seen to be the primary cause of fatalities among those aged between 15 and 29 in 2012 [2]. Developing countries that are experiencing rapid urban sprawl and rise in motorised traffic are those who suffers the escalating number of RTAs the most [3]. Comparatively slow pace of development in road infrastructure, levels of enforcement and policies became one of the main factors that are contributing to the rapid increase of the RTAs [4].

Brunei Darussalam is a small country that is located along the Northern coastline of Borneo Island on the Southern region of the South China Sea with a total land area of 5,770 square kilometres. It has a tropical climate and high humidity temperature and heavy rainfall uniformly throughout the year. In 2014, it was recorded that Brunei has a total road network of 3,191.2 km and a total population of 411,900 with 1.4% annual growth rate at which 69.6% of the entire population are concentrated within the Brunei-Muara District [5]. Community within Brunei depend heavily on cars rather than other transportation at which 92% of the traffic fleet are private cars. With this percentage, Brunei has the highest vehicles to population ratio in the world [6]. Predominance of private cars in Brunei is mainly due to the inadequate availability of public transport, dispersal of urban developments and even low price of car and fuel [7]. With the high number of private cars, RTA rate is higher than various developed and developing countries which results in significant social and economic impacts on the country [8].

One tool that can be used effectively to achieve a sustainable transportation planning and sustainable traffic management in urban areas is Geographic Information System (GIS) [4]. GIS also provide powerful tools in developing reliable geospatial databases that can be used when analysing RTA data [9].

In India, GIS was used to identify RTA hotspots along a road section that is known for accidents [10]. With the help of GIS, they visually identified locations along the road section that had a high rate of RTA. A crash factor, based on RTA rate, was computed for each of these locations. The crash factors were then compared to an empirical critical crash rate that was computed for the entire study road. If the crash rate is seen to exceed the critical crash rate, location is classified as an accident hotspot. However, their approach did not include the use of GIS analysis nor did they determine the spatial extent or even to rank the identified accident hotspots.

For a national highway in Algeria, Ref. [11] conducted hotspot analysis in GIS using Kernal Density Estimation (KDE) method to identify and prioritise those RTA hotspots. KDE method was used to transform RTA data, in discrete point form, into a continuous raster coverage that displayed the density distribution of RTA along the selected highway. The acquired coverage applied an arbitrary search with a radius of 0.5 km, and the output cell size was not specified. No proper justification was made as to why 0.5km search radius was selected even though it has a significant impact on the prediction accuracy of the kernel density estimation method in GIS [12, 13]. Aside from that, this method can only be used to rank accident hotspots based on the frequency of RTA.

The following sections of this paper demonstrate an approach using GIS to identify and delineate the spatial extension of RTA hotspots along a case-study road in Brunei Darussalam. The GIS approach also prioritises the identified hotspots using four different ranking schemes that are based on frequency, risk factor, economic impact of RTA within each hotspot zone and the combination of these three criteria.

2 Digitisation and geocoding of past RTA data

As a result of preliminary analysis of RTA cases, major accident-prone roads in Brunei Darussalam were identified. Jalan Gadong was selected as a case study for this research paper due its high number of reported accidents per km per year.

Jalan Gadong is a 10.2 km long dual carriageway connecting Jalan Kumbang Pasang in the east and Jalan Tutong to the west of the capital city of Brunei, Bandar Seri Begawan. As a major urban arterial carrying traffic from the west of the capital while providing access to various commercial buildings, residential houses and schools along its length, it has a relatively very high usage rate, estimated to be over 47,000 vehicles per day [14]. There are 9 signalised junctions, 1 signalised roundabout, 15 mid-block U-turns and many more uncontrolled junctions along the whole stretch of road. The posted speed limit is 65 km/hr throughout, except for an approximately 1 km long section with a 50 km/hr speed limit due to the presence of several uncontrolled junctions.

During the initial stage of data digitisation, RTA data for Jalan Gadong was exported manually from standard police report form into a spreadsheet program. The data was further filtered, cleaned and refined to ensure only reports with sufficient location details were used for GIS analysis. Suitable data field and coding systems were also developed to indicate the crash movement type and contributory factors of each case. As for the coordinates of the RTA location, they were estimated using both digital maps and handheld GPS units.

In all, RTA data of Jalan Gadong within the period of 2011 to 2015 were processed, comprising of 747 accident cases with 4 fatalities, 18 serious injuries, 52 slight injuries and 1262 no injuries for both traffic directions of the dual carriageway, as shown in Figure 1. The most common type of crashes was rear-end crashes, and involved 2 or more vehicles.

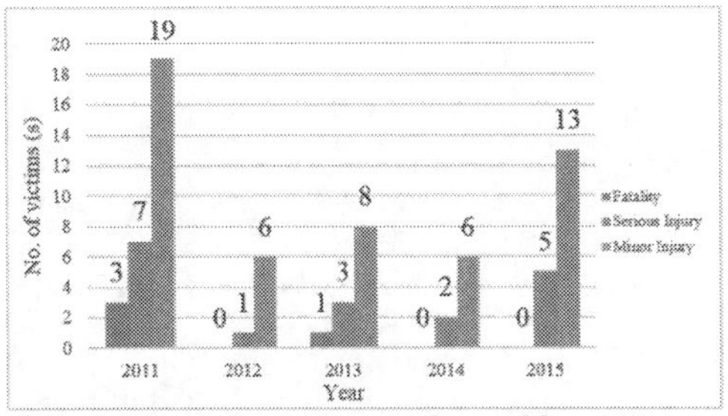

Figure 1. Jalan Gadong's RTA casualty trend from 2011 to 2015.

3 Identification and ranking of RTA hotspots

A GIS analysis method was developed to identify and rank RTA hotspots along Jalan Gadong. Initially, RTA data for Jalan Gadong was exported into ESRI ArcGIS 10.2 as a point shape file. A "Select by location" function in ArcGIS filtered out RTA points that were located outside of Jalan Gadong. For each RTA point, a proximity analysis was conducted by creating a buffer zone of a fixed radius centred at the RTA coordinates.

Previous studies have used arbitrary fixed distances to segment the road network for spatial analysis, irrespective of traffic speeds. However, the radius of the buffer zones in this study was taken to be equal to the Stopping Sight Distance (SSD) on Jalan Gadong. SSD is typically defined as the distance traversed by a vehicle from the time its driver perceives a hazard in the vehicle's path to the time the vehicle brakes to a stop [15]. It should be noted that the SSD is a proxy measure as vehicles do necessarily stop after that distance post-collision. However, noting that a large proportion of RTAs in Brunei are single-vehicle accidents [8], SSD has the advantage of reflecting the effect of speeds on the final resting locations of the vehicles, as the latter are more likely to be accurately recorded compared to say, the location of loss of control. To compute the SSD, the speed was chosen as the posted speed limit for the majority of Jalan Gadong (65km/hr), as no data on operating speeds was available. The perception-reaction time was taken as 1.5 seconds, while the longitudinal friction coefficient between vehicle tyres and road surface was taken as 0.33 [15], yielding an SSD value of 80 m.

Closeness and buffering of RTA points resulted in many overlapping circles which was not helpful in identifying RTA hotspots. Hence, every group of overlapping circles were merged into a hotspot zone, as RTA locations within SSD distance of each other could reflect a particular hazardous zone. This process resulted in 12 RTA hotspot zones as shown in Figure 2.

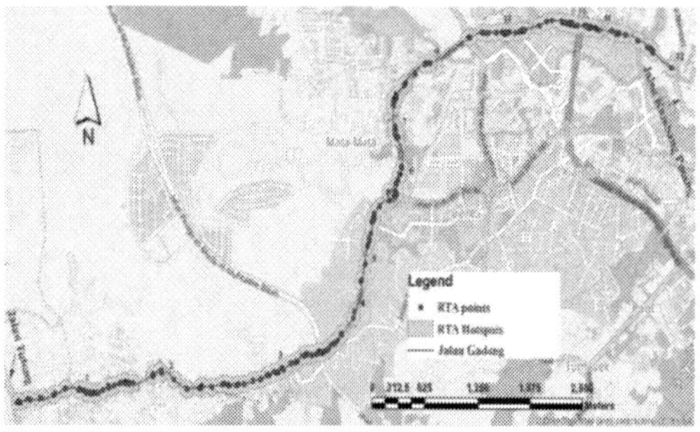

Figure 2. RTA hotspots along Jalan Gadong.

4 Prioritisation of RTA hotspots

With the help of a "Join" function in ArcGIS, RTA points located within each RTA hotspot zones, as shown in Figure 2, were automatically combined and counted. Moreover, the same join function helped to sum up the number of fatalities, serious injuries, minor injuries and no injuries related to RTA points located within each zone. For this research, four ranking schemes were used to prioritise RTA hotspots. The first ranking scheme was based on frequency of RTA points inside each hotspot zone. Equation (1) was developed for this scheme which involves the density of RTA within each zone and the number of years of data. It also involves SSD to account for the operating speed, longitudinal friction coefficient and gradient of the road, as well as the driver's perception-reaction time. These four parameters contribute significantly to the risk of having a traffic accident on the road.

$$F = RTA \times (SSD/L) \times (1/N) \quad (1)$$

where:
F = average number of yearly RTA within each hotspot zone per SSD,
RTA = RTA count within the hotspot zone,
SSD = Stopping Sight Distance,
L = length of hotspot zone along the road centreline, and
N = number of years of data.

The second ranking scheme was developed based on the severity within each RTA zone, which was adopted from an approach developed by Ref. [16]. Equation (2) was used to compute the risk factor within each zone.

$$S = X + 3Y + 5Z \quad (2)$$

where:
S = severity
X = total number of minor injury within each RTA hotspot zone,
Y = total number of major injury within each RTA hotspot zone, and
Z = total number of fatalities within each RTA hotspot zone.

The normalised severity (NS) is calculated from S using Equation (3).

$$NS = S \times (SSD/L) \times (1/N) \quad (3)$$

where:
NS = normalised severity (severity per SSD length per year),
S = severity,
SSD = Stopping Sight Distance,
L = length of hotspot zone along the road centreline, and
N = number of years of data.

The third ranking scheme was based on the socio-economic impact of RTA within each hotspot zone. A formula, developed by Ref. [17], was adopted for the estimation of the socio-economic impact in BND as follows:

$$SEI = (\text{Fatality} \times \$1{,}419{,}639) + (\text{Serious injury} \times \$70{,}205) + (\text{Slight injury} \times \$9{,}119) + (\text{No injury} \times \$3{,}300) \quad (4)$$

where:
SEI = socio-economic impact.

The normalised socio-economic impact ($NSEI$) is calculated from SEI using Equation (5).

$$NSEI = SEI \times (SSD/L) \times (1/N) \quad (5)$$

where:
$NSEI$ = normalised socio-economic impact (Socio-economic impact per SSD length per year),
SEI = socio-economic impact,
SSD = Stopping Sight Distance,
L = length of hotspot zone along the road centreline, and
N = number of years of data.

The RTA hotspot zones were split into four hierarchical classes with equal number of zones. A colour scheme was then applied to all hotspot zones. This was to display visually the hierarchy of frequency, normalised severity and normalised socio-economic impact of RTA within each zone, as shown in Figure 3a, b and c, respectively.

It is noticeable that the outputs from the three ranking schemes demonstrated some resemblance with each other. The results of NSEI and NS based schemes were more similar to each other than to those of the F based scheme. The reason for the difference in the results between the NS and NSEI based schemes is due to the consideration of non-injury cases in the latter while lacking this in the former scheme. In addition, the relative weight of fatality is higher in the latter than that in the former scheme. Therefore, the fourth ranking scheme was developed by combining once the frequency scheme and the normalised severity scheme, and another time the frequency scheme and the normalised socio-economic impact scheme, into two composite ranking schemes.

The acquired values of F, NS and $NSEI$ for Jalan Gadong were classified into four equal-percentile levels, resulting in four hierarchical levels for F, NS and $NSEI$. The two composite ranking schemes exhibit the risk factor of each RTA hotspot zone using two risk assessment matrices that employ the four hierarchical levels of F, NS and $NSEI$ as shown in Figure 4a and b. The risk assessment matrix has been modified from [18] based on road safety audit practices in New Zealand, which emphasise on prioritising accidents with higher severity of injuries. Then a colour scheme was assigned

 DOI:10.1051/matecconf/201712404003

to hotspot zones, in order to display the hierarchy of risk factors of RTA hotspots based on *NS* and *NSEI*, as shown in Figures 5 and 6 respectively.

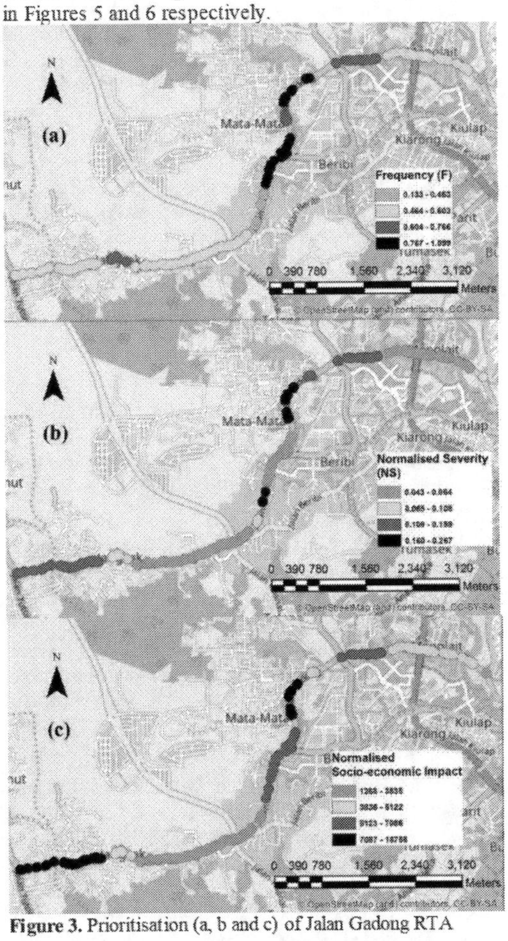

Figure 3. Prioritisation (a, b and c) of Jalan Gadong RTA hotspots.

(a)

Normalised severity (NS)	F > 0.77	0.6 < F ≤ 0.77	0.46 < F ≤ 0.6	0 ≤ F ≤ 0.46
NS > 0.16	Serious	Serious	Significant	Moderate
0.11 < NS ≤ 0.16	Serious	Significant	Moderate	Moderate
0.06 < NS ≤ 0.11	Significant	Moderate	Minor	Minor
0 ≤ NS ≤ 0.06	Moderate	Minor	Minor	Minor

(b)

Normalised Socioeconomic Cost (NSEI)	F > 0.77	0.6 < F ≤ 0.77	0.46 < F ≤ 0.6	0 ≤ F ≤ 0.46
NC > 7086	Serious	Serious	Significant	Moderate
5122 < NC ≤ 7086	Serious	Significant	Moderate	Moderate
3835 < NC ≤ 5122	Significant	Moderate	Minor	Minor
0 ≤ NC ≤ 3835	Moderate	Minor	Minor	Minor

Figure 4. (a and b) Risk Assessment Matrices.

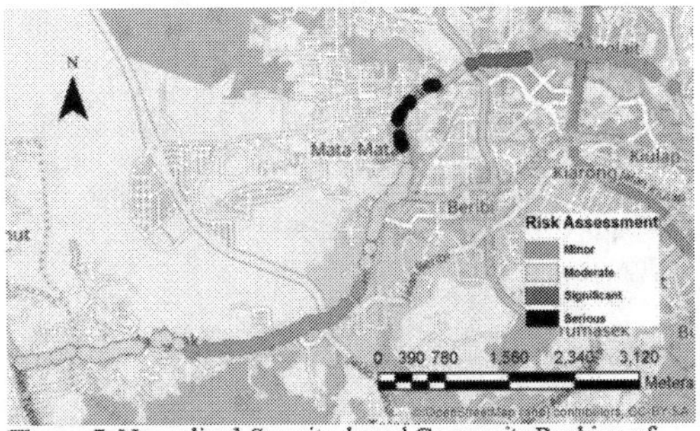

Figure 5. Normalised Severity-based Composite Ranking of Jalan Gadong RTA hotspots.

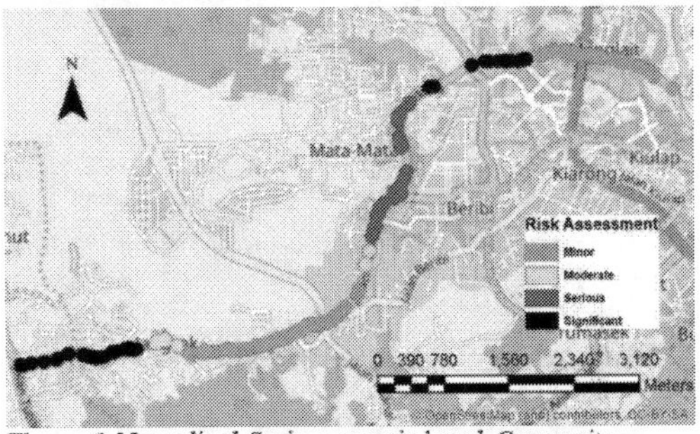

Figure 6. Normalised Socio-economic-based Composite Ranking of Jalan Gadong RTA hotspots.

5 Micro analysis of RTA hotspots

As can be seen in Figure 2, the initial stage of hotspot identification above resulted in a few continuous hotspot zones of up to 2 kilometres in length. A micro analysis method to achieve an improved visualisation of the spatial distribution of RTA within each identified hotspot zone was thus also developed. This method of data visualisation was used to determine road segments within the hotspot zone which had a relatively high density of RTA points. It could then facilitate subsequent road safety assessment and prioritisation of countermeasures, in light of limited resources for road safety improvement. The micro analysis method using GIS is described in the following.

The centreline of Jalan Gadong road was clipped by the RTA hotspot zones such that only the centreline portions within the RTA zones remained while those located outside the zones were eliminated. After that, the remaining centreline portions were split into equidistant segments, each of a length equal to the SSD. A "Near" tool in ArcGIS was used to project the RTA points onto each of the segments of Jalan Gadong. The "Join" function was then used to project RTA points onto the segments in all hotspot zones. To display the number of RTA points related to each segment, a symbology was assigned in the form of a column chart. A Python pre-logic script code was applied to split the number of RTA points into 5 hierarchal classes. A colour scheme was used to show the hierarchy of these classes. In addition to the colour scheme, the column's length in the column chart was used to symbolise the count of RTA points related to each road-centreline segment, as shown in Figure 5.

Through this method, particular sections of Jalan Gadong with higher incidence of RTAs were identified. They were then brought forward for detailed road safety auditing and proposal of mitigation measures, with the aim of improving the safety of these sections of Jalan Gadong.

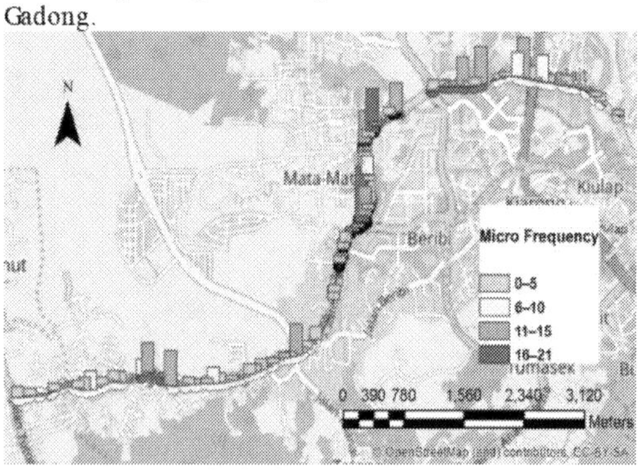

Figure 7. Micro Analysis of Jalan Jerudong RTA hotspots.

5 Conclusions and recommendations

Since Jalan Gadong is known to have a high number of accidents per kilometre per year, the selection of this road as a case study for this research was helpful, especially when comparing the results of the three ranking schemes that are based on frequency, normalised severity and normalised socio-economic impact of RTA. The developed hotspot identification approach used SSD value as a radius to create buffer zone around each RTA points. This approach has a better justification than using arbitrary values of search radius and output cell size used in KDE method. However, merging overlapping buffer zones along Jalan Gadong created long RTA hotspots with a very high number of RTA points. This may impede the ability to prioritise the intervention and maintenance measures within the long stretch of the same RTA hotspot. The micro analysis method, explained in Section V and illustrated in Figure 7, was subsequently developed to ensure that the road segments with relatively high number of RTA could easily be identified for the purpose of road safety assessment and subsequent prioritisation.

Frequency, normalised severity and normalised socioeconomic impact of RTA within each zone were used to develop the three different ranking schemes in Figure 3a, b and c. The division by the length of the hotspot zone in (1), (3) and (5) was found helpful to prioritise RTA hotspots according to the density of RTA points, severity and socio-economic impact within each hotspot. For the same number of RTA points, severity and socio-economic impact per hotspot, the longer the hotspot zone, the less the density of RTA points within the zone, and the less the value of F, NS and NSEI in (1), (3) and (5), indicating less priority. The division by both the length of the hotspot zone and the number of years in (1), (3) and (5) may allow the prioritisation of RTA hotspots across many roads with different lengths of hotspot zones and different number of years of data. More research is recommended to calibrate (2) and (3) to the local conditions of Brunei Darussalam.

Due to the similarity between the results of the three ranking schemes based on frequency, normalised severity and normalised socioeconomic impact of RTA, two composite ranking schemes were developed to combine both the frequency and normalised severity, and the frequency and normalised socio-economic impact, of RTA within each identified hotspot into a composite risk factor. The composite ranking schemes can help prioritise the development of intervention and maintenance programmes for road sections identified as RTA hotspots. Further research is needed to validate the hierarchy of risk factor identified at the two composite ranking schemes, and decide which one of them can provide a better prioritization of RTA hotspots.

It is anticipated that the methods and results developed in this study will facilitate a more efficient approach to improving the safety of the road network, by optimising the use of finite resources on priority hotspots to achieve more overall road safety gains.

Acknowledgment

The authors are grateful to the Brunei Research Council for their funding support and to the Brunei National Road Safety Council and the Royal Brunei Police Force for providing the necessary road safety data and information.

References

1. ICMR, Development of feasibility module for road traffic injuries surveillance. Indian Council of Medical Research Bulletin, **39**, 41-50 (2009).
2. World Health Organization, Global status report on road safety 2015. (2015).
3. R. Ewing, S. Hamidi, and J.B. Grace, Urban sprawl as a risk factor in motor vehicle crashes. Urban Studies, (2014).
4. G. Yalcin and H.S. Duzgun, Spatial analysis of two-wheeled vehicles traffic crashes: Osmaniye in Turkey. KSCE Journal of Civil Engineering, **19**, 2225-2232 (2015).
5. Department of Statistics, Brunei Darussalam Statistical Yearbook. (2014).
6. Q. Bandial, Number of new vehicles increased by 19% between 2011 and 2013, in The Brunei Times, (2015)
7. Centre for Strategic and Policy Studies, Review to Formulate a Roadmap and Draft National Masterplan for a Sustainable Land Transportation System for Brunei Darussalam. **5**, (2014).
8. S.J. Tan, S. Ladi, E.M.M. Zahran, Y.H. Yap, and E.K.A. Rahman, A Review of Road Traffic Accident Statistics and Trends in Brunei Darussalam: presented at BICET 2016, Brunei Darussalam, (2016)
9. S. Ladi, C.D. Wijeyesekera, A.J. Brimicombe, and Y. Li, *Road traffic accident hotspot identification using modified Voronoi Process.* Advances in

Computing and Technology (AC&T), 189-198 (2009).

10. G. Apparao, P. Mallikarjunareddy, and R.G. SSSV, Identification Of Accident Black Spots For National Highway Using GIS. International Journal of Scientific & Technology Research, **2**, 154-157 (2013).
11. R. Oulha, K. Brahimi, A. Boumediene, F. Dali, and M.A. Madouche, GIS Contribution to Identify Accident Black Spots on National Highway: Case Study of Wilaya of Mascara (Algeria). International Journal of Chemical, Environmental & Biological Sciences (IJCEBS), **1**, (2013).
12. S. Chainey, Examining the influence of cell size and bandwidth size on kernel density estimation crime hotspot maps for predicting spatial patterns of crime. Bulletin of the Geographical Society of Liege, **60**, 7-19 (2013).
13. L. Thakali, T.J. Kwon, and L. Fu, Identification of crash hotspots using kernel density estimation and kriging methods: a comparison. Journal of Modern Transportation, **23**, 93-106 (2015).
14. Centre for Strategic and Policy Studies, Review to Formulate a Roadmap and Draft National Masterplan for a Sustainable Land Transportation System for Brunei Darussalam. **3**, 31 & 35 (2012).
15. C.A. O'Flaherty, *Highways: Volume 1 Traffic Planning and Engineering*. 3 ed, 369-370, (1996)
16. K. Geurts and G. Wets, Black Spot Analysis Methods: Literature Review. (2003).
17. Asian Development Bank, The Cost of Road Traffic Accidents in Brunei Darussalam. (2005).
18. NZ Transport Agency, Road safety audit procedures for projects. Interim release, (2013).

Impacts of Large Vehicles on Traffic Safety in Freeway Interchange Merging Areas and Improvement Measures

Yifan Guo[1], Qiaojun Xiang[1], Shen Li[1], Ting Zhang[1] and Rao Yao[1]

[1]*School of Transportation, Southeast University, Si Pai Lou #2, Nanjing 210096, China*

Abstract. In order to work out the characteristics of traffic operation on freeway interchange merging areas under different ratio of large vehicles, aerial photography technique based on unmanned aircraft and other observation device were used to investigate the merge section traffic data firstly. Based on survey data and actual traffic characteristics, the paper then established simulation models with VISSIM with calibrated car-following and lane changing behaviour models. Next, the paper analysed the influence of traffic volume and acceleration lane length under various traffic compositions to evaluate safety status of interchange merge sections by traffic conflict technique and speed consistency index. Finally, two kinds of safety promotion strategies, speed limit and setting forbidden line, were evaluated and the results indicated these measures can raise safety level by about 10%~15% under certain conditions, which were determined as the best applicable conditions.

1 Introduction

Freeway interchange merging area is the bottleneck and accident-prone spot. In general, at merging section, due to aggressive lane changing behaviours, vehicles on mainline may slow down or pull out to overtake, which increases the risk of traffic accident. Therefore, a well-functioning merging area has a significant effect on improving traffic safety [1].

Currently, large vehicles have increasingly become one of the research hotspots with construction of freeway and development of modern logistics industry. Owing to poor performance of large vehicles, there is serious interference in traffic flow, especially at interchange merging areas, which leads to a decline in stability of traffic flow. Agent et al. evaluated geometric alignment design of expressway with Heavy goods vehicle (HGV) as the objective [2]. Gazis and Herman first proposed the concept of "Moving Bottleneck" and attempted to explain it according to traffic flow wave theory when they found that a HGV may cause a slow motorcade in 1992 [3]. Later, numerous researches on formation mechanism of moving bottleneck have been conducted, and some traffic phenomena and experimental data were analysed from the perspective of moving bottleneck theory [4-8]. However, most of these findings focus on basic sections of highway, not on interchange areas.

Regarding merging area, many previous studies have focused on influence factors of traffic safety and capacity [9-11]. S. Li et al. established crash risk prediction model and defined HCRI index based on traffic conflict technique for freeway interchange merging area [12]. L. Lei analysed the gear-alternating control regulation based on FI cellular automaton model and concluded that gear-alternating control can improve traffic situation significantly under large traffic volume [13]. Sun et al. simulated the operation characteristics of merging area under open boundary condition by calibrated car-following model and lane changing model [14].

From the literature review, there seems to be three main issues in China: 1) few studies have linked large vehicles to interchange merging areas; 2) the models used to analyse merging segment in China are based on research results from other countries, but the differences of traffic composition and other conditions between China and other countries, isn't taken into consideration; 3) the research on merging areas is mainly focused on the capacity rather than safety.

In this paper, with regard to a directional on-ramp, simulations are conducted with the calibrated car following model and lane changing model in VISSIM. Then traffic conflict technique and velocity consistency index are selected to evaluate the impact of HGV on traffic safety at confluence area. Finally, some measures are proposed to improve safety status of merging section and their applicable conditions are given.

2 Methodology

2.1 Data collection

The traffic survey was conducted at a semi-cloverleaf interchange with a typical merging area, which is composed of a three-lane mainline and a one-lane ramp if emergency lane is neglected. The width of every lane is

3.75m. The horizontal curve radius of mainline is 1500m and ramp radius is 650m. Unmanned aerial vehicle (UAV) photography was applied to shoot video in good conditions. Finally three hours high-precision video was obtained. Moreover, radar velocimeters were used to measure speed of gore nose, midpoint and end point of merge area, as shown in the Figure 1.

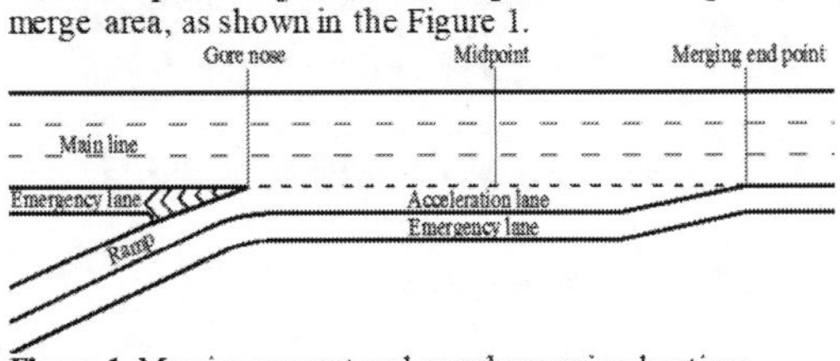

Figure 1. Merging segment and speed measuring locations.

2.2 Safety evaluation

Common traffic safety evaluation methods include three categories, namely accident-based method, conflict-based method and traffic flow characteristic based method. In this paper, the latter two methods are considered. The traffic conflict and traffic characteristic indicator are defined as follows.

2.2.1 Traffic conflict

Traffic conflict technique, as an approach that overcomes the lack of reliable accident records, is widely used for safety evaluation in China [15-17]. However, most previous studies on traffic conflict pay attention to the intersection rather than merging section. In fact, there exists great difference between them because of distinct traffic and road conditions. Therefore, the key of this method is how to identify the effective conflict at merging segments.

TTC (time to collision) is the critical parameter to determine a traffic conflict. In this paper, two typical types of conflicts are defined according to the angle between the vehicle travelling directions: 1) rear-end conflict, the angle is in [0°, 15°); 2) sideswipe conflict, the angle belongs to [15°, 85°]. Relevant studies pointed that the TTC thresholds of these two conflicts are 4.7s and 4.2s, respectively [16]. To make study convenient, this paper selects 4.2s as the TTC threshold for all conflicts.

2.2.2 Speed consistency

Based on traffic characteristics, some studies found that speed consistency can well reflect the traffic flow stability and safety situation of highway [18, 19]. Speed consistency indicators commonly used include the following two categories.

• SDI, speed difference index, can reflect directly safety of adjacent sections of highway, mainly including ΔV_{85} and $85(\Delta V)$.

• SSI, speed statistic index, reflect traffic safety by analysing distribution of velocity statistics such as σ and σ/μ, which is more applicable to a single segment.

As to confluence area, which is different from the basic section in traffic characteristics, SDI index may be more suitable. The equation of ΔV_{85} and $85(\Delta V)$ are listed as follows:

$$\Delta V_{85}=\left||V_1|_{0.85}-|V_2|_{0.85}\right| \tag{1}$$

$$85(\Delta V)=|V_1-V_2|_{0.85} \tag{2}$$

where V_1 and V_2 are respectively the speed of the same vehicle on different segments. In fact, ΔV_{85} overestimates the safety while $85(\Delta V)$ is more precise because the relationship and interaction of adjacent sections is taken into account [20]. Meanwhile, a novel indicator *SRC* based on $85(\Delta V)$ is proposed to reflect the reduction coefficient of speed.

$$SRC=\left|\frac{V_2-V_1}{V_1}\right|_{0.85}\times 100\% \tag{3}$$

3 VISSIM-based merging area model

3.1. Model assumption

In this paper, the purpose is to investigate the safety situation under different ratio of HGV. Therefore, a freeway merge section model was established. Specific to the characteristics of merge sections and the functions of the software VISSIM, some assumptions are listed as follows.

• Assume that both the mainline and on-ramp are straight and the slope of merge section is zero.

• Assume that the angle of mainline and ramp is 30° and parallel type acceleration lane is adopted.

• Neglect the effect of weather conditions and lateral clear width on the traffic flow characteristics.

• Assume that the design speed of main line and ramp respectively are 100km/h and 60km/h.

• Assume that the traffic composition on the entrance ramp is the same as the main line.

3.2 Model calibration

3.2.1 Vehicle parameter calibration

We divide the vehicles into two types, cars and large vehicles (HGV), and their various parameters are shown in Table 1.

Table 1. Parameters of car and HGV in model.

Type	Length (m)	Width (m)	Maximum Acceleration	Maximum Deceleration
Car	4.11~4.34	1.5	3.5 m/s^2	6.0 m/s^2
HGV	7.0~14.0	2.5	2.5 m/s^2	5.5 m/s^2

3.2.2 Car following model calibration

In this paper, Wiedemann99 car-following model, which is widely used in highway, is adopted to simulate the car following behavior at on-ramp bottleneck. Table 2 gives the calibrated parameters of car following model.

Table 2. Parameters of car following model.

Parameter	Value	Parameter	Value
CC0	1.5m	CC5	0.35
CC1	1.50s	CC6	12
CC2	5.5m	CC7	0.25
CC3	-8.00	CC8	3.5m/s^2
CC4	-0.35	CC9	1.5m/s^2

3.2.2 Lane changing model calibration

In general, at a merging section, the lane changing behaviours have the most significant influence on the traffic flow. Due to aggressive lane changing behaviours, vehicles on the mainline may slow down or enter into another lane. Lane changing model in VISSIM includes mandatory and discretionary lane changing behaviours and thus it is adopted in this study. The parameters of model is calibrated as Table 3.

Table 3. Parameters of car following model.

Parameter		Value	Parameter		Value
Max. deceleration	own	-4.43 m/s^2	Accepted deceleration	own	-1.00 m/s^2
	Trailing vehicle	-3.00 m/s^2		Trailing vehicle	-0.50 m/s^2
-1m/s^2 per distance	own	200m	Waiting time before diffusion		60.0s
	Trailing vehcle	200m	Min. HeadWay (front/rear)		0.5m

3.3 Model verification

Table 4. Result of speed and conflict test.

	Vehicle type and Location		Survey value (km/h)	Mean of simulation (km/h)	Error
Speed test	Car	Gore nose	88.25	89.61	1.54%
		Midpoint	82.64	82.17	0.57%
		End point	77.38	75.51	2.42%
	HGV	Gore nose	70.44	72.16	2.44%
		Midpoint	64.37	62.65	2.67%
		End point	55.42	53.51	3.45%
	Conflict type and Time interval		**Survey value**	**Mean of simulation**	**Error**
Conflict test	Rear-end	0~1h	18	20.3	12.78%
		1~2h	42	38.6	8.10%
		2~3h	31	32.7	5.31%
	Sides-wipe	0~1h	22	23.6	7.27%
		1~2h	53	55.1	3.96%
		2~3h	40	35.3	11.75%

The merging segment model is built in VISSIM and the investigation and statistical results (traffic volume, ratio of HGV, etc.) were input into simulation model. For more stable output results and smaller random error, each simulation lasts for 3600s and simulation in the same condition is carried out 10 times, thus the average value of the total 10 simulations is adopted as final result.

To test the validity of the model, this paper compared the simulation results with the observations by two indicators, average speed of section and number of traffic conflicts. The verification results are shown in Table 4. It can be seen that Error of section speed and traffic conflict respectively are less than 5% and 15%, which are within acceptable range in terms of engineering application. Therefore the traffic simulation model is valid.

4 Simulation and results analysis

4.1 Simulation scheme

This paper aims to explore the influence of large vehicle mixing rate on traffic safety in freeway interchange merging areas. Nevertheless, the security situation is not only affected by large vehicles, but also by the traffic conditions (traffic volume, merging ratio, etc.), road conditions (acceleration lane length, curve radius, number of lanes, etc.) and control conditions (speed control, channelization, traffic markings, etc.).

This paper establishes kinds of sub-models of simulation by inputting different traffic conditions and road conditions, and thus the elements of different simulation scenarios are shown in Table 5.

Table 5. Conditions of traffic simulation.

Variable	Value
Ratio of HGV (%)	20, 30, 40, 50, 60
Mainline traffic volume (veh/h/ln)	500, 750, 1000, 1250, 1500, 1750
Merging ratio	0.25
No. of main lanes	One way and three lanes
No. of ramp lanes	1
Width of lanes (m)	3.75
Acceleration lane length (m)	200, 250, 300, 350, 400, 450

4.2 Relationship of ratio of HGV vs. mainline traffic volume

The relationship between the ratio of HGV and traffic volume at the merging area is studied in this section. The confluence area with the acceleration lane length of 250m and merging ratio of 0.25 was taken as an example. The simulation results are shown in Figure 2 and Figure 3.

From Figure 2, it can be observed that with the increase of mainline traffic volume, the conflict quantity at merge segment also increases obviously when the traffic volume is less than a critical value. When ratio of HGV is not greater than 30%, total amount of collisions increases roughly in proportion to traffic volume. Once the ratio of HGV reaches or exceeds 40%, the collisions will encounter a decline after a rapid growth. The difference is that the peak value of curve occurs at 1500veh/h/ln when proportion of HGV is 40% while it occurs at 1250veh/h/ln under 50%~60% of HGV. The decline is in line with expectations because the interaction and interference will increase when volume is large enough, which may cause larger traffic density, slower velocity and smaller possibility of collision. In addition, traffic conflicts also increases significantly with

DOI:10.1051/matecconf/201712404004

the increase of the rate of large vehicles under the traffic volume of 1000veh/h/ln~1500veh/h/ln, which means the large vehicles, as the moving bottleneck of merging area, exactly have great influence on safe operation of traffic flow, particularly when ratio of HGV arrived at 40%.

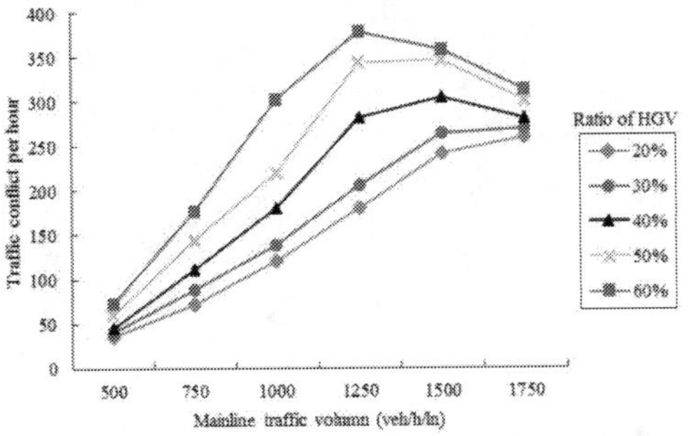

Figure 2. Relationship of traffic conflict vs. traffic volume.

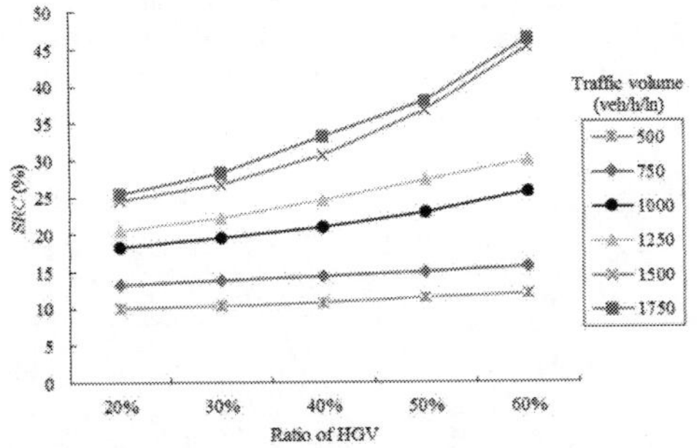

Figure 3. Relationship of *SRC* vs. traffic volume.

According to Figure 3, under the same traffic volume, the *SRC* increases along with the increment of the ratio of HGV as a whole, which implies the decline in security. Nevertheless, the curve showed different growth trend under different traffic volume. When volume is less than 750veh/h/ln, the SRC grows slowly, which suggests that the moving bottleneck caused by HGV has not yet formed. The SRC increases approximately in a line with the raise of ratio of HGV when the volume is between 1000~1250veh/h/ln. Specially, when volume continues to mount up, the SRC will increase exponentially which indicates HGV will have a greater impact on traffic stability under larger volume. Meanwhile, the curves under 1500veh/h/ln and 1750veh/h/ln have a similar shape and small spacing, indicating that the influence of traffic volume is weakened.

4.3 Relationship of ratio of HGV vs. acceleration lane length

To explore the relationship of HGV and acceleration lane length, the simulations under the situation that the mainline volume is 1250veh/h/ln, merging ratio is 0.25 and the acceleration lane length is between 200m and 450m were carried out with the results shown in the Figure 4 and Figure 5.

In Figure 4, it can be seen that the length of acceleration lane has significant influence over the traffic operation in the confluence area. In general, with the increase of the acceleration lane length, the number of traffic conflicts tends to decline when the length is less than a critical value and then it remains nearly stable. When the ratio of HGV is 20%, the conflicts are no longer reduced when acceleration lane length reaches 300m, while this critical value will climb to 350m when HGV account for 30%~40% and 400m when HGV exceed 50%. From another perspective, the acceleration lane exists a border, not the longer the better.

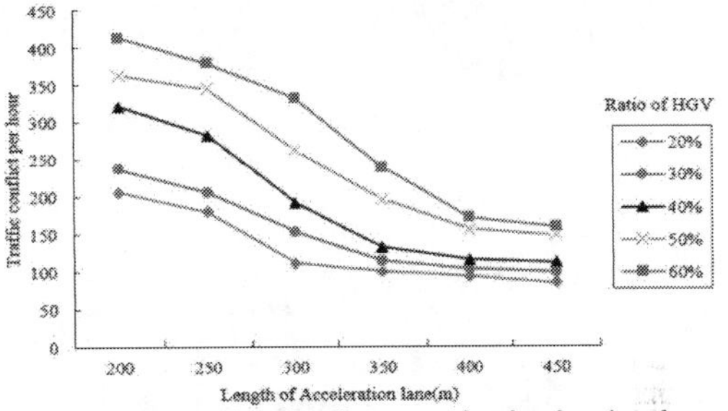

Figure 4. Relationship of conflict vs. acceleration lane length.

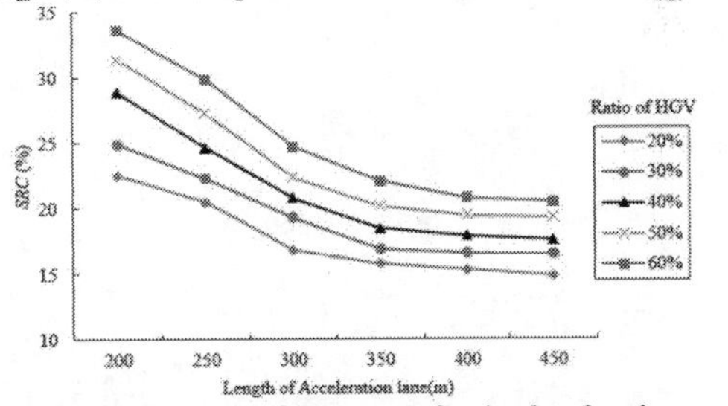

Figure 5. Relationship of *SRC* vs. acceleration lane length.

From Figure 5, it is found that the speed stability is improved with acceleration lane length increasing. Also, under the same length of acceleration lane and traffic volume, the more large vehicles, the lower the speed consistency and stability. This because the fact that large vehicles will cause more interference when changing lane at merging area.

4.4 The impact of speed limit measure

Under the same condition, a higher proportion of large vehicles may cause lower security on interchange on-ramp joints. By the analysis of section 4.3, setting reasonable length of acceleration lane can significantly improve traffic conditions. However, there are great difficulties in reconstructing an acceleration lane for an existing freeway interchange.

In addition to big size and poor performance of HGV, the main reason for the safety problems is that the vehicle enter into the main line too early, which cause large speed difference between ramp vehicles and mainline vehicles due to inadequate acceleration. Thus, a possible measure is limiting the mainline speed to reduce speed difference.

Because the design speed of mainline and ramp are respectively 100km/h and 60km/h, the speed of inner, middle and outer lane of mainline are respectively limited as 90km/h, 80km/h and 70km/h in this paper. Then the simulations under different conditions were conducted when merging ratio and acceleration lane length were fixed to 0.25 and 300m. The reduction rate of traffic conflicts is selected as the indicator to evaluate effcts and the result is shown in Figure 6.

From Figure 6, at the same traffic volume, a better improvement effect of sub-lane speed limit measure is observed at the condition of enhanced HGV proportion. And this measure can be implemented for optimum results when traffic volume is 1000veh/h/ln~1500veh/h/ln and ratio of HGV isn't less than 40%.

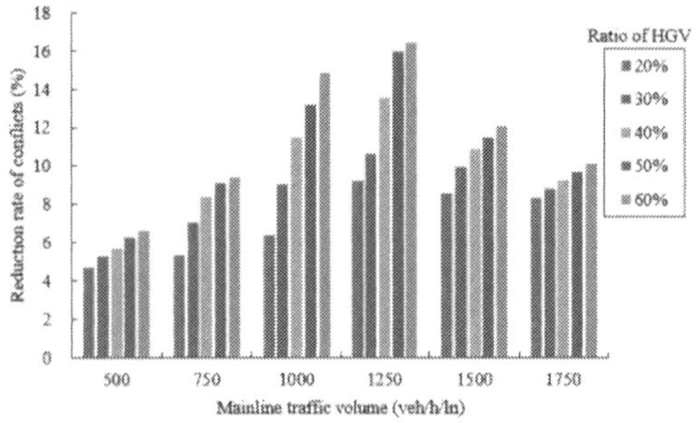

Figure 6. The effect of speed limit at merging area.

4.5 The impact of forbidden line

Regarding the parallel type acceleration lane, too large merging angle also may increase driving risk. Therefore MUTCD points out that the solid white line, namely forbidden line, setting at the end of optional diagonal or chevron approach markings on gore nose (Figure 7), can avoid premature merging into mainline [21]. However, the length of forbidden line and its adaptability in China and implementation effect have not been proven and evaluated.

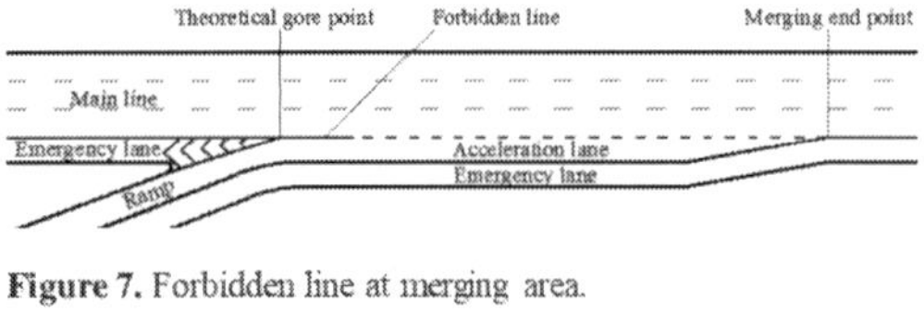

Figure 7. Forbidden line at merging area.

Taking the merge section under the condition that mainline volume arrives at 1250veh/h/ln, merging ratio is 0.25, acceleration lane is 300m long and HGVs account for 40% as an example, the simulation was carried out to study the effectiveness of forbidden line, with the results shown in Figure 8. Obviously, when forbidden line is 15m long, the best improvement has been achieved. And the effectiveness will decline when continuing to increase the length because this actually reduces acceleration lane length and the choice of merging space.

The merge section with forbidden line length of 15m, acceleration lane length of 300m, merging rate of 0.25, traffic volume of 500veh/h/ln~1750veh/h and HGV ratio of 20%~60% was taken as an example to compute the reduction rate of traffic conflicts. And the K-means cluster analysis method was applied for the effectiveness evaluation of the data obtained. The final clustering result obtained from ten iterations is shown in Table 6.

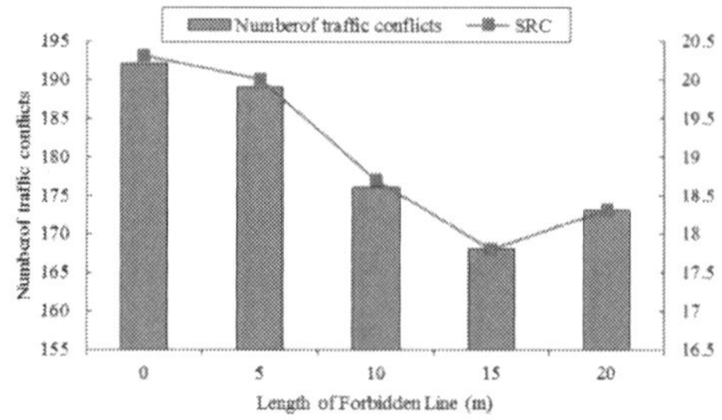

Figure 8. The effect of forbidden line at merging area.

According to the cluster results, the effectiveness of forbidden line is divided into four levels, excellent, good, medium and poor. The effectiveness under various conditions is shown in Table 7.

Table 6. Final cluster centres in simulation.

Level	Poor	Medium	Good	Excellent
Reduction Rate of Conflicts （%）	2.85	5.56	8.08	11.15

Table 7. Effect analysis for the setting of the forbidden line.

Traffic Volume (veh/h/ln)	Ratio of HGV	Reduction Rate of Conflicts	Level
500	20%	2.08%	Poor
	30%	5.33%	Medium
	40%	7.10%	Good
	50%	6.24%	Medium
	60%	4.27%	Medium
750	20%	3.34%	Poor
	30%	5.67%	Medium
	40%	8.36%	Good
	50%	6.91%	Good
	60%	5.02%	Medium
1000	20%	6.11%	Medium
	30%	8.43%	Good
	40%	10.05%	Excellent
	50%	8.81%	Good
	60%	6.67%	Medium
1250	20%	9.03%	Good
	30%	11.02%	Excellent
	40%	12.50%	Excellent
	50%	10.57%	Excellent
	60%	8.61%	Good
1500	20%	7.59%	Good
	30%	9.43%	Good
	40%	11.62%	Excellent
	50%	9.26%	Good
	60%	7.11%	Good
1750	20%	5.15%	Medium
	30%	7.08%	Good
	40%	7.33%	Good
	50%	3.83%	Poor
	60%	2.16%	Poor

From Table 7, when the large vehicle ratio is 30%~50% and mainline traffic volume is 1000veh/h/ln~1500veh/h/ln, setting forbidden line has excellent or good effect. When the traffic volume is relatively small or large, setting forbidden line has poorer effect along with the increment of the large vehicle ratio. The reasons are following: when the traffic volume is relatively small, the traffic flow is under free state and large vehicles slightly influence small vehicles. However, when the traffic volume is relatively large, the traffic flow is under saturated state, and forbidden line cannot effectively reduce traffic conflicts.

5 Conclusion

On the freeway interchange merging area, the moving bottleneck caused by large cars is not independent, which may affect traffic safety when lateral lane changing behavior and longitudinal car-following behavior occur. The impact of large vehicles and the improvement measures were systematically researched in this paper with the goal of traffic safety.

The influence of the moving bottleneck caused by large vehicles on the traffic flow is significant. With the increment of large vehicle ratio, traffic safety decreases for the traffic conflicts and *SRC* increases significantly. When traffic volume is less than 750veh/h/ln or more than 1500veh/h/ln, the impact of HGV is relatively small. This indicates that moving bottleneck has not yet formed at merging area at that time.

Setting appropriate length of acceleration lane can enhance traffic safety level effectively. Under the condition of traffic volume of 1250veh/h/ln, the length of acceleration lane should be set to 300m, 350m and 400m when large vehicle ratio are 20%, 30%~40% and 50%~60%, respectively.

Besides, speed limit is an effective method to improve traffic operetion at merge sections, especially when traffic volume of mainline is 1000veh/h/ln~1500veh/h/ln and ratio of HGV is less than 40%. Forbidden line proposed by MUTCD is also considered in this paper and we conclude that 15m long forbidden line may have better improvement effect under medium traffic volume condition. Meanwhile, by K-means cluster analysis, when traffic volume is 1000veh/h/ln~1500veh/h/ln and the ratio of HGV is 30%~50%, setting forbidden line on freeway interchange merging area can effectively promote the traffic safety; especially, when the large vehicle ratio is about 40%, the positive effect is the most obvious.

However, this paper was limited to analyzing data from the perspective of part of road and traffic conditions. In reality, merging ratio, number of lanes, ramp radius, type of acceleration lane and other traffic control or environment conditions should be taken into account. Future studies may look into these influencing factors as variables for modeling.

Acknowledgments

This research is funded by China's National Science and Technology Support Programme Project (Project No.2014BAG01B01).

References

1. Z. Yuan, The traffic simulation and evaluation of merging-area on freeway. Hefei University of Technology (2004)
2. K. R. Agent, and J. G. Pigman, Evaluation of Highway Geometries Related to Large Trucks. (1991)
3. D.C. Gazis, and R. Herman, The Moving and "phantom" bottlenecks. Transportation Science, **26**, 3, 223-229, (1992)
4. G.F. Newell, A moving bottleneck. Transportation Research Part B Methodological, **32**, 8, 531-537, (1998)
5. J. C. Muñoz, and C. F. Daganzo, The bottleneck mechanism of a freeway diverge. Transportation Research Part A Policy & Practice, **36**, 6, 483-505, (2002)
6. C. F. Daganzo, and J. A. Laval, On the numerical treatment of moving bottlenecks. Transportation Research Part B Methodological, **39**,1, 31-46, (2005)
7. Y. Wu, D. Yu, W. Yu, T. Qin, and Y. Li, Double-layer ramp-metering model for incident congestion on expressway. Journal of Traffic & Transportation Engineering, **1**, 2, 129-137, (2014)
8. J. Wang, and H. Niu, Graded-information feedback strategy in two-route systems under ATIS. Journal of Traffic & Transportation Engineering, **1**, 2, 138-145, (2014)
9. M. Sarvi, and M. Kuwahara, Using ITS to improve the capacity of freeway merging sections by transferring freight vehicles. IEEE Transactions on Intelligent Transportation Systems, **9**, 4, 580-588, (2008)
10. L. Leclercq, V. L. Knoop, F. Marczak, and S. P. Hoogendoorn, Capacity drops at merges: New analytical investigations. IEEE International Conference on Intelligent Transportation Systems, **62**, 1129-1134, (2014)
11. Z. Li, P. Liu, W. Wang, and C. Xu, Development of control strategy of variable speed limits for improving traffic operations at freeway bottlenecks. Journal of Central South University, **21**, 6, 2526-2538, (2014)
12. S. Li, Q. Xiang, Y. Ma, X. Gu, and H. Li, International Journal of Environmental Research & Public Health, **13**, 11, 1157, (2016).
13. L. Lei, L. Dong, and H. Ge, Study of gear-alternating control regulation in the interfluent location of on-ramps with cellular automaton model. Acta Physoca Sinica, 56, 12, 6874-6880, (2007)
14. J. Sun, Z. Li, and J. Sun, Study on traffic characteristics for a typical expressway on-ramp bottleneck considering various merging behaviors. Physica A Statistical Mechanics & Its Applications, **440**, 57-67, (2015)
15. X. Qu, S. Jin, and J. Weng, Analysis of the relationship between aggregated traffic volume and

traffic conflicts on expressways. Transportmetrica, **11**, 7, 648-658, (2015)
16. S. Li, Q. Xiang, X. Gu, J. Dai, C. Tang, and S. Zhao, MATEC Web of Conferences, **81**, 02006, (2016)
17. L. Zheng, K. Ismail, and X. Meng, Traffic conflict techniques for road safety analysis: open questions and some insights. Canadian Journal of Civil Engineering, **41**, 7, 633-641, (2014)
18. Y. Wu, and Z. Wu, A safety analysis method for highway based on average speed and speed standard deviation. Journal of Highway and Transportation Research and Development, **25**, 3, 139-142, (2008)
19. Y. Pei, and G. Cheng, Research on the relationship between discrete character of speed and traffic accident and speed management of freeway. China Journal of Highway and Transport, **17**, 1, 74-78, (2004)
20. Z. Zhang, Effects of truck on traffic flow in multi-lane freeway diverge segment. Southeast University, (2015)
21. Federal Highway Administration, MUTCD, (2009)

Safety Assessment of Auxiliary Lanes in Freeway Interchange Weaving Areas based on Traffic Conflict Technique

Rao Yao[1], Qiaojun Xiang[1], Xin Gu[1] , Yifan Guo[1] and Ting Zhang[1]

[1]School of Transportation, Southeast University, Si Pai Lou #2, Nanjing 210096, China

Abstract. Auxiliary lanes can provide an improved weaving environment rather than a forced or direct merge or diverse for vehicles entering and departing freeway weaving segments. Considering China's traffic composition and operational status, this paper proposed safety assessment of auxiliary lanes in freeway interchange weaving areas based on traffic conflict technique in order to estimate its safety impacts and related influential factors considered in design. Based on freeway interchange weaving areas data and actual traffic characteristics, simulation models were built with VISSIM. This paper used Surrogate Safety Assessment Model (SSAM) to calculate conflicts by inputting vehicle trajectories from VISSIM, analysed the influence of ramp spacing, lane number of mainline, traffic volume, weaving ratio and percentage of heavy vehicles on safety performance of auxiliary lanes in freeway weaving areas. Finally, Traffic Conflict Modification Factor was presented to evaluate effectiveness of implementing auxiliary lanes. The paper found the safety impacts and related influential factors of auxiliary lanes. The findings can provide support to effective decision making with regards to constructing future auxiliary lanes in freeway interchange weaving areas.

1 Introduction

As an important part of freeway, interchange and its operation performance is of great importance to the entire road network. Traffic flow features in weaving areas are much more complicated than other function areas, therefore, weaving areas have a great influence on interchange safety. Auxiliary lane (AL) is a common facility in interchange weaving areas and is often used to improve weaving environment by adding lanes between entrance-and exit-ramp pairs when the ramps are close. There are many standards specifications about AL setting and designing methods: A Policy on Geometric Design of Highway and Streets (AASHTO Green Book, 2004) [1], Manual on Uniform Traffic Control Devices (MUTCD, 2009) [2], Japan Highway Design Essentials (1991) [3], Design Specification for Highway Alignment (2006) [4] and Guidelines for Design of Highway Grade-separated Intersections [5]. However, the setting methods and geometric design guidelines listed in these specifications are mainly based on capacity analysing and lack detail, which could lead to uncertainty to traffic engineers when considering designing ALs. Besides, a large proportion of AL-related researches focus on the operational impacts of ALs, very few on safety impacts.

As for the researches on safety impacts of ALs, Beverly Kuhn.et al analysed roads setting conditions and traffic crash rates in ramp influential areas and found that implementing ALs in merge and weaving areas can effectively reduce traffic crash and conflicts [6]. According to principles of lane balance, Hongjun Tian took Wang Guantun interchange on Ji Nan-He Ze highway as an example to evaluate impacts of ALs and found that ALs can reduce the interruptions of main line vehicles, make traffic running smoother and safer [7]. Worku Mergia identified crash injury severity influential factors were diver-related, traffic, environmental and geometric design factors. A generalized ordinal logit model and partial proportional odds model were applied to identify the factors that increase the likelihood of one of five levels of injury severity: no injuries, possible/invisible injuries, non-incapacitating injuries, incapacitating injuries, or fatal injuries. It was found that the use of continuous auxiliary lanes between an entrance ramp and an exit ramp tends to increase the likelihood of severe injuries near the diverge areas [8]. Sarhan.et al studied 34 weaving segments in 26 interchanges to investigate the effects of ramp terminal spacing and traffic volumes on safety performance. Some findings were found included: 1) the historical crash records show that the implement of ALs did not improve the safety performances significantly at these sites. 2) Weaving Type A was associated with relatively lower collision frequencies when compared with weaving Type B. 3) the number of collisions will decrease with increasing length of speed-change lane [9]. Richard Glad analysed datasets and found that two-thirds of the rear-ends and one-third of the sideswipe occurred in ALs in weaving areas, although the crashes were mainly attributed to weaving traffic instead of the presence of ALs [10]. Yi Qi.et al presented Traffic Conflict Modification Factor (TCMF) as an indicator to analyse the changes in conflict frequency before and after implementing ALs under different conditions by simulation software. It was found that the reduction of traffic conflicts could be especially significant when high weaving volumes were present [11].

From the literature review, we can conclude that the safety impacts of ALs and related influential factors haven't been studied clearly. Besides, the analysis on safety impacts of ALs are mainly based on historical crash records. Crash records have issues of small simple size, difficulty of obtaining, low data reliability and large time span. What is more, the method of analysing crash records is not available for facilities or managements to be or are being planned. It has been proved that traffic conflict is in correlation with traffic crash and is an effective surrogate measure of evaluating traffic safety. Traffic Conflict Technique (TCT) has the advantages of short periods, large simple size and high data reliability, etc. Safety performance can be evaluated under different conditions by combining TCT with simulation.

Therefore, TCT is applied in this paper to assess safety impacts of ALs in freeway interchange weaving areas. Take the basic setting form of ALs as an example, different influential factors were studied in order to provide detailed guidelines on ALs planning and designing. This paper also provide support to effective decision making with regards to weaving areas geometric design, management and prevention of accidents.

2 Method

2.1 Data collection

The traffic survey was conducted at a semi-cloverleaf interchange with a typical weaving areas, which is composed of a one-lane mainline and a one-lane ramp (one-lane entrance and one-lane exit with one AL between). (Fig.1) The width of every lane is 3.75m. The ramp spacing is about 165m and the radius of the ramps is 60m. Note that the mainline is in fact a collector-distributor road, so the weaving volume was high in this area, satisfying the requirement of sample size.

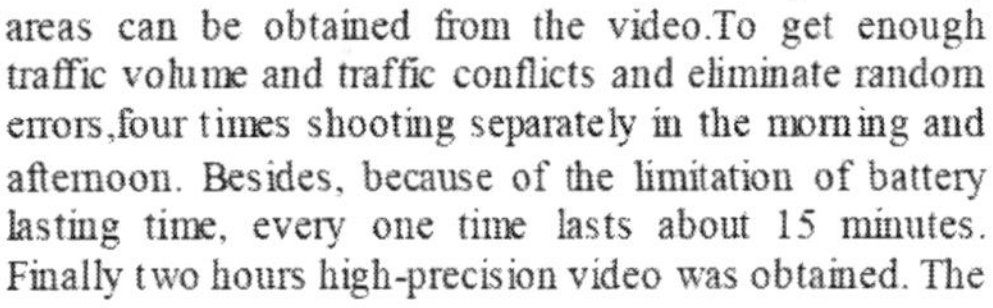

Figure 1. Weaving area in survey

In order to observe and analyse the traffic conflicts clearly, unmanned aerial vehicle (UAV) photography was allowed to shoot video in good conditions. High definition of the whole view of the interchange weaving areas can be obtained from the video.To get enough traffic volume and traffic conflicts and eliminate random errors,four times shooting separately in the morning and afternoon. Besides, because of the limitation of battery lasting time, every one time lasts about 15 minutes. Finally two hours high-precision video was obtained. The aerial photo was below. (Fig.2)

Figure 2. Weaving area in aerial photo

2.2 VISSIM-based simulation model of the auxiliary lanes

2.2.1 Model Assumption

Specific to the characteristics of the weaving area and the function of the simulation software, some assumptions are present as below.

• According to Chapter 12 of HCM 2010[12], there is at least one AL in freeway weaving area. It also states that a freeway segment between an on-ramp and an off-ramp but without an AL should be analysed separately as isolated merge and diverge areas. However, for continuity and simplicity in analysis for the rest of this paper, we assume that a weaving area can be designed with or without ALs.

• Assume that the mainline and the ramps are straight, the influence of horizontal curve is neglected. Just consider the geometrical parameters such as length, width of the lane in the simulation experiment.

• The influence of gradient is considered to be 0.

• The angle between the ramp and mainline is 25° and the typical auxiliary lanes analysed in this paper is parallel type introduced in HCM 2010.

• The influences of weather conditions and lateral clear width on the traffic flow are neglected.

• Only small cars (Car) and heavy vehicles (HGV) are considered in the discussion.

• Assume that the design speed of the mainline is 80km/h and the ramp is 40 km/h.

2.2.2 Parameter Calibration

Calibration in simulation mainly includes the representative vehicle parameters and three types of driver behaviours models: the car-following model, the lane-changing model and the transverse behaviour model [13]. Refer to the previous related research and take this paper's research object into consideration, parameters are set as follows:

• Representative Vehicle Parameters Calibration

Cars and HGVs are considered in this paper and the calibration of vehicle characteristic parameters include the length, maximum/minimum and expectation of acceleration/deceleration. Specific settings are shown in Table 1.

• The Car-following model

The Wiedemann99 Car-following model was adopted and the specific parameters are shown in Table 2.

• The lane-changing model

DOI:10.1051/matecconf/201712404005

Line-changing is the most significant behavior in weaving areas. Frequent lane changes may lead to disruption and congestion in traffic flow. The lane-changing model parameters include Min.Headway, Lane change Behavior and to slower lane if collision time above. Specific parameters setting are shown in Table3.

• The transverse behavior model

Desired Positon at Free Flow, Min.Lateral Distance and Overtake on same lane are the main parameters in the transverse behaviour model.

• Simulation time period and number of times

Considering the field investigation data, sample size requirement and in order to eliminate the influence of randomness of both traffic conflicts and simulation, every simulation under each different combination of conditions conduct 10 times, every time lasts 4500s and the average is adopted as the final result.

Table 1 Vehicle Parameters in simulation software

Vehicle Type	Length of Vehicle (m)	Width of Vehicle (m)	Acceleration (m/s^2)		Deceleration (m/s^2)	
			Max.	Exp.	Max.	Exp.
Car	4.2-4.9	1.5	3.5	3.1	5.0	4.6
HGV	7.0-13.0	2.5	2.5	2.0	4.8	4.2

Table 2 Parameters in Wiedemann99 Car-following model

Parameter	Value	Parameter	Value	Parameter	Value
Standstill distance (CC0)	1.50m	Negative following thresholds (CC4)	-0.35	Standstill acceleration (CC8)	$3.5m/s^2$
Headway time (CC1)	1.60s	Positive following thresholds (CC5)	0.35	Acceleration at 80km/h (CC9)	$1.6m/s^2$
Following variation (CC2)	3.60m	Speed dependency of oscillation (CC6)	11.44	Look Ahead Distance(m)	0~200
Threshold of entering following (CC3)	-7.67	Oscillation acceleration (CC7)	0.25 m/s^2	Observed Vehicles	4

Table 3. Parameters in lane-changing model

Parameter		Value	Parameter		Value
Max. deceleration	own	-5.86 m/s^2	Accepted deceleration	own	-1.00 m/s^2
	Trailing vehicle	-3.00 m/s^2		Trailing vehicle	-0.50 m/s^2
$-1m/s^2$ per distance	own	150m	Waiting time before diffusion		60.0s
	Trailing vehcle	100m	Min. HeadWay (front/rear)		0.6m
Safety distance reduction factor		0.60	Maximum deceleration for cooperative braking		-3.50 m/s^2

2.2.3 Model Verification

We carry out the simulation by setting up a simulation network of a freeway weaving area with an AL and take the practical survey data as input parameters, such as traffic volume, percentage of HGV, etc. Mean travel time and traffic conflict counts are chosen as indicators of simulation test, and mean absolute percentage error is used to calculate the errors. Mean travel time can be obtained by setting time sampler on the roads and traffic conflict counts can be obtained by analyzing the vehicle trajectories through SSAM.

The verification result is shown in TABLE 4. It is shown that the errors of mean travel time and traffic conflict counts are 1.648% and 9.69% respectively. So the simulation model is valid.

Table 4. Model verification test results

	Test Group	Observation value (s)	Simulation value (s)	MAPE %
Travel Time Test	1	11.2	11.4	1.786
	2	10.93	11	0.640
	3	11.3	11.2	0.885
	4	11.49	11.1	3.394
	5	12.39	12.2	1.533
	Ave.	11.462	11.38	**1.648**
Conflict Test	1	67	57	14.93
	2	42	39	7.14
	3	64	58	9.38
	4	77	68	11.69
	5	94	89	5.32
	Ave.	68.8	62.2	**9.69**

2.2.4 Simulation Scheme

This paper's purpose is to evaluate safety impacts of auxiliary lanes in freeway interchange weaving areas. Safety influential factors can be mainly divided into geographic conditions (ramp spacing, lane numbers of ramp, lane numbers of mainline, lane width, etc.), traffic conditions (through volume, weaving ratio, ramp-to-ramp volume, percentage of HGV, etc.) and other management conditions.

To avoid complication of processing data and to get representative results at the same time, kinds of combinations of conditions are established in the simulation, the specific conditions of simulation schemes are shown in TABLE5. We can get a total of 72 simulation scenarios to compare safety conditions with an AL or not.

Table 5. Conditions of traffic simulation.

Road Conditions	Ramp Spacing (m) L_s		200, 300, 400
	Lane Number of Mainline N		1, 2, 3
Traffic Conditions	Traffic Volume	Through Volume (veh/h/ln) V_{FF}	1000,1400
		Weaving Volume Ratio R_w	0.2, 0.4
		Ramp-to-Ramp Volume(veh/h/ln) V_{RR}	10

	Percentage of large vehicles (%) R_{HGV}	25,50

2.3 Traffic Conflicts Model and Assessment Index

SSAM is used to process data describing the trajectories of vehicles provided in simulation software, VISSIM in this paper. SSAM calculates surrogate measures of safety among every two vehicles and discern conflicts if the interaction satisfies the criteria for conflicts. [14] In this paper, TTC is chosen as the criteria for conflicts and it is defined as the minimum time to collision value observed during the interaction of two vehicles on collision course. In this paper, we use 4s as the threshold and if at any time step the TTC is below this value, the interaction is tagged as a conflict [15].

Traffic Conflict Rate is used as assessment index, and we use TCMFs (Traffic Conflict Modification Factor) to analyze the safety impacts of ALs. Similar to the CMFs presented in the AASHTO HSM, the TCMFs were provided for estimating the expected changes of traffic conflict rate at a location after implementing specific geometric treatments associated with an AL. The TCMFs were calculated as follows [11]:

$$\text{TCMF}=\frac{\textit{Traffic Conflict Rate after Treatment}}{\textit{Traffic Conflict Rate before Treatment}}\times 100\% \qquad (1)$$

If a TCMF value is less than 1.0, it means the treatment may potentially reduce the traffic conflicts and improve the safety performance.

3 Results

Decreases in traffic conflict rate are found after implementing an AL by analyzing the statistical results. The safety performance of ALs in weaving areas is related to ramp spacing, lane number of mainline, traffic volume , weaving ratio and percentage of HGV, as shown in Table6~8.

Table 6 TCMFs for weaving areas (N =1)

L_s (m)	Traffic Volume		R_{HGV} (%)	Conflict Rate Before (%)	Conflict Rate After (%)	TCMF
	V_{FF} (veh/h/ln)	R_w				
200	1000	0.2	25	22.10	4.42	0.2
			50	25.05	4.76	0.19
		0.4	25	26.83	4.83	0.18
			50	29.88	5.08	0.17
	1400	0.2	25	34.27	5.14	0.15
			50	39.00	5.46	0.14
		0.4	25	40.07	5.61	0.14
			50	44.08	5.73	0.13
300	1000	0.2	25	8.07	3.55	0.44
			50	8.53	3.67	0.43
		0.4	25	9.21	3.87	0.42
			50	10.63	4.25	0.40
	1400	0.2	25	11.47	4.36	0.38
			50	12.67	4.56	0.36
		0.4	25	13.57	4.75	0.35
			50	13.91	4.87	0.35
400	1000	0.2	25	3.06	1.53	0.50
			50	3.20	1.60	0.50
		0.4	25	3.41	1.67	0.49
			50	3.81	1.79	0.47
	1400	0.2	25	4.00	1.84	0.46
			50	4.40	1.98	0.45
		0.4	25	4.77	2.05	0.43
			50	5.05	2.12	0.42
					Average	0.34

Table 7 TCMFs for weaving areas (N =2)

L_s (m)	Traffic Volume		R_{HGV} (%)	Conflict Rate Before (%)	Conflict Rate After (%)	TCMF
	V_{FF} (veh/h/ln)	R_w				
200	1000	0.2	25	8.91	3.12	0.35
			50	9.59	3.26	0.34
		0.4	25	12.96	3.63	0.28
			50	15.69	4.08	0.26
	1400	0.2	25	18.43	4.24	0.23
			50	21.33	4.48	0.21
		0.4	25	25.61	4.61	0.18
			50	26.44	4.76	0.18
300	1000	0.2	25	4.36	3.01	0.69
			50	4.72	3.07	0.65
		0.4	25	4.92	3.10	0.63
			50	5.43	3.31	0.61
	1400	0.2	25	6.33	3.61	0.57
			50	6.76	3.65	0.54
		0.4	25	7.21	3.75	0.52
			50	8.00	3.84	0.48
400	1000	0.2	25	1.77	1.33	0.75
			50	2.01	1.47	0.73
		0.4	25	2.24	1.52	0.68
			50	2.60	1.61	0.62
	1400	0.2	25	2.86	1.69	0.59
			50	2.98	1.73	0.58
		0.4	25	3.33	1.80	0.54
			50	3.85	2.0	0.52
					Average	0.49

Table 8 TCMFs for weaving areas (N =3)

L_s (m)	Traffic Volume		R_{HGV} (%)	Conflict Rate Before (%)	Conflict Rate After (%)	TCMF
	V_{FF} (veh/h/ln)	R_w				
200	1000	0.2	25	5.60	2.52	0.45
			50	6.21	2.61	0.42
		0.4	25	6.90	2.69	0.39
			50	7.83	2.74	0.35
	1400	0.2	25	9.06	2.81	0.31
			50	10.21	2.86	0.28
		0.4	25	11.68	2.92	0.25
			50	13.17	3.16	0.24
300	1000	0.2	25	3.04	2.31	0.76
			50	3.40	2.45	0.72
		0.4	25	3.89	2.57	0.66
			50	4.14	2.69	0.65
	1400	0.2	25	4.54	2.77	0.61
			50	4.80	2.83	0.59
		0.4	25	5.18	2.90	0.56
			50	6.10	3.11	0.51
400	1000	0.2	25	1.54	1.28	0.83

DOI:10.1051/matecconf/201712404005

			50	1.67	1.35	0.81
		0.4	25	1.95	1.48	0.76
			50	2.08	1.52	0.73
	1400	0.2	25	2.41	1.64	0.68
			50	2.71	1.71	0.63
		0.4	25	3.02	1.78	0.59
			50	3.48	1.95	0.56
					Average	0.56

Overall, installing auxiliary lanes at weaving areas can significantly reduce traffic conflicts rate as opposed to the cases without auxiliary lanes.

3.1 Safety Impacts with geographic conditions

3.1.1 Lane number of mainline

Traffic conflict rate and TCMFs when L_s is 200m, R_w is 0.2 and R_{HGV} is 25% are shown in Fig.3~5. TCMFs values get lower when there are fewer lanes in mainline, in other words, the safety impact is more effective. Because when lanes in mainline are fewer, there is a more complicated situation that through and weaving vehicles mixing in the same line. For the lack of space for changing speed and waiting for a safety gap, more conflicts will occur. So when there are fewer lanes in mainline of weaving areas, auxiliary lanes can be taken into consideration.

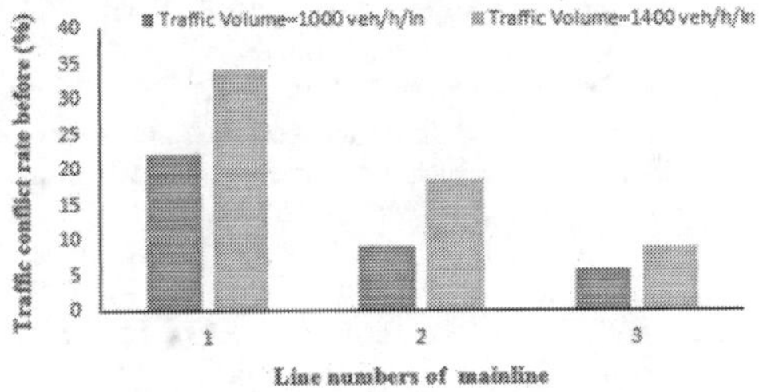

Figure 3. Traffic conflict rate before implementing ALs

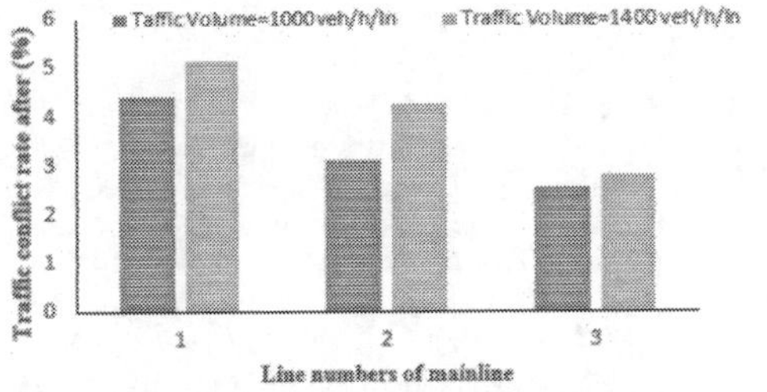

Figure 4. Traffic conflict rate after implementing ALs

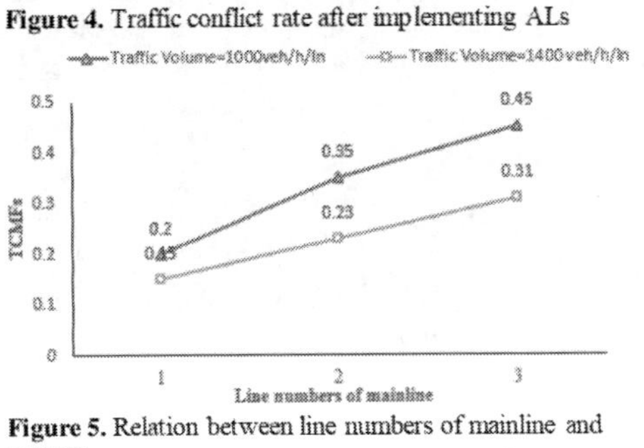

Figure 5. Relation between line numbers of mainline and TCMF

3.1.2 Ramp spacing

TCMFs values get lower when the ramp spacing are shorter (Fig.6). Because when the length between on-ramp and off-ramp is short, there are not enough time and space for those weaving vehicles to perform a series of behaviors including acceleration/deceleration, lane-changing, etc. Large differences of vehicle speeds can lead to disturbance and crash risk, especially when the speed of vehicles on freeway are high. The merge and diverge influence areas are normally defined at a length of 450m in the HCM (2000, 2010), so when ramp spacing is shorter than this value, it's essential to design ALs. This is particularly true for weaving areas at cloverleaf type interchanges. ALs can be a cost effective method to increase capacity and safety particularly in cases where the facility experiences high entrance and exits from adjacent interchanges. The threshold value can be smaller when traffic volume and weaving volume are high.

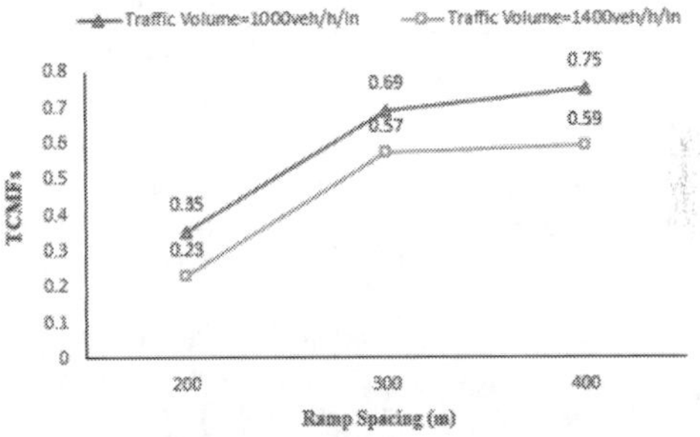

Figure 6. Relation between ramp spacing and TCMF

3.2 Safety Impacts with traffic conditions

3.2.1 Traffic volume

The reduction of traffic conflicts with ALs could be especially significant when high traffic volume, especially weaving volumes were present. (Fig.7) (e.g., with a through volume of 1400 veh/h/ln and weaving ratio of 0.4). Weaving is the most complicated behaviour in weaving areas, it consists of groups of vehicles performing speed-changing, lane-changing and other actions. Besides, driver factors is much more complicated and could also be important in this situation. ALs can provide sufficient space for those weaving vehicles and could also be taken as a buffer zone for those careless drivers.

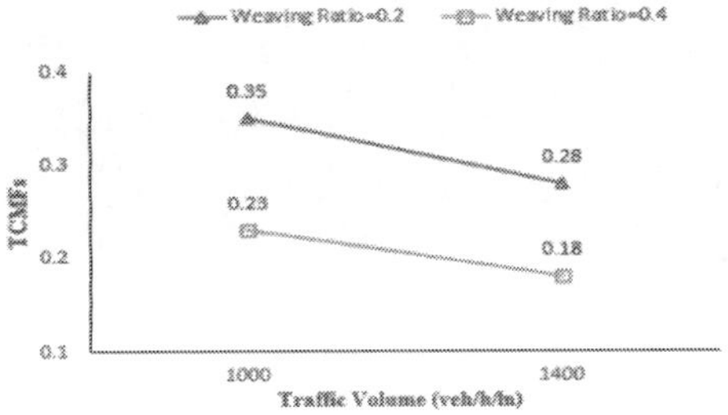

Figure 7. Relation between traffic volume and TCMF

3.2.2 Percentage of HGV

TCMFs values get lower when the percentage of HGV are higher.(Fig.8) Because the differences of speed, acceleration/deceleration and other characteristics between HGV and cars, HGV can form a blocking effect on lateral roads to cars lane-changing. So it can be a problem to have HGV mixed in traffic flow.

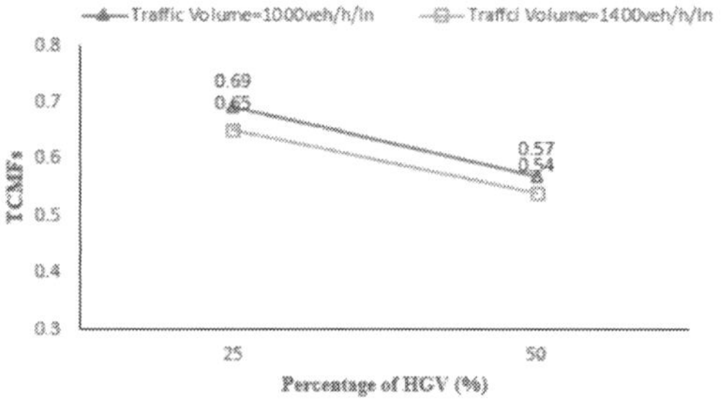

Figure 8. Relation between percentage of HGV and TCMF

4 Conclusion

This paper investigated the safety impacts of adding auxiliary lanes by analyzing the traffic conflicts derived from the traffic simulation studies. The results showed that:

• Adding auxiliary lanes can significantly reduce the ratio of traffic conflicts in freeway weaving areas.

• Safety influential factors considered in ALs design include geographic conditions, traffic conditions and other management conditions. In particular, this paper analyze the factors of ramp spacing, lane numbers of mainline, through volume, weaving ratio and percentage of HGV. The results show that these factors have great influence on safety impacts of ALs.

The findings can provide support to effective decision making with regards to constructing future auxiliary lanes in freeway interchange weaving areas.

Acknowledgements

This work is supported by the Fundamental Research Funds for the Central Universities and the Graduate Scientific Innovation Research Foundation of Jiangsu Province (KYLX16_0274). The authors also would like to thank the graduate research assistant at the School of Transportation at the Southeast University for their assistance in field data collection.

References

1. American Association of State Highway and Transportation officials (AASHTO). *A policy on geometric design of highways and streets*. Washington D.C., (2004)
2. DoT, U.S. *Manual on uniform traffic control devices* (MUTCD). Federal Highway Administration, Washington, DC, (2003)
3. Japan Highway Public Corporation. *Japan Highway Design Essentials*. Shaanxi Travel & Tourism Press, China, (1991).
4. *Design Specification for Highway Alignment*. Publication JTG D20-2006. Ministry of Transport of the People's Republic of China, (2006).
5. *Guidelines for Design of Highway Grade-separated Intersections*. Publication JTGT D21-2014. Ministry of Transport of the People's Republic of China, (2014).
6. Kuhn B, Balke K, Chaudhary N A, et al. Managed Lanes Strategies Feasible for Freeway Ramp Applications. (2007).
7. Hongjun Tian, Chengxu Wang, Yanhua Han, Applications of auxiliary lanes and collector-distributor lanes in interchanges. Ji Lin Transportation Technology. Vol **4**, pp: 11-13. (2006)
8. Mergia W Y. Exploring factors contributing to injury severity at freeway merging and diverging areas. University of Dayton, (2010).
9. Sarhan M, Hassan Y, Abd El Halim A O. Safety performance of freeway sections and relation to length of speed-change lanes. Canadian Journal of Civil Engineering, Vol.**35**, No.5, pp:531-541, (2008).
10. Glad R W. Weave analysis and performance: The Washington state case study. (2001).
11. Qi Y, Chen X, Cheu R K, et al. Design and Scope of Impact of Auxiliary Lanes. (2014).
12. Transportation Research Board, National Research Council, *Highway Capacity Manual*. Washington, DC (2010).
13. PTV VISSIM User's Manual 5. 0. Planung Transport Verkehr AG, D-76131 Karlsruhe, Germany, 2007.
14. Gettman, D., L. Pu, T. Sayed, and S. G. Shelby. Surrogate Safety Assessment Model and Validation: Final Report. Publication FHWA-HRT-08-051. FHWA, U.S.Department of Transportation, 2008
15. Li S, Xiang Q, GU X, et al. The Types Division and Severity Threshold of Traffic Conflict on Freeway Interchange. *MATEC Web of Conferences*. EDP Sciences, Vol.**81**, No. 02006, (2016).

DOI:10.1051/matecconf/201712404006

Research on three - dimensional Cooperative Formation of Sea and Air for Cruise and Rescue

Fan Wei[1]

[1] *School of Navigation, Wuhan University of Technology, Wuhan, China*

Abstract. Aiming at the problems of maritime traffic safety, a new model of jointing USV and UAV for Three-dimensional cruise and rescue is studied. Using the Information Fusion Technology, through the leader-follower formation mode, The collaborative fleet remote control. At the same time, it discusses how to carry out effective supervision on the responsible sea area under the complicated and changing sea conditions, especially the quick and efficient completion of the maritime search and rescue mission.

1 Introduction

The role of the oceans as an important player in the world's economic activities is becoming increasingly important, the maritime traffic environment is more complex, highlighting a lot of maritime traffic safety problems. Therefore, the search and rescue of maritime cruise and distress accidents has also become the focus of maritime research in order to achieve "all-weather operation, all-round coverage, emergency rapid response" for the sea traffic safety management objectives [1-3].

Unmanned Surface Vessel (USV) is a kind of maritime intelligent sports platform capable of self-safe navigation under complex sea conditions, and has the characteristics of low cost, high efficiency and no casualties. Combined with unmanned aerial vehicles (UAV) widely used, the advantages of strong mobility, USV, UAV can cooperate with each other to form a convenient control of the formation of the formation of complex sea conditions in the face of cruise and rescue missions, the formation of cruise and rescue Of the sea and air three-dimensional collaborative formation.

2 Formation diversity

2.1 Formation pattern

The determinants of formation include the nature and characteristics of the task, as well as the prevailing sea conditions and weather conditions.

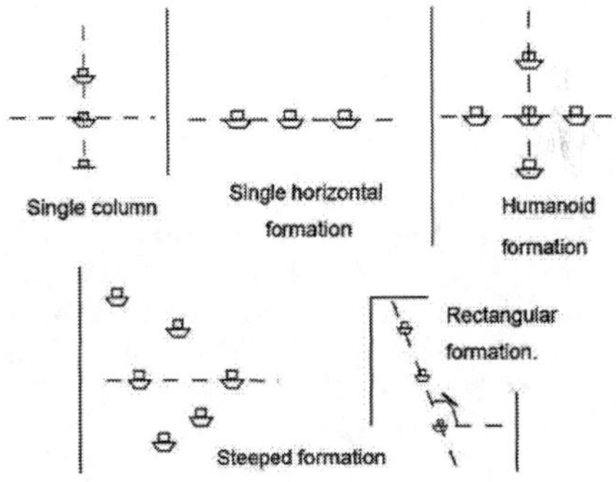

Figure 1. Formation

Table 1. Formation pattern classification

Formation pattern	Formation elements	Formation characteristics
Single column	β=180°,α=0°	Formation is simple, easy to maintain, more flexible
Single horizontal formation	β=90°,α=90°	Steering difficulties, manipulation inconvenience
Stepped formation	β≥90°	Better depth, Larger width
Rectangular formations	β=180°or 90° α=0°or 90°	Better depth, Large width
Humanoid formation	γ1=γ2	Good depth, Larger width

2.2 UAV in formation

UAV's position in the formation is flexible, applicable to any USV formation mode. UAV has the characteristics of fast flight speed, strong maneuverability, wide monitoring range and long battery life. It can reduce the emergency response time of the formation, expand the effective area of the cruise and rescue, makes the function more Three-dimensional, makes up for the shortcomings of the USV formation, provides new regulatory perspectives.

3 Cooperative control of formation

3.1 Information fusion technology

The basis of control are communications, collaborative communication to form a specific network within formations, the members can achieve resource sharing. The information fusion technology will deal with the multi-source information of each sensor, combine the information according to certain optimization criteria, thus eliminate the possible redundant and contradictory data, and complement the information, reduce the uncertainty, and finally get the same complete conclusion, for the formation of decision-making to provide a strong guarantee .Includes Data conversion, Data related, Situation database,Fusion calculation [4].

3.2 Centralized control mode

Centralized control mode [5] is to use a higher configuration, better performance of the USV as a carrier, the carrier to complete the collaborative formation of the data analysis and processing, and sent to other members to carry out the task deployment. Here the centralized control system uses the leader - followed model [6]. The L-F model is shown in Fig.2:

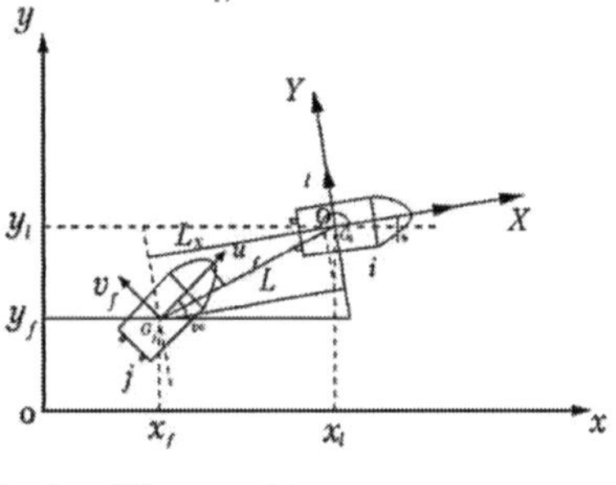

Figure 2. Leader - follower model

3.3 Remote control system

Remote control system [7-9] is divided into land-based control system and ship control system. The former mainly completes the real-time display of the status and velocity information of the USV and can send control commands to the ship-borne system. The latter can access the ship's position and velocity information, and can analyze the commands conveyed by the land-based control system and control the USV, including start up, turn, speed change, berthing, etc. In order to realize the positioning function of USV, adopts GPS positioning mode. GPS can not only determine the exact location but also calculate the speed of USVs. The land-based system uses the electronic chart system and establishes the GUI control system. The electronic chart can accurately display the real-time position of the USV. The GUI panel can send control commands to the ship borne system by setting different command buttons.

Land-based system access network , the ship-borne system access GPRS network, using the traditional Internet and GPRS wireless communication combination of the way through the TCP / IP protocol to achieve the two systems between the data communication function. USV use ultra- short wave communication for information exchange and task communication, also known as the use of 1 to 10 meters wavelength of electromagnetic waves for line-of-sight transmission. Ultra-shortwave wave shortwave wave propagation mode is highly stable, subject to seasonal and day and night changes in the small impact.

4 The maritime cruise and rescue of formation

Due to the limitations of the USV itself and its equipment and sensors, it is often difficult to complete a given task in a complex environment. But the formation of collaborative search tour can effectively overcome the restrictions, so the goal of a full range of search and observation. In short, USV, UAV collaborative formation makes the search system has better reconnaissance ability and higher fault tolerance, while reducing the waste of resources while improving the efficiency of search mission.

4.1 Formation search mode

The formation of the cooperative search mode [10] can be roughly divided into coverage mode and non-coverage mode. Coverage mode includes Extend Square mode, Parallel Line mode and Fan-shaped mode. Non-coverage mode includes Track mode.

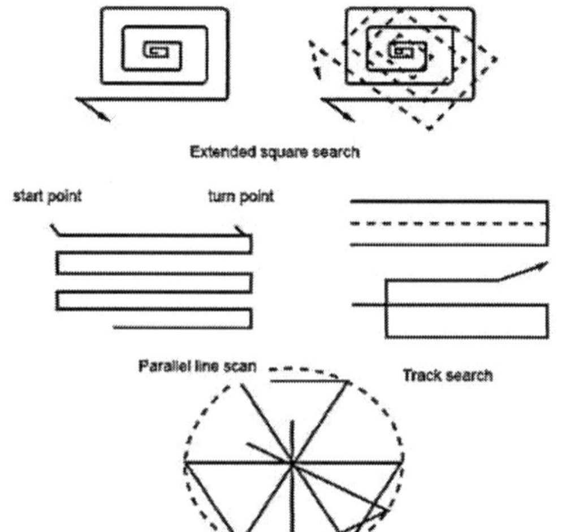

Figure 3. Formation search mode

Parallel line scan search and rescue [11] is the most common, the calculation width is "w", USV speed is "v", search area is "S". Assuming that the search area is rectangular, the target position is unknown, its probability is evenly distributed in the search area, and there is no coverage area, the search route function can be expressed as:

$$b(t) = \frac{wvt}{S}, (t \geq 0)$$

USVs do not find the target within time "t", and the probability of finding the target within the increment time "Δ t" is $[1 - b(t)]wv\Delta t / S$,Whereby the probability that the USV finds the target at time "t + Δ t":

$$b(t + \Delta t) = b(t) + [1 - b(t)]wv\Delta t / S$$

By the above formula :

$$b'(t) = \lim_{\Delta t \to 0} \frac{b(t + \Delta t) - b(t)}{\Delta t} = [1 - b(t)]wv / S$$

B (0) =0, $b(t) = 1 - e^{-twv/S}, (t \geq 0)$

From the nature of the b (t) function, the longer the formation search time, the slower the change in the probability of the target being found. When the search length "v" is close to "S / w", the probability of finding the target is close to "1". So for this search and rescue mode, should be selected perpendicular to the minimum span direction as USV search and rescue navigation direction, that is, the optimal path.

4.2 Search and rescue equipment combination applications

In the case of good visibility, the photoelectric tracking and monitoring equipment imaging system[12] [13] can clearly monitor the situation of the long distance sea surface with the maritime search lamp, and can effectively carry out the pollutant supervision and the night emergency rescue, Improve the success rate of finding targets (especially drowning man).

4.3 Cruise and rescue in the UAV

In the same distance, UAV's patrol radiation area is 12 times the USV [14], patrol effect is much better than the USV. Therefore, in the cruise and rescue process using UAV can effectively improve the cruise efficiency, rapid detection of regulatory waters within the latent risk factors. Through the machine, the ship can greatly enhance the flexibility of the formation of the task. UAV as the air flight platform [15], its control mode and formation function in the final analysis through its airborne system to complete the relevant equipment, include airborne digital camera, multi - spectral imager

.

Table 2. UAV maritime applications

Maritime business	UAV function	Functional description
Daily cruise	Regular cruise	To inspect the waters and vessels in the area, and to supervise the traffic control areas, to curb the sewage discharge
	Used as a VHF aerial relay	As the USV ship borne VHF and land-based communication distance is short, need UAV equipped with VHF relay equipment, with the USV cruise
	With USV supervision	USV is slow at cruise, cannot be observed long-range unknown vessels, the use of ship carrying UAV can expand the scope of reconnaissance
Anti – pollution supervision	Collect evidence of pollution	Use UAV for oil spills and chemical accident emergency evidence, reach the scene of the accident will be real-time transmission to the control center
	Collect evidence of dangerous accident	UAV-USV joint investigation is required for marine hazardous sites, and UAV can expand the survey area
Emergency search and rescue	Search targeting	Borrow the UAV self-positioning system and airborne AIS to determine the location of the accident, and the information back to the control center
	Video transmission	Through the video equipment to shoot the scene of the accident, the scene of the real-time situation back to the land-based control center, easy to co-ordinate the command
	Material delivery	Through the UAV can complete the food, GPS buoys and small life-saving equipment and other materials put
	Act as a float	UAV installed on the automatic inflatable equipment, set the location identification signal, if necessary, can fall into the sea as a life-saving buoy

5 Experimental verification

5.1 Design of the ship

In order to verify the feasibility and robustness of the cooperative formation control system, the model ship experiment platform is used to verify. The experimental platform includes the leader system equipped with the leader ship, followed ship and UAV, for remote control of the computer. Results as shown:

Figure 4. Model diagram

Microcontroller is control center. It can be written into the slave computer program, the program's role is to receive the instructions of the host computer, and code analysis, to understand what kind of action issued by the user.

5.2 Algorithm design

5.2.1 Sports model

First assume that the formation of the ith USV is the leader ship, the jth USV is followed ship, the formation of two ships composed of the leader- followed[16] model, as Fig 2 shown. Modeling the leader ship, definition: "L" is the relative distance between the followed ship and the leader ship; "Ψ" is the relative angle between the followed ship and the leader ship; "G" is the center point of the USV; "Φ f "is the heading angle of the leader ship and the followed ship; Pl (x l, y l), Pf (x f, y f) are the coordinates of the position of the leader ship and the followed ship. The mathematical expression is:

$$\left\| L_y - L_{yr} \right\| \leq \delta_1$$
$$\left\| \Psi - \Psi_r \right\| \leq \delta_2 \quad (1)$$

Where "δ1" and "δ2" are bounded positive numbers.

Define the tracking error of the heading angle:

$$Z\varphi = \varphi f - \varphi l \quad (2)$$

Define the errors in the structure:

$$\begin{cases} \xi_x = L_{XR} - L_X \\ \xi_y = L_{YR} - L_Y \\ \xi_\phi = \phi_f - \phi_l \end{cases} \quad (3)$$

The kinematic model of USVs is:

$$\begin{cases} L_X = -u_l + u_f\cos\xi_\phi - v_f\sin\xi_\phi + L_{rl} \\ L_Y = -v_l + u_f\cos\xi_\phi + v_f\sin\xi_\phi + L_{rl} \\ \xi_\phi = r_f - r_l \end{cases} \quad (4)$$

Can be derived from the formation of the error system model:

$$\begin{cases} \xi_x = -u\cos\xi_\phi + v_r\sin\xi_\phi + \xi_y r_i + u_i - L_R\psi_R\sin\psi_R \\ \quad -L_R r_1 \sin\psi_R \\ \xi_x = -u\sin\xi_\phi - v_r\cos\xi_\phi - \xi_y r_i + v_i + L_R\psi_R\cos\psi_R \\ \quad +L_R r_1 \cos\psi_R \\ \xi_\phi = r_f - r_i \\ m_1 u - m_2 vr + d_1 u = \tau_1 \\ m_2 v - m_1 vr + d_2 v = 0 \\ m_3 u + (m_2 - m_1)ur + d_3 r = \tau \end{cases} \quad (5)$$

In the actual formation voyage, the distance L between USVs (measured by laser ranging) and the relative angle "Ψ" (which can be calculated from the relative position) are calculated by comparing the actual distance "L" between the two ships angle "Ψ" according to the external environment within the allowable range of error within the allowable range can achieve the purpose of formation control, that is, to complete the formation, maintain, re-generation, formation to maintain the formation of obstacle avoidance and other tasks.

5.2.2 Bow control

During the process of ship moving horizontally, the hydrodynamic damping force is the main force, and the traverse velocity "Vl"is bounded. Assuming that the interference is bounded:

$$|w_1(t)| \leq w_0$$
$$|w_2(t)| \leq w_0$$
$$|w_3(t)| \leq w_0$$

Design the controller to make the formation of the members of the track tracking error ψ_1、μ_1、Z_1 final bounded.

The Define state variables "φ_{ie}", Using the filter to get its estimate:

$$\varphi_{ie}=\psi_{1e}+\gamma\psi_{1e} \tag{6}$$

(6) On both sides of the derivative get:

$$\mathrm{m}_i\varphi_{ie} = f_{i1}(x) - \tau_{ir} - w_{i\varphi} \tag{7}$$

f (x) for the unknown dynamic function, with the network to get its theoretical estimates:

$$f(\mathrm{x}) = W_i^T\sigma(V_i^T\xi_i) + \varepsilon_i \tag{8}$$

To stabilize the system (7), we need to eliminate f (x), so we use the adaptive term to approximate it:

$$f(\mathrm{x}) = W_i^T\sigma(V_i^T\xi_i) \tag{9}$$

The controller is:

$$\tau_{ir} = (k_{i1} + h_{i1})\varphi_{ie} + W_i^T\sigma(V_i^T\xi_i) \tag{10}$$

$$h_{ir} = k_{i4}(\|\xi_i W_i^T\sigma'\|_F^2 + \|\sigma' V_i^T\xi_i\|^2 + 1) \tag{11}$$

Substituting control (10) into (7):

$$\mathrm{m}_{33}\varphi_{ie} = -f(x) - (k_{i1} + h_{i1})\varphi_{ie} - w_{i\varphi} \tag{12}$$

Into the f (x) items, finishing to get:

$$\mathrm{m}_{33i}\varphi_{ie} = -W_{i1}^T(\sigma - \sigma' V_{i1}^T\xi_{i1}) - W_{i1}^T\sigma'(V_{i1}^T\xi_{i1}) + \varepsilon_{i1} - (k_{i1} + h_{i1})\varphi_{ie} - d_{ie} - w_{i\varphi} \tag{13}$$

Longitudinal control and Bow control are similar.

5.3 Experimental results

According to the design idea of the formation, the L-F motion model of the bow control and longitudinal control of its experimental verification. In the remote control under the USV to follow the ideal path forward, control effect as shown:

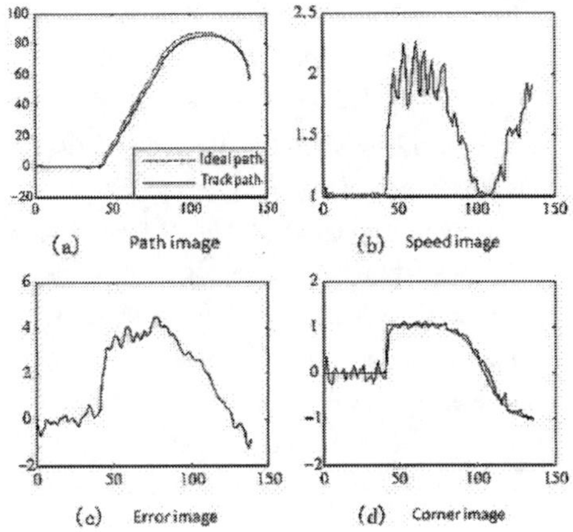

Figure 5. Control effect picture

Among them, a for the USV to track the ideal path forward curve, in considering the storm interference under the premise of USVs with good robustness, can follow the ideal path forward; Fig. b for the tracking speed chart, In the process of unmanned sailing, the stability of the input speed, has been the right to follow the ideal path forward; Fig. c is the tracking error graph, the tracking error can be stabilized within 4 unit distance, can be more accurately follow the ideal path forward; Fig. d is no progress; Fig. d is sailing steering angle, the USV straight sailing first and then turns at 60 ° steering angle, and finally moves circumferentially along the 60 ° direction.

In the L-F formation, the initial position of the three USVs is set to (0, 0), (0, 10), (0, -10), the leader ships are experimentally designed to keep the speed of the ship at any time, follower ship at any time to adjust the speed of navigation, angular speed to follow the leader ship. While setting the safety distance L = 10, in the navigation to ensure the safety of USV.

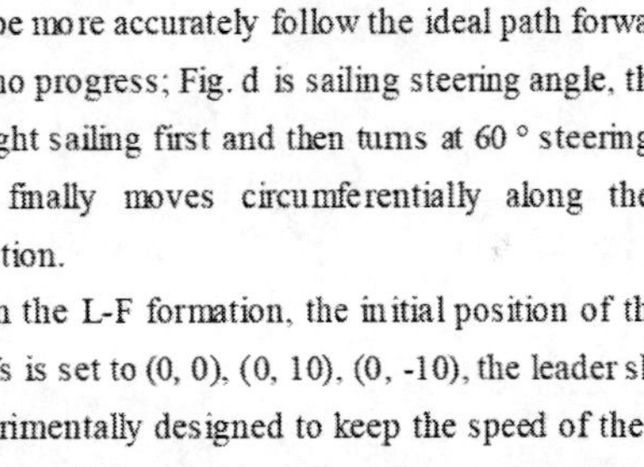

Figure 6. Rule formation effect chart

Fig.9 shows the leader-followed formation in accordance with the ideal path of the experimental results, followership to keep in a safe distance within the sailing process can basically follow the route of the leadership, the error remains within the ideal range.

Set the initial coordinates of the leader ship (0,0), the safety distance L = 5, the coordinate position of the two follower ships (-10,10) 、(-10, -10),when the initial formation is irregular. The leader ship in the storm interference to do a straight line before the movement, followed by circular motion, asked follower ships to follow the leader ship, so that the formation to maintain stability.

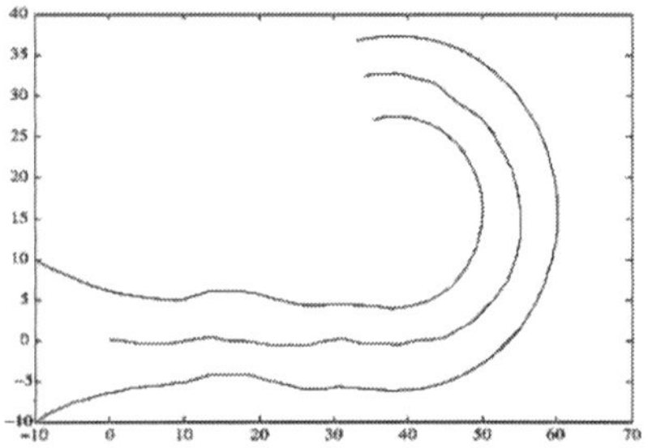

Figure 7. Irregular formation effect picture

Through the experimental results can be seen, the initial formation for the irregular state, after L-F control mode, the formation tends to be stable, the control effect is ideal.

6 Conclusion

The USV and UAV are coupled with each other to form a cooperative fleet for maritime cruise and rescue. Compared with the traditional formation mode, there is an unparalleled advantage in terms of supervision scope, search and rescue efficiency and real-time information.

The L-F model is used to control the USV and UAV cooperative formation, and the information fusion technology is applied to the cooperative information communication. The feasibility of the principle is verified by establishing the mathematical function.

Explores the various search and rescue modes in the process of maritime search and rescue, and emphasizes the role of UAV in maritime cruise.

The feasibility of the cooperative formation control is verified by the simulation test of the control platform and the shell. However, due to conditions, the UAV experimental link is relatively weak, should be further strengthened.

References

1. Kristina Dundović, Nilda Lazzo-Kurtin , Karlo Colić ,"The Role And The Importance Of The Maritime Rescue Coordination Centre", Journal of the Polytechnic of Rijeka (2015)
2. Ma Minglu ,"Research on the Key Problems of Maritime Cruise Search and Rescue Integration", China Water Transport(2016)
3. Liu Li,Gu Qun,Peng Weiming,Deng Xiaoming, "Research on the System Framework of Maritime Stereoscopic Supervision System Based on Unmanned Aerial Vehicle", China Water Transport,ISSN：1006-7973 (2014)
4. Wang Wei, "Research on Remote Control System of Unmanned Monitoring Ship", M.S. thesis, Zhe Jiang University (2014)
5. Jiang Feng,Xiao Jianmei,Wang Xihuai,"Ship Formation Based on Generalized Predictive Control",Computer and Modernization,ISSN ：1006-2475(2014)
6. H.Niu.H.Oh, A. Tsourdos,"Development of Collision Avoidance Algorithms for the C-Enduro USV", IFAC Proceedings Volumes(2014)
7. Joseba Menoyo Larrazabal, Matilde Santos Peñas "Intelligent rudder control of an unmanned surface vessel",Expert Systems With Applications(2016)
8. Cao Shijie,Zeng Fanming,Chen Yutao,"Method of Synchronous Control of Unmanned Surface Boat", Chinese Journal of Ship Research,ISSN：1673-3185 (2015)
9. Gao Fei,"Design and Implementation of Software System for Unmanned Surface Boat Based on",M.S. thesis,Ocean University of China (2015)
10. Yang Wangfeng,Yang Songlin, "Multi - unmanned cohort synergy strategy and search and rescue application" ,The 18th China International Boat & Technology & Equipment Exhibition & High Performance Boat Symposium(2013)

11. Chen yi, "A Preliminary Study on the Cooperative Strategy Based on Optimization and Its Realization",M.S. thesis,Jiangsu University Of Science And Technology (2016)
12. Keller, John, "Northrop Grumman to provide BAMS maritime surveillance UAV with ability to sense and avoid other aircraft", Military & Aerospace Electronics (2011)
13. Lin Jianhui, "Photoelectric tracking and monitoring, satellite communications, maritime search lights and other equipment in the maritime cruise search and rescue combination of applications", Ship Communication and Navigation Proceedings. TN915.85(2009)
14. Anonymous, "RESPEKT: The German Maritime Search and Rescue Service at 150" , Ausmarine (2015)
15. Wei Lei,Li Yunbin, "Research on the Optimization of Maritime Cruise Path of Unmanned Aerial Vehicle",Tian Jin navigation(2015)
16. Shen DongBin, Sun ZhenDong, Sun WeiJie, Leader-follower formation without leader's velocity information ,Science Chine(Information Sciences, 2014)

Speed-consistency Based Safety Evaluation for Freeway Diverging Segments

Ting Zhang[1], Qiaojun Xiang[1], Shen Li[1], Rao Yao[1] and Yifan Guo[1]

[1]*Traffic safety engineering center, School of transportation, Southeast University, 210096, Si Pai Lou#2, Nanjing, China*

Abstract. Diverging segments are potential bottlenecks because of frequent lane change and deceleration maneuvers. Speed study has been proven to be an effective way to understand the safety performance of diverging segments. This paper addresses the safety level of diverging segments in terms of speed consistency. Speed data is extracted by Tracker based on aerial video imager collected at the diverging segment of MaQun interchange of G42 in Nanjing, China. The operating speed of vehicles is assumed to follow the Linear Regression Model with two explanatory variables—initial velocity and running distance, and the estimated operating speed is amended under actual traffic conditions. Furthermore, average speed, speed difference and the coefficient of variation are used as safety indicators to establish the speed evaluation criteria. Last but not the least, based on vague set of material-units, the safety level of diverging segments in terms of the velocity arrives at level III, which means that consideration must be given about the safety of diverging segments. The proposed methodology can be used to estimate the safety of diverging segments.

1 Introduction

Freeways play a vital role in serving mega-cities, which could significantly raise the speed for mid and long distance traffic. Nevertheless, the efficiency and safety of freeways has been dramatically constrained by bottlenecks. Diverging segments may easily become bottlenecks because of frequent lane change and deceleration maneuvers. Therefore, understanding the efficiency and safety of diverging segments are important and necessary.

Freeway diverging segments occur primarily at off-ramp junctions with the freeway mainline [1]. Normally, there are two movements in diverging segments, namely mainline-to-mainline and mainline-to-ramp, which can affect the operations of each other. Influence area of diverging segments is defined as the deceleration lane(s) and Lanes 1 and Lanes 2 of the freeway for a distance of 450m upstream of the diverge point [1]. Figure 1 illustrates the influence area of diverging segments.

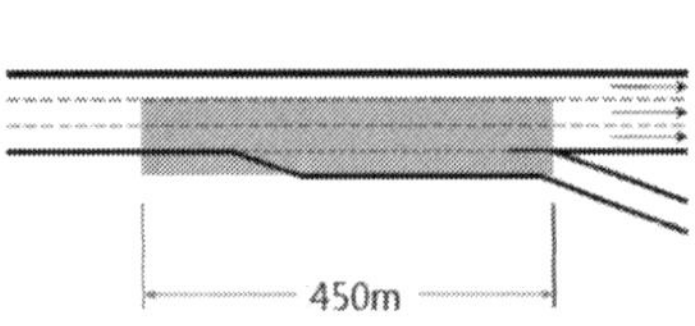

Figure 1. Influence area of diverging segment

Because of its importance and efficiency, diverging segments have already gained considerable attention from researchers. The capacity studies of diverging segments have been a focus. Researchers have analysed the capacity of diverging segments using empirical and simulation methods [2-3]. Other studies have also been conducted on level of service (LOS), traffic delay and operational characteristic [4-6]. In addition to the studies above, there were also researches on diverging segments' safety. Some researchers explored the contributing factors to crashes at freeway diverging locations [7-8], they discovered that AADT, lane-type configuration and speed are always tend to increase the severity of crashes at freeway diverging areas. The operating speed of diverging segments were also been studied [9-10], and linear regression model were adopted to predict the speed of diverging area. Studies using speed have been proven to be an effective way to understand the safety performance of road. Some researchers identified the relationship between crashes and speed [11-13]. Their researches demonstrated that the larger the variance of speed, the higher the crash rates. As for the speed safety of diverging segments, the linear regression model of speed was established and the safety level of speed was arrived at based on the 85th percentile individual speed difference in diverging segments [14].

From what have been mentioned above, it is not hard to conclude that there were numerous studies on speed and traffic safety at diverging segments respectively. However, with respect to speed safety for diverging segments, there were not enough studies. Hence, this study conduct speed safety analysis for diverging segments. The initial speed and the travel distance are considered in the regression model. Meanwhile, volume and weather conditions are used to revise the speed model developed. Exploring the connection between speed and safety would be helpful in finding hazardous diverging segments.

The sections are organized as follows. Section 2 first introduces the research methodology which is used in building the model and assessing the safety level of diverging segments. Section 3 presents the statistics analysis of speed collected. Section 4 demonstrates the regression model of speed at diverging segments. The speed safety evaluation result is showed based on material-unit in section 5. Finally, Section 6 summarizes the findings, conclusions and limitations of the paper.

2 Methodology

This study built a multivariable linear regression model to estimate the speed of vehicles at freeway diverging segments according to the statistics analysis of speed data collected. Linear regression allows multi-variables to be properly calibrated, and outputs are relatively easy to interpret. What is more important is that the assumptions of linear regression models are often suitably satisfied in practical applications [15].

In this study, vehicles are classified into heavy (wheel base is larger than 7m or power density is small or equal to 15kw/t) and small ones (the others). The operating speed model are established for heavy and small vehicles individually, the operating speed of vehicles at diverging segments is then arrived at according to the ratio of the speed for heavy vehicles and the small vehicles estimated.

The research assumes that the operating speed are jointly influenced by initial speed, travel distance, volume and weather factors. To be more specific, initial speed v_0 and travel distance L are put in the regression model, the others are used to revise the estimated model. Speed can be expressed as (1):

$$v=\beta_0+\beta_1 v_0+\beta_2 L+\varepsilon \quad (1)$$

where ε follow a normal distribution with mean and standard deviation equal to 0 and σ, respectively. β_0 β_1 and β_2 are estimated constants. Linearity, homoscedastic disturbances and normally distributed disturbances should be checked to make sure that linear regression is appropriate.

In our study, speed v_0, speed difference d and coefficient of variation for speed cv are calculated according to the speed regression model built, which are assumed to be the indicators related to speed safety of diverging segments. Vague sets [16] surpass fuzzy sets when deal with uncertain information [17]. Thus vague sets are applied in this study to estimate the safety level of speed at diverging segments. Compared with traditional multi-objective decision methods, vague sets can effectively deal with uncertainty conditions. The comprehensive material-element can be calculated as (2):

$$v_{ij}=[t_{ij},1-f_{ij}] \quad (2)$$

where v_{ij} is the vague value of object i with indicator j, t_{ij} represents for the true membership function and f_{ij} denotes the false membership function. In our study, all the indicators are supposed to be as smaller as they can, thus t_{ij} and f_{ij} are calculated as (3) and (4), respectively.

$$t_{ij}=(x_j^{\max}-x_{ij})/(x_j^{\max}-x_j^{\min}) \quad (3)$$

$$f_{ij}=(x_{ij}-x_j^{mim})/(x_j^{\max}-x_j^{\min}) \quad (4)$$

where x_{ij} denotes the value of object i with indicator j, $x_j^{\max}$ represents for the maximum value of indicator j of all objects, and $x_j^{\min}$ is the minimum value of indicator j of all objects. The essence of this process is normalization.

The weights of indicators are calculated by means of AHP. And similarities between the matter-element under test and the indicator of standard are computed as (5):

$$M_z(x,y_j)=1-\frac{|t_x-f_x-t_{yj}+f_{yj}|}{8}-\frac{|t_x+f_x-t_{yj}-f_{yj}|}{4}+\frac{|t_x-t_{yj}|+|f_x-f_{yj}|}{8} \quad (5)$$

where x represents for the indicator value for the matter-element under test, y_j is the indicator value for different levels. The comprehensive similarity is computed as (6):

$$d_j=\sum_{i=1}^{n} w_j M_{ij}(x_{in},y_{ij}) \quad (6)$$

where d_j is used to illustrates the similarity between the indicator under test and the standard value, w_j is the weight for indicator j, $M_{ij}(x_{in},y_{ij})$ denotes the similarity between the matter-element under test and the indicator of standard. It is obvious that a larger similarity value indicates the safety level of speed.

3 The statistic characteristics

The research study the diverging segment of MaQun interchange of G42 in Nanjing, China. The speed data was extracted by Tracker based on aerial video imager collected. Three cross-sections (the diverge point C, the midpoint B and terminal point A) of datasets were screened from data extracted at the influence area of diverging segment. In this research, the segment is set as straight line for convenience, and the speeds are analysed and modelled for heavy vehicle and small vehicles(the others), respectively.

Descriptive statistics are useful way to understand the primary characteristics of speed. Take heavy vehicles as example, Table 1 provides speed information about the basic statistics for heavy vehicles. The statistics indicate that the speed of each cross section follows normal distribution. There seems little difference among the statistics for section A, B and C, which means central tendency of heavy vehicles' speed.

Table 1. The summary statistics of speed for heavy vehicles

Cross-section	A	B	C
Mean	19.172465	18.209627	17.412200
Std. deviation	2.7175877	2.7816214	2.9391556
skewness	-0.232	0.041	0.289
kurtosis	-0.494	0.702	-0.423
minimum	12.6778	12.3483	11.9281
maximum	24.0144	23.8396	24.4921
15th percentile	16.156235	15.127065	14.483093
50th percentile	19.217385	17.852405	16.986620
85th percentile	22.350810	21.313448	21.089265

The curves of speed with heavy vehicles and small vehicles are illustrated by Figure 2. The characters of the speed curve are as follows. First, it is obvious that the vehicle will decelerate when reaching the diverge point. Second, there appears dissimilarity in terms of speed range with heavy and small vehicles. Furthermore, it can be easily seen from Figure 2 that the speeds present a decreasing trend for all the vehicles. What is more important is that deceleration of different vehicles demonstrate different and that the deceleration distance are longer for the heavy vehicles in comparison to small vehicles.

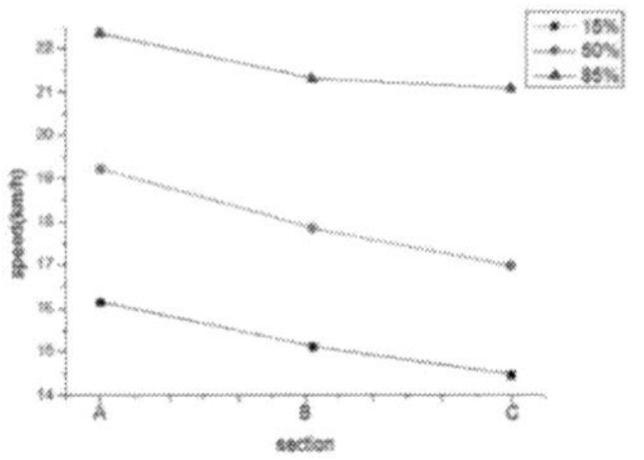

a) Heavy vehicles

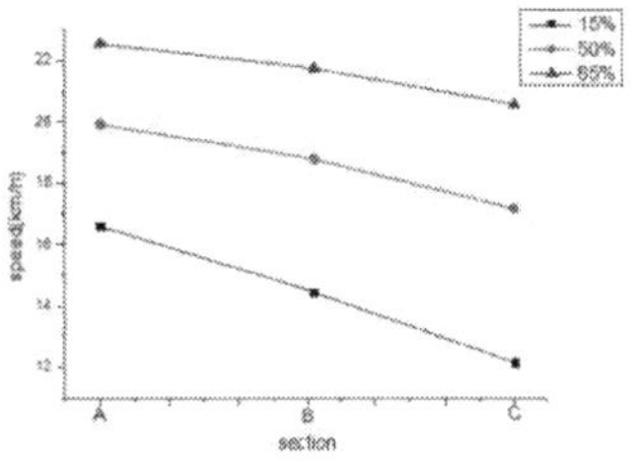

b) Small vehicles

Figure 2. The speed curve of vehicles

4 The regression model

4.1 Model estimation

A multivariable linear regression model is built to estimate the speed for heavy and small vehicles at diverging segment. Table 2 shows the parameter estimates of the regression estimated on the collected data for heavy vehicles.

It can be easily seen from Table 2 that this model contains a constant and two explanatory variables-the initial speed and travel distance. The signs of the estimated parameters are in line with expectation. Travel distance is with a negative coefficient, indicating that long distance might decreases the speed of heavy vehicles, which is easy to be understood. Meanwhile, long distance (*L*) also indicates of worse travel conditions, and then results in low speed of heavy vehicles. Initial speed v_0 is with positive coefficient which determines the operating speed. The two independent variables are statistically significant with the value of $|t| \geq 1.96$ under level of significance 5%. The value of R_{adj}^2 is 0.7325, which indicates that 73.25 percent of total variance can be explained by independent variables.

Table 2. The linear regression parameters of heavy vehicles

Variables	Coefficients	Std.	t-value	P>t
L	-0.0213	0.0016	-13.27	0
v_0	0.8233	0.0097	84.69	0
Cons	4.4756	0.1782	25.11	0

*Variable significant at 95% interval.

Linearity, homoscedastic disturbances and normally distributed disturbances are checked to make sure that linear regression is appropriate. Thus the regression model of speed can be expressed as (7):

$$v_h = 4.4756 + 0.8233 v_0 - 0.0213 L \qquad R_{adj}^2 = 0.7325 \quad (7)$$

where v_h denotes the operating speed of heavy vehicles, v_0 and L are of the same meaning defined before.

The regression models of small vehicles can be expressed as (8) by the same method.

$$v_s = 3.2417 + 0.8474 v_0 - 0.0268 L \qquad R_{adj}^2 = 0.6806 \quad (8)$$

where v_s represents for the operating speed for small vehicles. The explanatory variables are of the same meaning as noted before.

4.2 Model calibration

Volume and weather factors are found to have significant impact on the actual operating speed of vehicles. Volume is with negative impact, indicating that high volume might decrease the speed of vehicles. Wet road surface condition reduces the speed of vehicles. It leads to smaller friction and results in longer braking distances. Meanwhile, vehicles are more likely to lose control under wet road condition, thus the driver tends to decelerate for safety consideration.

The speed model can be rewritten as (9) with consideration of volume and weather conditions:

$$v = v_0 + v_v + v_0 f \qquad (9)$$

where v represents for the actual operating speed of vehicles, and v_0 is the estimated speed by regression model, v_v denotes the speed reduction because of high volume and f denotes the coefficients of speed reduction due to road surface conditions.

The speed reductions for freeways with volume and weather condition are shown as Table 3 to Table 4, respectively.

Table 3. The speed reduction of different LOS

LOS	v/c	Speed reduction	
		Heavy vehicles	Small vehicles
A	<0.25	0	0
B	<0.65	3	4
C	<0.75	5	8
D	1	-	-

Table 4. The speed reduction of weather condition

Road surface condition	Dry	Wet
f	0	0.15

5 Safety evaluation

5.1 Evaluation criteria

Transportation is a complex, diverse and stochastic system involving people, vehicles, roads and traffic management. And operating speed of vehicles is influenced by all these factors. Research shows that average speed (v), speed difference (d) and the coefficient of variation (cv) are assumed to have straightforward influence on speed safety of vehicles [18]. Speed difference is the difference between the average speed and the design speed. The coefficient of variation is the ratio of the standard deviation and the average speed.

In this paper, three indicators are used to evaluate the safety level for vehicles according to Specifications for Highway safety audit [19] and the speed safety criteria of highway [20], the criteria of vehicles developed with the consideration of traffic circumstances, which is shown as Table 5.

Table 5. The criteria for speed safety

Indicators	Level I	Level II	Level III
v	v £ 20	20 < v ? 40	v >40
d	d £ 10	10 < d ? 16	d >16
cv	cv £ 10	10 < cv ? 15	cv >15

5.2 Safety evaluation

In our study, vague sets are applied to estimate the safety level of speed for vehicles. There are several steps to follow to arrive at the safety level of vehicles speed. First, the comprehensive material-element can be calculated by (2). Second, the weights of indicators are calculated by means of AHP. Furthermore, similarities between the matter-element under test and the indicator of standard are computed by (5). Finally, the comprehensive similarity is computed by (6) and the safety level of speed is confirmed.

5.2.1 Vague sets

The research studies the speed safety level of the diverging segment of MaQun interchange of G42 in Nanjing, China. The diverging segment is of three lanes and one auxiliary lane with mainline design speed 60 km/h.

According to the regression model and speed reductions, the average operating speed is calculated at 20.0175km/h (the ration of heavy vehicles and small ones is calculated at 17.1372%), the speed difference is 39.9825 km/h and the coefficient of variation is 14.1745%. The comprehensive material-units and the vague sets are calculated by (2), as is shown in Table 6 and Table 7, respectively.

Table 6. The comprehensive material-units

Indicators	Level I	Level II	Level III	Indicators under test
v	(0,20)	(20,40)	(40,60)	20.0175
d	(0,10)	(10,20)	(20,60)	39.9825
cv	(0,10)	(10,15)	(15,60)	14.1745

Table 7. The comprehensive vague sets of material-units

Indicators	Level I	Level II	Level III	Indicators Under Test
v	(0,0.333)	(0.333,0.667)	(0.667, 1)	(0.3336, 0.3336)
d	(0,0.167)	(0.167, 0.333)	(0.333,1)	(0.6664, 0.6664)
cv	(0,0.167)	(0.167, 0.25)	(0.25, 1)	(0.2362, 0.2362)

5.2.2 The weights of indicators

The weights of indicators are calculated by means of AHP, which is presented as Table 8.

Table 8. The indicators weights

Indicators	v	d	cv
weights	0.139	0.332	0.528

5.2.3 The similarities

The similarities between the matter-element under test and the indicator of standard are computed by (5). The results are presented as Table 9. It is obvious that a larger similarity value indicates the safety level of speed. The safety levels of average speed, speed difference and the coefficient of variation are Level II, Level III and Level I, respectively.

Table 9. The indicators similarity

Indicators	M_1	M_2	M_3
v	0.874813	0.874906	0.708438

d	0.687688	0.771063	0.833219
cv	0.902754	0.771063	0.833219

5.2.4 The safety level

Table 10 provides information about the comprehensive similarity computed by (6). The similarity is 0.8150, 0.7847 and 0.8266 for Level I, Level II and Level III, respectively. Which indicates the speed safety level of diverging segment is level III. Hence, the result found in this study is that the safety level of diverging segment in terms of speed is not the optimal. And measures like speed limit should be taken to manage the speed of diverging segments.

Table 10. The comprehensive similarity for different speed safety levels

Level	Level I	Level II	Level III
Comprehensive Similarity	0.8150	0.7847	0.8266

6 Conclusion

Freeways diverging segments may easily become bottlenecks for traffic safety. One effective way to improve the safety of diverging segments is to study the speed safety with the objective of identifying the hazardous diverging segments and reducing the risk of diverging segments by traffic control and management.

In order to provide effective predictors for diverging segments' speed, this research study the diverging segment of MaQun interchange of G42 in Nanjing, China. The speed data was extracted by Tracker based on aerial video imager collected. The regression linear model shows that initial speed and travel distance are important factors with negative and positive coefficient, respectively. Besides the two factors, volume and weather factors are used to calibrate the operating speed. Based on the proposed model, the safety level of diverging segments can be identified using vague sets. The results show that the safety level of diverging segments in terms of speed is not the optimal, and measures like speed limit should be taken to manage the speed safety of diverging segments.

There are limitations of this study. The regression model considered only two variables regardless of significant GOF. The studied segments are straight lines, other geometric designs do not vary too much. More geometric designs can be explored in the future.

Acknowledgements

The authors thank the China's National Science and Technology Support Programme Project (Project No.2014BAG01B01) for funding this research.

References

1. *Highway Capacity Manual* (TRB, Washington, D.C. , 2010)
2. Bertini, R. L., Hansen, S., and Bogenberger, K. *Empirical analysis of traffic sensor data surrounding a bottleneck on a german autobahn*, TRR. 97(2005)
3. Leong, L.V., Mahdi, M. B., and Chin, K. K. *Microscopic simulation on the design and operational performance of diverging diamond interchange*. TRP, **6**,198 (2015)
4. Wu, N., and Lemke, K. *A new model for level of service of freeway merge, diverge, and weaving segments*. Procedia - Social and Behavioral Sciences, **16**, 151 (2011)
5. Newell, G. F. *Delays caused by a queue at a freeway exit ramp*, **33**, 337 (1999)
6. R. Qiu. *Research on traffic stream characteristics of off-ramp junction in the interchange*. (Doctoral dissertation, Southeast University,2006)
7. Chen, H., Liu, P., Lu, J. J., and Behzadi, B. *Evaluating the safety impacts of the number and arrangement of lanes on freeway exit ramps*. AAP. J., **41**,543(2009)
8. Mergia, W. Y., Eustace, D., Chimba, D., and Qumsiyeh, M. *Exploring factors contributing to injury severity at freeway merging and diverging locations in ohio*. AAP. J., **55**,202, (2012)
9. Z. Zhang, X. Hao, and W. Wu. *Research on the running speed prediction model of interchange ramp*. Procedia - Social and Behavioral Sciences, **138**,340 (2014)
10. W. Wu. *Research on the operating speed prediction model of interchange ramp* (Doctoral dissertation, Beijing Technology University, 2009)
11. Solomon, D. *Accidents on main rural highways related to speed*. (1964)
12. Liu, G., and Popoff, A. *Provincial-wide travel speed and traffic safety study in saskatchewan*. TRR, **1595**, 8 (1997)
13. Baruya, B. *Speed-accident relationships on european roads*. Trl Annual Research Review. (1999)
14. H. Xu. *Research on level of service of safety for freeway diverging area based on speed consistency*. (Doctoral dissertation, Harbin Institute of Technology, 2011)
15. *Statistical and Econometric Methods for Transportation Data Analysis*. (A CRC Press Company, Washington, D.C. 2003)
16. W.Gau. *Vague sets*. IEEE Transactions on Systems Man and Cybernetics, **23**,610(1993)
17. W. Yan. *The application of entropy multipurpose decision method at evaluation ofbiding*. **36**,98(2004)
18. B. Luo. *Safety appraisal and improvement measures of two-lane roads based on operating speed*. (Doctoral dissertation, Beijing, 2008)
19. *Specifications for Highway safety audit*. (China communication press, 2015)
20. Y. Yang. *Study on the speed of public transport in mountainous cities*. (Doctoral dissertation, Chongqing Jiaotong university, 2014)

Analysis of drivers' errors at roundabouts in Morocco

Fatiha Moutchou[1], and Abdelghani Cherkaoui[1]

[1]*Research team "Engineering, management and optimization of Systems" (EMOSYS), Industrial Engineering Department, Mohammadia School of Engineering, Rabat, Morocco, BP 765 Rabat Agdal, Code postal 10100, Morocco*

Abstract. Roundabouts have increased in a significant way in recent years in Moroccan cities as they have been seen as a solution to congestion problems and traffic safety issues. However, roundabouts are relatively new to Moroccan drivers and the rules of priority to which the users are accustomed are reversed in roundabouts. This can lead to confusion and misleading. The overall objective of this study is to identify roundabout features that appear to play a role in the difficulties that drivers encounter when crossing roundabouts. This paper will present a comprehensive understanding of what is required of drivers when crossing roundabouts. We then predict failures that can occur at roundabout using our driver error taxonomy. A case study of two roundabouts in Morocco is analyzed as illustration. The paper concludes with implications of the findings to future design.

1 Introduction

Today, Moroccan cities are experiencing, as elsewhere in the world, congestion problems characterized by increased travel time, and even bottlenecks. This congestion is therefore a result of high volume of traffic, and little available space.

In the field of urban planning, the concept of "roundabout" thus appears as a solution to these problems and is now gaining wide use across Morocco.

Indeed numerous studies [1-4] have found that roundabouts are quite advantageous regarding traffic safety compared to conventional intersections. In addition extensive studies have been carried out on benefits of roundabouts in the following areas: traffic calming effects due to the creation of a low speed environment and the elimination of conflict points, reduced delay and concomitant emissions [5-7], capacity and quality of traffic flow [8].

Although roundabouts provide many benefits, there are some disadvantages as well. The primary drawback is that the rules of priority to which the users are accustomed are reversed in roundabouts. The priority is giving to circulating traffic rather than entering traffic. This means that the driver entering the roundabout must give priority to driver that comes on his left.

However, the term "roundabout" should not be confused with other traffic circles such as rotaries and circular roadways. Roundabouts are circular intersections with specific design and traffic control features. These features include yield control of all entering traffic, channelized approaches utilizing splitter island to divide entering and exiting traffic on each leg, and appropriate curvature to ensure that travel speeds on the circulatory roadway are low [9].

In Morocco, the Highway Code defines the roundabout as "an intersection where all traffic merges into and emerges from a one way road around a central island of circular shape, the circulation on this roadway is in a counterclockwise direction". However, it should be mentioned that the Moroccan Highway Code does not explicitly states who has priority between users in roundabouts but stipulates that roads signs must be used to inform road users of obligations, restrictions or prohibitions with which they must comply.

It is therefore, through signage that roundabouts are regulated in Morocco. Since drivers are used to traffic lights telling them what to do, roundabout as a new concept, can be confusing and understanding of its operating rules can take time.

Drawing on this, the aim of our study is to gain a better understanding of drivers' behaviours at roundabouts. In order to properly address this question, we analyze circumstances leading to failures during roundabout negotiation and the cognitive processes that underlie them.

In this article, we first present a detailed analysis of tasks performed by drivers when crossing a roundabout. We also approach the human errors in driving context. Then, we propose our driver error taxonomy while crossing roundabouts. In situ observations are used to collect data relating to the roundabout crossing task. To conclude, we use our driver error taxonomy to analyze and interpret the drivers' errors and to propose measures that can help drivers and therefore enhance road safety.

2 Prescribed task of crossing roundabouts

To drive in a roundabout properly and safely, the driver has to take into account the layout of the site, its geometry, and interactions with other users. From a cognitive analysis perspective the interaction between the car drivers 'capabilities and the demands of the driving task determines the outcome in terms of a more or less safe driving behaviour. [10]

This can be called situation awareness, formally defined by Endsley [11] as the "perception of the elements within a volume of time and space (Level 1), the comprehension of their meaning (Level 2), and the projection of their status in the near future (Level 3)".

Roundabout can be visually complex requiring that drivers scan several different areas and keep track of several different elements to get the information need to safely pass [12]. So, while crossing roundabouts, drivers have to perform many subtasks in an environment with many rules and complex interactions. Indeed, beyond the formal rules of using the road as defined by the highway codes, the crossing task mobilizes cognitive skills.

In this sense, the approach of our study is descriptive and our analysis underlies the assumption that to properly negotiate a roundabout, drivers need competences more diversified than only the respect of the Highway Code rules.

To determine a formal framework for modeling the roundabout crossing task, the focus is placed on cognitive analysis of the crossing task from Endsley's work [11] combined with the rules of the road at roundabouts.

first as "component", who is sometimes subject to functional failures due to his own adaptations limits, second, as "regulator" who regulates his activity and adjusts it to overcome the problems arising from the different components of the system including him.

We relied also on the work of Nachtergaele and Fleury[13] to define the prescribed tasks in crossing roundabouts (For more details see our previous paper) [14].

Our approach aims at analyzing the prescribed task (what a driver is expected to do to negotiate roundabouts safely) and the effective task (What a driver effectively does) [15], to explore the possible discrepancies between these tasks and their origins.

The technique of Hierarchical Task Analysis (HTA) (See [16]) is carried out to collect information on the tasks and to determine key functions performed by drivers as they approach and navigate through roundabout.

The first step was to divide the roundabout crossing task into subtasks. Within each segment (approach, entry, within the roundabout, exit), we have identified subtasks that drivers must consider to safely cross the roundabout. Each task accomplished by the driver involves information that needed to be obtained , clues that needed to be understood, decisions that needed to be made, and actions that needed to be taken.

The Figure1 below, shows the subtasks that divers must perform when negotiating roundabouts.

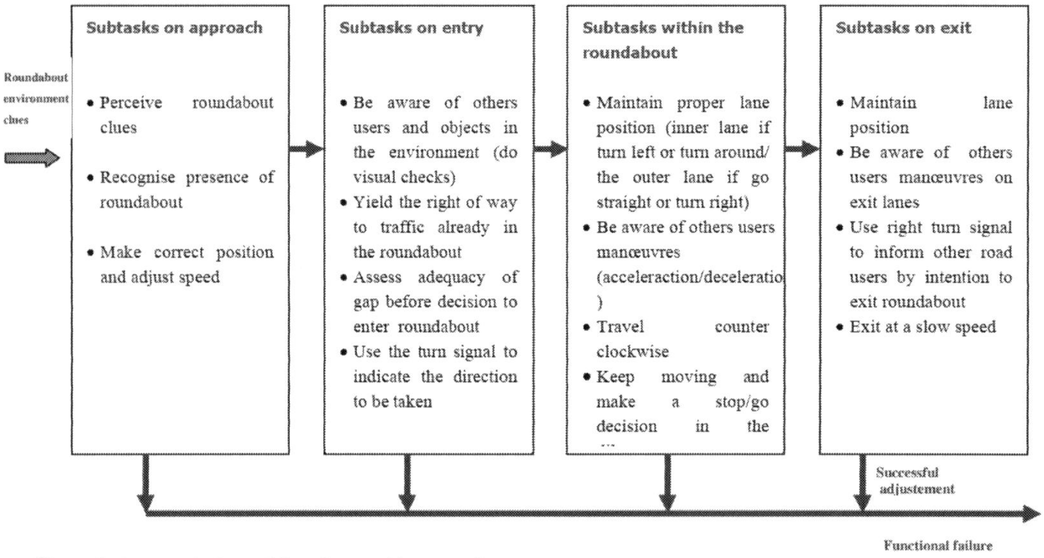

Figure 1: Framework for modeling the roundabout crossing task

3 Human functional failures in driving context

In driving context, as stated by numerous authors [17-19] it is important to consider the driver in a double status:

Indeed, the road environment and temporary factors such as the presence of others road users and weather conditions can affect driver behaviour, also factors involving the driver's state (fatigue, stress, illness...) will affect driver's perception of the road and the behaviour

DOI:10.1051/matecconf/201712404008

that he thinks is appropriate for the road. In order to avoid any ambiguity, some consensus about terminology seems to be useful.

For the purposes of this paper, "human error" refers to "human dysfunctions" and will be reported with the label of "human functional failures" described by Van Eslande and al [17] such as following: "the impairment of one (at least) of the functions that usually allows the operator to adapt to the difficulties he meets when fulfilling his task".

According to Leplat and Hoc [15], "error" reveals the discrepancies between the prescribed task and the real task, it must be considered as the symptom of the difficulties encountered by the operator in carrying out its activities within a given context.

The explanatory factors of errors correspond to the factors coming from the context, the person, and so on that have favored the emergence of a functional failure.

In the scientific literature, different taxonomies of human errors have been proposed in several areas to serve as descriptive or theoretical tools.

a comprehensive set of errors that may lead to an incident (such as a near miss) or accident (crash) at roundabouts.

Our study was based on the taxonomy of contributing factors of errors developed by Wierwille and al [20]. One of the innovative features of this taxonomy is that it focuses on the causes of driver errors. It provides an explanation of not only *what* happened, but also *why* it happened. The fundamental concept underlying the taxonomy is that failures in driver performance are a result of a combination of contributing factors. Four major categories of contributing factors have been identified: 1-Inadequate knowledge, training, and skill, 2-Impairment, 3-Willful inappropriate behavior, and 4-Infrastructure and environment. This taxonomy is presented in figure 2 below:

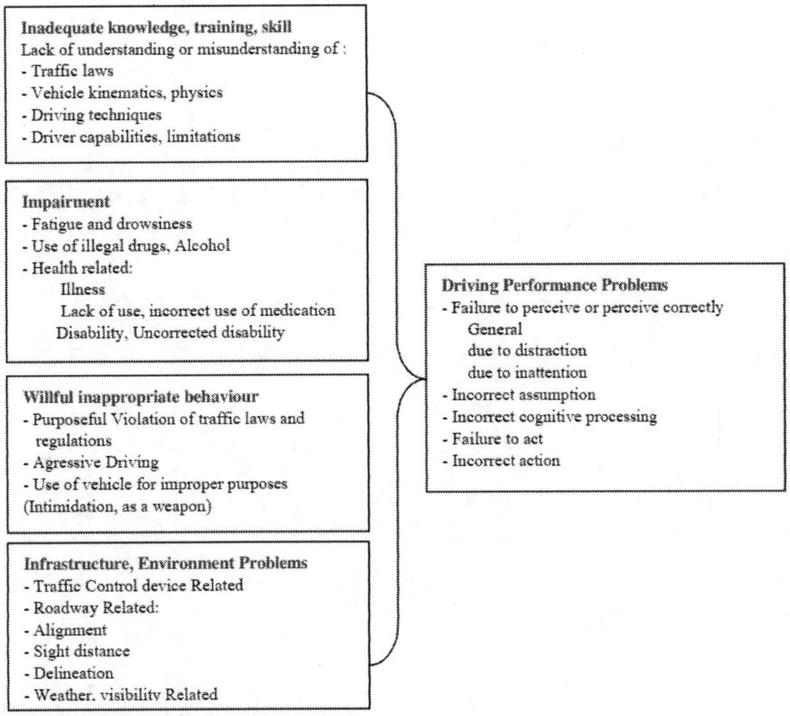

Figure 2: Contributing factors taxonomy (source [20])

4 Proposal of driver error taxonomy when crossing roundabouts

The taxonomy of errors can be useful to deeper understanding of situations that are prone to errors and to find corrections to these situations. In our driver error taxonomy, emphasis has been placed on the multiplicity of contributing factors. The subtasks chain presented in "Figure. 1" was used in our taxonomy in order to identify

The errors analysis was carried out using a standard human HAZOP (Hazard and Operability) technique to provide a description of potential errors that can arise at each subtasks.

To each of these subtasks we have associated the cognitive functions necessary to achieve it, and which are more or less preponderant in the subtask concerned. Once the failure has been identified, the sources and types of errors have been described based on the literature review of research that focused on sources of errors [13, 20, 21].

The table 1 below presents our driver error taxonomy when crossing roundabout and their explanatory elements.

Table 1. Our driver error taxonomy at roundabouts

N	task	Domain	Internal error mode	Deviation	Errors Type	Endogenous Factors	Exogenous Factors
1	**Subtasks on approach**						
1.1	Perceive roundabout clues	Perception	Miss see No detection (visual) See too late	Failing to perceive roundabout	Perception error	Inattention Lack of Vigilance Distraction Overload Impairment due to dysfunctions	Roundabout signs missing, obscured or not clearly visible. Poor weather conditions Poor Visibility of roundabout
1.2	Recognise presence of roundabout	Memory	Forget information Miss recall information Wrong information obtained	Failing to recognise roundabout	Recognition errors	Memory confusion Overload Misinterpretation Knowledge problem (user is unfamiliar with roundabout	Poor weather conditions Inconsistency of signage location and size between roundabouts Poor legibility of roundabout
1.3	Make correct position Adjust speed	Decision Action	Decision omitted Action too late Action too early	Failing to formulate safe stopping strategy Failing to control speed on approach	Decision errors Performance errors	Distraction Time pressure Inadequate model mental Knowledge problem (user is unfamiliar with roundabout) Impairment due to dysfunctions	Angle of approach affects the user's judgements Long sighting distance available Poor weather conditions Crowded and complex traffic flow
2	**Subtasks on entry**						
2.1	Be aware of users and objects in the environment (visual checks)	perception	Miss see No detection (visual)	Failing to stop behind queued vehicle	Perception errors	Expectation Lack of Vigilance Distraction Time pressure Impairment due to dysfunctions	Crowded and complex traffic flow Road layout inciting drivers to travel at an inappropriate high speed
2.2	Yield the right of way to traffic already in the roundabout	Decision Action	Decision omitted Action too late Deliberate violation	Failing to yield the right of way to vehicles in the roundabout	Decision errors Recognition errors	Inadequate model mental Knowledge problem (user is unfamiliar with roundabout) Confusion Distraction Impairment due to dysfunctions	Poor Visibility of roundabout Poor weather conditions No appropriate roundabout signs Crowded traffic which may lead to unsafe crossing
2.3	Assess adequacy of gaps for entering	Decision	Lack of strategy poor strategy late strategy	Accepting an unsafe gap distance Conflict in roundabout	Decision errors	Low perception of risk Misinterpretation Knowledge problem Decision overload	High traffic flow Insufficient capacity of roundabout leading to the formation of queue and increasing risk taking behaviours Poor visibility Poor weather conditions
2.4	Use the turn signal to indicate the direction to be taken	Decision Action	Decision omitted Action too late Wrong action	Signal driving direction omitted Failing to use the turn signal to indicate the direction	Decision errors Performance errors	Distraction Time pressure Knowledge problem overload	Poor legibility of roundabout
3	**Subtasks within the roundabout**						
3.1	Maintain proper lane position	Decision Action	Decision omitted Action too late Wrong action	Failing to detect proper lane when entering	Decisions errors Performance errors	Inadequate model mental Knowledge problem (user is unfamiliar with roundabout) Impairment due to dysfunctions	Insufficient capacity of roundabout High Traffic flow
3.2	Be aware of other users acceleration/deceleration	perception	Miss see No detection (visual)	Failing to avoid conflicts with others users	Perception errors Recognition errors	Inattention Lack of Vigilance Distraction Time pressure Impairment due to dysfunctions	Insufficient capacity of roundabout High Traffic flow Poor Visibility Poor weather conditions
3.3	Travel counter-clockwise	Action	No performed action Deliberate violation	travelling clockwise	Performance errors	Confusion Time pressure Knowledge problem	Insufficient capacity of roundabout leading to the formation of queue Poor legibility of roundabouts
3.4	Keep moving make a stop/go decision in the dilemma zone	Action Decision	Action too late Wrong action decision omitted decision too late	Stop at roundabout and refusal of priority	Performance errors Decision errors	Confusion Distraction Inadequate model mental Knowledge problem (user is unfamiliar with rules of priority) Impairment due to dysfunctions	Poor visibility Poor legibility of roundabouts High speed of non priority users
4	**Subtasks on exit**						
4.1	Maintain lane position	Decision Action	Decision omitted Action too late Deliberate violation	Changing lanes incorrectly when exiting	Decision errors Performance errors	Confusion Time pressure Knowledge problem Decision overload	High traffic flow Insufficient capacity of roundabout
4.2	Use right turn signal to signal intent to exit	Decision Action	Decision omitted Action too late Wrong action	Signal driving direction omitted Failing to use right turn signal to signal intent to exit	Decision errors Performance errors	Distraction Time pressure Knowledge problem (user is unfamiliar with roundabout)	Poor legibility of roundabouts
4.3	Be aware of others users on exit lanes	Perception	Miss see No detection	Failing to avoid conflict with others users	Perception errors Recognition errors	Inattention Lack of Vigilance Distraction Time pressure Impairment due to dysfunctions	Insufficient capacity of roundabout High Traffic flow Poor Visibility Poor weather conditions
4.4	Exit at a slow speed	Action	Action omitted Action too late Wrong action	Failing to control speed on exit	Performance errors	Confusion Time pressure Knowledge problem	Poor legibility of the exit Exit radius too tight

The proposed taxonomy table aims to gain an in-depth understanding of deviations that are likely to occur when crossing roundabout and their nature (errors type). Emphasis is placed on endogenous factors (driver state) and exogenous factors (environmental and contextual conditions) that can contribute to these errors.

One of the most important roles of this taxonomy is to provide a framework that can be used in retrospective analysis (accident or incident) to determine human causes and weakness in design, or in prospective analysis to predict possible errors. It constitutes a starting point for a more accurate and consistent classification of human functional failures at roundabouts.

To validate this taxonomy practically, we have applied it to the assessment of drivers' behaviours at roundabouts in Rabat, Morocco. A further goal was to make recommendations to attenuate drivers' errors, and thus to improve traffic safety.

5 Study Methodology and main findings

5.1 Research methodology

The research reported in this article was undertaken at two different roundabouts in Rabat, Morocco. In term of research methodology, our study was based on in situ observations. The roundabouts studied were observed for different random periods of time (30 minutes to 1 hour). The observations were based on an observation sheet developed to carry out the assessment, and to examine whether it would be possible to identify characteristics of roundabout that coincide with a higher likelihood of unwanted events. For each subtasks involved in crossing roundabout, the observer had to check if all subtasks required were accomplished by the driver and to mark missing, incorrectly, or unrequired one. The observer also counted the frequency of different driving manoeuvres, (e.g. lane-changing, overtaking). These measures served to calculate error rates for different driving tasks when negotiating roundabouts.

The sample established was composed by 184 vehicles observed (21 of taxi drivers and 163 drivers of other vehicles). Two-wheeled vehicles and pedestrians were excluded from our study.

5.2 Main findings

5.2.1 Overall numbers of deviations observed

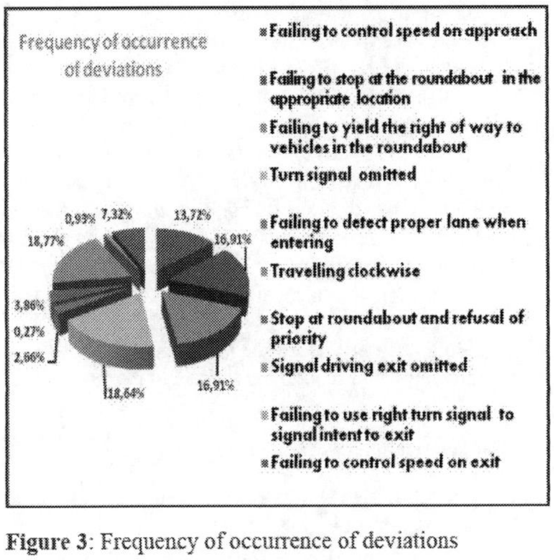

Figure 3: Frequency of occurrence of deviations

The results show that the most observed deviations, concern the omission of the turn signal either at the exit or to indicate the direction with a percentage of occurrence of more than 18.5% each, followed by the non-respect of the priority rule and failure to stop at the roundabout in the appropriate location, with a percentage of almost 17% for each failure. The percentage of vehicles that did not slow down when approaching the roundabout is also important with 13.7%.

These findings make it possible to state that the majority of failures are located in the approach and entry of the roundabout, and account for more than 66% of the observed deviations. These deviations are related primarily to perception and recognition errors, and can be explained using our driver errors taxonomy by endogenous and exogenous factors.

Concerning endogenous factors, these failures can be induced by the lack of knowledge about priority rules, problems of distraction and inattention, lack of knowledge of roundabout features for unfamiliar users, the use of mental model inappropriate, or even the time pressure inducing a bad assessment of gap for entering.

For exogenous factors, they are mainly related to the poor readability and visibility of the roundabouts studied (signs damaged and poorly maintained, appropriate markings were absent), the excessive generosity of their dimensioning and the disturbances related to the density of traffic in the transit roads in which these roundabouts are arranged.

5.2.2 Deviations by types of users

The following graph outlines the distribution of deviations by types of users.

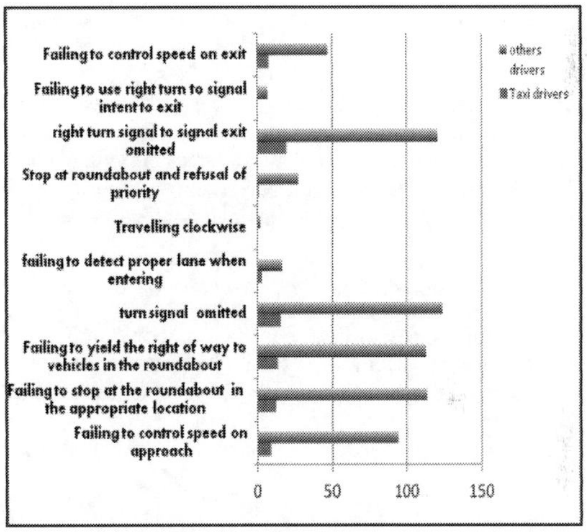

Figure 4: Deviations by types of users

Our analysis of problems encountered by drivers when crossing roundabouts reveals a difference between the two samples (taxi driver and other drivers). This indicates a particularity of errors that concern the two driver types.

It was observed that taxi drivers were more involved in the failures to detect proper lane (38% against 28% for the other vehicles) and the omission of the turn signal at the exit of the roundabout (95% against 74% for other users). The excessive self-confidence that may exist among Taxi drivers could be advanced here as an explanation.

The percentage of errors related to the failure to stop at the roundabout in the appropriate location, to yield the right of way to vehicles in the roundabout, and to indicate to other vehicles the direction to be taken were almost the same for both types of users. Overall, more than 76% of the drivers (cars and taxis) observed did not signal the direction they were going to drive, and more than 66% failed to give way to priority vehicles.

5.2.3 Signage at roundabouts

Signalling is the most problematic aspect of the roundabouts in Moroccan cities. Indeed, the Economic Commission for Africa highlighted in its report 'Road Safety in Morocco [22] that 'the urban environment suffers from multiple dysfunctions. Roundabouts having the same technical characteristics with similar traffic conditions have different signage from one city to another or even within the same city". This statement is consistent with our findings. Indeed some aspects of the current signage at roundabouts visited are not optimal, it was observed that:

- Roads signs are not always conspicuous due to their size, colour, and position;
- Warning sign which indicates the presence of a roundabout ahead wasn't always used at roundabouts visited;
- At some locations a mixture of road signs and other signs for the public could be seen as confusing or conflicting. An example is "advertising sign" posted with a "yield sign" at one of the roundabouts visited;
- It was noted that "yield signs" were worded differently at roundabouts visited. The sign at one roundabout reads "give way", whereas the sign at another roundabout reads "you don't have priority". This discrepancy increases the ambiguity of the instruction.

5.2.4 Suggestions to improve safety

Recommendations can be made to improve traffic safety at roundabouts in Moroccan cities. It should first be pointed out that addressing the urban planning problem in Morocco with regard to road safety requires a comprehensive and holistic approach at several distinct levels:

DRIVERS:

The ignorance of priority rules at roundabouts can affect the performance and traffic safety. Efforts must be done on educating drivers and introducing safety campaigns to raise awareness among road users in and around roundabouts. Also inclusion of more training and assessment on roundabouts in the driving test, both in the practical and theory tests would help to reduce the number of incidents involving new drivers at roundabouts.

- URBAN PLANNING

The safety of roundabouts can be improved through better analysis of users' needs at the design phase of the project.

- ROUNDABOUTS FEATURES

In order to design well functioning roundabouts, we need to take into account drivers' limitations, thus actions may notably consist in:

- Promoting visibility to meet the driver expectations, ensuring that the driver will not be surprised by the roundabout or the reactions of other drivers. It is suggested to increase visibility of warning signs, with reflectors or luminance to alert drivers and provide them with additional time to take the appropriate decision.
- Roundabouts with sufficient capacity to avoid queuing too long or too frequent. This question raises the need to situate any action in the implementation of new roundabouts on the basis of a good knowledge of the evolution of urban mobility.
- Improving readability of roundabouts through clear, uniform, and visible signs.

6 Conclusion

The taxonomy of driver error developed, aims to provide an operational framework to enhance understanding of human functional failures, consistent with cognitive concepts and specifically adapted to the driving task of crossing roundabouts. It could serve as a good basis for analysis of deviant behaviours that are likely to occur at roundabouts not only in Morocco but in other countries worldwide.

The results of our study address that there is a general lack of awareness and understanding on the part of most drivers regarding roundabouts. This is enhanced by inconsistencies in the road signs at roundabouts, which can be confusing and misleading.

More generally it can be emphasized that road scene elements and traffic condition play a role in the difficulties that drivers encounter when crossing these intersections.

The analysis leads to the general conclusion that roundabouts, with all kinds of traffic users, are too complex and extremely difficult to negotiate, hence the need to set up roundabouts which facilitate visibility and predictability for all users. This means that roundabouts must be designed in a consistent and standardised manner.

The research outcomes suggest that road authorities have to consider the roundabout safety impacts when planning and evaluating new roundabouts.

References

1. T. Thaweesak, Modern Roundabouts for Oregon. Technical Report Form DOT F 1700.7, June (1998).
2. G. Schermers, F.Wegman, P. Vliet, R. Horst and J. Boender, Country report. The Nederlands, 4th international symposium on highway geometric design, Valencia, June,(2010).
3. P.Adams, Traffic signals and roundabouts: Are they really safer?. Roads & Transport Research, ARRB, Victoria, December (1995).
4. E. Jorgensen and N.O Jorgensen, Traffic Safety at Danish Roundabouts Constructed after 1985. VTI Konferens, n° 4A, part5. Sweden, (1996).
5. R. Elvik, and T. Mysen., Traffic safety Handbook (Norwegian). Institute of transport economics. Norway, (1997).

6. P. Garder, The Modern Roundabout, the sensible alternative for Maine. Final report, Department of Civil and Environmental Engineering, University of Maine, Orono, (1997).
7. R.A. Retting, G. Luttrel and E. Russell., Public Opinion and Traffic Flow Impacts of Newly Installed Modern Roundabouts in the United States. In TRB 2002 Annual Meeting CD-ROM. CD-ROM. Transportation Research Board, Washing, D.C, (2002).
8. M. Rahman and T. Hicks, A critical look at roundabouts. Compendium of Technical Papers, ITE, (1994).
9. Public works roundabout technical advisory committee county of Los Angeles department of public Work, Roundabout policy and design practices for county of Los Angeles. November (2007).
10. W. Fastenmeier, H. Gstalter, Driving task analysis as a tool in traffic safety research and practice. Safety Science, pp 952-979, (2007).
11. M. Endsley and D. Kaber, Team Situation Awareness for Process Control Safety and Performance. Process Safety Progress, Vol. 17, No. 1, (1998).
12. M., Le Thu Huyen, Risk analysis, drivers' behaviour and traffic safety at intersections in Motocycle-Dominated Traffic flow. Tuprints.ulb.tu-darmastadt, 2009
13. C. Nachtergaêle and D. Fleury, In depth analysis of accidents at urban development. Convention Report INREST / RE-03-909-FR, June (2003). (Report in French)
14. F. Moutchou, A. Cherkaoui, and E. El Koursi, An approach to assessing driver's behaviour at roundabouts. Proceeding of the 3rd international conference on road and rail infrastructures, CETRA, Split, Croatia, April (2014).
15. J. Leplat and J.M. Hoc, Task and activity in psychological analysis of situations. Cahiers de Psychologie Cognitive, 3, 49-63, (1983). (Paper in French)
16. J. Annett, Hierarchical Task Analysis. The handbook of task analysis for human-computer interaction, edited by Diane Diaper and Neville Stanton, September, (2003).
17. P. Van Eslande and K. Fouquet, Analysing human functional failures in road accident. Report Project N° 027763, Trace, (2007).
18. P.Salmon, M. Regan and I. Johnston, Human error and road transport: phase one- literature review, Report n° 256, December (2005).
19. J. Reason, Human error. Cambridge University Press, New York, (1990).
20. W.W. Wierwille, R.J. Hanoswski, J.M.Hankey, C.A.Kieliszewski, S.E.Lee, A.Medina, A.S.Keiser and T.A.Dingus, Identification and evaluation of driver errors: overview and recommendations. US departement of transportation, Federal Highway Administration, Report No FHWA- RD-02-003, (2002).
21. M.Räsänen, and H. Summala, Car drivers' adjustments to cyclists at roundabouts. Transportation Human Factors, 2(1), 1-17, (2000).
22. B. Boulaajoul, Road safety in Morocco, assessment and prospect. Report of the Economic Commission for Africa, July (2009). (Report in French)

Factors Affecting Accidents Risks among Truck Drivers In Egypt

Ahmed Fathalla Elshamly[1], Ragaa Abd El-Hakim[2], and Hafez Abbas Afify [3]

[1]*Tanta University, Egypt*
[2]*Public Works Engineering Department, Faculty of Engineering, Tanta, Egypt*
[3]*Public Works Engineering Department, Faculty of Engineering, Tanta, Egypt*

Abstract. Egypt is ranked among the countries with the highest rates of road accidents. According to the American Chamber of Commerce more than 96% of Egypt's goods are transported by trucks and due to their large volume and excessive weight, the severity and number of truck accident fatalities are much higher than other vehicles in Egypt. The present study aims at identifying truck driver's behavior and its influence on crash involvement. Due to the shortage in recording accident data and the inaccurate road accident audit, data was collected from several governorates in Egypt through questionnaire. Questionnaire forms were filled out through personal interviews with truck drivers. The total number of respondents was 643. The final analysis was made on the 615 questionnaires with complete answers. The data was analyzed and logistic regression was applied to accident related data to examine the contributing factors affecting accident occurrence of truck drivers. Results showed that fatigue in terms of driving hours (continuous and total) and lack of sleep, drug use during driving, and driver obesity are the most influencing factors on the occurrence of truck accidents in Egypt. The findings of this research highlight the important role human factors have on the risk of crash involvement amongst Egypt's truck drivers and the need to improve their work conditions.

1 Introduction

Road safety is a very serious issue in Egypt and the Middle East in general. The crash numbers and road fatalities in Egypt are very high. The victims of road fatalities are mostly young and middle aged people which has a great impact on the emerging economy in Egypt. Therefore, the reduction or prevention of traffic accidents is one of the top priorities of researchers, community and the government in Egypt [1] .Truck accidents, in particular, are one of the most important issues facing Egypt's road safety at present due to the large number of fatalities and death toll [2]. A key aspect of trucks is their large size and weight which represents a great danger to other vehicles and their passengers, and results in the complexity of any accident involving trucks. The aforementioned feature of trucks requires that truck drivers and other drivers as well be cautious about the risk represented by large vehicles. Behavior is a fundamental part of people which affects their perception, attitudes and driving character. This behavior concerned characteristics have a direct impact on one's safety or vulnerability to traffic accidents. Central to the entire results of truck accidents is the tragic loss of other drivers' lives, pedestrians, and the truck drivers themselves. Behavior of truck drivers' effect on the ratio of truck accidents is the main concern of many researchers. Truck accidents have many reasons, in some cases, lack of attention, including general inattention, misdirected attention, falling asleep, and distraction [3].One of the most common sleep disorders is excessive daytime sleepiness (EDS) which affects about 4 to 12% of the population, and causes psychological stress and lack of productivity at work plus an increase in the risk of accidents [4]. Sleeping habits could be a major contributor to truck accidents due to lack of sleep, long shifts, continuous driving hours, long journeys and tight timetables that lead to fatigue of drivers [1]. Obesity is also a risk factor for automotive accidents. Obese truck drivers have a significantly higher risk of causing accidents than non obese truck drivers [5]. The National Transportation Safety Board concluded that more than 30 % of fatal-to-the-driver truck crashes are fatigue related [6]. Overweight imposes abnormal body movements, which can lead to the limited range of movement of major joints, decreased muscle strength and the inability to maintain fixed positions for long periods, and reduces visual observation [7]. Unfortunately, some truckers use illegal materials such as alcohol, cannabis and cocaine [8]. Traditionally, many researchers have subscribed to the belief that drivers take drugs on widely varying estimates. This is known by self reporting questionnaires or through a drug test but most reported was largely the use of anabolic substances [8]. Aggressive driving represents a serious problem in today's society with an increasing number of drivers identifying this issue as alarming[15]. Degrees of aggressive driving behaviors caused by violations of traffic rules result in hazards for other road users [16]. Many accidents are caused by errors and violations resulting from Aggressive driving [17].

2 Objectives

The main purpose of the current study is to identify the factors contributing to accidents among truck drivers in Egypt. It also aims at recommending the necessary solutions for the treatment of these factors.

3 Data collection and study area

Due to the lack of accurate accident data, a Driver Behavior Questionnaire (DBQ) was formulated to collect important information needed to assess and evaluate the relationship between truck drivers' behavior and traffic accident occurrence. Based on the studied literature and keeping in mind the Egyptian habits, 47 different questions were included in the DBQ and arranged in five different sections. The covered sections are demographic characteristics, accident exposure, distraction while driving, aggressive driving, and different driver behavior aspects. The aggressive driving part of the questionnaire inquired about speeding, racing with others, driving on the form of sutures, overconfidence that makes the drivers violate traffic laws, drug use traffic offense, and drug use in general. A total of 643 truck drivers were interviewed, 28 of whom were excluded because the drivers refused to complete the questionnaire. All of the interviewed drivers were male drivers (it is not common for females to work in this profession in Egypt). Questionnaire interviews regarded the place of residence of all drivers in order to cover all different parts of Delta region and cover all the international highways inside Egypt. Questionnaire forms were compiled in three governorates, namely; Kafr El-Sheikh, Gharbia, and Dakahlia. The survey was conducted between April and July 2015.

4 Analysis and discussion

The analysis of the remaining 615 respondents was done using the statistical package for social sciences (SPSS) program. The results of the most important questions of the questionnaire are summarized in table 1.

Table 1. Percentage of drivers by study variables

Demographic Characteristics				
Age (year)	21 - 29	30 - 39	40 - 50	
	22.8%	31.8%	33%	12.5%
Weight (kg)	< 70	70 - 89	90 - 110	> 110
	7.6%	48.6%	37.4%	6.3%
Height (cm)	< 160	160 - 175	176 - 190	> 190
	7.3%	48.8%	40.3%	3.6%
BMI	Normal	Over weight		Obese
	51.1%	27.3%		21.6%
Monthly income (Pound)	< 1000	1000 - 2500	2500 - 4000	> 4000
	11.4%	37.9%	35.9%	14.8%
Driving Experience(Year)	< 5	5 - 9	10 - 15	> 15
	18.4%	39.7%	33.8%	8.1%
Educational Level	Primary	Preparatory	Secondary	University
	9.9%	31.4%	45.4%	13.3%
Total Driving Hours	1 - 5	6 - 10	11 - 15	> 15
	12.5%	35.5%	35.6%	16.6%
Continuous Driving Hours	< 4	4 - 6	7 - 8	> 8
	16.7%	40.7%	24.1%	18.5%
Sleeping Hours	< 6	7 - 8		< 8
	69.6%	27.2%		3.3%
	Variable	**Not at All**	**Sometimes**	**Often**
Distraction Activities	Eat while driving	32.5%	44.2%	23.3%
	Smoking	16.1%	47%	36.9%
	Cell phone	15.9%	55.3%	28.8%
	Radio	23.6%	46%	30.4%
	DVD	45.9%	33.2%	21%
	Talking	21.3%	37.2%	41.5%
Aggressive Driving	Sutures	56.9%	31.1%	12%
	Racing	44.2%	44.6%	11.2%
	Speeding	40.5%	42.9%	16.4%
	Self Overconfidence	54.5%	36.4%	9.1%
	Drug use Traffic Offense	56.3%	34.1%	9.6%
Self Reported Drug Use	Before long Journeys	37.4%	27.6%	14.5%
	20.5%			
Driver Behavior	Sleep while Driving	37.2%	49.6%	13.2%
	Lack of Sleep	30.1%	52%	17.9%
Number of Accidents	1 – 2	2 - 3	4 - 5	< 5
Total Driving Hours	52%	38%	7%	3%
Fatalities Number	1	2	3	< 3
Continuous Driving Hours	51%	32%	13%	5%

DOI:10.1051/matecconf/201712404009

Strong evidence of aggressive driving behavior was found. 12% of drivers often drive on the form of sutures, 44.6% of drivers sometimes race with other cars, 16.4% of drivers exceed the speed limit on the roads and 9.1% of drivers often have overconfidence due to size of the truck, this represents a significant danger to other cars. From the literature, it was known that there is a strong relationship between driver fatigue, obesity and crash involvement. To assess the effect of fatigue on accident exposure, variety of questions were asked in the questionnaire. These questions inquired about the number of sleeping hours, and continuous and total driving hours.

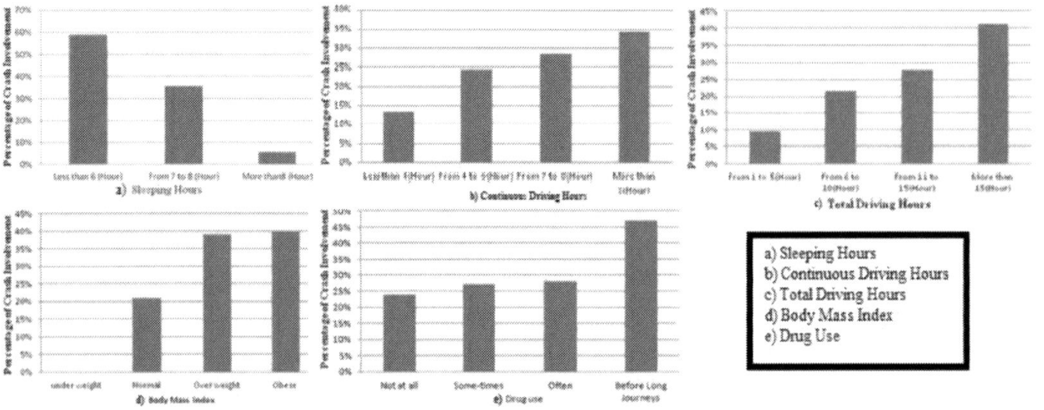

Figure 1. Percentage of Drivers Involved in Accidents by a- Sleeping Hours, b- Continuous Driving Hours, c- Total Driving Hours, d- Body Mass Index, e- Drug Use.

Figure 1,a shows that the less the truck driver sleeps the more exposure to traffic accident. This finding is in agreement with Souza findings [4]. figure 1,b illustrates that percentage of crash involvement increases when the number of continuous driving hours per day increases. figure 1,c presents the relationship between percentage of crash involvement and the number of total driving hours per day. It is apparent from this figure that the number of total driving hours per day is proportional to crash involvement. Unfortunately, in Egypt, there are no regulations that limit the number of continuous driving hours or the total driving hours per day for truckers. In fact, the current traffic laws encourage and forces truckers to drive long periods as every 24 hours they are charged for the goods they transport. In order to understand how driver obesity is related to crash involvement, the drivers were asked about their weight and height, and the body mass index (BMI) was calculated. The drivers were classified according to BMI as underweight, normal, over weight and obese[9]. Figure 1,d shows that obese and overweight truck drivers are the most involved in traffic accidents by 40% obese, 39% over weight, 21% normal drivers. Percentage of truck accidents was associated with higher rates of obesity. Percentage of obese and overweight drivers are so close that indicates most of truck drivers in a risk category, Surveys such as that conducted by James et al [6] showed that obese drivers represent more than 50% of commercial truck drivers. To measure the effect of drug use on crash involvement, a question inquiring about how frequent the driver uses drugs was asked and the answers were associated with crash involvement. From figure 1,e it is apparent that truck drivers who take drugs before long journeys are the most involved in accidents by 47%. Unfortunately, this result confirms that many truck drivers are using drugs before long journeys. Their belief is that this helps them to focus and continue to drive long distances. Table 2 illustrates the effect of long driving hours (continuous and total) on people who sleep less. From the table, it can be seen that drivers who sleep less than 6 hours, and drive more than 8 hours continuously are the most exposed to accident occurrence (31%). The table also shows that drivers who sleep less than 6 hours, and drive more than 15 hours a day are more involved in crash accidents (42%). It can be seen that percentage of accidents to drivers who sleep less than 6 hours decreased whenever total driving hours were less. This explains the significant impact of fatigue due to long driving hours on the occurrence of accidents.

Table 2. Percentage of Accident occurrence to Drivers who sleep less than 6 hours by total and continuous driving hours

Drivers sleep Less than 6 (Hour)				
Continuous driving hours	Less than 4(Hour)	From 4 to 6(Hour)	From 7 to 8(Hour)	More than 8(Hour)
	14%	30%	25%	31%
Total driving hours	From 1 to 5(Hour)	From 6 to 10(Hour)	From 11 to 15(Hour)	More than 15(Hour)
	11%	19%	28%	42%

Table 3 summarizes the percentage of drivers for different accident related variables (sleeping hours, continuous driving, total driving hours, body mass index and drug use). It can be seen from this table that 59% of drivers involved in accidents sleep less than 6 hours. 79% of drivers who are exposed to accidents are 40% overweight and obese. Also it can be noted that 28% of drivers involved in accident often take drugs and 47% do that before long journeys.

Table 3. Percentage of Drivers involved in accidents related Variables(NA= Not Applicable)

Sleeping Hours	Less Than 6 Hours	7-8 Hours	More Than 8 Hours	
	59%	36%	5%	
Continuous Driving	Less than 4(Hour)	From 4 to 6(Hour)	From 7 to 8(Hour)	More than 8(Hour)
	13%	24%	29%	34%
Total Driving Hours	From 1 to 5(Hour)	From 6 to 10(Hour)	From 11 to 15(Hour)	More than 15(Hour)
	9%	21%	28%	41%
Body Mass Index	Under Weight	Normal	Over Weight	Obese
	NA	21%	39%	40%
Drug Use	Not at all	Sometimes	Often	Before Long Journeys
	24%	27%	28%	47%

5 Severity of accidents

Accidents are unplanned and sudden events but they cause damage to person or property. In an accident, there can be more than one type of casualty. Thus, the severity of accident was divided into damage only, injuries only, deaths, damage + injuries, damage + deaths, injuries + deaths, damage + injuries + deaths and other losses are illustrated in Figure 2. It can be seen from this figure that the most frequent type of truck accident severity is (damage = 49%), (injuries = 19%), (deaths = 13%) and (damage + injuries + deaths= 6%).

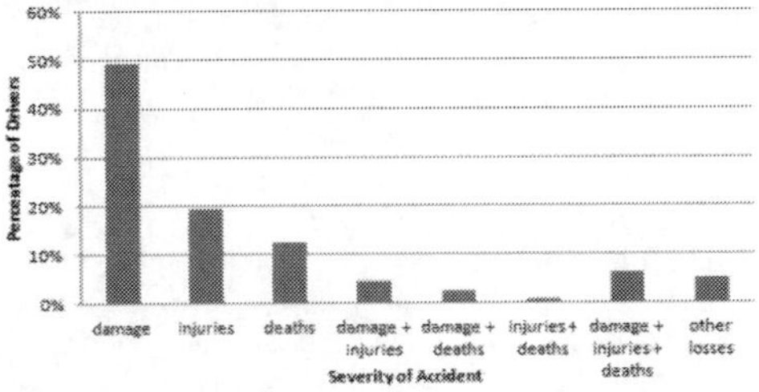

Figure 2. Percentage of Casualties as Reported by Truck Drivers Related to Severity of Accidents

Truck drivers who had an accident or more were asked about the cause of the accident. Figure 3 shows self reported accident causes, namely speeding 19%, lack of attention 15%, bad weather 13%, drug use 12%, sleepiness 11% and overloading 6%. Drivers admitted several reasons of accidents by their confessions and compared with the data analysis, that explained that sleep, drug use and obesity are common causes of truck accidents. Fatigue, long driving hours and lack of sleep caused a lack of attention and resulted in accident occurrence.

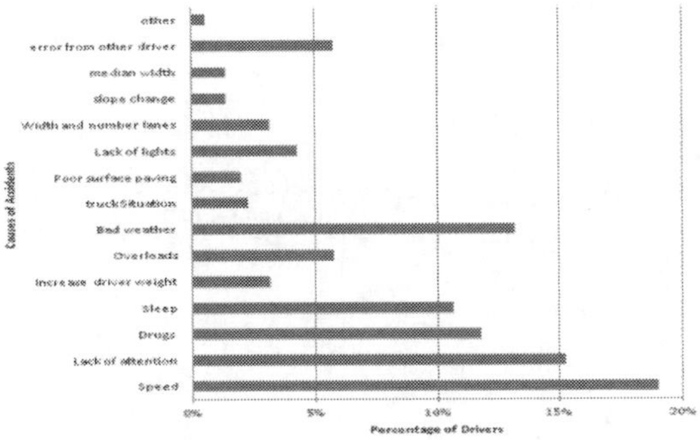

Figure 3. Percentage of Drivers' Self Reported Causes of Accidents

Figure 4 shows that number of death causalities is one killed person in 51% of fatality accidents, two killed persons in 32% of fatality accidents, three persons killed in 13% of fatality accidents and more than three persons in 4% of fatality accidents as reported by the truck drivers in the questionnaire.

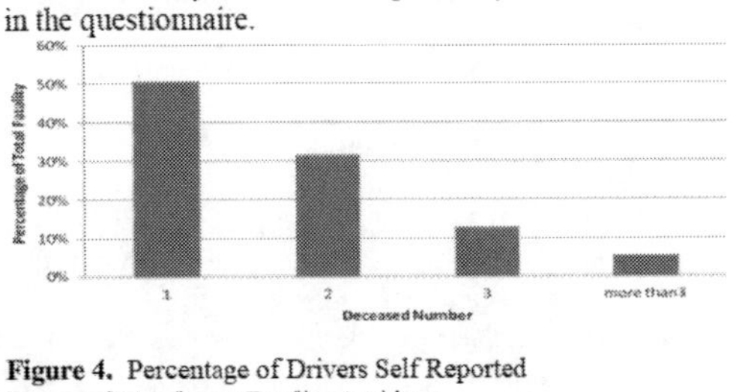

Figure 4. Percentage of Drivers Self Reported Deceased Number at Fatality Accident

7 Development of driver behaviour models

The characterization of truck drivers' behavior, statistical analysis and modeling of accident probability were conducted using SPSS. Due to of the nature of the collected data, the binary logistic regression technique was used to develop truck driver behavior models. The method was designed to find factors that were most effective in predicting the probability of accident occurrence. It adds variables to the model in a stepwise base, provided that these variables have significance value less than 0.05 and check for the significance of variables already in the model to consider if any variables should be removed. Several accident probability models were developed considering the influencing factors of fatigue in terms of sleeping hours, continuous and total driving hours, obesity, and drug use. Table 5 contains the two pseudo R2 measures (Cox and Snell & Nagelkerke) that judge the accuracy of the models. Cox and Snell attempted to calculate the coefficient of determination based on maximum likelihood [12]. It is better to use the Nagelkerke R^2, because the Cox and Snell can never quite get to 1, and is therefore less accurate [13]. In this study drivers slept less than 6 hours, total driving more than 15 hours, continuous driving more than 8 hours, obese drivers, and drivers who often use drugs accident models showed R^2 values of 0.70, 0.64, 0.56, 0.61, and 0.6.

Table 5. R^2 Test for model accuracy

Accident Model	-2 Log likelihood	Cox & Snell R^2	Nagelkerke R^2
Drivers slept less than 6 hours	206.441a	0.47	0.70
Total driving more than 15 hours	58.765a	0.44	0.64
Continuous driving more than 8 hours	64.113a	0.37	0.56
Obese drivers	111.124a	0.45	0.61
Use drugs before Long Journeys	98.609a	.453	0.60

8 Conclusions and recommendations

The present paper aimed at investigating the behavior of truck drivers and its impact on crash involvement. Based on the results and analyses of this research the following conclusions were drawn:

- Driver fatigue, driver obesity and drug use emerged as reliable predictors of truck traffic accidents. These results were also confirmed using the odds ratio analysis.
- The casualties of truck accidents were 49% damage only, 19% injuries, 13% deaths and 6% (damage + injuries + deaths).
- The number of death casualties at fatality accidents were 51% one person killed, 32% two persons killed, 13% three persons killed, 4% more than three persons killed.
- The developed models demonstrated that the highest influence on probability of accident occurrence is aggressive driving and distraction activities.
- A number of recommendations are suggested to reduce both the incidence and severity of crashes. These include:
- Unless the Egyptian government adopts legal legislation for regular detection of truck drivers' drug usage periodically and apply points system which has different levels of penalties, reduction of accident number and severity will not be attained.
- There is a definite need for updating current traffic laws and regulations especially the penalties for repeated traffic violations.
- Moreover, more real accident data should be made available to researchers for analysis and simulation. This requires establishing electronic database for the accidents including all data.
- Develop a national handbook on the rules and principles of safety driving rules.
- Another important practical implication is that truck drivers should be given extensive and professional training with emphasis on safe driving qualities.

References

1. Ahmed Mohamed Ismail, H.Y.A., Mohamed Ahmed owais, Analysis and Modeling of Traffic Accidents Causes for Main Rural Roads in Egypt. 2010.
2. World Health Organization. Violence, I.P., and World Health Organization. Global status report on road safety 2013: supporting a decade of action. World Health Organization, 2013.
3. McKnight, A.J., and George T. Bahouth. "Analysis of large truck rollover crashes." Annals of Advances in Automotive Medicine/Annual Scientific Conference. Vol. 52. Association for the Advancement of Automotive Medicine, 2008..
4. Souza, J.C., Teresa Paiva, and Rubens Reimão. "Sleep habits, sleepiness and accidents among truck drivers." Arquivos de neuro-psiquiatria63.4 (2005): 925-930..
5. Anderson, J.E., et al. "Obesity is associated with the future risk of heavy truck crashes among newly recruited commercial drivers." Accident Analysis & Prevention 49 (2012): 378-384..
6. James evan, James evan, J.Goglia, National Trasportation Safty Board, Annual report to congress,1999, 490 L'Enfant Plaza, S.W., NTSB/SPC-00/03 Washington, D.C. 20594.
7. Capodaglio, P., et al. "Functional limitations and occupational issues in obesity: a review." International Journal of Occupational Safety and Ergonomics 16.4 (2010): 507-523..
8. Doyle, K., Drug use high among commercial truck drivers. Oct 25,2013.
9. Anderson, J.E., et al., Obesity is associated with the future risk of heavy truck crashes among newly recruited commercial drivers. Accident Analysis & Prevention, 2012. 49: p. 378-384.
10. Hosmer Jr, D.W., Stanley Lemeshow, and Rodney X. Sturdivant. Applied logistic regression. Vol. 398. John Wiley & Sons, 2013.
11. Stutts, J.C., Jean W. Wilkins, and Bradley V. Vaughn. "Why do people have drowsy driving crashes." Input from drivers who just did 202/638 (1999): 5944..
12. Cox, D.R. and E.J. Snell, Analysis of binary data. Vol. 32. 1989: CRC Press.
13. Nagelkerke, Nico JD. "A note on a general definition of the coefficient of determination." Biometrika 78.3 (1991): 691-692.
14. Hosmer Jr, David W., Stanley Lemeshow, and Rodney X. Sturdivant. Applied logistic regression. Vol. 398. John Wiley & Sons, 1989.
15. Car crashes rank among the leading causes of death in the United States, Traffic Safety Culture Index April, AAA Foundation of Traffic Safety, 2008
16. Cerrelli, Ezio C. Trends in large truck crashes. No. HS-808 690. 1998.
17. Luke, R., and Gert J. Heyns. "Reducing risky driver behaviour through the implementation of a driver risk management system." Journal of Transport and Supply Chain Management 8.1 (2014): 10-pages.

DFT analysis of Ship pilotage risk based on A.D. Hall models

Wei hongbin [1]

[1] China waterborne transport research institute, Beijing, China

Abstract. Based on A pilot station more than 10 years of pilotage safety accident database, from three angles, including time, link and space, the statistical of ship pilotage safety accident risk is analysis, from which obtain the pilotage safety accident characteristics, and provide the basic data for the risk assessment and risk analysis model. By using the improved A.D. Hall model, based on the study of accident database, a 3D model of ship pilotage risk assessment is established, which is from the time dimension, logic and resource dimension angle. From the point of view of resource dimension, the dynamic fault tree model is established considering the dynamic relationship between events, and the model is analysed by dynamic fault tree analysis (DFT). Based on the analysis results, measures to reduce pilotage risk are proposed.

1 Introduction

Pilotage, defined in the Regulations refers to the guidance of ship navigation, berthing and shifting activities.

In the whole process of ship pilotage, it is analysed in terms of human, material and environment, can be decomposed into the pilot, the cited ship, and the external environment of ship pilotage three key factors, which need to be considered.

Vessel pilotage, defined in the law, refers to the activities that lead the ship to sail, berth, berth and berth.

In the whole process of ship pilotage, it is analysed in terms of human, material and environment.

With the development of large-scale ships, the tonnage and main dimensions of vessels are becoming larger and larger, and the difficulty of piloting ships is increasing gradually;

The rapid development of the port, the number of daily arriving ship will continue to increase, the pilot ships will increase year by year, the pilot work intensity is increasing year by year;

With the increasing volume of old port operations, the gradual development of new ports and the pilot operation environment become strange.

The above three aspects increase the potential safety risk of pilotage.

The risk assessment of ship pilotage has been analysed for many years. Fang Quan Gen (2006) proposed risk matrix and risk criteria, which based on ship pilotage accident [1].Wu Yong Jun ,Ma Fei and Xuan Shao Yong use FTA methods to analyse system failures and risks [2-4]. Fang Cheng (2008) used the Bayesian method to forecast the risk of ship pilotage risk, which is based on the analysis of the actual situation and the accident [5]. Zhou Lili (2008) used the grey comprehensive evaluation model to sort the correlation degree of the ship pilotage risk [6]. Fan Hong (2004) used the evidence theory to evaluate the system risk, when the underlying factors are subjective judgments due to the lack of data [7]. Chen Zheng Hua (2005) made a more comprehensive and accurate assessment of the risk situation in the pilot area and the ship's navigation [8]. Xue Yi dong (2005) made a detailed analysis of human error, which based on the case of ship pilot accident [9]. XI Yong Tao established the index system and evaluated the safety risk of the ship's pilotage by the unknown measure model and the confidence identification criterion [10]. Lin Tie Liang used the FTA method to analyse the risk of the bridge collision [11]. Zhang Qing and Bai Xu use the FMEA and FTA methods to analyse system failures and risks, and propose solutions based on the results for the corresponding risks [12-13]. Ji Changming use the A.D. Hall models to do the system analysis [14].HU Shen ping and Q Fang use Synergy-based mode to study the management risk of marine traffic [15] [16].In this paper the A.D. Hall models is improved to assessment of ship pilotage risk, which is from the time dimension, logic and resource dimension angle. DFTA is used to make a comprehensive and accurate assessment of ship pilotage.

2 Pilotage safety accident database

The statistics of pilotage risk accidents is based on a pilot station for more than 10 years. The accident time, historical information, link information and spatial location information is statistically analysed.

Through time history analysis of accidents, the average of every 10,000 pilot ships occurred in various types of security incidents 7.6 times (including cars, pilot

 DOI:10.1051/matecconf/201712404010

boats, ships). The average probability of occurrence of pilot accidents is 0.08%. According to the occurrence of the link, the accidents contain two links, which are the process of the pilot reaching the pilot site and beginning pilot operation.

In this two links, there are three categories including ship accidents, car accidents, pilot boats accidents and ship piloting accidents. All of the pilotage risk accidents, ships accidents accounted for 40%, followed by the pilot boats accidents accounted for 31%, cars accidents accounted for 28%. Ship piloting operation is the main link of the pilot accident, ship piloting accident occurred in the average probability of 0.03%.

According to the location of the ships, the accidents are divided into three categories including port, channel and anchor. According to the analysis of the piloting t accident report, 62% of the ship accidents occurred in the port area; 25% of accidents occurred on the channel; 13% occurred in the anchorage.

3 Identification of risk

On the basis of full understanding of risk characteristics, identify potential risks and specific risk factors of these risks. Risk identification can generally be carried out by imagination and by the use of standard analysis techniques. Imagination must not be based solely on the identification of hindsight or existing dangers, but should also take full account of the anticipated dangers that may or may not be present.

The risk factors, according to the boundary division of risk, can be divided into internal and external risks, the internal risk refers to the events in the internal organization, and external risk refers to the risk of occurrence in the main external organization. According to this classification method, pilotage risk is divided into two categories, internal risk and external risk.

The internal risk includes the pilot process by pilot and pilot organization internal security management risk management. The pilot's own analysis of the situation, including health, fatigue, emotional control, safety awareness and quality of mind. The pilot and pilot experience, including the laws and regulations, familiar with the pilot program and emergency plan making, the state of the environment, cited the ship and tug the cognitive level and the pilot operating conditions of degree. The risk management of the internal safety management of pilotage institutions includes the risk of the managers involved in the pilot accident.

For external risks, it includes the risk factors of the ship and the risk factors of the external environment. Among them, the risk factors of ships can be divided into the risks of the cited vessels and the risks of tugs, as well as the risks of other ship equipment on the pilotage accidents, as well as the risks of the crew members and captains of the ship being cited. The risk factors of the external environment and environment factors, detailed for the waterway anchorage risk, dock risk, the risk of traffic (that is caused by the density of ships on the route of accident factors), risk information sharing, the other ships risk (the ship around the ship non normal sailing accident caused by the risk. Especially the accident caused by dredging and fishing boats crossing route) and pilotage accident natural factors influence risk factors.

4 The risk assessment model

Risk is the impact of uncertainty. Usually the effect is to deviate from expectations, can be positive or negative. Uncertainty is the state of the lack of understanding or knowledge of an event, or even a partial result or possibility. Typically, the risk is presented as a potential event and a consequence or a combination of both. Generally, the risk is expressed in terms of the combination of the consequences of an event, including the change of the situation and the likelihood of occurrence.

How to control the risk? According to Systematic perspective, prefer "risk management" than "accident response". Risk management should focus on the prevention of accidents;

Combined with Synergy-based mode risk copying, an improved A.D. Hall 3D structure chart is established, the risk assessment and analysis is carried out. The model consists of three coordinates, which are logical dimension, time dimension and resource dimension.

Thus from logic dimension, time dimension and resource dimension, the pilotage A.D. Hall risk management structure model is constructed.

4.1 The logic dimension

In accordance with the logic dimension, risk assessment contain risk identification, risk analysis, risk control and risk response.

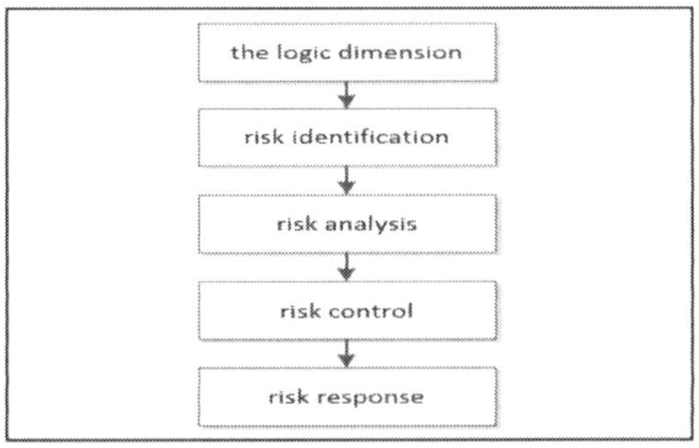

Figure 1. The logic dimension

In Figure 1, system risk management must first determine the risk of affecting the ability to achieve goals. Identify risk sources through risk identification; Conduct risk analysis, monitoring, measurement, analysis and evaluation of risks and opportunities; Take steps to address risks and opportunities to ensure that activities are implemented, increase the ability to cope with risks.

Risk identification: Risk identification is the risk factor that may exist in the defined system, and what impact these factors have on the system.

Risk analysis：Risk analysis is on the basis of risk identification, the corresponding index system and evaluation standard, to classify the degree of risk, reveal the key risk factors affecting the security of the system, according to the key risk factors, take preventive measures, to reduce the risk and improve the existing security situation. At the same time, through this assessment, relevant risks can be kept within the acceptable range as much as possible.

Risk control：The risk control scheme is based on the risk identification and risk assessment, the paper puts forward the corresponding measures of risk reduction, risk control scheme and according to these measures feasible, including the formulation and revision of some regulations. In the risk control plan should carefully consider the known risks and in danger of "recognition" and "risk analysis" risk identification in step two, we should be fully aware of the risk as a result of new technology or update the operation method of the cause, so as to fully implement the risk control scheme for all risk.

Risk response：Risk response should be based on the risk control scheme, take corresponding emergency measures and take specific emergency measures for the risks that have occurred or have occurred.

4.2 The time dimension

Pilotage service according to the time dimension contains accepting the pilot application, making the pilot plan and implementation of the pilot scheme.

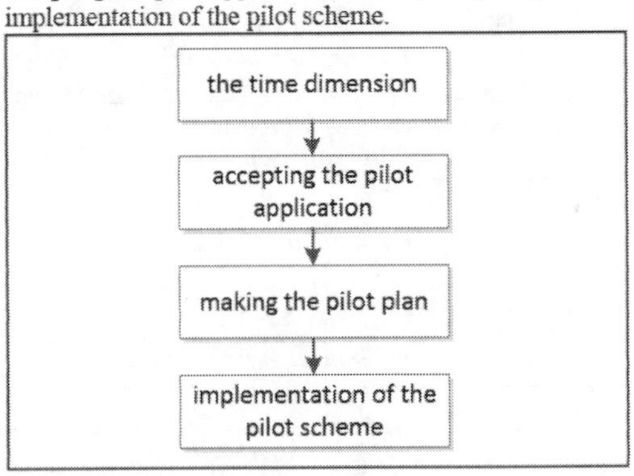

Figure 2. The time dimension

In Figure 2, system risk management penetrates the full process of pilot activities. Carry out pilot production, understand the pilot external and internal conditions, fully consider the ship navigation environment, meteorological and hydrological conditions, wharf and navigation facilities and other potential risk factors, and formulate the pilot plan. Time dimension refers to the system engineering, from acceptance to execution can be divided into 3 stages: After accept the pilot task, formulate pilot program, and implement the pilot scheme.

Acceptance of pilotage application: the pilot ship applies for pilotage to the pilotage institution, and the pilotage institution accepts and provides pilotage service in accordance with the regulations.

Pilotage plan formulation and Implementation: pilotage organization according to the application formulation pilot plan and implementation plan.

The formulation and implementation of the pilot scheme: according to the plan of pilotage scheme, and assigned to the pilot operation, the specific implementation of pilot scheme.

4.3 The resource dimension

Pilotage service according to the resource dimension, contains: Software, hardware (natural objects, man-made objects, such as facilities, equipment), environment-conditions, human.

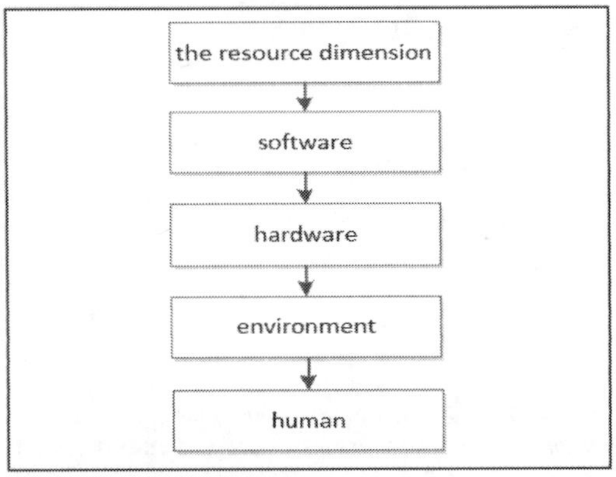

Figure 3. The resource dimension

In Figure 3, the relationship between human and objects-state, these three aspects are organically combined to form a dynamic risk system.

Human factors mainly refer to the pilots and other personnel, including human behaviour and cognitive judgment operation. Factors of objects mainly include, ship, natural conditions, navigation traffic conditions. Factors of the relationship between human and object include information communication and organization management.

Software: Usually refers to system management, including information exchange between organizations, between the object, the planning policy, emergency response program. In this paper, including information exchange with the piloting ship、contact with the other ship、contact with tug and emergency plan。

Hardware: Usually refers to objects (natural objects, man-made objects, such as facilities, equipment). In this paper, including the cited ship and passage past ship.

Environment: Usually refers to External natural conditions, including environmental and climatic factors. In this paper, including visibility 、wind and flow conditions.

Human: Usually refers to human factor，including Human-behaviour and perception.

The realization of security needs to prevent, control and eliminate the risk from three aspects of above three aspects content.

DOI:10.1051/matecconf/201712404010

5 Dynamic fault tree construction method

The establishment of fault tree can rely on joint research of experts, which can also be combined with the database and refer to other industry establishment methods. In this paper, the ship pilotage failure case database is used to construct the DFT.

Consider the fault tree consisting of n basic events, and the basic events are independent of each other. The state parameter X_i represents the basic event, and Φ represents the state parameter of the top event. Fault tree structure function

$$\Phi = \Phi(X_1, X_2, \cdots X_n) \tag{1}$$

$$\overline{X} = (X_1, X_2, \cdots X_n) \tag{2}$$

DFT is developed on the basis of static fault tree, which is an extension of static fault tree. In DFT, due to the dynamic characteristics of the system, researchers need to take into account the interaction between the components and the order of events, which is the content of the static fault tree cannot be achieved. Because the failure mode has sequence problem, therefore the cut sets cannot be used, but the sequence cut sets can be used.

The resolution of a DFT is much different than the one of the SFT, because temporal and cross dependencies cannot be modelled and solved through Boolean algebra [17].

The DFT are using dynamic logic gates, namely Priority And Gate(PAND),The Spare Gate(SP) and The Functional Dependency gate(FDEP),Sequence enforcing gate(SEQ) [18].

In this paper the failure of a Priority-AND(PAND) ,Functional Dependency gate(FDEP) and Sequence enforcing gate(SEQ) have been used as the dynamic logic gates.

In these paper the dynamic fault tree is converted to Markov chain, which can be used to solve the dynamic the sequence cut sets of the fault tree .

All the failure modes and propagation paths can get from the state transition graph of the dynamic fault tree, which is represented by a sequence cut set. In a Markov state transition diagram, each step corresponds to transfer a part of the fault system, if a final state Markov chain for system failure, the sequential chain of all transfer plus the ordinal relation becomes a system failure mode [19].

Therefore, in order to express the order relation, a new method for representing the failure mode is introduced. Such as: failure mode A first fault, B after the failure, system failure can be expressed in <u>AB</u>. The C is the top event, A and B are the bottom events, and the sequence cut set in figure Ⅱ can be represented as {<u>A, B</u>} [19].

6 Dynamic fault tree analysis

According to the logical dimension of pilotage system, the main purpose of fault tree analysis is to identify the cause of the pilot failure, by the risk analysis, find out the weak link of pilotage activity. And then carry out the control measures.

The fault tree analysis was used to determine the top event as collision. The cause of the collision was analysed by the method of system analysis, from risk control of resource dimension. Taking the collision as the top event, the fault tree was constructed .Detailed see Table 5 Fault tree table and Appendix Figure 7 Fault tree figure.

6.1 Dynamic sub trees analysis

According to the dynamic logic gate, the fault tree diagram can be divided into three dynamic sub trees. From left to right, followed by FDEP、SEQ and PAND, three pairs of trees, respectively, using Markov chain method to solve the sequence cut set.

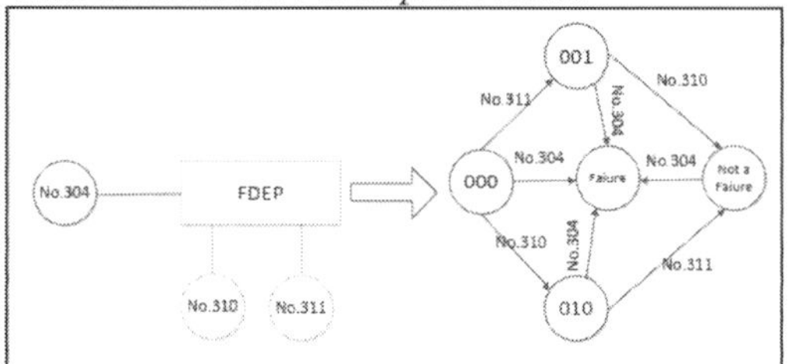

Figure 4. The "FDEP" gate transfer to Markov chain

In Figure 4, the logic gates have 8 state chains in the process of transforming into Markov chains, and only 4 chains correspond to failure, Through Boolean transport of logical gates or gates, a sequential cut set in Table 1 is obtained by combining the Markov chain transformation of Fig. 4.

Table 1. The "FDEP" gate sequence cut set of fault tree

Set name	The cut set of basic events
Set 1	<u>{No.304}</u>

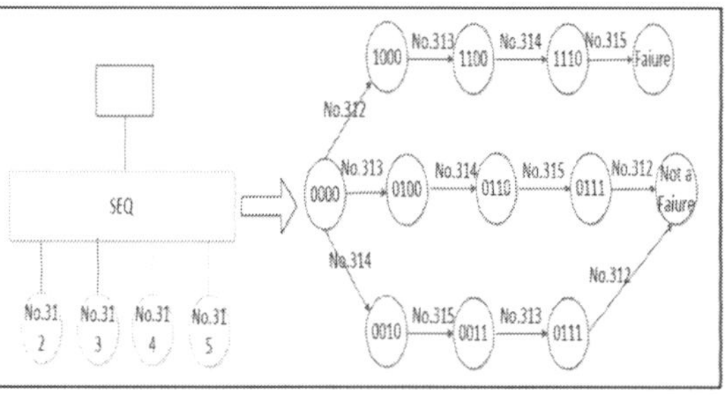

Figure 5. THE "SEQ" GATE TRANSFER TO MARKOV CHAIN

In Figure 5, the logic gates have 24 state chains in the process of transforming into Markov chains, and only 1 chains correspond to failure, As shown in Figure 5, the top 1, The other 23 is not failure, As shown in Figure 5, the middle 1 and the next 1, the other 21 are not drawn in Figure 5, but the results are not failure. Through Boolean transport of logical gates or gates, a sequential cut set in Table 2 is obtained by combining the Markov chain transformation of Fig. 5.

Table 2. The "SEQ" gate sequence cut set of fault tree

Set name	The cut set of basic events
Set 1	{No.312、No.313、No.314、No.315}

In Figure 6, The logic gates have 2 State chains in the process of transforming into Markov chains, the input events A and B in the 2 chains represent two input events in Figure 7 Fault tree figure, respectively, corresponding to logical gates or gates, and No.23. Through Boolean transport of logical gates or gates, a sequential cut set in Table 3 is obtained by combining the Markov chain transformation of Fig. 6.

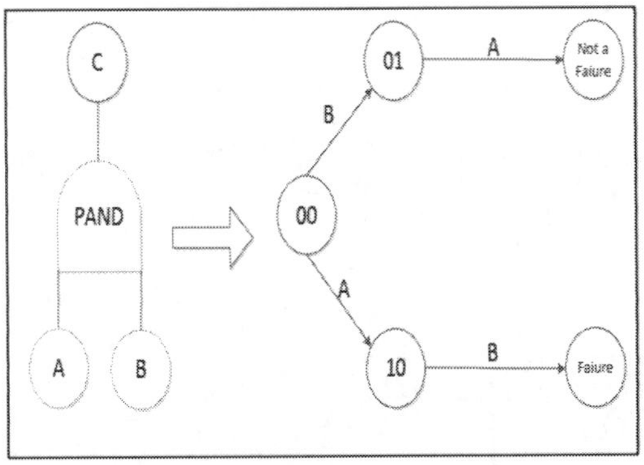

Figure 6. The "PAND" gate transfer to Markov chain

Table 3. The "PAND" gate sequence cut set of fault tree

Set name	The cut set of basic events
Set 1	{No.307、No.305}
Set 2	{No.307、No.306}
Set 3	{No.307、No.301}
Set 4	{No.307、No.302}
Set 5	{No.307、No.303}
Set 6	{No.308、No.301}
Set 7	{No.308、No.302}
Set 8	{No.308、No.303}
Set 9	{No.308、No.305}
Set 10	{No.308、No.306}
Set 11	{No.309、No.301}
Set 12	{No.309、No.302}
Set 13	{No.309、No.303}
Set 14	{No.309、No.305}
Set 15	{No.309、No.306}

6.2 Subtree synthesis analysis

Based on the above results, the synthesis of three sub trees can be used to get the risk mode of the top event of the ship pilot fault tree:

Table 4. The synthesis sequence cut set of fault tree

Set name	The cut set of basic events
Set 1	{No.304}
Set 2	{No.312、No.313、No.314、No.315}
Set 3	{No.307、No.305}
Set 4	{No.307、No.306}
Set 5	{No.307、No.301}
Set 6	{No.307、No.302}
Set 7	{No.307、No.303}
Set 8	{No.308、No.301}
Set 9	{No.308、No.302}
Set 10	{No.308、No.303}
Set 11	{No.308、No.305}
Set 12	{No.308、No.306}
Set 13	{No.309、No.301}
Set 14	{No.309、No.302}
Set 15	{No.309、No.303}
Set 16	{No.309、No.305}
Set 17	{No.309、No.306}

From the above results, in Table 4,we can see that the dynamic fault tree have 17 sequence cut sets, and there are 17 ways to lead to the top fault event. 17 sequence cut sets concentrated {No.304} ,which contains the least number from the qualitative analysis, the structure importance of bottom events will be the most important, which should be strengthen security measures to avoid incidents.

Table 5. Fault tree table.

Top event	The first floor	the second floor
No.11 Collision	No.21 software	No.301 Insufficient information exchange with the piloting ship
		No.302 Contact with the other ship failure
		No.303 Contact with tug failure
		No.304 No emergency plan
	No.22 hardware	No.305 Piloting ship - ship fault
		No.306 Passage past ship - ship fault
	No.23 Environment	No.307 Poor visibility
		No.308 Wind
		No.309 Flow
	No.24 Human	No.310 Emergency handling fault
		No.311 Emergency action is not enough
		No.312 Careless alert
		No.313 Collision risk Judgment error
		No.314 Shipping position Control error
		No.315 Collision avoidance action fault

7 Conclusion

In this paper, based on the collected ship pilotage failure case database, analysis of statistical characteristics of ship pilotage accident, identify potential risk sources causing ship pilotage failure.

Through the internal and external risk identification, using improved A.D.Hall model in logic dimension, time

DOI:10.1051/matecconf/201712404010

dimension and the dimension of resources, to evaluate the risk of ship pilotage, identification of the risk classification.

Dynamic fault tree analysis obtains the relevant risk results. Relevant results show that this method can be used for the analysis of ship pilotage risk.

However, the application of complex fault tree model is still very difficult, in the future, a simplified calculation method can be used to solve the dynamic fault tree model.

References

1. FANG Quan-gen. Journal of Harbin Engineering University, **03**, 329-334 (2006).
2. Wu Yong Jun. Jimei University, (2013).
3. Ma Fei. Dalian Maritime University, (2008).
4. Xuan Shao Yong. Shanghai Maritime University, (2005).
5. Fang Cheng, NAVIGATION OF CHINA, 31, 388-391(2008).
6. Zhou Lili, Journal of Shanghai Maritime University, 29, 21-25 (2008).
7. Fan Hong. Wuhan University of Technology, (2004).
8. CHEN Zheng Hua.NAVIGATION OF CHINA,**01**,3-8+15(2005).
9. Xue Yi dong, NAVIGATION OF CHINA, 3, 28-32 (2005).
10. XI Yong-tao, SHIP & OCEAN ENGINEERING, 28, 56-58 (2009).
11. Lin Tie liang, JOURNAL OF TONG JI UNIVERSITY(NATURAL SCIENCE), 34, 467-471 (2006).
12. Zhang Qing, CHINESE JOURNAL OF CONSTRUCTION MACHINERY, 14, 109-113 (2016)
13. Bai Xu .SHIP BUILDING OF CHINA,**04**,171-179(2012).
14. Ji Changming, Asia Pacific Power and Energy Engineering Conference, 28-31,(2010).
15. HU Shenping. Procedia Engineering,45, 31 – 37 (2012)
16. Q Fang. Journal of Shanghai Maritime University,3,50-55 (2009)
17. Manno, Gabriele, Expert Systems with Applications, 39, 10334–10342 (2012)
18. K. Durga Rao, E Reliability Engineering and System Safety, 94, 872–883(2009)
19. Liu wenbin, Nanjing University of Science and Technology,(2006

Appendix

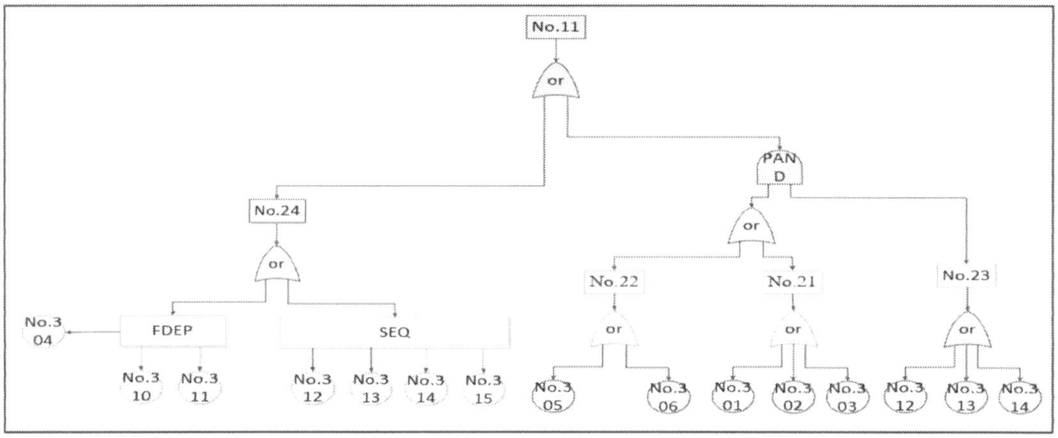

Figure 7. Fault tree figure

DOI:10.1051/matecconf/201712404011

A Logic Architecture for 360 ADAS-Alerts for Hazards Detection Based in Driver Actions

Javier Izquierdo-Reyes[1], Luis A. Curiel-Ramirez[1], Ricardo A. Ramirez-Mendoza[1] and M. Rogelio Bustamante-Bello[1]

[1]*Escuela de Ingeniería y Ciencias, Tecnológico de Monterrey*
Calle del Puente 222, Tlalpan, Ejidos de Huipulco, 14380 Ciudad de México, México.

Abstract. In this work is presented a novel approach for passive safety in vehicles by Advanced Driver Assistance Systems (ADAS) alert emission in 360° around driver to notify about hazards nearby the vehicle depending on the actions taken by driver per the context. This proposal would create a more robust system compared to current passive ADAS systems since the feedback to driver is in the same direction that hazard is detected (Punctual Sound Source Alert), compared with most assistance systems that emits sounds from the monitor or from the dashboard provoking distractions when emits alerts unnecessarily. The increase of security by this method will allow the driver to be aware of their surroundings even in a very quiet cabin or in a noisy environment. Also, it would detect the steering wheel angle, speed of movement and the activation of turning lights among other alerts, which would allow us to define a critical action during driving; apart from using sensors and cameras aimed at the driver to detect patterns of movement during these critical actions and have a prediction of a possible turn or manoeuvre when driving, refer to Figure 1. It will be necessary a reconfiguration of the alert in frequency, time of action depending upon the level of risk to prevent an accident or to reduce the consequences in an imminent accident.

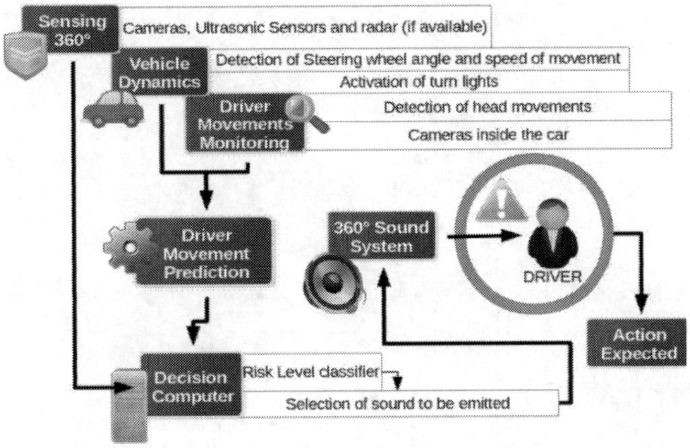

Figure 1. Graphical Abstract.

1 Introduction

Advanced driver assistance systems are becoming more common in mid-segment vehicles and are fully required in the high segment due to high added value. But despite all the new developments, the issuance of alerts has not evolved beyond a screen.

Emission alert is necessary to avoid traffic accidents, however, not all alerts are well perceived by all the people, or sometimes the screen is overloaded and therefore induces a state of distraction.

Nowadays there are different kind of alerts such as:

- Sound
- Visual
- Audio-visual
- Haptics

In Table 1 is a list of brands and models that incorporates some passive ADAS that emit this type of alerts.

In previous studies [1], [2], have been demonstrated that a poorly designed alert can be worse than if no alert is emitted because of the distraction it will produce, in [3] we used electroencephalography to detect driver concentration, and we found that some drivers got distracted due to ADAS Alerts.

Also, the National Highway Traffic Safety Administration (NHTSA) of the United States of America (USA) has conducted some studies to analyse the effects that auditory alerts have over the driver and how they are perceived (Warnings, Alerts, Status Notification or Social Notification) [4]. This research has been done to promote a standardized set of sound that allows easily and clearly identify urgent alert from other messages.

On the other hand, the proposed vibration and audio alerts have been shown to reduce reaction time significantly, even if the driver is involved in a second task, such as a telephone call, a high working memory load task [2], [5], [6]

Table 1. Visual or sound alerts in ADAS in Latin-American

Brand	**Models**	**Included Since (Year)**	**Driver drowsines s alert**	**Lane keep assist**	**Traffic Alert**	**Speed limit alert**	**Blind spot alert**	**Pre-collision System**	**Parking Sensor alerts**
Toyota Motor Corporation	All	2015	Optional	Optional	Optional		✓	Optional	✓
Hyundai Motor Company	Stinger	2018	✓	✓	Optional		✓	✓	✓
	All	2015					✓		Optional
VW Group	Passat	2013	✓				✓		✓
	Touareg	2015	✓	Optional				Optional	✓
	Golf	2013	✓						✓
	A3-6	2013	✓	Optional	Optional		✓	✓	✓
	Leon	2013	Optional	Optional					✓
	All	2013	Optional						Optional
General Motors Corporation	Equinox	2012		Optional		Optional	✓	Optional	✓
	Terrain	2012		Optional		Optional	✓	Optional	✓
	Astra	2015				✓			✓
	Malibu	2016	Optional	✓			✓	✓	✓
	All	2008							Optional
Ford Motor Company	Mustang	2013	Optional	✓		Optional	✓	✓	✓
	Focus	2017	Optional	✓			✓	✓	✓
	Escape	2016	Optional	✓			✓	✓	✓
	All	2006							Optional
Nissan Motor	Leaf	2013		✓			✓		✓
	Infiniti Q/Qx	2015	Optional	✓			✓	Optional	✓
	All	2013					Optional		Optional
Fiat Chrysler Automobiles	Pacifica	2017	✓	✓			✓	✓	✓
	300	2016	✓	✓			✓	✓	✓
	All	2010							Optional
Honda Motor Company	Accord	2016	✓	Optional	Optional		✓		✓
	RDX	2016	Optional	✓			✓	✓	✓
	RLX	2013	Optional	✓			✓		✓
	All	2010		Optional			Optional		Optional
Suzuki Motor Corporation	Vitara	2015	Optional	Optional			✓	✓	✓
	All	2013					Optional	Optional	Optional
Grupo PSA(Peugeot)	3008		Optional	✓			✓	✓	✓
	All	2011					Optional	Optional	Optional
Renault	Clio	2017	Optional	Optional			✓	Optional	✓
	Koleos	2017	✓	Optional			✓	Optional	
	All	2012					Optional		Optional
BMW AG	4,5,6	2014	Optional	✓			✓	✓	✓
	All	2010	Optional	Optional	Optional	Optional	✓	Optional	✓
Mazda Motor Corporation	6	2017	Optional				✓		✓
	MX-5	2016		Optional		✓	✓	✓	✓

	All	2012					Optional		✓
Mitsubishi Motors Corporation	outlander	2015	Optional	Optional			✓		✓
	All	2012					Optional		Optional
Tata Motors Limited (Jaguar, Land Rover)	xj/xf	2014	Optional	✓		Optional	✓	✓	✓
	All	2012		Optional	Optional		✓	Optional	✓
Geely Automobile (Volvo)	S90		✓	✓		Optional	✓	✓	✓
	All	2011	Optional	Optional			Optional	Optional	✓
Daimler AG(Mercedes-Benz)	GLA gamma	2012	✓	Optional		Optional	✓	✓	✓
	All	2011					Optional		✓

2 Alert implications

Distraction: As stated above, if the alert system is not precisely designed, it may involve distraction during driving due to confusion about the type of alert or risk detected, which generates a high mental workload and usually means a change in the attention of the road to dashboard or infotainment system [7].

Modern vehicle security systems display information on a central screen that is sometimes overloaded due to the amount of information being displayed to the user (music, GPS, social media, application system status, alerts, etc.).

It is because of the alerts are difficult to understand and represent a risk due to the need to change the look from the road to the screen to confirm the source of the alert.

False Alarm: Another problem with ADAS systems is when a false alarm is emitted, it is more common in systems based on artificial vision, which have a certain percentage of precision, which allows errors. Or when the alert is too repetitive due to an incorrect configuration and sounds at all times. All these false alarms are annoying to the driver and create a lack of confidence in the system [8]–[10].

In addition, a false negative is a big problem especially for vehicles on the scale 2 or higher automation by NHTSA [10], as these vehicles can be 'autonomous' under certain conditions. The best example was in 2015, when, unfortunately, a false negative cost the life of at least one person in the USA. Because the system was not able to detect a truck, the car did not stop and kill the passenger.

So, to avoid potential dangers created by the warnings and alerts in non-autonomous cars, it is important to design the alerts in the best way or making them adaptable to the driver.

3 Perception of alerts

According to the research made by Westat and the NHTSA [4], the perception of alerts and the means of urgency depends strongly of the noise condition and the urgency of the alert.

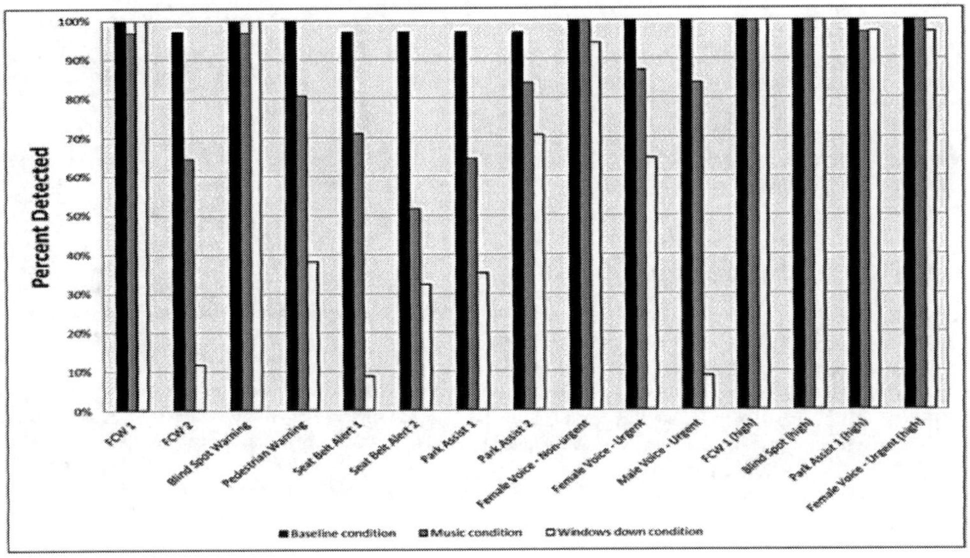

Figure 2. Percentage of drivers who detects alerts [4].

The configuration for each alert is indicated and explained in detail in the experimental protocol of the Study 2 'Warning perception in ambient noise environments' [4]

In figure 2 and 3 are shown the results obtained in an experiment [4] with 34 drivers, in which can be observed that low frequencies are less detectable than high in noise conditions, for example, in Figure 1. The FCW2, Seat Belt alert 1 and male voice has 11% or less in percentage detected for the windows down condition, while the FCW, Blind spot Warning, FCW1 (high), and Blind Spot (high) reached the 100% of detection.

And the same happens with the mean urgency rating for FCW2 and Seat Belt Alert 1.

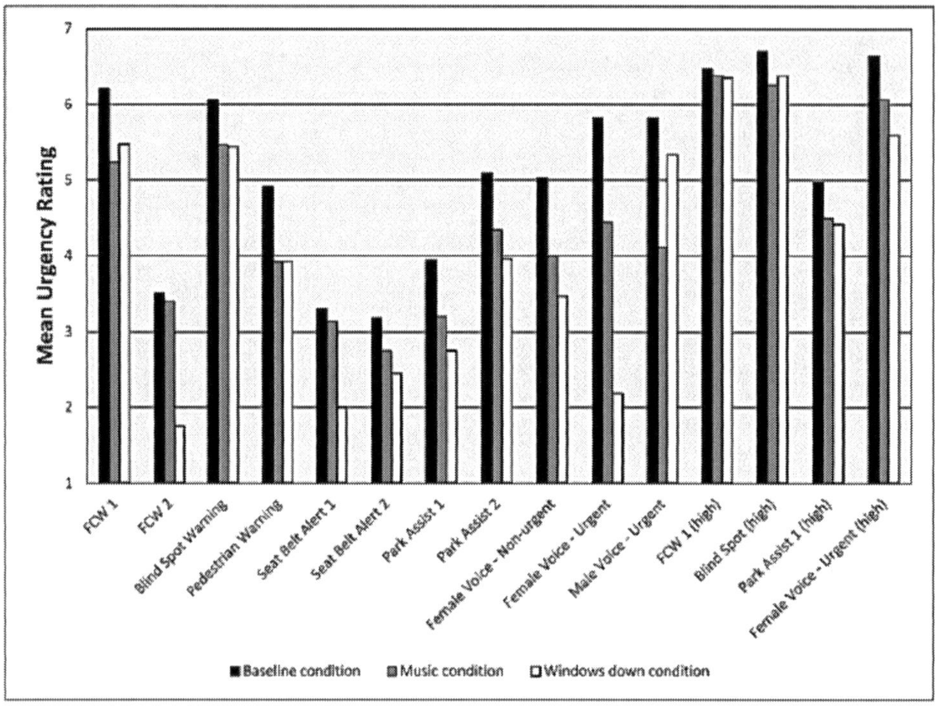

Figure 3. Mean Urgency Rating [4], [11].

For the male voice is different and drivers rated relative high the mean urgency, it is possible, because speech warnings are fast to be classified and better understood than bursts or beeps.

To explain the effect Zhang et. al. [1] developed a mathematical model in which is analysed the response time due to speech alerts and all aspects involved in the process.

4 360° ADAS alert proposal

After analysing the state of the art of the alerts, the variety of them and the experience obtained in the experiment developed in [3] and [12], we propose the use of a surround system 360° inside the vehicle as is shown in Figure 4, to give the user the ability to know the existence of hazards around him even without the need of see a monitor or infotainment screen.

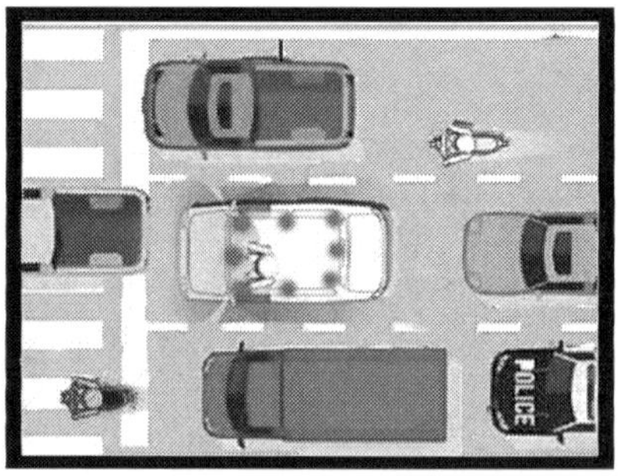

Figure 4. Audio source location in the 360° proposal

Since 2016, there is a system in the market called Arkamys, which pretends to enhance security by changing the way alerts are emitted. The system uses as sources of sound the original car speakers to create the '3D-ADAS'.

DOI:10.1051/matecconf/201712404011

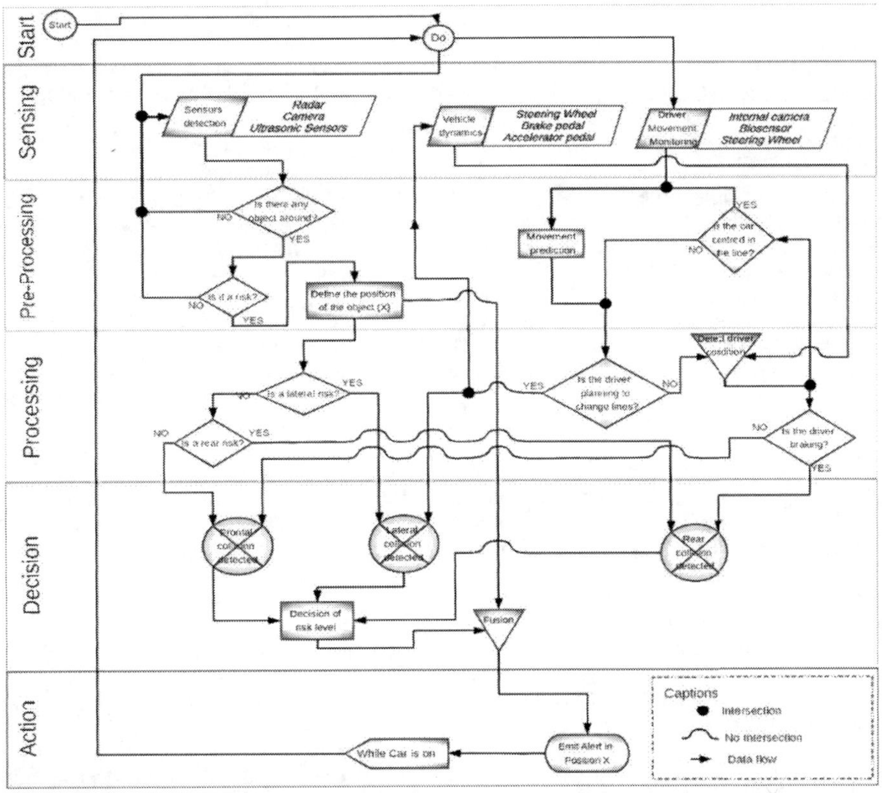

Figure 5. The general data flow diagram of proposed system

Despite it is a good concept, it has not been installed in any car yet, so, it is not clear how the system will be connected to ADAS and if necessary a major hardware and software integration.

Our proposal is to create a system like Arkamys, but smarter, means that we propose to use a 360-sound system inside the vehicle, which emits punctual sound source alerts based on prediction of driver movements and detection of hazards around the vehicle.

4.1 360° Sensing

Nowadays, the automobiles begin to carry more sensors that allow us to know the environment of the vehicle, providing the user with greater safety when driving.

There is a great variety of these, for example, with the use of ultrasonic sensors, radars of near and far range. Now is possible to determine the distance of objects to the car by using cameras with Artificial Intelligence, it is possible to detect and recognize objects to classify and issue alerts accurately.

Everything above is technologically possible and available in a car, so in our proposal we are increasing the 'intelligence' by predicting the driver's movements and

avoiding annoyances to the driver by emitting sound alerts only when necessary. Depending on the equipment available in the car, this will be used to improve the accuracy of alerts to the driver.

The main problem in a 360 ADAS alerts is the lateral monitoring, in a crowd via, the car will be surrounded by many cars, so predicting driver's movements will allow us to detect an intention to change lane and assisting the driver efficiently.

In figure 4, is the data flow diagram to understand the basic logic of this proposed system, it is important to indicate that we have omitted the exhaustive detail of each part of the system since it will be stablished according to the available equipment.

4.2 Movement predictions

To predict driver movements, there are some approaches propose to detect it, in this case as the main objective is to assist the driver in the lateral control, the most common methods are:

- Cameras
- Vehicle dynamic parameters
- Neuromuscular Dynamics

The use of cameras by Kondyli in [13] shows an approach for the use of range cameras to detect upper body posture and driver movement in the car cabin with a skeletal model that follows the driver's movements, this principle could be useful for our approach and it is non-invasive to the driver.

In the other hand is the method proposed by Angkititrakul et al. [8] to create a line departure prediction system (LDPS) that is based in driving signals calculating the time to line crossing (TLC), The method of predicting the TLC is to estimate vehicle path and road curve and then calculate the time before the intersection of the two trajectories using the following relationship:

$$TLC = \frac{y_l}{v sin\phi} \quad (1)$$

Where y_l is the lateral distance from the lane boundary to the vehicle's reference (in meters), v is the vehicle speed (in meters per second), and $sin\phi$ is the vehicle relative yaw angle (in radians).

Finally, another proposal implies neuromuscular data to predict driver movements with a high success rate using a Queuing Network based driver model [14]. The results from 16 subjects showed that the proposed model performs very well, and its performance is closer to driver's actions when a vehicle runs at a relatively high speed.

By using the information obtained by the ECU through its OBD II port, it would detect the steering wheel angle, speed of movement and the activation of turning lights among other alerts, that would help the cameras system, which would allow us to define a critical action during driving.

Based on the review, available technology and methods of lane assistance shown above, it is proposed to create an intelligent driver assistance system by emitting sound alerts only when necessary, avoiding annoyance, in a 360-sound configuration.

5 The proposed approach

To start with the development, a sound system will be installed inside a car around the driver as was shown in figure 4, a 360° audio set would be prepared to test different kinds of alerts that will help drivers detect the location of hazards around him even without seeing them.

The sound set in 360 would have alerts based on the configuration presented in the previous section by the NHTSA investigation [4], this would help us determine the effectiveness of this type of alerts in a 360-sound environment selecting only a few audio types.

In addition, the vehicle dynamics would be recorded to generate a mathematical model to predict the driver's movements, as one of the presented in the previous section and it will depend of the instrumentation installed in the car when test is executed.

We firmly believe that as autonomous cars reach everyday life, it is important to develop systems to help humans in driving tasks and avoid traffic accidents that means high monetary costs and mainly lost lives.

References

1. Y. Zhang, C. Wu, and J. Wan, "Mathematical Modeling of the Effects of Speech Warning Characteristics on Human Performance and Its Application in Transportation Cyberphysical Systems," pp. 1–13, 2016.
2. F. Biondi, D. L. Strayer, R. Rossi, M. Gastaldi, and C. Mulatti, "Advanced driver assistance systems: Using multimodal redundant warnings to enhance road safety," Appl. Ergon., vol. 58, pp. 238–244, Jan. 2017.
3. J. Izquierdo-Reyes, R. A. Ramirez-Mendoza, and M. R. Bustamante-Bello, "A study of the effects of advanced driver assistance systems alerts on driver performance," Int. J. Interact. Des. Manuf., Jan. 2017.
4. J. Singer, N. Lerner, C. Baldwin, and E. Traube, "Auditory Alerts in Vehicles: Effects of Alert Characteristics and Ambient Noise Conditions on Perceived Meaning and Detectability," … Conference on the …, vol. 1, pp. 1–14, 2015.
5. C. L. Lewis, B. A., Penaranda, B. N., Roberts, D. M., & Baldwin, "Effectiveness of bimodal versus unimodal alerts for distracted drivers," Hum. Factors Driv. Assessment, Train. Veh. Des., pp. 376–382, 2013.
6. R. Kawamura, M. S. Bhuiyan, H. Kawanaka, and K. Oguri, "Simultaneous stimuli of vibration and audio for in-vehicle driver activation," IEEE Conf. Intell. Transp. Syst. Proceedings, ITSC, pp. 1710–1715, 2011.
7. E. E. Wiese and J. D. Lee, "Auditory alerts for in-vehicle information systems: the effects of temporal conflict and sound parameters on driver attitudes and performance.," Ergonomics, vol. 47, no. 9, pp. 965–986, 2004.
8. P. Angkititrakul, R. Terashima, and T. Wakita, "On the Use of Stochastic Driver Behavior Model in Lane Departure Warning," IEEE Trans. Intell. Transp. Syst., vol. 12, no. 1, pp. 174–183, Mar. 2011.
9. F. Meng, S. Li, L. Cao, Q. Peng, M. Li, C. Wang, and W. Zhang, "Designing Fatigue Warning Systems: The perspective of professional drivers," Appl. Ergon., vol. 53, pp. 122–130, 2016.
10. F. Biondi and L. Skrypchuk, "Use Your Brain (and Light) for Innovative Human-Machine Interfaces," pp. 99–105, 2017.
11. V. Design, "In-vehicle noise alters the perceived meaning of auditory signals," pp. 401–407.
12. J. Izquierdo-Reyes, R. A. Ramirez-Mendoza, M. R. Bustamante-Bello, S. Navarro-Tuch, and R. Avila-Vazquez, "Advanced driver monitoring for assistance system (ADMAS)," Int. J. Interact. Des. Manuf., Sep. 2016.
13. A. Kondyli, V. P. Sisiopiku, L. Zhao, and A. Barmpoutis, "Computer Assisted Analysis of Drivers' Body Activity Using a Range Camera," Intell. Transp. Syst. Mag. IEEE, vol. 7, no. 3, pp. 18–28, 2015.
14. L. Bi, M. Wang, C. Wang, and Y. Liu, "Development of a Driver Lateral Control Model by Integrating Neuromuscular Dynamics Into the Queuing Network-Based Driver Model," IEEE Trans. Intell. Transp. Syst., vol. 16, no. 5, pp. 2479–2486, Oct. 2015.

The Application of Uncertainty Theory in Vehicle Collision Accident Reconstruction

Su-yang Ji[1], Jing-jing Xia[2]

[1]*School of Mechanical and Power Engineering, Nanjing Tech University, Nanjing 210000, China;*
[2]*School of Traffic Science Specialty, Huaiyin Institute of Technology, Huaian 223003, China*

Abstract. The theory of uncertainty analysis technique is presented. A kinect model is established, the uncertainty factors such as adhesion coefficient and brake trace are considered. The method of uncertainty theory is applied to analyze the uncertainty of velocity calculation in accident reconstruction, and then the practicability of the method was also demonstrated by a true accident case.

1 Introduction

According to the 2015 China Automobile Safety Development Report, in 2013, a total of 198 thousand and 400 traffic accidents involving casualties, resulting in 58 thousand and 500 deaths, 213 thousand and 700 people were injured [1].Accident reconstruction not only can accurately carry out accident investigation and responsibility identification, but also has a great help to accident prevention. Therefore, it is very important to realize the accurate representation of vehicle collision accident. But because of the complexity of vehicle collision accidents, There is a lot of uncertainty in speed identification. Uncertainty theory is applied in the paper, and discussed the uncertain parameters in accidents, according to the calculated uncertainty, calculate the range before the collision speed, make more accurate determination of the cause of the accident.

2 Uncertainty theory

Uncertainty is usually used to characterize the dispersion of the measured values. In general terms, the measurement uncertainty is defined as the degree to which the measurement results are correct. There are all kinds of measurement errors in the actual accident measurement, the results always have a certain degree of uncertainty, we can only get the scope which can be used to describe the degree of uncertainty [2].

2.1 Uncertainty description

Measurement uncertainty is the degree to describe uncertainty of the measurement results, Suppose that we need to measure the value of Y related to the quantity of input $x_i(i = 1, 2 \dots)$ with uncertainty, x_i is usually a random variable. recorded the multiple functions as:

$$Y = f(x_1, x_2, x_3, \dots, x_n). \quad (1)$$

if and only if the approximate solution y to be obtained when the optimal value of each input x_i is taken, write down y as:

$$y = f(x_1, x_2, x_3, \dots, x_n). \quad (2)$$

Therefore, the uncertainty of the results (y) is derived from$x_i(i = 1, 2, \dots, n)$, In other words, the uncertainty u_y of y depends on the degree of uncertainty $u_{(x_i)}$ of each input $x_i(i = 1, 2, \dots, n)$.

2.2 Method for evaluating uncertainty

There are two kinds of evaluation method of uncertainty, the Type A and Type B evaluation of uncertainty, they are show in Figure 1.

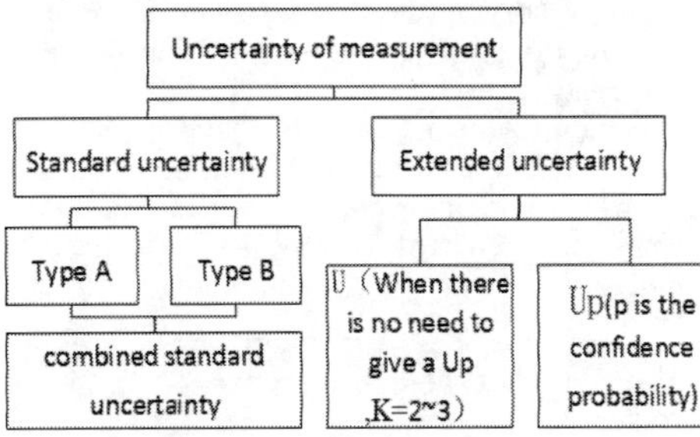

Figure 1. Structure of measurement uncertainty

The Type A evaluation of uncertainty assessed the uncertainty based on taking statistical analysis on

DOI:10.1051/matecconf/201712404012

observation while the Type B evaluation of uncertainty taking other methods. Such as long-term practical experience, common sense or reference books, etc.

In this paper, the Type B evaluation of uncertainty was been adopted. The calculation method is given.

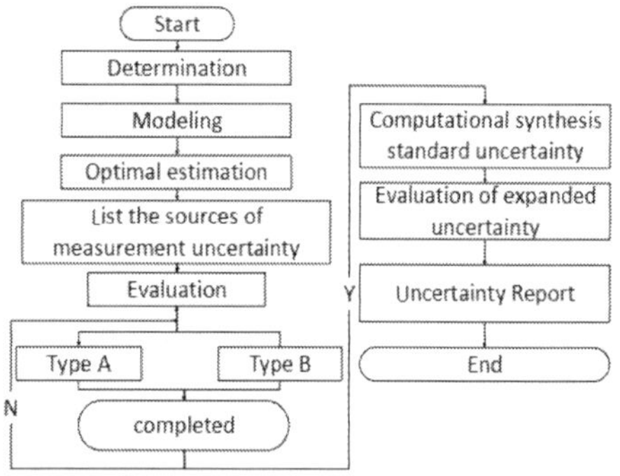

Figure 2. Steps for the evaluation of uncertainty in measurement

2.3 Standard uncertainty calculation

If the half width dispersion interval x_i of the known estimate a_i is given. And the probability of falling in the interval $[x_i - a_i, x_i + a_i]$ was 100%. Then estimate the distribution, we can get the uncertainty of x_i was $u_{x_i} = a_i/k_i$, k_i means coverage factor, generally taken equal to 2-3[3].

2.4 The calculation of combined standard uncertainty

The standard uncertainty component of a function y caused by any standard uncertainty x_i is recorded as:

$$u_{y_i = \left|\frac{\partial y}{\partial x_i}\right| u_{x_i}} = |c_i| u_{x_i} \tag{3}$$

c_i means transmission coefficient, which used to describe the change in the estimated value of the output y caused by the unit variation of the estimated value of the input x_i. Played a role in the spread of uncertainty.

Some measurements are direct. The standard uncertainty of each component is uncorrelated. Synthesis by combination of square root. For most situations the inputs are uncorrelated. Then synthesis with the following formula:

$$u_c(y) = \sqrt{\sum_{i=1}^{n}\left(\frac{\partial f}{\partial x_i}\right)^2 {u_{x_i}}^2} \tag{4}$$

$$= \sqrt{\sum_{i=1}^{n} u_i^2(y)} \tag{5}$$

From the above formula we can know, Combined standard uncertainty means the uncertainty caused by input in the form of superposition of square .When one of the input value has the most uncertainty. The uncertainty of the parameters will be significantly greater than other parameters, means this parameter uncertainty has the biggest impact. Others confirmed to belong to small uncertainty.

2.5 The calculation of the expanded uncertainty

In fact, the confidence probability of the standard uncertainties often in a low position. For example, the confidence probability of normal distribution of combined standard uncertainty is 68%.In order to make the measurement more reliable, we will need to expand the interval of confidence probability, we can use coverage factorkmultiplied bycombined standard uncertainty$u_c(y)$, so that we can get the expanded uncertainty.

$$U(y) = ku_c(y) \tag{6}$$

The value of the multivariate function Y obtained at this time can be written as

$$Y = y \pm U \tag{7}$$

It means the best value of multivariate function Y with maximum probability will falls in the range$[Y - U, Y + U]$

3 Calculation method of vehicle speed and evaluation of uncertainty based on momentum conservation

3.1 The basic theory of momentum conservation in vehicle collision

The law of conservation of momentum is the most common rule in nature. In the actual traffic accident, The collision is often made up of two objects, to establish impact models:

$$m_1 v_{10} + m_2 v_{20} = m_1 v_1 + m_2 v_2 \tag{8}$$

Taking into account the fact that the collision is often similar to the plane collision, $0 - xy$ Coordinate system can be selected, the formula（2-1）can be written as:

$$m_1 v_{10x} + m_2 v_{20x} = m_1 v_{1x} + m_2 v_{2x} \tag{9}$$

$$m_1 v_{10y} + m_2 v_{20y} = m_1 v_{1y} + m_2 v_{2y} \tag{10}$$

The speed can be calculated according to the brake print formula

$$v_j = \sqrt{2\varphi g S_j} \tag{11}$$

Simultaneous formula（9）and formula（10）, we can get v_{10} and v_{20}

3.2 The uncertain parameters in vehicle accident

In the foregoing content, we assumed the length of the brake marks, the friction coefficient and the velocity component angle are already known .However in the

DOI:10.1051/matecconf/201712404012

actual traffic accident, these parameters can not be measured accurately, there is great uncertainty .Observe the brake formula (11), Acceleration of gravity g is a constant, but the coefficient of friction φ and the length of the brake marksall have uncertainty. Their values are $\varphi_{min} \le \varphi \le \varphi_{max}$, $S_{min} \le S \le S_{max}$. Therefore, the instantaneous velocity of the vehicle can be divided as follows:

$$\sqrt{2\varphi_{min}gS_{min}} \le v \le \sqrt{2\varphi_{max}gS_{max}} \quad (12)$$

4 Accident case analysis

A van along the road from west to east to DP Rd and XA Rd .intersection, hit a garbage truck which was along the XA Rd from north to south .Field inspection map shows, Before the collision the truck had brake marks on the road, began to appear the break seal$S_1 = 15.6m$ of right side wheels, then produced the break seal$S_2 = 4.4m$ of both side, At this time the two vehicles collided, there was no break seal of garbage truck before the collision, than two cars move forward together and have brake seal $S_3 = 3 \pm 0.5m$ after the collision, garbage truck

In the above formula, $m_1$means quality of car A, $m_2$means quality of car B, v_{10}means speed of car A before collision, v_{20} means speed of car B before collision, v_1 means speed of car A after collision, v_2 means speed of car B after collision.

overturned eventually .both of their driving direction are south by east after the collision, by supplementary measure, the angle between the brake traces and the north south direction after collision is$\alpha = 20^{\circ}$, The accident occurred on the road is a newly built asphalt pavement.

Table 1. Vehicle parameters.

Vehicle	Curb weight	Total Passenger	Total mass after collision
Van	1920kg	2	2050kg
Truck	10490kg	1	10555kg

4.1 Determination of related parameters and calculation of vehicle speed

This is the scene graph derived from police who deal with traffic accidents, the driving direction and end position of the accident and their respective brake marks are given.

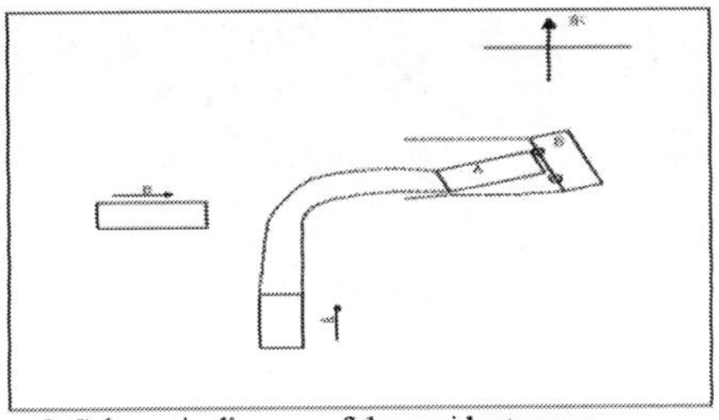

Figure 3. Schematic diagram of the accident scene

According to literature, we can know the range of sliding attachment coefficient is $\varphi_{S1} = 0.60\sim0.70$, $\varphi_{S2} = 0.80\sim0.90 = \varphi_{S3}$ from Table 2[3].

Table 2. Table of reference value of longitudinal slip adhesion coefficient of automobile.

Pavement behaviour		Dry		Damp	
		≤ 48km /h	≥ 48km /h	≤ 48km /h	≥ 48km /h
concrete	new	0.80-1.00	0.70-0.85	0.50-0.80	0.40-0.75
	old	0.60-0.80	0.60-0.75	0.45-0.70	0.45-0.65
Asphalt	new	0.80-0.90	0.60-0.70	0.50-0.80	0.45-0.75
	old	0.60-0.80	0.55-0.70	0.45-0.70	0.40-0.65

According to the law of conservation of momentum to calculate the speed of the collision of two cars, before the collision, truck A from west to East, the garbage truck B from north to south, then the instantaneous speed of van in collision v_1 is

$$m_1v_1 = (m_1 + m_2)v_c\sin\alpha \quad (14)$$

$$m_2v_2 = (m_1 + m_2)v_c\cos\alpha \quad (15)$$

$$v_1 = \frac{m_1+m_2}{m_1}v_c\sin\alpha = 14.87m/s \quad (16)$$

According to the law of conservation of energy, the speed of van before braking v_{10} is

$$\frac{1}{2}m_1v_{10}^2 = \frac{1}{2}m_1g\varphi_{s_1}S_1 + m_1g\varphi_{s_2} + \frac{1}{2}m_1v_1^2 \quad (17)$$

4.2 Evaluation of uncertainty in vehicle speed identification.

Based on the introduction of uncertainty in the paper, take a confidence level at 0.94, At this point the corresponding K = 2[6]. the calculation results of the above parameters are presented in Table 3.

Table 3. Van speed uncertainty component evaluation form

Parameter	φ_{S_2}	S_3
Range	0.05	0.3
coverage factor (k)	2	2
$u(v_i) = a_i/k_i$	0.025	0.15
$u'(v_i)$	0.0312	0.0109
Propagation coefficient	8.7456	2.4779
Relative standard uncertainty component	0.0156	0.0054

According to the above parameters, the relative standard uncertainty of v_{10} is:

$$u_{v10} = \sqrt{\sum_{i=1}^{2} u'^2(v_{BOI})} = 1.65\%$$

Expanded uncertainty is:

$$U' = Ku_{v10} = 2 \times 1.65\% = 3.3\%$$

The range of v_{10} is:

$$v_{10} = v_1 \pm v_1 U' = (53.53 \pm 1.77)km/h = 51.76 \sim 55.30km/h$$

Similarly,

$$v_{20} = (28.56 \pm 0.94)km/h = 27.62 \sim 29.5km/h$$

5 Summary

Through the above analysis, selecting appropriate uncertain parameters for uncertainty evaluation, the identification results of vehicle speed and the relevant judicial authorities are consistent, , and has higher accuracy. The research shows that the uncertainty theory has certain applicability in the automobile accident reconstruction, and it can be used in the actual traffic accident. However, the calculation of uncertainty is more complicated and needs further study.

Acknowledgments

This work was financially supported by The Jiangsu provincial production and research cooperation project, Research on the key technology of vehicle crash accident reconstruction, BY2015051-10.

References

1. China Automotive Technology Research Center.China automotive safety development report:2015[M].Social science literature press, (2015).
2. Yuan Quan, LI Yi-bing.Uncertainty evaluation on the energy method of vehicle traffic accident reconstruction[J].China Journal of Highway and Transport, 15(1):110-112, (2002).
3. GA/T643-2006, typical traffic accident vehicle speed technology in the form of identification[S].
4. Wang Ji-zhong.Study on the Interaction of Tire Tread Element with Road Surface and Tread Element Design[D].Changchun:Jilin University of Technology, (2000).
5. Zhuang Ji-de.Principles of Automobile Tyre[M].Beijing:Beijing Institute of Technology Press, (1996).
6. Lu Zhi-xiong, Yang Rui.Automobile accident identification[M].China Machine Press, (2013).

Music Recommendation System for Human Attention Modulation by Facial Recognition on a driving task: A Proof of Concept

Roberto Avila - Vázquez[1], Sergio Navarro – Tuch[1], Rogelio Bustamante – Bello[1], Ricardo A. Ramírez Mendoza[1] and Javier Izquierdo - Reyes[1]

[1]*Tecnológico de Monterrey, School of Science and Engineering, Calle del Puente 222, Ejidos de Huipulco, 14380, Tlalpan, Ciudad de México, Mexico*

Abstract. The role of music on driving process had been discussed in the context of driver assistance as an element of security and comfort. Throughout this document, we present the development of an audio recommender system for the use by drivers, based on facial expression analysis. This recommendation system has the objective of increasing the attention of the driver by the election of specific music pieces. For this pilot study, we start presenting an introduction to audio recommender systems and a brief explanation of the function of our facial expression analysis system. During the driving course the subjects (seven participants between 19 and 25 years old) are stimulated with a chosen group of audio compositions and their facial expressions are captured via a camera mounted in the car's dashboard. Once the videos were captured and recollected, we proceeded to analyse them using the FACET™ module of the biometric capture platform iMotions™. This software provides us with the expression analysis of the subjects. Analysed data is postprocessed and the data obtained were modelled on a quadratic surface that was optimized based on the known cestrum and tempo of the songs and the average evidence of emotion. The results showed very different optimal points for each subject, that indicates different type of music for optimizing driving attention. This work is a first step for obtaining a music recommendation system capable to modulate subject attention while driving.

1 Introduction

The role of music listening in the vehicle while driving represents an important element of the design of the user experience. Ünal [1] suggests that the presence of music selected by the participant while driving increases the accuracy of lateral control and better time response latency on speed change. Other researchers like Brodsky [2] argues that music has a positive influence in the emotions of the driver, but the effect on driver accuracy is still unclear. Mizoguchi [3] stablishes that music for the driver with slow tempo and medium volume level could work as an effective safety driver assistance. Also, the use of music could reduce the cardiovascular damage of driving by the induction of positive emotions [4]. Unfortunately, the election of soundtracks lacks of a formal methodology, thus is necessary an improvement in order to obtain better and reproducible results in this kind of experiments.

Some procedures have been proposed on the election of music as a safety drive assistance or a comfort element. Kinoshita [5] describes a system based on the evaluation of a visual simulated scenario for the election of a playlist with a specific musical genre. This idea was improved by Krishnan [6], that include the metadata of the song, the musical features and demographic data of the driver. This process of recommendation could be done by different approaches, like a distance of subject punctuation in the article cited above, a self – organizing map [7] or an artificial neural network [8]. The usability and specificity of the variables measured depends on the state of the art in the instrumentation and research on physiological and psychological values.

This work proposes a system that relates the soundtrack listened by the driver with the attention of the driver in the road. Previous work [9] suggests that the number of emotions expressed by a subject is related with the attention level of the subject in the task. Thus, a decrease in the number of emotions detected while listening to music could be used an approximation of the effectiveness of the soundtrack heard by the user as an attention modulator. For this purpose, iMotions FACET module [10] is used for emotion recognition by facial expression.

2 Methodology

For this study, a group of young drivers was recruited. The selected age range was chosen cause it's one of the groups with most accidents related with euphoric driving and distraction in which the emotions of the subject have an important role. Seven volunteers between 19 and 25 years old (2 women and 5 men) all of them with valid driver's license and with knowledge in driving automatic transmission vehicles. All participants were invited to be part of this study and none of them was against him/her will. The subjects were labelled from S1 to S7 for analysis maters. The circuit selected for the driving task is presented in Figure 1. The circuit had a length of approximate 3.5 kilometres with a normal traffic of a scholar zone.

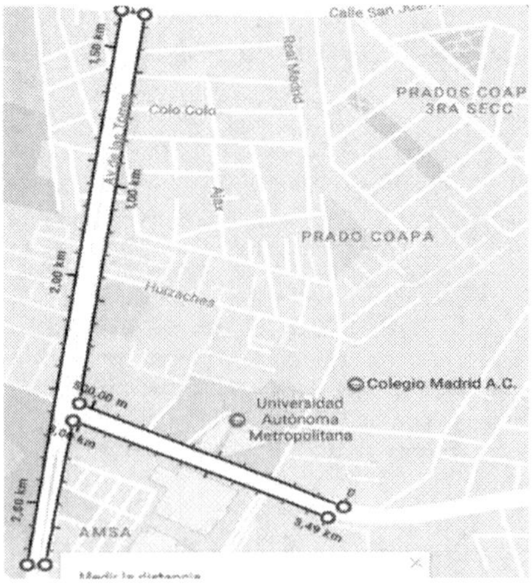

Figure 1. Driving circuit used in the experiment

The experiment was designed with a fixed duration of six minutes and 45 seconds. Six songs of 45 seconds were selected from a free music database [11] characterized by the emotional reaction. This duration was elected for an average speed of 35 km/h and an adequate duration of the test without an effect of fatigue that could affect the results.

Mean tempo (Te) [12] and cepstrum (Ce) were selected as significant characteristics for music classification and relevant impact while driving. The characteristics of these two features in the whole database are presented in Table 1 and the properties ones of the songs selected appeared in Table 2. These songs allow to have an adequate sample of the database with an approximate distribution of tempo and cepstrum that could be used in a response surface model described in Equation 1 [13].

Table 1. Ranges of Cepstrum and Tempo of free music database

Variable	Minimum	Maximum
Cepstrum	3876.44	28718.95
Tempo	52.14	159.94

The ranges of cepstrum and tempo indicates to us a database with high variability of genre in the content of the database. This variability could be used to have an adequate surface response representative of subject behaviour.

The playlist was designed in the same order as it is presented in the table, with a silence of 1 minute as a baseline and 15 seconds of silence between each song.

Table 2. Ranges of Cepstrum and Tempo of songs selected

Song	Cepstrum	Tempo
167.mp3	11126.61	103.19
389.mp3	4453.76	80.98
562.mp3	6972.56	159.67
649.mp3	7301.37	60.79
797.mp3	9307.82	123.96
852.mp3	11864.08	159.94

The driver's face was recorded with a webcam connected to a computer controlled by one researcher that supervises the proof. Also, the computer reproduces the playlist with its loudspeakers. The subject receives oral instructions about the objective of the experiment and the course of the driving circuit. The record was stopped after six minutes and 45 seconds.

iMotions system through the implementation of the FACET module. Such module analyses the face images in order to detect the movement by tension or relaxation of the muscles, identifying the Action Units (AU) after the detection of such units the correlation of the AU by the use of the Facial Action Coding System (FACS) deploys a value related with the probability of the emotion being displayed by the subject. In this test we worked with the evidence records of 9 emotions (joy, anger, disgust, contempt, sadness, surprise, fear, confusion and frustration). These emotions were measured with facial recognition with a rate of 15 samples per second. Emotions presented were counted and the signal obtained were smoothed with a mean filter of 200 samples. The maximum of every period of song were registered and a quadratic model was modelled as is presented in Equation 1.

$$S = p_{00} + p_{10} * Ce + p_{01} * Te + p_{20} * Ce^2 + p_{11} * Ce * Te + p_{02} * Te^2 \quad \text{Eq. 1}$$

Symbol	Variable
S	Average Emotion Predicted
P_{nm}	Weigths
Ce	Cepstrum
Te	Tempo

This model is a common assumption in experimental design without a known response, thus is a preliminary model that could be used to predict attention related to audio features. This model is relevant because includes the interaction between both features, a hypothesis that not had been reported in literature. After the determination of this surface each surface, was optimized in order to obtain the optimal sound characteristics that produces the maximum attention for each subject. This maximization process was obtained with a constrain of tempo between 60 and 160 beats per minute and a mean cepstrum between 5000 and 12000.

3 Results

Figure 2 shows the signal obtained from its number of emotion in the whole proof. The segment while the songs were reproduced are in darker grey and silence in lighter grey.

As we see in Figure 2, the behaviour of the signal along time is not homogeneous. These differences indicate also a variability in the number of emotions expressed across the driving task. This difference could be associated to environmental conditions, but the differences founded in every subject indicates a personal preference to music because every subject have similar driving conditions.

The higher value of the signal post processed (that indicates a major number of emotions expressed in that point) for every subject is summarized in Table 3.

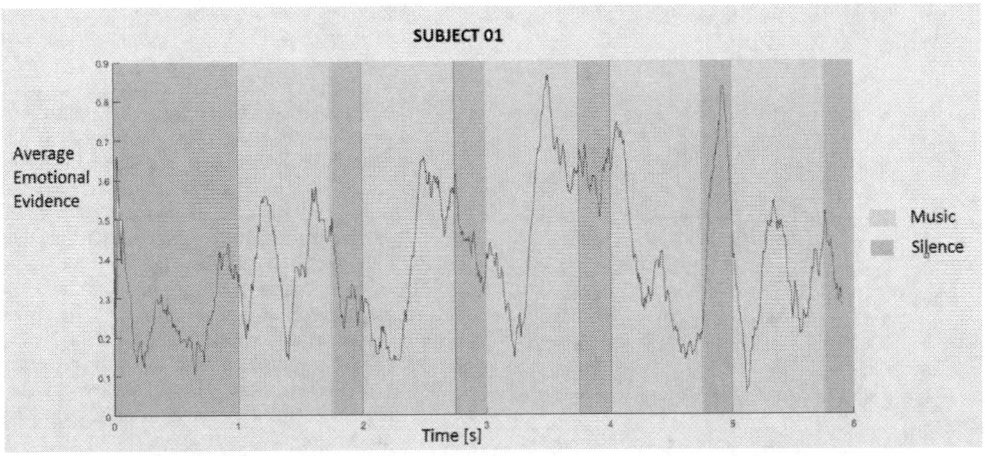

Figure 2. Example of a subject signal of time vs. number of emotions.

Table 3. Maximum number of emotions per subject per song.

Song	S1	S2	S3	S4	S5	S6	S7
167.mp3	0.58	0.75	1.27	1.347	1.77	1.29	0.26
389.mp3	0.65	1.76	1.35	1.28	2.135	1.35	0.35
562.mp3	0.865	1.91	1.51	1.21	2.28	0.995	0.48
649.mp3	0.75	2.3	0.95	1.855	1.525	1.2	0.49
797.mp3	0.83	2.27	1.45	1.38	1.595	1.15	0.62
852.mp3	0.515	1.32	0.31	1.35	1.3	1.08	0.53

The surfaces described in the methodology were obtained for each subject. The model elected in the methodology allows to find a maximum value that indicates an optimal induction of attention. An example of this kind of surfaces is presented in Figure 3. Unfortunately, the parameters for each surface, presented in Table 4, indicates strong differences in each model and optimal values. These results could indicate important differences from each subject, but also problems in the acquisition of each driving test. The parameters obtained for each surface are summarized in Table 4. Some values of the model are small. That situation could indicate a simplification of the equation proposed, but also affects the optimal values for each subject. Thus, its necessary a research on the meaning of each parameter of the model in order to obtain adequate musical features for the creation of playlists.

Information in Table 5 reflexes the high diversity of optimal characteristics as we expected in the diversity of parameters founded in Table 4. The optimal parameters are inside the values of database ranks; thus its values are constrained in typical values of songs. This optimization could be used in the election of individualized playlists, but its efficacy must be proven in future experiments.

Table 4. Parameters of the quadratic forms of tempo and cepstrum vs Average number of emotions

Parameter	S1	S2	S3	S4	S5	S6	S7
p00	-6.04E-01	-6.84E-01	-3.10E+00	1.991	1.568	1.542	-2.57E-01
p10	2.92E-04	1.20E-03	3.01E-04	3.03E-04	-3.30E-04	-1.61E-04	2.75E-04
p01	5.95E-03	-2.47E-02	6.79E-02	-3.06E-02	3.15E-02	7.75E-03	-5.85E-03
p20	-2.14E-08	-1.29E-07	-7.73E-09	-2.25E-08	3.96E-08	1.10E-08	-3.09E-08
p11	2.49E-07	6.90E-06	-2.50E-06	9.31E-07	-3.86E-06	-1.74E-07	2.00E-06
p02	-2.96E-05	-1.27E-04	-2.02E-04	8.02E-05	1.37E-05	-3.93E-05	-3.79E-05

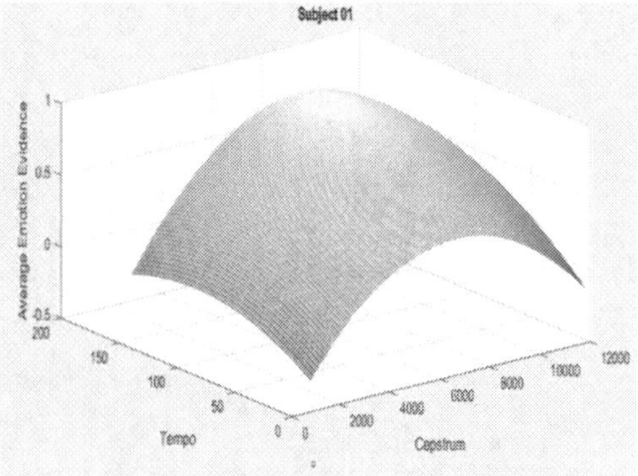

Figure 3. Subject response modelled as a quadratic model

Table 5. Optimal characteristics of songs that maximizes attention

Subject	Optimal Cepstrum	Optimal Tempo
S1	6.47E+03	6.00E+01
S2	7.46E+03	1.05E+02
S3	5.00E+03	1.37E+02
S4	7.97E+03	6.00E+01
S5	1.20E+04	1.60E+02
S6	5.00E+03	8.76E+01
S7	9.62E+03	1.60E+02

4 Discussion and Analysis

This experiment shows an important area of exploration in driving ergonomics. Music playlist could be understood as an active element in the driver's environment, that allows to improve the driver's performance. This election must be transformed from a subjective election of songs to an objective evaluation of the subject based on the acoustic characteristics of the samples. As we see in the present work, it is possible to reduce the experiment to some songs elected in a way that allows to make a model of the subject response and finally a prediction about the kind of songs that could be adequate for every subject.

Table 5 is an interesting example about how individual preferences affects the behaviour of the driver. We were not able to identify a pattern that indicates that users with similar demographics also shares similar music preferences. This methodology allows us to identify this differences and approximates a first results about how to identify optimal characteristics of the songs that must be reproduced in a driving task. A preliminary analysis of the results obtained confirms previous results [2] about the preference of young people of high tempo and aggressive music, that could be associated with lower cepstrum and electronic music compared with acoustic music with higher cepstrum. Not all the electronic music had low cepstrum and other genres could have these characteristics, so the model must be refined in order to obtain an adequate music playlist for driver's attention optimization.

This experiment also was performed on realistic conditions. That implies a problem in face recognition because the variation of sun light, road conditions and obstacles in the way. Ideal conditions could imply a best convergence of the results in recognizable patterns of optimal sound condition, but also could affect the analysis of real driving tasks. This equilibrium and the differences in the subject in both experimental environments must be deeply analysed to adequately extrapolate conclusions between real and simulated driving.

This work presents a first approximation about the relationship between attention, emotion expression and music influence on driving task. The use of a quadratic surface in the optimization method is a valid first approximation, but a real approximation about the significance of the surface must be explored in future work. The mathematical relationship between audio characteristics and the human response could indicate behavioural mechanisms that could not be described yet. Also the inclusion of other audio parameters could make a more complicate and complete model about these responses and give ideas about hidden mechanism of daily life activities.

5 Conclusions

This work shows a method for music playlist creation based on the individual characteristics of the driver and the driving task. Cesptrum and tempo proved to be valuable indicators for music classification and playlist optimization, but other parameters must be included on future experiments. Quadratic model of music features vs. subject response works as a valuable starting point for the evaluation of the subject state, particularly attention while driving. These discoveries must be confirmed by other experiments that include other physiological values. In future work, other sensors can be included to analyse the correlation between the variables mentioned in this work and the impact on the dynamics of the car. An example of future works could include accelerometers measurements in the circuit to correlate the behaviour of the driver and the emotions affected by the auditory stimuli.

References

1. A.B. Ünal, D. de Waard, K. Epstude & L. Steg, *Transportation Research Part F: Traffic Psychology and Behavior* **21**, 52-65 (2013).
2. W. Brodsky & M. Kizner, *Transportation Research Part F: Traffic Psychology and Behavior* **15 (2)**, 162-173, (2012).
3. K. Mizoguchi & S. Tsugawa *Vehicular Electronics and Safety (ICVES), 2012 IEEE International Conference on,* IEEE pp. 117-121 (2012,July)
4. S.H. Faiclough, M. van der Zwaag, E. Spiridon & J. Westerink *Physiology & Behavior* **129**, 173-180 (2014)
5. Y. Kinoshita, Y. Masaki, T. Muto, K. Ozawa & T. Ise *Consumer Electronics, 2009. ICE'09. IEEE 13th International Symposium on* , IEEE, pp. 94-98 (2009, May).
6. A.S. Krishnan, X. Hu, J.Q. Deng, R. Wang, M. Liang, C. Zhu, Y.K. Kwok *IEEE 7th International Conference on Cloud Computing Technology and Science (CloudCom)* (2015)
7. N.H. Liu *EURASIP Journal on Audio, Speech and Music Processing* **20 (1)** 52-65 (2013)
8. N.H. Liu *Multimedia tools and applications* **72 (2)**, 1341-1361
9. R. Avila – Vázquez, R. Bustamante – Bello, R.A. Ramírez – Mendoza, A. Beltrán Fernández, J. Izquierdo Reyes, S.A. Navarro – Tuch *10th International Conference on Advanced Computational Engineering and Experimenting ACE-X* (2016, July).
10. iMotions. iMotions Emotient https://imotions.com/blog/facial-action-coding-system/
11. M. Soleymani, M.N. Caro, E.M. Schmidt, C.Y. Sha, Y.H. Yang *Proceedings of the 2nd ACM International Workshop on Crowdsourcing for Multimedia* pp. 1-6 (2013, October)
12. W. Brodsky *Transportation Research part F: Traffic Psychology and Behavior* **4 (4)**, 219 – 241 (2001)
13. K. Hinkelman, O. Kempthorne *Design and Analysis of Experiments* p.504 (2007)

Optimal location of dry ports in the Pan-Pearl River Delta

Qitao Wu[1], Hongou Zhang[1] and Yang Wang[1]

[1]*Guangzhou Institute of Geography/Guangdong Public Laboratory of Geospatial Information Technology and Application/ Guangdong Institute of Innovation and Development, Guangzhou 510070, China*

Abstract. Dry ports have become a very important part of globalization and a cornerstone in port competitiveness. Based on the analytic hierarchy process, this paper established the dry port index system to evaluate the potential location of dry ports linked to the port system in the Pearl River Delta, in a case study covering 115 cities in the Pan-Pearl River Delta. The results show that it is optimal to locate 21 dry ports in the research region, and each dry port serves different areas. There are 4 optimal transport corridors for the dry port logistic network, including Guangzhou-Kunming transport corridor, Guangzhou-Chengdu transport corridor, Guangzhou-Changsha transport corridor and Shenzhen-Nanchang transport corridor.

1 Introduction

With continuous economic development, coastal port expansion is under growing pressure brought by traffic congestion, land constraints, environmental pollution, etc., hence coastal ports have begun to build inland dry ports. A dry port refers to an inland logistics and distribution center which is located away from and closely linked to the seaport by rail, road, or inland waterway, providing warehousing, customs clearance and other value-added services [1]. The dry port originated in Italy and then spread throughout Europe and North America [2]. It has the advantages of increasing cargo transport, reducing pollution, facilitating transport, enhancing safety and security, and lowering transport costs, and can also promote the economic development of the hinterland. In recent years, coastal ports have started to strengthen logistics cooperation and consolidation with the hinterland by locating and building dry ports there. Dry ports have become a new part of port competiveness in addition to dock facilities and shipping lines as well as a priority in the future development of ports in China and abroad [3].

In their systematic research on the dry port, overseas scholars have summarized the concept, functions, and classification of the dry port[4], analysed its advantages[5], assessed its environmental effects and impacts[6], and analysed the roles of different stakeholders in dry port construction[7]. They have also conducted case studies on numerous dry ports in different parts of the world, including West Europe and North America [3], North Europe [8], India[9-11], Malaysia[12], Russia[13], and China[14]. In China, a growing number of studies on the dry port have been conducted, but most of them are confined to the introduction of the dry port and qualitative research of its economic and social benefits [15-18]. The site selection and location of dry ports is an important aspect of dry port studies, on which domestic and foreign scholars have made many attempts with numerous methods. Studies on optimal location of dry ports mainly use two techniques. The first one is use a systematic network model and data about cargo flow between the coastal port and the hinterland to calculate the location and network structure of dry ports under the condition of minimum logistics costs; models and methods used include the hub-and-spoke network model [19], ant colony algorithm [20], mixed integer arithmetic model [9,21], and bi-level programming model[22]. The second is to establish an evaluation indicator system of dry port location, using methods such as the analytic hierarchy process (AHP)[23-24] and cluster analysis[25-26] to calculate the dry port location indicators of alternative regions for the selection and optimal location of dry ports. Both approaches have their own pros and cons: The first method features dry port location and optimization based on logistics cost control, yet it requires gaining access to complete, detailed data of cargo flow between the coastal port and all alternative dry port locations (origin-destination data), which are difficult to obtain; what's more, such data are not completely recorded and vary from year to year, and may thus affect the simulation results. The second one is frequently adopted nowadays and easy to use, but it hardly takes into consideration actual cargo flow and transport network. Hence it is necessary to conduct an optimization analysis taking the current logistics situation into account. Meantime, how a scientifically reasonable indicator system is established also have some impact on research findings.

As the maritime Silk Road construction keeps advancing, Guangdong Province is speeding up the construction of the Pearl River Delta (PRD) as an international logistics and distribution center, actively locating and building dry ports in the Pan-Pearl River Delta (Pan-PRD) to enhance interconnection with inland regions and create smooth intermodal transport corridors. With the Pan-PRD region as the object of study, this study is intended to, by establishing an indicator system, conduct a systematic evaluation of the scientific reasonability of cities in the region as a potential dry port location, optimize the PRD port system's dry port location and transport corridor network in the Pan-PRD, with a view to providing scientific basis for dry port location and construction of the PRD port system.

2 Research methodology and region

2.1 Research region

The PRD port system, consisting of hub ports, such as Guangzhou Port and Shenzhen Port, and important ports, including the ports of Humen, Zhuhai, Zhongshan and Huizhou, is one of the top three port systems in China. Amid the trend that large-sized container vessels and international liner alliances become prevalent, PRD ports have become vital nodes of global ocean shipping, with a total of more than 320 international shipping lines and 50-plus shipping lines to Taiwan, connecting with over 200 ports in more than 100 countries and regions. In response to increasingly stiff competition in recent years, all the ports have begun to strengthen construction of inland dry ports to expand their service coverage of the hinterland. In 2014, the container throughput of the PRD port system exceeded 46 million TEUs, including 24.03 million TEUs handled by Shenzhen Port and 16.60 million TEUs handled by Guangzhou Port, and the two ports combined to account for 89% of the region's total. The ports of Shenzhen and Guangzhou, both having extensive international shipping lines and complete inland transport network, are the leading ports of external ocean shipping for the hinterland. Therefore, this paper analyzed the location of inland dry ports linked to the PRD port system with the ports of Shenzhen and Guangzhou as the key objects of study.

The Pan-PRD concept was proposed by the Guangdong provincial government in 2003, referring to a regional economic circle consisting of nine provinces and autonomous regions, namely Guangdong, Fujian, Jiangxi, Hunan, Guangxi, Hainan, Sichuan, Yunnan, and Guizhou, plus the Hong Kong and Macao special administrative regions, which is called "9+2" for short. The Pan-PRD has a land area of 1.99 million km2 and a population of 446 million, making up 20.78% and 34.76% respectively of the country's total area and population. The study focused on the location of dry ports linked to PRD ports by rail. Hainan Province is excluded from the study as it can connect with PRD ports by convenient sea transport. The Special Administrative Regions of Hong Kong and Macau are also not included in the research region due to their specialty. The real scope of the study covers eight provinces and autonomous regions in the Pan-PRD, namely the provinces of Guangdong, Fujian, Jiangxi, Hunan, Sichuan, Yunnan, Guizhou and Guangxi Zhuang Autonomous Region, consisting of 115 prefecture-level cities and autonomous prefectures.

2.2 Research methodology

The optimal location of dry ports involves many elements and decision-making comparison. This study used the AHP to explore the most optimal location. The AHP was proposed by Thomas L. Saaty in the 1970s. It models a complex multi-goal decision-making problem as a system and breaks it down into multiple goals or criteria, and then selects an optimal decision from multiple decision alternatives through quantitative analysis of qualitative indicators.

The first step was establishing an evaluation indicator system of dry port location. The site selection of the dry port as a link between the coastal port and the hinterland involves regional transport conditions and infrastructure, economic development and polity support, among other things. The study established the composite index and indicator system of dry port location mainly based on the above three elements (Table 1).

Regional transport conditions and infrastructure are important elements affecting logistics and distribution costs and efficiency. Regional transport capacity is a significant element influencing dry port logistics and distribution; the geographical advantages of regional transport determined by the conditions of transport infrastructure are major elements affecting dry port site selection. (1) The size of regional transport volume. Reflecting the demand and supply of regional transport and logistics market and the development level of the regional transport sector, it is measured by regional freight volume, consisting of rail, road and shipping freight volumes. (2) Regional transport infrastructure. It reflects the conditions of infrastructure connecting regional urban center with the surrounding areas. Good external linkage is a vital foundation for logistics and distribution. It is measured by the number of freeways, national and provincial highways, and railways in the region.

Regional economic development and foreign trade constitute an important foundation of regional external logistics and a determinant of dry port location. (1) Regional economic development has an impact on the region's output of goods suitable for container shipment and the volume of goods needed to be transported into the region for production and consumption purposes. It is measured by regional GDP, regional gross output value of industrial enterprises above designated size, and regional total investment in fixed assets. (2) Container generation and transport is an important product of external trade. Transnational corporations and investment are a driving force of the growth of foreign trade and a cornerstone of the efficient operation of dry ports. This indicator is measured by the actual utilization of foreign investment, total amount of imports, and total amount of exports.

The dry port construction must meet the requirements of national macro-economic and transport development planning. Location priority shall be given to regions with national policy support to create better external conditions for container logistics. (1) Policy conditions. The main consideration is whether the candidate city is a national comprehensive transport hub and a national road transport hub. (2) Rail transport conditions. The main consideration is whether the candidate city is a rail container handling center or a rail container handling station.

Table 1. The indicator system of dry port location.

Objective layer (weight)	Theme layer (weight)	Indicator item (weight)
Regional transport infrastructure (0.2)	Size of regional transport (0.1)	Rail freight volume (0.0249)
		Road freight volume (0.0594)
		Shipping freight volume (0.0157)
	Transport infrastructure (0.1)	Number of freeways (0.0167)
		Number of national highways (0.0167)
		Number of provincial highways (0.0167)
		Number of railways (0.050)
Regional economic development & foreign trade (0.6)	Regional economic development (0.3)	GDP (0.100)
		Gross output value of industrial enterprises above designated size (0.100)
		Total investment in fixed assets (0.100)
	Regional foreign trade (0.3)	Actual utilization of foreign investment (0.150)
		Total amount of imports (0.075)
		Total amount of exports (0.075)
Development policies & external conditions (0.2)	Policy conditions (0.1)	National comprehensive transport hub (0.075)
		National highway transport hub (0.025)
	Rail transport conditions (0.1)	Container handling centre (0.075)
		Container handling station (0.025)

Second, each of the indicators was assigned a weight by expert scoring. In this study, 43 questionnaires were sent to scholars and experts of domestic scientific research institutions and Guangdong provincial transport and port authorities for them to rate the importance of each pair of indicator items on a scale of 1 – 9, and a total of 21 validly completed questionnaires were recovered. The average values of all indicator items derived from the statistics of the questionnaires were calculated by the Data Processing System (DPS) to yield the weights of all parametric indicator items (Table 1). The DPS calculation results indicated that all the weights passed the consistence checks of single hierarchical ranking and overall hierarchical ranking.

Third, the data were normalized to be all positive. The formula of normalization is:

$$Z_i = \frac{A_i - A_{\min}}{A_{\max} - A_{\min}} \quad (1)$$

In which, Zi denotes the normalized datum, and Zi∈[0,1]; Ai denotes the actual value of the indicator; Amin denotes the minimum value of the indicator; Amax denotes the maximum value of the indicator.

Finally, the composite indexes of all cities as dry port location alternatives were calculated. The composite index represents a combination of all elements of each candidate city. The bigger the composite index of a city, the more suitable the city is to be selected as the dry port location.

The calculation formula of the composite index is:

$$C_i = \sum_{i=1}^{n} Z_i \times G_i \quad (2)$$

In which, Ci denotes the composite index of dry port location for the ith city ; Zi denotes the normalized value of the ith element; Gi denotes the weight of the ith element.

2.3 Data Sources

The data used in modeling are sourced from officially published statistical yearbooks and relevant plans. Among them, the quantitative data including regional GDP, gross output value of industrial enterprises above designated size, total investment in fixed assets, actual utilization of foreign investment, rail freight volume, highway freight volume, and shipping freight volume are sourced from China City Statistical Yearbooks, and the total amounts of imports and exports are sourced from statistical yearbooks of the relevant provinces and autonomous region. The data about freeways, national and provincial highways, and railways are sourced from the 2015 edition of China Atlas published by SinoMaps Press. Data about national comprehensive transport hubs, regional road transport hubs, container handling centers, and container handling stations are sourced from the relevant plans published on the website of the PRC Ministry of Transport.

3. Research findings

3.1 Calculation results of the composite index of dry port location

As shown by the calculation results of three indicator items, the three highest rated cities in the calculation results of regional transport infrastructure (Figure 1a) are Guangzhou, Chengdu, and Kunming. In terms of spatial distribution, the index is generally higher for provincial capitals and their surrounding areas and lower for cities far away from the provincial capital. From a regional perspective, the PDR is rated the highest, followed by the Changsha-Zhuzhou-Xiangtang (CZT) region and the region on the West Coast of the Taiwan Straits. In terms of regional development and foreign trade (Figure 1b), the calculation results of the index vary greatly, with most cities rated lower. In spatial distribution, except for the PRD, the index is generally higher for provincial capital cities and lower for other regions. As shown by the calculation results of development policies and external conditions (Figure 1c), the index is evenly distributed among all cities and generally higher for most regions, except some remote cities and regions.

Weighted summing of the calculation results of the three indicator items yielded the dry port location composite indexes of all cities (Figure 1d). The index ranges among the cities from the maximum value of 76.64 to the minimum value of 0.04, with an average of 11.77. As for the distribution of values, the index is higher for a small number of cities; only 31 cities, accounting for 26.96% of all the cities in the research region, are rated higher than the average value. With regard to spatial distribution of the composite index, for the PDR, the region on the West Coast of the Taiwan Straits, and the CZT region, cities with a higher composite index are relatively concentrated; by contrast, in Guizhou, Yunnan, Jiangxi and Guangxi, regions with a higher composite index are scattered, and there is a sound spatial combination of cities at different levels.

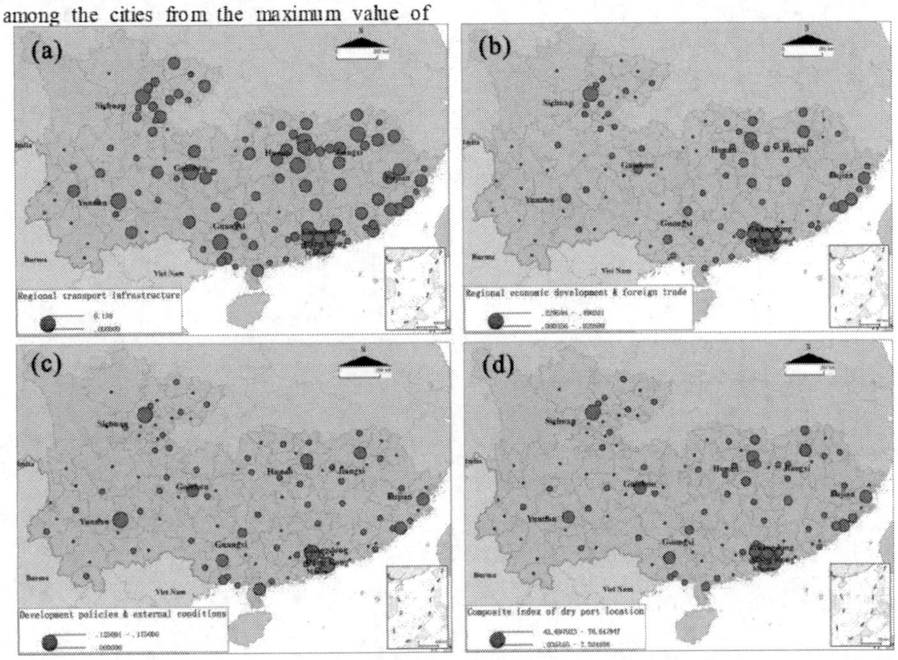

Figure 1. The computing result of (a) regional transport infrastructure, (b) regional economic development and foreign trade, (c) development policies and external conditions, (d) composite index of dry port location.

3.2 Optimal dry port location

As large-sized logistics infrastructure, a dry port mainly serves the city where it is located and the surrounding areas, hence its location shall take into account the conditions of alternative cities themselves and, more importantly, scientifically reasonable spatial distribution. The composite index of dry port location reflects the order of priority of all regions in dry port location. The higher the index of a city, the more probable it will be selected; and the lower, the less probable. However, the composite index only indicates the potential of a city being selected as the dry port location. The spatial distribution of dry ports is not necessarily dependent on the composite index. As illustrated by the spatial distribution of dry port location composite indexes of the cities, cities with a higher index are largely distributed in eastern regions and relatively concentrated. If the number and location of dry ports completely depend on the composite index ranking, it will result in imbalance among the regions and duplicated construction. Optimal dry port allocation requires optimal selection based on considering spatial distribution. Each province shall be considered separately with service coverage divided spatially. On the assumption that each province is allocated with three dry ports and considering that coastal prefectures and cities already have ports, coastal cities in Guangdong, Guangxi, and Fujian are excluded from being an alternative dry port location. The optimal allocation in the Pan-PRD of dry ports linked to the PRD port system was yielded according to the composite indexes of dry port allocation of alternative cities in the provinces and autonomous regions.

It is optimal to locate 21 dry ports in the Pan-PRD (Figure 2), among which, 3 ones are located in Chengdu, Dazhou, and Neijiang of Sichuan Province; 3 ones are located in Kunming, Dali, and Honghe of Yunnan Province; 3 are located in Guiyang, Zunyi and Liupanshui of Guizhou Province; 3 ones are located in Nanning, Liuzhou, and Guilin of Guangxi Zhuang Autonomous Region; 3 ones are located in Changsha,

Hengyang and Yueyang of Hunan Province; 3 ones are located in Nanchang, Jingjiang and Ganzhou of Jiangxi Province; 3 ones are located in Zhaoqing, Shaoguan and Meizhou of Guangdong Province. Each of these dry ports serves different parts of the province. Let's take Guizhou Province for example. The dry port in Guiyang mainly serves central and southeast Guaizhou, the one in Zhunyi serves north Guizhou, and the one in Liupangshui serves west Guizhou, south Sichuan, and northeast Yunnan.

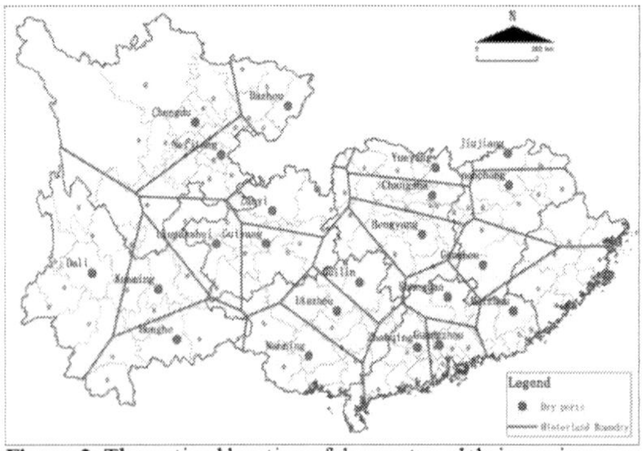

Figure 2. The optimal location of dry ports and their service coverage.

3.3 Optimal transport corridor of dry ports

Dry ports are linked to PRD ports mainly by existing trunk railways, optional depending on distance. All dry ports are connected with PRD ports by rail. From a dry port, containers are transported by rail to PRD ports or inland regions. In the selection of regions to be served by a dry port, the surrounding areas of the dry port are selected as its service area due to their closest proximity. The optimization yielded four transport corridors between dry ports and PRD ports (Figure 3), with all inland dry ports located on the four axial transport corridors.

The Guangzhou-Kunming transport corridor runs from Guangzhou through Zhaoqing and Nanning to Kunming, and can further extend to the prefectures of Dali and Honghe. Linking a total of five cities where dry ports are located, the transport corridor is underpinned by the Nanning-Kunming, Litang-Zhanjiang, Guangzhou-Maomin railways, and in the future transport distance and time through the transport corridor will be further shortened with the construction and operation of the Luoding-Cenxi railway. The corridor is mainly used for container transport between Guangzhou Port and south Guangxi and Yunnan, with a huge potential for development. Its construction has now taken shape. Guangzhou Port has built its "Dry Port in Kunming", shortening the time of container transport between Kunming and Guangzhou Port form 7 days to 54 hours, and realizing one-stop customs clearance in Kunming for local goods. The Guangzhou-Chengdu transport corridor runs from Guangzhou through Zhaoqing, Liuzhou, Guiyang, and Chongqing to Chengdu, also extending to the two dry ports in Panzhihua and Guilin. This corridor involves a total of nine cities where dry ports are located, and among all the corridors, it has the most extensive service scope and links the largest number of cities. This corridor is underpinned by the Chengdu-Chongqing, Chongqing-Guizhou, Guizhou-Guangxi, Jiaozuo-Liuzhou, Guangzhou-Maoming railways, and with the construction and operation of the Liuzhou-Zhaoqing railway in the future, the container transport distance and time through it will be reduced to save costs. Mainly serving north Guangxi, Guizhou, Sichuan, and possibly Chongqing, it is the most important corridor for PRD ports to expand their service scope to southwest China. The corridor is very long with a huge service space, covering many economically developed cities in the region, and therefore has the biggest potential for value growth. The Guangzhou-Changsha transport corridor runs from Guangzhou through Shaoguan and Hengyang to Changsha, and can extend to Yueyang, with the expansion potential for connection with Hubei Province. It runs through four cities where dry ports are located, mainly serving the markets of north Guagndong and Hunan, and is a vital container transport corridor which can indirectly serve central China and even north China. This corridor mainly relies on Beijing-Guangzhou railway for container transport. The Beijing-Guangzhou passenger rail line has been built along this corridor, greatly enhancing its cargo transport capacity, hence the regions along this corridor are key destinations for PRD ports to expand into inland container transport markets. So far the Guagnzhou-Changsha container regular train has been in operation along this corridor. Shenzhen-Nanchang transport corridor runs from Shenzhen through Ganzhou to Nanchang, and can extend to Jiuiang and Meizhou. This corridor involves four cities where dry ports are located and mainly serve Jiangxi Province, indirectly serving Anhui and even north China. It mainly relies on the Beijing-Kowloon railway for container transport and can serve the areas along the railway.

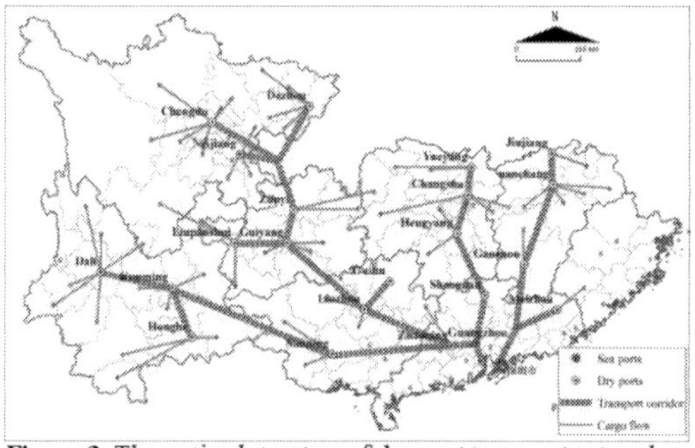

Figure 3. The optimal structure of dry port transport network

4. Conclusion and discussion

As an inland cargo distribution point linked to the coastal port, the dry port is a vital node of the linkage between the coastal port and the hinterland. The dry port construction will increase the coastal port's freight volume, expand its service coverage of the hinterland, facilitate and ease outward cargo flow from the hinterland, and therefore represents an important development direction for the construction and expansion of port service networks at home and abroad. As a case study on the PRD port system, this paper used the AHP to establish a composite indicator system of dry port location and calculate the composite indexes of dry port location for cities in the Pan-PRD region, and then conducted an optimization analysis of dry port location and transport network based on the calculation results.

(1) As shown by the calculation results of the composite index of dry port location, in the PRD region, the region on the West Coast of the Taiwan Straits, and the CZT region, cities with a higher composite index are relatively concentrated, while in Sichuan, Guizhou, Yunnan, Jiangxi and Guangxi, cities with a higher composite index are scattered, and there is a sound spatial combination of cities at different levels.

(2) According to the dry port composite index, it is optimal to locate 21 dry ports throughout the Pan-PRD region, in capital cities such as Kunming, Chengdu, Guiyang, Changsha, Fuzhou and Nanning, as well as economically developed and strategically situated cities like Guilin, Chenzhou, Hengyang, and Shaoguan.

(3) In line with the location of dry ports, it is advisable to build four transport corridors for dry ports, namely Guangzhou-Kunming transport corridor, Guangzhou-Chengdu transport corridor, Guangzhou-Changsha transport corridor, and Shenzhen-Nanchang transport corridor. Each of the transport corridors can use existing rail networks to connect different dry ports, forming an intermodal transport corridor between the PRD port systemand inland dry ports.

The location and construction of dry ports is a systems engineering project, which requires taking into account numerous issues and elements including the economy, environment, investment and industry. The composite index and indicator system of dry port location reflects the reasonability of the location of dry ports in Pan-PRD cities. To build dry ports in Pan-PRD provinces and cities requires systematic planning. In particular, given that the dry port construction needs huge investments, it is necessary for dry ports located in the Pan-PRD to be constructed by phases. In the phase-by-phase construction, all the cities' composite indexes of dry port location shall be taken into consideration. Priority may be given to cities with a higher index, while cities with a lower index can wait until conditions are mature. The dry port construction also needs active cooperation from inland cities. Therefore, to achieve an optimal dry port location requires the PRD port system to, in view of the proactivity and support policy framework of inland cities for dry port construction, cooperate with inland cities to build dry ports in flexible and diverse modes of dry port construction, in a bid to advance, accelerate and improve the location and construction of dry ports. Members of the PRD port system can also team up to build inland dry ports shared by them, thereby reducing the costs of inland dry port construction. Bold attempts shall be made to pilot application of favorable policies in the PRD region to inland dry ports to help them open wider to the outside world and enable the PRD and Pan-PRD to play a still greater role in China's "One Belt and One Road" campaign.

Acknowledgements

This research is supported by the national Natural Science Foundation of China (No. 41301132), the China Postdoctoral Science Foundation (No.2013M530065, No. 2014T70116)., The High-level Leading Talent Introduction Program of GDAS(No.2016GDASRC-0101)

References

1. Q.T. Wu, H. Zhang, Shipping Management **32**, 22-24 (2010)
2. P. Rodrigue, T. Notteboom, Maritime Policy & Management **36**, 165-183 (2009).
3. P. Rodrigue, T Notteboom, Research in Transport Business & Management 5, 4-15 (2012)
4. V. Roso, J. Woxenius, K. Lumsden, Journal of Transport Geography **17**, 338-345 (2009)
5. A Jarzemskis,A V Vasiliauskas, Transport **22**, 207-213 (2007).
6. V. Roso. Transport Research Part D: Transport and Environment **12**, 523-527 (2007).
7. P. Rodrigue, J DebrieJ, A Fremont, E Gouvernal, Journal of Transport Geography **18**, 519-529 (2010).
8. A. Bask, V. Roso, ANDERSSON D, Hämäläinen E. Journal of Transport Geography **39**, 85–95 (2014).
9. K. Y. Ng, G. C. Gujar, Journal of Transport Geography **17**, 346–356 (2009)
10. K. Y. Ng, G. C. Gujar, Transport Policy **16**, 232–239 (2009)
11. K. Y. Ng, J L Tongzon, Eurasian Geography and Economics, **51**(5), 669–682 (2015).
12. J Jeeval, S L Chen, E S Lee, The Asian Journal of Shipping and logistics **31**,109-134 (2015).
13. E. Korovyakovsky,Y Panova, Research in Transport Economics **33**, 25-34 (2011)
14. Y. H. Li, Q. L. Dong, S. W. Sun, Journal of Coastal Research **73**, 641-646 (2015).
15. S. Lu, Y Dong, Shipping Management **29**, 20-22 (2007)
16. G Wang, Port Economy 3, 27-30 (2009)
17. J. L. Yang, X. Li, Journal of Ocean University of China **3**, 40-43 (2010)
18. H. Liang, China Ports, **5**, 26-28 (2014)
19. Q. Wu, H. Zhang, Research on Waterborne Transportation, **4**, 34-37 (2010)
20. X. Q. Song, P. Zhang, Y. J. Guo, Journal of Dalian University of Technology **47**,853-857 (2007)
21. G Crainic, P Dellolmo, N Ricciardi, Transportation Research Part C **55**, 518-534(2015).
22. X. J. Feng, W. Wang, Y. K. Zhang, Journal of Hohai University(Natural Sciences) **41**, 525-530 (2013)
23. K. A. Bian, The Asian Journal of Shipping and Logistics **27**, 331-354 (2011).
24. M. WU, Transportaion Scinece Technology **2**,176-178 (2014)
25. Z. M. Zhang, Journal of Shanghai Maritime University **29**, 34-38 (2008)
26. Y. Wang, J Wang, Railway Transport Economy **4**, 65-70 (2010)

Port Logistics Policies Evaluation Based on Difference-in-Difference Model

Husong Xing[1], Zhehui Zhang[1], Yuanhui Deng[2]

[1]China Waterborne Transport Research Institute Beijing, China
[2]National Academy of Innovation Strategy CAST Beijing, China

Abstract. In order to study port logistics policies' impact on related entities of port logistics, this paper try to use difference-in-difference model (DID model) in port logistics policy evaluation. Reviews related port logistics policies announced recent years in China, evaluates the policies impact on public port enterprises, and demonstrates the feasibility of using DID model to evaluate the port logistics policies. Based on whether a public port enterprises develop port logistics business, defines the public port enterprises which develop port logistics business as the experimental group, others as the control group, and establishes DID model and evaluates the policy influence. Evaluation results show that the port logistics policies have a negative impact on the gross profit of the experimental group, and DID model can be used in port logistics policies evaluation. Perhaps the strengthening of infrastructure construction, diversification and expansion of development channels enhance the operating costs of port enterprises in short-term and result in short-term earnings decline.

1 Introduction

During more than 30 years of Chinese Reform and Opening-up, the development of port logistics has gained great achievements. In order to accelerate the development of port logistics, a series of policies (including laws and regulations, planning, guiding opinions and demonstration projects) have been put forward to promote the development of port logistics in China, such as " Guiding opinions on accelerating the healthy development of international freight forwarders logistics industry ", "Guiding opinions on promoting port transformation and upgrading", "The medium and long-term planning for logistics industry development (2014-2020)", etc. As the official document of the competent government departments related made for the port logistics industry, port logistics policies could regulate and affect the behaviors of relevant port logistics organizations. The policy evaluation can let the policy-makers know whether the effect of policies achieved the expected goals or promoted the target groups' development. So it is necessary to work out a scientific assessment mothed for the port logistics policies evaluation.

Therefore, this paper attempts to analyze the impact of port logistics policies on port logistics enterprises by using DID model. On the one hand, it is of great significance to explore the establishment of an objective and quantitative study tool for evaluating port logistics policies, and on the other hand, it has important guiding significance to improve the implementation of port logistics policies.

At present, many scholars have evaluated the impact of port and logistics policies on port logistics entities. Some scholars researched the impact of port management system's reform on the port operations efficiency and benefits by quantitative or qualitative research methods (mainly by qualitative method), such as Tomás Serebrisky[1] (2011), Sophia Everett[2] (2011), MC Ircha [3] (2011) and José I. Castillo-Manzano[4] (2008) studied the port reforms in Argentina, Australia, Canada and Spain by quantitative method respectively. But there were few studies on the influence of port logistics policies in China, and only a small number of researchers evaluated the policies impact on port logistics enterprises preliminary, such as the influences of the port law[5], port investment and financing policies[6], value-added tax policy[7], the Shanghai Free Trade Zone on[8], and lacked the impact of port logistics policies studies by quantitative methods. In terms of DID model, it has been applied to evaluate the policies impacting on fields of the medical and health care[9], the rural tax and fee reform[10, 11], the real estate policy reform[12] and many others, since Zhou Li'an first introduced the DID method in 2005. Specifically in the field of transportation and logistics, only Wang Yao[13] (2014) and Song Xiaoli[14] (2015) used the DID model to analyze the impact of transportation infrastructure on cities. So there is no research on the port logistics policies evaluation on the real economy by the application of DID model.

2 Port Logistics Policies Review in China

Although the port industry and the logistics industry have been developing for many years, it is difficult to define the accurate and high-level recognition for port logistics because it involves a large number of industries and more logistics activities of business forms. Because there is no unified definition of port logistics, there is no specific policy that can be called port logistics policy in policy-making. Therefore, this paper evaluates a series of policies for port logistics which comes from port and logistics policies. We consider the port loading and unloading and stockpiling associated inside the port area as port industry, and consider activities related to port industry outside the port area as port logistics business[15], such as dredging port transportation, warehousing, freight forwarding, customs clearance agent, logistics financial, etc. Therefore, this paper will consider the policies of dredging port transportation, warehousing, freight forwarding, customs clearance agent, logistics finance and other related to port industry businesses as port logistics policies.

Since the official document of “Planning for the logistics industry restructuring and revitalization” was announced in 2009, the logistics industry has been developed rapidly in China. There were a number of policies announced for the logistics industry development. “The guidance of transportation to promote the healthy development of the logistics industry” in 2013 clearly put forward to enhance the logistics service capabilities of the traditional transport hub, support the traditional transportation and loading-and-unloading services to extend to modern logistics services, and rely on ports, inland water ports and other port resources to enhance the international logistics service capabilities. “Guidance on accelerating the healthy development of international freight forwarders logistics industry” in 2013 put forward to build a number of international logistics enterprises with strong competitiveness of large-scale characterized by prominent main business, advanced management mode and overseas network sound. “Guiding opinions on promoting port transformation and upgrading” in 2014 clearly put forward to develop the port logistics, support port enterprises to develop logistics real estate, supply logistics services of all-round multi-level, encourage port enterprises to extend the port logistics industry chain by diversified operation, such as cooperation with shipping, trade and other related businesses based on port industry. “The medium and long-term planning for logistics industry development (2014-2020)” in 2014 clearly put forward to promote the development of international logistics actively, strengthen infrastructure construction of the hub ports, airports, railways, highways, , form a group of international freight hub, enhance the capacity of import-and-export goods distribution, improve the facilitation level of international logistics, support the logistics enterprises joint with each other, build an international logistics service network and competitive internationally transnational logistics enterprises

The port and logistics policies mentioned above, proposed relevant policies to promote the development of port logistics, such as supporting port logistics enterprises to strengthen infrastructure construction and operate diversification, supporting international freight forwarding enterprises transformed to international large-scale logistics enterprises, supporting the diversification, innovation and developing multi-level logistics services for the port enterprises, etc. These policies pointed out the development direction for the relevant port logistics enterprises, and provided opportunities and supports for the related enterprises to extend logistics businesses to some extent.

3 Establish DID Model for Port Logistics Policy Evaluation

3.1 The basic principles of DID model

The DID model has been widely used to evaluate the effect of policy or project implementation in by quantitative econometrics recent years. Usually studying a wide range of policies is different from the general scientific research, because it is difficult to guarantee completely random for distributing the samples between the experimental group and the control group. Experiments of the non-stochastic allocation between experimental group and control group can be called natural trial, which has a notable feature that there may be differences between different groups of samples before the implementation of the policy. It may ignore these differences only through a single before-and-after comparison or lateral comparison, which will lead to biased estimates for the effects of policy implementation. DID model based on the data from natural trial can control the differences among the study objects of different group by effectively modeling, and separate the effective impact of the policy.

In the aspect of policy evaluation, the DID model can control the impact of some factors other than the intervention factors by combining the “before-and-after” and “horizontal-difference” effectively. At the same time, adding other covariates which may affect the outcome variables can further control the suspected factors between experimental group and control group, which may make up the defect of natural trial that the distribution of samples can't separate completely random. Then we could get a true policy evaluation.

There are three hypotheses needed to be fulfilled in DID model building: (1) The experimental project carried out can affect variables of the experimental group and can’t affect the relevant research variables of the control group, namely the experimental project carried out can only change the relevant research variables of the experimental group; (2) During the experiment, the experimental group and the control group have the same characteristics and macro environment which has the same influence expect the control variable;(3) Some important characteristics of the experimental and control groups are stable and don’t change with time, which remain stable throughout the whole experimental project.

The core of the DID model is the construction of DID estimator. By combining the simple before-after comparison (before and after intervention) and the simple cross-sectional comparison (the experimental group vs. the control group), we can get the following formula:

$$d = \Delta\bar{Y}_{treatment} - \Delta\bar{Y}_{control} = (\bar{Y}_{treatment,t_1} - \bar{Y}_{treatment,t_0}) - (\bar{Y}_{control,t_1} - \bar{Y}_{control,t_0}) \quad (1)$$

d is the DID estimator, Y is outcome variables, the subscript $treatment$ and represent the experimental group and control group respectively, t_0 and t_1 represent the experiment before and after the experiment respectively.

3.2 The applicability analysis of DID model

The objects of port logistics policies evaluation is usually ports, logistics, shipping and other related enterprises. This paper attempts to use DID model to evaluate the impact of port logistics policies on port enterprises. According to statistics, there are 16 listed port enterprises in China. Nine of them regard the port logistics sector as a separate module, and the others don't regard the port logistics sector as a separate module when they make main business cost and income statistics. Therefore, we can divide the 16 listed port enterprises into two groups according to the listed port enterprises whether regard the port logistics sector as a separate module, one group is port enterprises engaged in port logistics business, the other group is not engaged in port logistics business. At the same time, the port enterprises are facing the same macro policy environment and the economic environment, and the important business indicators of the port enterprises are relatively stable. So we can make such a treatment for the port logistics policy set: the nine listed port enterprises engaged in the port logistics business can be regarded as experimental group which affected by the port logistics policy set, and the other seven port companies can be regarded as control group. We could find the port logistics policies dense announced in 2013 and 2014 according to the port logistics policies review, so we could consider it as accepted experiment after 2014, and consider it as control before 2013.

We find evaluate the impact of port logistics policy set on the operation of port enterprises quantitatively by using the DID model can basic satisfy the three basic hypotheses of the DID model building. Therefore, we can use DID model to evaluate the impact of port logistics policy set on the operation of port enterprises quantitative.

3.3 Construct DID model for port logistics policies evaluation

This paper selects 16 listed Chinese port enterprises as samples. Regards Lianyungang port, Rizhao port, Yantian port, Yingkou port, Shenzhen Chiwan port, Jinzhou port and Nanjing port enterprises which don't regard the port logistics sector as a separate module when they make main business cost and income statistics as the control group, and regards Beibu Bay port, Dalian port, Shanghai port, Xiamen port, Tangshan port, Tianjin port, Zhuhai port, Chongqing port and Ningbo port which regard the port logistics sector as a separate module when they make main business cost and income statistics as the experimental group. At the same time, make the data of 2012 as pre-test data, and make the data of 2015 as the experimental data after the port logistics policies announced. Then we could establish the port logistics policies evaluation model as follows.

$$Y_{it} = b_0 + b_1 \bullet year_t + b_2 \bullet policy_i + b_3 \bullet year_t \bullet policy_i + \theta X_{it} + \varepsilon_{it} \quad (2)$$

In formula (2), Y_{it} is the dependent variable, $year_t$ and $policy_i$ are dummy variables representing the time and the group respectively, namely the variable $year_t$ for 2012 is 0, the variable $year_t$ for 2015 is 1. The variable $policy_i$ of the experimental group is 1, while the control group is 0. $year_t \bullet policy_i$ is the interaction of dummy variables of time and group. This paper selects gross profit of the 16 listed port enterprise as the dependent variable. At the same time, in order to better explain the dependent variable, the paper also introduces some control variables, including total assets and net assets of the port enterprise, total import-and-export volume of the port city, port throughput, fixed assets of whole society in the port city. Therefore, introduce the vector X_{it} as the combination of these control variables, make θ as the coefficient vector come from the regression of the control variables. ε_{it} is the residual, subscript i represents each individual, subscript t represents different time points.

According to the basic principles of DID model, $policy_i = 0$ when the port belonging to the control group, the mathematical expectation of time effect before and after the introduction of port logistics policies announced as follows.

$$E(Y_{policy=0,year=1}) - E(Y_{policy=0,year=0}) = b_1 \quad (3)$$

$policy_i = 1$ when the port belonging to the experimental group , the mathematical expectation of time effect before and after the introduction of port logistics policies announced as follows.

$$E(Y_{policy=1,year=1}) - E(Y_{policy=1,year=0}) = b_1 + b_3 \quad (4)$$

Therefore, the policy effect of port logistics policies is as follow.

$$d = [E(Y_{policy=1,year=1}) - E(Y_{policy=1,year=0})] - [E(Y_{policy=0,year=1}) - E(Y_{policy=0,year=0})] = b_3 \quad (5)$$

4 Empirical analysis of The Port Logistics Policies Effect

4.1 The results of Port logistics policy evaluation

This article selects total assets and net assets of the port enterprise (TAP and NAP), total import and export volume of the port city (IEVPC), port throughput (PT), fixed assets of whole society in the port city (FASPC) as the independent variables, selects gross profit of the port enterprise (GPPE) as the dependent variable, and then analyzes the port logistics policy set impact on the gross profit of the port logistics enterprise. Related data could gained from the Statistical Communique on Social and Economic Development of the Port City and Annual Report of Listed Port Enterprises. Using DID model established above for port logistics policies evaluation, we can calculate and obtain the results as shown in Table 1.

Table 1. Results of DID model for port logistics policies evaluation

Variables	Dependent Variable: Gross Profit of Port Enterprise		
	Coefficient	*T-value*	*P-value*
Policy	6.655379	1.76	0.092
Year	-6.45891	-1.74	0.095
Policy × year	-9.055916	-1.82	0.082
TAP	-0.0698608	-1.73	0.098
NAP	0.2466289	3.76	0.001
IEVPC	0.0026239	1.95	0.064
PT	2.851475	2.02	0.055
FASPC	0.0011856	2.7	0.013
c	-6.453745	-1.98	0.059
R^2		0.945	

In table 1, we can see that the relevant parameters passed the T-test and F-test, which indicates that the results of regression model has robustness and significates. Among them, the parameter policy × year characterized the policy effect has a value of -9.055916, which shows that the port logistics policy has a negative impact on the gross profit of port enterprises of the experimental group which carry out port logistics business and regard the port logistics sector as a separate module.

4.2 Analysis of the evaluation results

According to port logistics policies review above, we can see that supporting port logistics enterprises to strengthen infrastructure construction and operate diversification, supporting international freight forwarding enterprises transformed to international large-scale logistics enterprises, supporting the diversification, innovation and developing multi-level logistics services for the port enterprises are the main port logistics policies and measures carried out in China. These policies and measures will greatly increase port enterprises investment in a short time. For example, in order to build inland dry ports, port logistics enterprises should expend a lot of money on infrastructure construction. In order to start a new business and manage diversified, the port logistics enterprises need expend a lot of money on market development and training costs, which may have an adverse impact the short-term profit of port logistics enterprises. Meanwhile, the port logistics policy announced mainly puts forward the development direction of port logistics enterprises, and don't have corresponding solutions for the development cost and the existing problems, which the port logistics enterprises have to bear alone. Those could cause cost increase and earnings decline in the development of the port logistics enterprises.

5 Conclusions

This paper introduces the DID model into the port logistics policies evaluation, and establish port logistics policies evaluation model for the object of port enterprises. In the empirical case, we set 2013 and 2014 as period of policies action, set 2012 as pre-experiment and 2015 as after the experiment, and then evaluate. Because promoting infrastructure ability, diversifying business, expanding the development channels will increase the burden of enterprises in the short term, the results show that the port logistics policy set have a negative impact on gross profit of the port enterprises which develop the port logistics business and regard the port logistics sector as a separate module, the results are credible. We can't have the long-term evaluation results because the implementation of the policy is relatively short, the long-term impact on port enterprises remains to be studied, but it is proved that DID model can be used in port logistics policies evaluation.

References

1. Tomás Serebrisky. "*An assessment of port reform in Argentina: outcomes and challenges ahead*", Maritime Policy & Management, vol. **32,** 2011, pp.191-207.
2. Sophia Everett. "*Port reform in Australia issues in the ownership debate*", Maritime Policy & Management, vol. **34**, 2011, pp.107-119.
3. M. C. Ircha. "*Port strategic planning: Canadian port reform*", Maritime Policy & Management, vol. **34**, 2011, pp.125-140.
4. José I. Castillo-Manzano. "*Economic analysis of the Spanish port sector reform during the 1990s*", Transportation Research Part A, vol. **42**, 2008, pp.1056-1063.
5. ZHANG G J. "*Study on the impact of port law on corporation performance of Guangzhou port group*", Southwest Jiaotong University, 2003.
6. LI Y N. "*Evaluation on the port investment and financing policy*", Shanghai Maritime University, 2007.
7. GUO Y P. "*The influence of value-added tax policy transformation on logistics enterprise*", Shang Ye Jing Ji, vol. **5**, 2009, pp.109-110.
8. FU X. "*Research of China(Shanghai) Pilot Free Trade Zone's significance of transforming , development and value of port logistics corporate*", Shanghai Jiao Tong University, 2014.
9. Torche F. "*The effect of maternal stress on birth outcomes: exploiting a natural experiment* " Demography, 2011.
10. ZHOU L A, CHEN Y. "*The policy effect of tax-and-fees reforms in rural China: a Difference-in-Differences estimation*", Economic Research Journal, vol. **8**, 2005, pp.44-53.
11. WANG W, AI CH R, CAO H. "*Research the tax and fee reform influence on the rural residents' consumption*", Management World(Monthly), vol. **1**,2013, pp.89-100.

12. LIU J Y. "*The evaluation of reform of property tag and effect on the household saving*", Huazhong University of Science& Technology, 2014.
13. WANG Y, NIAN M. "*High speed railway and urban scale expansion - based on empirical research of China*", Finance & Economics , vol. **10**, 2014, pp.113-122.
14. SONG X L, LI K W. "*The effects of improving the quality of traffic infrastructure on urban population scale—An empirical analysis based-on railway speedup*", Modern Economic Science, vol. **5**, 2015, pp.19-26.
15. XING H S, ZHANG ZH H. "*The effect of developing port logistics on the port industry*", Shipping Management, vol. **3**, 2015, pp.14-17.

MATEC Web of Conferences **124**, 05003 (2017) DOI:10.1051/matecconf/201712405003
ICTTE 2017

Challenges and Strategies of Abidjan Port-Hinterland Connectivity

Pascal Kany Prud'ome Gamassa[1] and Yan Chen[2]

[1-2]*Dalian Maritime University, Transportation and Management College, No. 1 Linghai Road, 116026, Dalian, China*

Abstract. Abidjan port-hinterland is composed of three countries which include Ivory Coast, Mali and Burkina Faso. The fast economic growth of the above countries resulted to a greater demand for the foreign goods from their population, causing a congestion problem at Abidjan port. Because of this problem, the connectivity movement between the port and its hinterland has considerably slowed down. In this article a qualitative methodology is used to analyse the Abidjan port-hinterland connectivity. The constant delays of the train services, the road congestion and the lack of coordination and communication between Abidjan port and its hinterland customers are the main presented challenges. The results of this research demonstrate that the poor quality roads and railways and the lack of a new dry port in Ivory Coast greatly affects the fluidity of the Abidjan port-hinterland connectivity. This article suggests some strategies on how to solve the current challenges and will be helpful in deciding on how the Abidjan port should be developed better and improve its connectivity with its hinterland.

1 Introduction

Nowadays, every major port of West Africa aims to develop and expand its hinterland, thus there is a high level of competition among the ports in this region. Abidjan port is a port which looks forward to becoming the hub port of the entire west coast of Africa. To reach this objective it is important to maintain the current customers of the port and also attract new customers and investors. Indeed, the more customers the Abidjan port can get, the more revenue it can obtain. As regard to maintaining its position of a leading port in the region, it is important to facilitate the accessibility of the Abidjan port to its hinterland. Port is the foundation industries of the national economy and critical infrastructure of transport system [1]. The hinterland of a port is basically made up of areas where the port serves and can be served from [2]. Despite the fact that in West Africa, customers of the hinterland pay greater attention to the distance between their current location to the nearest ports and the facilities, the rates and the services proposed by these ports, but nowadays their connectivity to the chosen port is one of the first priority they think about in making their final decision. Analysing Abidjan port-hinterland connectivity is of great importance for the Abidjan port Authorities as it will help to develop a better plan for the development of Abidjan port. Construction and development of ports logistics can support the regional and even the country's economy [3]. In this article, a qualitative methodology is applied and face to face interviews were conducted with representatives of Abidjan port Authorities, some regular customers of Abidjan port, some officials of the Sitarail Company that is in charge of linking Abidjan port to its hinterland by railway, some representatives of private companies operating in Abidjan port, some officials of Ivory Coast's customs administration, some representatives of the Ivorian Shippers' Office, the Burkinabe shippers' Council, the Malian shippers' council and other stakeholders. The data being used were collected from the official websites of the World Bank database, the distance calculator and the interviews. This research aims at analysing the challenges currently faced by the Abidjan port-hinterland connectivity and providing some solutions to those problems.

2 Literature review

The transportation system between a port and its hinterland plays an important role in their connectivity. The port and the hinterland bear a relationship of facilitating and conditioning each other [4]. Because of the current high competition among ports of the same regions in the world, topics related to port-hinterland supply chain management and connectivity, are of great interests for many researchers. Researchers who have already written papers on similar topics have in majority noted that multimodal transport should be developed and a port and its hinterland can never be dissociated, thus thinking about developing a port automatically means developing the port connectivity with its hinterland. They wrote about different aspects of port-hinterland connectivity. This article is the first to present a complete analysis of the current challenges in the overall Abidjan port-hinterland corridor connectivity, using two routes, Abidjan-Bamako route and Abidjan-Ouagadougou route, and to provide concrete solutions to this particular case.

M. Zhang, B. Wiegmans and L. A. Tavasszy presented up to date insights of intermodal transport development in port of Shanghai, port of Rotterdam and their hinterlands via a comparative study. Their paper introduced the recently completed or underway projects carried out in both of the hinterland transport systems. They concluded that at this moment, construction of rail facilities near the Shanghai port is not effective, because the capacity for container transport in the national rail network is very limited and the newly launched sea port container terminals cause increased congestion on motorways as a result of the providing more handling capacity of the sea side. The different projects helped to improve the accessibility of the port of Rotterdam, but still further improvements are needed as especially road transport access of the port of Rotterdam is limited, difficult to extend, and vulnerable for accidents [5]. Davide Infante, Giuseppe Paletta and Francesca Vocaturo in their article focused on an intermodal freight transport service in which containers represent the moved loading units. With a view to minimizing the total cost of the service, they formulated a ship-truck intermodal transportation problem as a Travelling Purchaser Problem (TPP), in this way broadening the real-world applications field of the TPP. To tackle this problem, a heuristic algorithm was presented and experimentally tested. Computational results indicated that the algorithm was very efficient on a set of benchmark instances, quickly achieving optimal or near-optimal solutions [6]. On the basis of analyzing customer's choice behavior, Nannan Hu, Jie Wang and Dong Mu in their paper constructed a utility function of port and obtained an equilibrium pricing strategy before and after competition and cooperation by applying Nash equilibrium. They also analyzed the impact of hinterland traffic condition on the competition and cooperation between ports. The analyzed results demonstrated that with the improvement of hinterland traffic condition, part of ports use cooperation strategy, which can not only improve the overall profit and share of the cooperative ports, but also can enhance the interest and market share of non-competitive ports so that the improvement of overall interest and market share of regional port group can be realized [7]. ShuHai Zhao, Wei Liang and Dan Han in their paper put forward a logistics information sharing platform framework facing e-commerce for an international Seaport of China abbreviated QDIP. They think that the platform changed all aspects of port services information passive situation, and not only improved the customs clearance and regional logistics operation efficiency and reduce logistics costs and increase trade opportunities, but also contributed greatly to the improvement of port business environment. Through the platform, enterprises can reduce intermediate links and costs and obtain economic benefits from goods circulation and cash flow [8]. Benyi Xie, Huiyuan Jiang, Jin Zha and Xiang Li in their article established a Logistic exponential regression model, applied marginal analysis and elastic analysis to analyze the interaction effects between Wuhan port and its hinterland's economy. They proposed that the construction of Wuhan Yangtze should be enhanced to better serve for regional economy development and guarantee the middle boost strategy. Their findings showed that the marginal effect of Wuhan port cargo handling capacity on Wuhan GDP increases every year [9]. Fedele Iannone's article presented an optimization model for the economic analysis and strategic planning of port-hinterland container logistics systems. The model was employed to investigate the inland multimodal distribution of import/export containers handled at the seaports located in the Campania region of Southern Italy. A numerical prototype was formulated and solved using a high-level programming language for large-scale mathematical optimization problems. The results found demonstrated how the competitiveness of the regional container seaport cluster can be boosted by an inter-port based extended gateway system with adequate customs facilities and improved railway connections [10]. Charles Kipkoech Kotut and Dr. Fred Mwirigi Mugambi in their paper focused into the influence of hinterland logistics and transport connectivity on the port of Mombasa with the objective of determining the northern corridor inefficiencies and how they influence performance of port of Mombasa. They employed a descriptive research to describe the state of affairs as it exists and a stratified sampling method to reach at most representative respondents in the study universe of fifty-eight respondents covering port users and clearing and forwarding agents, port officials and leading transport and logistics organizations: KSC and KMA. Questionnaires and interviews were used by the researchers as instruments to collect data and finally analyze and synthesize for presentation using Microsoft Excel data analysis software for purposes of achieving objectives. The findings of the research effectively revealed that the influence of hinterland transport inefficiencies on performance by Kenyan Ports Authorities was glaring. Northern corridor inefficiencies was found to contribute to slow uptake of cargo into the hinterland leading to high truck turn round time and therefore high cargo dwell time at the port, leading to a conclusion that indeed hinterland transport connectivity plays a significant role in the success of ports [11]. Elvira Haezendoncka, Michael Doomsa and Alain Verbeke developed a new governance perspective on port–hinterland linkages and related port impacts, by using a governance lens to assess this farsighted contracting challenge. They used the PHI matrix in their empirical, governance-based analysis of contractual relationships between the port authorities in Antwerp and Zeebrugge, and their respective stakeholders. Their findings revealed that the used matrix represents a critical complement of traditional value added and employment-related impact analyses, which are typically restricted to a narrow geographic zone and are devoid of insight into the nature of contractual relationships with actors located outside of the port and immediately adjacent geographic areas [12].

However, An Caris, Cathy Macharis and Gerrit K. Janssens in their paper, developed a service network design model for intermodal barge transport which was applied to the hinterland network of the port of Antwerp in Belgium. Selected cooperation schemes were simulated by means of a discrete event simulation model for intermodal barge transport and compared with simulation results of bundling in the port area. They concluded that cooperation between inland terminals offers an opportunity to attain economies of scale, but may not be perceived as a sole solution for reducing waiting times of inland barges at sea terminals. Thus a combination of bundling measures in the port area and in the hinterland may be necessary to improve the intermodal transport chain [13]. In his article, Dr. Olivier Podevins highlighted the decisive role of terminals in the hinterland and the changing structure of the terminal network in Western and Central and Eastern Europe (CEE) based on the empirical evidence and market observations. In this concern he firstly considered the main seaports of the Northern Range and their connected terminal network system in Western Europe, as well as the development on the Mediterranean/Black Sea Range and the terminal network in Central/Eastern Europe. He mentioned in his findings that strategic partnerships between seaports and in land waterways ports such as Duisburg, and the ARA-Ports or between Constanta and Vienna has to be seen as a long term process in order to reduce port congestion and increase scales. And he added that intermodal terminals located on existing transport corridors by rail, road and inland navigation between the heart of the EU, the Mediterranean, East and Central Europe provide important new impulses for strategies in an enlarged European Union [14]. Ying Wang and Jian Wang after having studied the regional economy and transportation situation of the western side of the Taiwan straits port group, firstly set up an evaluation index including factors that influence the dry port, secondly used Fuzzy Cluster Analysis to select and rank the candidate locations, and finally classified three grades of locations, each requiring a different construction strategy. Their findings revealed that from the view of geological location and promotion of regional economy, Longyan, Sanming, and Nanping cities should be considered as first grade locations to build dry port, and the cities of Changsha and Nanchang should be considered as second grade locations, while the other cities should be considered as last grade locations [15].

3 Overview of Abidjan port and its hinterland

Inaugurated in 1951, Abidjan port is the major port of Ivory Coast. It has access to the Atlantic Ocean through the Vridi canal and also plays a great role in the economic development of the country. It serves as a gateway for Mali and Burkina Faso. The Abidjan port hinterland is thus considered to have a total population of about 52.28 million people and a surface area of about 1,836,872 square kilometres. Ivory Coast has cobalt, nickel, copper, iron, diamonds and natural gas as main mineral resources. Mali has salt, phosphates, uranium and gold as the main natural resources. Burkina Faso has manganese, gold, marble as the main mineral resources and its economy greatly relies on the agricultural production. Because mineral resources are great source of revenue for these countries, therefore it is important for their authorities that these resources can be easily exported. On the other hand, the large profits made by these countries from selling their natural resources have contributed to the improvement of the living conditions of their population. This resulted into a net growth of these countries' Gross Domestic Product which has created a demand problem. Because of the higher consumption power, population of these regions consumes more imported goods. Most of these goods transit through Abidjan port. The Table 1 illustrates the Abidjan container throughput from the year 2007 to the year 2014. It is important to note that by the time this paper was written, the Abidjan port container throughput of the year 2014 was the latest data available in the World Bank database [16].

Table 1. Abidjan port container throughput annual, 2007-2014; Measure: TEU (Twenty-foot Equivalent Unit); Source: World Bank data 2015.

Years	Abidjan Port Container Throughput
2007	590,306
2008	713,625
2009	677,029
2010	607,730
2011	642,371
2012	690,548
2013	745,102
2014	783,102

The Abidjan port is expected to reach 15,952,521 TEUs by 2020 [17]. From Table 1 it can be seen that the container throughput of Abidjan port kept increasing from 2007 to 2009. Due to the political crisis that Ivory Coast went through in 2010, its container throughput decreased in that year but restarted to increase after the crisis in 2011. Since then a constant growth has been noticed. It is important to mention than since the year 2011, Ivory Coast became the second economic power in West Africa after Nigeria which is the first, and the Abidjan port authorities hope to attract more customers, investors and ships in the coming years. Mali and Burkina Faso economies positively grew notably due to the increase of the selling prices of the mineral resources that these countries export. For the year 2017, Ivory Coast's doing business rank is 142, Mali's is 141 while Burkina Faso's is 146 over 190 countries all over the world [18]. Their ranking positions are among the best in the West African region. These data show that these countries are normally attractive for potential investors.

4 Abidjan port-hinterland connectivity challenges

The connectivity between Abidjan port and its hinterland is mainly done by railways and roads. This was facilitated by the fact that the three countries belong to the Economic Community of West African States (ECOWAS) which guarantees citizens of this region to freely move their goods from one country to another. The different challenges of Abidjan port-hinterland connectivity are thus introduced in two routes, Abidjan-Ouagadougou route and Abidjan-Bamako route. It is difficult to underestimate the transportation disadvantages that landlocked countries face in international trade logistics and transportation [19]. In Africa, the efficiency of the transport chain and of the maritime services is hampered by poor connectivity to markets, resulting from the poor quality and high cost of land transport infrastructure [20]. The Fig. 1 below represents the railway from Abidjan in Ivory Coast to Ouagadougou which is the capital city of Burkina Faso. In Fig. 1, the red colour line represents the existing railway while the green colour line represents the railway that is about to be built. As it can be noticed, there is a total absence of railway from Ivory Coast to Mali. Consumers thus use two logistics methods: they can firstly transport their goods from Abidjan directly to Bamako, which is the capital city of Mali, by road or secondly transport them from Abidjan to Bouake by train and then from Bouake to Bamako by road.

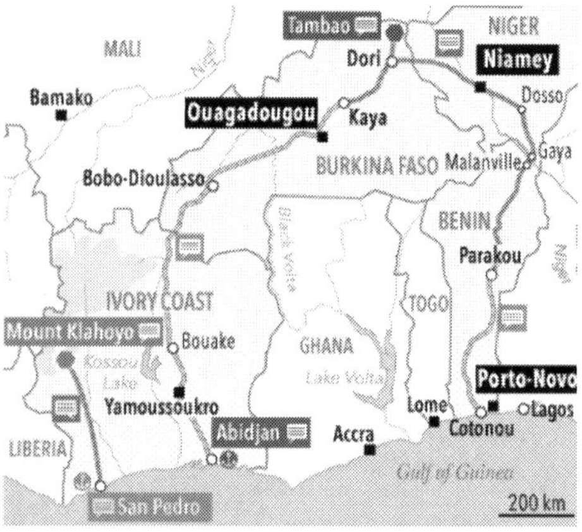

Figure 1. Railway from Abidjan to Burkina Faso

The distance from Abidjan to Ouagadougou is 830 kilometres and the distance from Abidjan to Bamako is 926 kilometres [21]. It takes 5 days for a truck to go from Abidjan to Ouagadougou [22]. There are about 40 checkpoints along the route from Ouagadougou to Abidjan. This is not normal because at each check point a truck driver has to pay a certain amount of money to pass, thus this is a sign of bribery. Every stop at checkpoints creates a delay. In Ivory Coast, while the truck drivers are driving from the hinterland to Abidjan port, they can also be stopped at any moment by the local police for their trucks' papers to be checked. They also mostly have to give the policemen a certain amount of money as bribery to continue their trip without any problem. These briberies directly result in the increase of the final price of the goods that are sold on the hinterland markets, making it expensive for hinterland local buyers. There is a railway line of 1,260 kilometres between Abidjan and Bamako [23]. This line was partly damaged at the beginning of this century when Ivory Coast had a socio-political crisis and after the presidential elections in 2010. There is a train service from Abidjan to Ouagadougou three times per week, taking thirty-six hours to arrive at destination. Unfortunately, these train services have mostly delays of several hours, which is not very convenient.

Hinterland customers do not usually receive any update on whether there are new regulations at Abidjan port especially which concern the rates and the documents to present for importing or exporting goods. There is constant road congestion on the road leading to Abidjan port in Abidjan and on the roads to Bamako and Ouagadougou. This is so inconveniencing and it is mainly due to the poor quality roads, the lack of interchange roads, and the small size of roads. Accidents are nowadays also noticed a lot on these roads. Moreover, Abidjan port does not communicate with its hinterland customers, which is something inacceptable. Though Ivory Coast has become a political stable country, but truck drivers also face armed robberies problem on these roads. These are done by ex rebels, who still have guns and whose main business has become to steal goods and money from trucks drivers. Nowadays, almost once in a month Ivory Coast's soldiers launch mutinies demanding higher salaries to the government and when doing that, they put stop barriers on all the roads heading to Abidjan, preventing any cars including truck drivers from reaching Abidjan [24]. This causes delays. Areas around ports are usually congested and worst of all they have pitiable road infrastructures [25]. Once they arrive at the Abidjan port main entrance gate, truck drivers usually park on the two sides of the road, which creates a terrible traffic jam in that zone. They usually park this way to wait for the loading or the unloading of the goods because there is no parking reserved for them. Trucks are mostly overloaded which not only affects the conditions of the vehicles but can also create accidents. It even happens that some drivers of trucks run away with the goods that they were supposed to deliver to a customer of the Abidjan port who is from the hinterland. The Abidjan port Authorities seem not to pay much attention to the hinterland logistics management as it mostly happens out of the Abidjan port.

5 Suggested strategies

Based on the findings of this article, for the first time some strategies on how to solve the challenges mentioned above are suggested. These strategies can be easily implemented with the great will of the Ivory Coast's government, Abidjan port Authorities and their different partners.

5.1. Build a new dry port in Ivory Coast

After analysing the Abidjan port-hinterland connectivity current situation, building a new dry port in Korhogo is found as a good solution. It will not only help to solve the road congestion problem and to decongest Abidjan port but also to promote the city, develop the city, create more jobs for locals and attract potential investors. The revenue generated by this new dry port can contribute to improve the quality of life of the local citizens by building hospitals, schools, etc. In the Fig. 2 below, which was adapted by the authors from Wordtravels, the city of Korhogo is indicated with a black arrow. As it can be seen, it is really close to Mali and Burkina Faso [26].

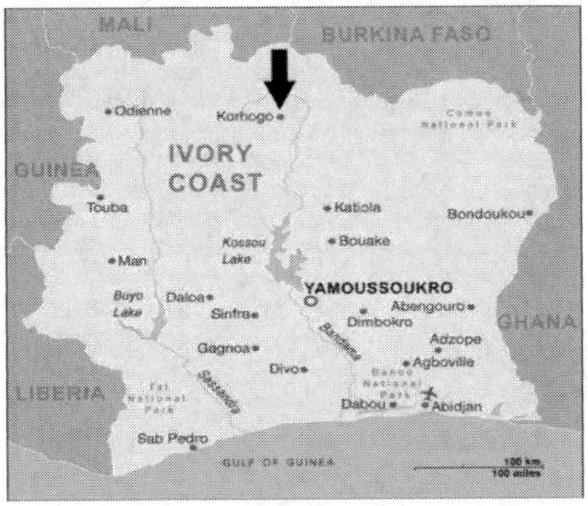

Figure 2. Korhogo new dry port location

The choice of Korhogo can be explained by the fact that it is the most populated city in the northern part of Ivory Coast and the distance to Bamako is 439 kilometres, to Ougadougou is 552 kilometres and to Abidjan is 495 kilometres. To give the customers from the hinterland a fast access to this new dry port, two highways can be built, one linking Bamako directly to Korhogo and the other linking Ouagadougou to Korhogo. This is a cost effective strategy. Based on the data of the population number and the GDP of the countries of Abidjan port-hinterland which show a constant growth since the beginning of this new century, it can be said that this strategy will serve on a long run basis.

5.2. Set up a communication platform

A communication platform needs to be set up for Abidjan port to exchange information with customers and trucks drivers. It can be created on an online platform through a website or in a physical platform through the establishment of an information centre which will facilitate communication to the customers of the different information related to the port while noting down their different request and complaints.

5.3. Open up a new department in charge of hinterland connectivity at Abidjan port

A department in charge of Abidjan port-hinterland needs to be opened, as people working in this department will be fully in charge of the connectivity route of the port to its neighbouring countries. This department will work in partnership with Mali's and Burkina Faso's Infrastructure Bureau, the Sitarail Company, the Ivorian Shippers' Office and the customs administrations of the three countries located in Abidjan port hinterland.

5.4. Build a new railway

If a railway which will link Abidjan directly to Bamako is not built, the Abidjan port will lose this part of its hinterland to Dakar port in Senegal whose government is currently renovating the direct railway to Mali with the help of the Chinese government. Ivory Coast and Mali governments can ask private investors or other foreign governments' help, to fund this project, like other countries do. An agreement can be made among the different parties, but the two countries in the long run will be the great beneficiaries of this investment as the railway will be built in their country, which will boost their countries' economies and attract more potential investors.

5.5. Renovate the existing railway and regulate train services time

Many customers in West Africa usually prefer to transport goods by train instead of using trucks as it is faster and cheaper. It helps to avoid the numerous checking points that the roads have. It is thus important to regulate the train services time so that train can always leave on time. It will help the customers to save time and will also boost the traffic hence creating higher profits.

5.6. Build new roads connecting Abidjan to Abidjan port-hinterland

With the help of potential partners, new roads that respect international standards can be built. These roads will link Abidjan port directly to Bamako and Ouagadougou. The current roads are of poor quality and very small. To accelerate the delivery time of goods from the port to its hinterland, having large roads will enable drivers to avoid losing time in traffic jam as it can be seen currently.

5.7. Create a special parking and management system for trucks

Trucks that arrive at Abidjan port should be better managed by creating a system that can manage them. A special parking can be built for them to avoid the traffic jam at the entrance of Abidjan port. Trucks drivers should be able before arriving at Abidjan port, to check online the status of the goods they will collect, to know approximately at what time they should arrive at the port. This will help hinterland truck drivers to save time, by reducing their time spent in Abidjan. By being faster they can be able to make more turns and earn more money.

5.8. Reduce checkpoints on the roads

Based on the regulations of ECOWAS, the importance of having many checking points on the Abidjan port-hinterland route is not justified. Thus, reducing the number of checkpoints to two checkpoints will drastically reduce the checking fees paid by the truck drivers. The amount of the checking fees should also be lowered down in order to attract more customers. The strategy will be to reduce prices and have a lot of customers instead of having higher prices but just few customers.

5.9. Reinforce security in the corridor

Security on the Abidjan-Bamako-Ouagadougou corridor should be reinforced by Ivory Coast authorities as it is a very strategic corridor for Ivory Coast's economy. A part of the large profits made by the authorities in the taxes paid by the truck drivers can be used to pay very well police officers that will be ensuring security along the corridor.

5.10. Exchange information among Custom services

Information and data should be exchanged between Abidjan port customs and customs services of its hinterland in order to have a better control of the goods that are exported from or imported goods in the hinterland. This will contribute to harmonise the decisions of each country governments on what are the kind of goods that can be imported in the Abidjan port-hinterland area or exported from this area. The customs administrations and the local police authorities of the three concerned countries can work together in seriously punishing overloaded trucks. As an example the drivers of those trucks can be asked to pay a very high amount of money as a fine.

5.11. Set up a tracking system for trucks

A tracking system should be set up by shippers' council of each of the three concerned countries in partnership with local authorities and Abidjan port Authorities in order to identify the different trucks drivers operating in the Abidjan-Bamako-Ouagadougou corridor, their trucks conditions and quality, and if necessary give the information on a truck to local hinterland customer. This will help the countries authorities to have some control on the trucks drivers operating in the region. They will be able to easily locate a truck and in case of complaints from customers, to easily find the driver of a truck.

6 Conclusion

Abidjan port that has the largest hinterland in West Africa was presented in this article and its connectivity system with its hinterland reviewed and assessed. It has been noted that without the establishment of a department in charge of managing, coordinating the hinterland connectivity at Abidjan port and the construction of a dry port, it will be difficult to offer an efficient service to customers. The governments of the countries located in the hinterland should also greatly contribute to the improvement of the port-hinterland connectivity by building new roads and railways with the help of some private investors or some foreign governments or institutions. As Ivory Coast aims to become an emerging country by 2020, tackling the current challenges mentioned above by using the proposed strategies that embody the two routes, Abidjan-Bamako and Abidjan-Ouagadougou, can be beneficial to the country and to the West African region in general. This article reveals that a great collaboration between the different entities composing the Abidjan port hinterland will contribute to the improvement of the Abidjan port-hinterland connectivity. An important insight is given on how having an excellent connectivity between the port and its hinterland can not only attract a lot of potential partners, investors and customers but also generate higher revenue to the Abidjan port. If the proposed recommendations are not applied there is a great risk that Abidjan port loses its hinterland market to Tema port in Ghana and Dakar port in Senegal in the coming years. This article can serve as a reference for future researchers, Abidjan port Authorities, maritime experts and all the institutions and companies that are involved in the Abidjan port-hinterland connectivity.

Acknowledgment

This work was supported by the National Natural Science Foundation of China (Grant No. 71271034), the National Social Science Foundation of China (Grant No.15CGL031) and the Fundamental Research Funds for the Central Universities (Grant No. 3132016306 and 3132017085).

References

1. Wang Jie, Zhang Huaqiang, "Analysis on hinterland segmentation in port project", International Conference on Engineering Computation, pp. 84-88 (2009)
2. Tetteh Evans, Hualong Yang, "Hinterland Demarcation Between Two West African Ports", International Conference on "Multidisciplinary Academic Research and Global Innovation", (2015)
3. Wang Yang, Huang Wenwen, "Economic Analysis of Logistics Financial Management of Supply Chain System in Ports-Hinterland", International Conference on Electronic Commerce and Business Intelligence, pp. 207-212, (2009)
4. Wang Yang, Ma Xing-rui, Chen Yuan-zhi, "Study on the Harmonious Operation of Logistics Finance Management of Supply Chain in Ports-Hinterland Based on System Dynamics", International Conference on Risk Management and Engineering Management, pp. 82-86, (2008)
5. M. Zhang, B. Wiegmans, L. A. Tavasszy, "A Comparative Study on Port Hinterland Intermodal Container Transport: Shanghai and Rotterdam", The Fifth Advanced Forum on Transportation of China, pp. 15-25, (2009)
6. Davide Infante, Giuseppe Paletta and Francesca Vocaturo, "A ship-truck intermodal transportation problem", Maritime Economics and Logistics, vol. **11**, no. 3, pp. 247-259 (2009)
7. Nannan Hu, Jie Wang, Dong Mu, "Research on the Cooperation and Competition Strategies of Port

Based on Hinterland Traffic Condition", International Conference on Logistics, Informatics and Service Sciences, (2016)
8. ShuHai Zhao, Wei Liang, Dan Han, "Seaport Logistics Information Sharing Platform in E-commerce: A Case Study of QDIP in China", Asian Himalayas International Conference on Internet, pp. 1-4, (2012)
9. Benyi Xie, Huiyuan Jiang, Jin Zha, Xiang Li, "Analysis of interactive effect between the port and the hinterland based on logistic model", Third International Conference on Intelligent System Design and Engineering Applications, pp. 304-307, (2013)
10. Fedele Iannone, "A model optimizing the port-hinterland logistics of containers: The case of the Campania region in Southern Italy", Maritime Economics and Logistics, vol. **14**, no. 1, pp. 33-72, (2012)
11. Charles Kipkoech Kotut, Dr. Fred Mwirigi Mugambi, "The Influence of Hinterland Transport Inefficiencies on the Performance of Ports-A Case Study of Kenya Ports Authority", International Journal of Science and Research, Vol. **3**, no. 8, pp. 405- 417, (2014)
12. Elvira Haezendonck, Michael Doomsa, Alain Verbeke, "A new governance perspective on port–hinterland relationships: The Port Hinterland Impact (PHI) matrix", Maritime Economics and Logistics, vol. **16**, no. 3, pp. 229-249, (2014)
13. An Caris, Cathy Macharis, Gerrit K. Janssens, "Corridor network design in hinterland transportation systems", Flexible Services and Manufacturing Journal, vol. **24**, no. 3, pp. 294-319, (2012)
14. Dr. Olivier Podevins, "Sea Port system and the inland terminals network in the enlarged European Union", International Symposium on Logistics and Industrial Informatics, pp. 151-155, (2007)
15. Ying Wang, Jian Wang, "The Optimal Location of Dry Port: A Case Study of the Hinterland of Western Side of the Taiwan Straits Port Group", International Conference on Industrial Engineering and Engineering Management, pp. 1864-1868, (2010)
16. http://data.worldbank.org/indicator/IS.SHP.GOOD.TU
17. Gamassa Pascal Kany Prud'ome and Chen Yan, "Application of Several Models for the Forecasting of the container Throughput of the Abidjan port in Ivory Coast", International Journal of Engineering Research in Africa, vol. 28, pp. 157-168, (2017)
18. http://www.doingbusiness.org/data/exploreeconomies/burkina-faso
19. Carana corporation, "Impact of transport and logistics on Mali's trade competitiveness", United States Agency for International Development, pp. 1-84, (2004)
20. African development report, "Connecting Ports to the Markets", AfricanBank, pp. 107- 153, (2010)
21. https://www.distancecalculator.net/from-bamako-to-abidjan
22. Borderless, "Abidjan-Ouagadougou Corridor Road Governance Report", West African Economic and Monetary Union, pp. 1-17, (2013)
23. http://www.railwaygazette.com/news/infrastructure/single-view/view/abidjan-ouagadougou-railway-upgrade-begins.html
24. http://foreignpolicy.com/2017/01/12/the-real-cost-of-ivory-coasts-military-mutiny-africa-fastest-growing-economy/
25. Regine Adele Ngono Fouda, Nana Darcis Romeo, Muhammad Azizi, S. Rick Fernandez, "Port Logistics in West and Central Africa: A Strategic Development under Globalization", Open Journal of Applied Sciences, Vol. **4**, pp. 76-84 , (2014)
26. http://old.wordtravels.com/Travelguide/Countries/Ivory+Coast/Map

The Impact of International Logistics to International Trade Research Based on an Empirical Analysis of the Data in 1995-2014 in Shandong Province

Zhenyu Jiang[1] and Yaohua Wu[1]

[1]*Modern Logistics Institute, Shandong University, China*

Abstract. By using time series model, this paper uses the data of Shandong province GDP, total value of import and export, the port cargo throughput and traffic volume from 1995 to 2014 in Shandong province, and empirical research on Shandong international logistics and international trade relations is based on the co-integration theory and Granger Causality Test. The empirical results show that the relationship has been long-term stable co-integration, and the one-way causation is between the port cargo throughput, traffic volume and the total amount of import and export, and deduced guiding mechanism of international logistics to international trade. So government ought to pay attention to strengthen international logistics influence to international trade, and complete open trade policy, strengthen the logistics personnel training, increase the intensity of port infrastructure construction, accelerate the process of bonded area logistics system.

1 Introduction

With the advent of modern times, international trade globalization and transport logistics integration have already become general direction and an important theme of the development of world economy. Also the development of the logistics industry has gained more and more support and attention, not only because it is a kind of advanced management technique or a way of organization on its own, but also it has gradually become an important index to measure the whole national comprehensive power and level of modernization. With the in-depth development of the reform and opening, and the increasingly rapid tendency of the economic globalization and trade internationalization, international trade, which is as the close tie between China and the rest of the world, has created a certain effect on the economic development of China and the world and also becomes a core and focus of the world economic competition, subtly affecting a country's world status and economic pattern. Joint development of the international logistics and international trade is both a kind of development status and a requirement in our own time. Many countries have realized international commercial trade and international business services through economic globalization and the integration of information technology to pursue the economic benefit maximization and efficiency of international trade market.

In recent years, Shandong province attaches great importance to the development of international trade, especially international logistics, both offering huge investment and cultivation, either in the construction of logistics infrastructure or in the logistics technical innovation. Also Shandong province has obtained some results in constantly consummating the logistics infrastructure, pushing forward the marketed-based logistics process and the electronization of logistics industry. Since the fully implement of five-year plan for 2011 to 2015, the reform of Shandong province's international trade has rapidly developed, adjusting structure, changing economic increasing methods and growing steadily, which also makes some achievements. In 2011, Shandong province's total volume of import and export breaks the barrier of 200 billion dollars at first time, totaling $235.87 billion, with year-on-year growth of 24.7 percent, hailing it as a "milestone" for the international trade of Shandong province. At the same time, Shandong province, whose international logistics industry maintains its stable and fast growth, is one of the China's developed provinces of trade. Since "Twelve Five-Year" plan, the logistics market of Shandong province is growing continuously. And this amount accounts for 6.5 percent of regional GDP and for 14.9 percent of proportion of the third industry. Its steady growth model protects the development of the international trade of Shandong province, playing an vital role in the support and safeguard. And with more and more attention payed on the logistics industry, Shandong province has made great progress either in the infrastructure construction and Industrial linkage development or in the process of international trade and logistics information development. During the "Twelfth Five-year Plan" period, logistics enterprises in Shandong province has reached more than 22000, among which there are 836 key enterprises. There are 339

DOI:10.1051/matecconf/201712405004

logistics parks in Shandong province, whose development not only promotes the social employment and strengthens the social security, but also boosts the regional economic development and promotes the scale effect. Shandong province has increased production berths of coastal ports to 540, and open 8 civil aviation and 368 air routes, among which there are 41 international flights, primarily establishing the framework of the international trade of Shandong province.Modern information technology, such as Internet and cloud computing gradually start to apply in the international trade and economy of circulation. Most of international logistics enterprises have established logistics information management system and determine the information of transportation, warehousing and forwarding of international goods by real-time controlling. And they promote the construction of logistics information sharing platform while they also perfect the standardization of international logistics, setting 49 standards for logistics successively, which exerts an enormous influence on the development of Shandong province's logistics. Thus it can be seen that the international trade and logistics in Shandong province, which is one of the China's developed province of trade, are developing in coordination. Driven by the national strategy Blue Economic Zone of Shandong Peninsula, the international logistics' demand and standard of Shandong province will get more improvement and more progress.

No matter from the development status of international trade and international logistics or the support of policy, it embodies the attention of our country paid to the both aspects. China's logistics industry after more than 30 years of reform, release and accumulation, is striding forward towards modernization, intensification, standardization and the investment in logistics infrastructure and the logistics efficiency is getting better and better, so the research for international trade and international logistics is valuable for theory and practice. But believe it or not, there are some problems in the development of international business and international logistics in China, such as high costs of logistics, insufficiency in logistics demand, unreasonable network layout, and imperfect technology standards and so on. So China's logistics industry must keep updating and innovating, realizing the service ability, improving the efficiency, in order to adapt to the rapid development of international business and lay a solid foundation for development of international business.

2 Literature review

Logistics, originated in western countries, is a highly-developed discipline in terms of its scope, process and methods. The studies conducted in this field are comprehensive and detailed with advanced approaches. However, the case is different in China. In fact, the concept was recently introduced and did not receive enough attentions until the globalisation has shown its irresistance. As a result, certain challenges still wait to beconquered. Yet the theoretical researches did achieve abundant results and provided a solid foundation for this article.

2.1 Relationship between international logistics and international trade

The International logistics and international trade are of inseparable and complementary relationship. The former is emerged due to the prosperity of international trade while it now pushes the latter into a better future. This relationship reflects their mutual support and influences during interaction. Besides, the concepts of low-carbon economy and environmental logistics also contribute to tightening these two fields and turning them into the new normal of economic growth after the financial crisis based on the theory of sustainable development [1]. Consequently, increasing number of experts and scholars at home and abroad have dedicated themselves to the empirical study of international logistics and international trade, some of which studied the data of bilateral trade and international logistics between China and Australia. The results showed that the bilateral trade has a positive and noticeable influence on the development of the international logistics. However, the influence is not significant due to the late start and slow process of logistics in China. Further studies analysed cases of international trade between Australia and other countries, such as China, America, Japan, etc., by VAR model. Besides, Tongzon (2009) recognised the accelerator role international trade has played in the development of international logistics industry [2]. Still, Nguyen and Tongzon (2009) also found the lack of infrastructure and investment causes the lagging growth of logistics and failed to promote the international trade [3]. While in China, experts at first also used Ganger Causality Test to examine this relationship by the volume of seaborne trade and the statistics of import and export between China and Europe, North America and Japan from 1983 to 2004. Yang (2007) proved the existence of strong feedback and causalityin international logistics and international trade [4]. Zhang (2011) explored the same phenomenonby taking the bilateral trade and cargo throughput between China and Thailand into consideration. Their conclusion is that certain interaction between international trade and logistics do exist in China-Thailand trade yet the promotion from international trade to international logistics are more significant than the other way around [5]. Same result has been reached by the appliance of logisticsin modelling analysis. The scientific modelling of Yue (2013) in empirical study of the total volume of international trade and cargo volume of China and Japanese from 1994 to 2011 demonstrated that the development of international logistics did encourage the business interaction. However, it pointed out that the future boost of international logistics has to rely on innovation instead of pure expansion [6]. Besides, appreciable achievements can be noticed based on data collected from provincial level. Taking the total value of international logistics as the reflection of its development state, Wang and Liu (2014) conducted empirical studies on the data of international trade and logistics in Tianjin municipality, which indicated a notable Ganger Causality and a long-term balance between the total value of international logistics and the trade volume of exportation and importation [7]. On top of that, similar conclusions

 DOI:10.1051/matecconf/201712405004

have also been reached in Wang's study analysing the volume of cargo traffic, port throughput and import and export in Shanghai from 1978 to 2008. The co-integration relationship among the three variables can be further explained as the following three aspects by Wang (2010): the trade value of import and export shows a positive effect onthe port throughput; no clear contribution is made when the value of throughput increased; and the value of cargo traffic has a time-leg effect on the trade value of import and export [8]. Their findings revealed the one-sidedness of expanding the construction and coverage of logisticsparks and provided suggestions on using innovative method to conduct a deep reform of the logistics industry. Furthermore, Wang and Wang (2011), and Gao and Meng (2012) arrived at the same conclusion with statistics from Taiwan and Guangdong province, respectively, proposing a significant influence of international logistics on international trade in a stable and lasting relationship while little from international trade to international logistics [9-10].

2.2 The influence of international logistics on international trade

The International trade and international logistics has mutual influence in the developing process. In other words, a strong relevancy can be recognised in these two fields. Therefore, a large number of experts and scholars carry on studies of theinfluential mechanism and factors focusing on logistics efficiency, costs, etc. It aims to produce more powerful momentum of international logistics for international trade through decreasing logistics costs, increasing service standard, improving profits and adding commercial value. Kong (2010) applied the impulse response function of VAR model on the time series data between 2002 and 2008 to launch a quantitate analysis. The result illustrated that the development of international logistics in China has generate a significant positive influence on international trade [11]. More particularly, Devlin and Yee (2002) took specific trade in the Middle-East and North Africa (such as tuna and coffee in Yemen) as research samples to conduct empirical studies on time cost which is an important section of international trade cost. As the results demonstrated, the international logistics efficiency affects the international trade level through the impact on the time costs [12]. Also, the analysis and evaluation of international trade behaviour and logistics level between America and Canada provided evidence for the promotion of international logistics efficiency by decreasing the logistics costs by Keane and Feinberg (2007). This phenomenon facilitated the trade and economic benefits of the two countries and accelerated the overall development of international trade [13]. Detailed analysis was conducted by Nordas et al. (2006) as well [14]. They adopted the international trade statistics of 191 countries and found out how exactly international logistics and international trade influenced each other. The result illustrated that the decrease of time cost is achieved by raising logistics efficiency in international trade. Besides, they noticed the inverse relation between time costs and the possibility of trading, in which the lower the cost becomes, the higher the possibility would be. Additionally, the possibility of international trade can promote logistics industry domestically and internationally. In this area, Li (2012) analysed data of international trade and logistics from 108 countries and proposed six main factors that influence the international logistics development on the national basis, which are customs efficiency, infrastructure, international shipping, service quality, the ability to track cargos, and punctuality [15]. Conforming to Li's findings, different factors affect the development of international logistics on different levels and there is clear existence of a positive relation between international trade and logistics. In terms of specific area in China, Chen (2007) conducted a modelling analysis of the relationship between the international trade and logistics in Shanghai [16]. The measurement result illustrates a parallel growth in international trade and logistics. Yet through the data of throughput and logistic growth curve, the international logistics has a clear motivation to international trade. Other researches such as Li and Sun (2013) creating modelling using statistics of Yunan province from 1981 to 2010 pointed out the stable Ganger causality on a long-term basis between international trade and logistics. Only the influence of logistics on international trade is more significant [17].

3 The empirical study of the relationship between the international logistics and trade in Shandong province

3.1 Selection for data index and data processing

Before the empirical analysis for correlation between the international logistics and international trade, we need to choose proper variables for the data description and the model building. At first, this paper select GDP as the index to measure economic development level; secondly, the gross import and export value (XM) is chosen as the index to measure the international trade level. Thirdly, for the reason that there lacks unified measure in the international logistics, to better reflect the relationship between the international logistics and international trade, this paper select cargo throughput(TTL) and cargo transportation(YSL)as the index to measure the international logistics.

The data in this paper are from China Statistical Yearbook and Shandong Statistical Yearbook, intercepting the gross domestic product (GDP), total export-import volume (XM),cargo throughput (TTL) and cargo throughput (YSL)in Shandong province from 1995 to 2014. By taking logarithm of GDP, XM, TTL and YSL to eliminate the possible heteroscedasticity, the results as follow: lnGDP, lnXM, lnTTL and lnYSL. The results are in the Table I .

Table 1. Sample data

Obs.	LNGDP	LNXM	LNTTL	LNYSL
1995	8.507819	14.148410	11.570628	11.105649

1996	8.679958	14.295708	11.701941	11.165692
1997	8.785244	14.377199	11.819336	11.195196
1998	8.856711	14.323376	11.810323	11.249129
1999	8.921837	14.418237	11.843365	11.292428
2000	9.028515	14.731400	11.984484	11.434780
2001	9.126419	14.878949	12.170342	11.507551
2002	9.237518	15.037571	12.303018	11.584818
2003	9.399153	15.311948	12.452503	11.675996
2004	9.617260	15.620209	12.629640	11.790830
2005	9.818304	15.855285	12.858424	11.904961
2006	9.994251	16.069831	13.060616	12.028804
2007	10.157234	16.321999	13.262946	12.198580
2008	10.339588	16.576437	13.396799	12.419121
2009	10.431071	16.444545	13.501786	12.558358
2010	10.575664	16.754412	13.669571	12.605033
2011	10.722427	16.976723	13.776645	12.660207
2012	10.820043	17.016405	13.879943	12.707666
2013	10.919267	17.100768	13.982185	12.749562
2014	10.992497	17.137360	14.066993	12.472209

Sourse:China Statistical Yearbook and Shandong Statistical Yearbook.

3.2 Data dependence analysis

This paper regards the correlation coefficient among the 4 variables, which is through complex computing, as the evaluation standard to distinguish the correlation and its degree by means of Eviews8.0 to get the correlation coefficient. The results are in the Table Ⅱ. It can be seen from the data that the variables are almost positively related to each other. So conclutions show that there may be a certain correlation among the 4 variables.

Table 2. Correlation among GDP,XM,TTL and YSL

	LNGDP	LNXM	LNTTL	LNYSL
LNGDP	1.0000	0.9955	0.9990	0.9869
LNXM	0.9955	1.0000	0.9969	0.9867
LNTTL	0.9990	0.9969	1.0000	0.9873
LNYSL	0.9869	0.9867	0.9873	1.0000

It is obvious that the correlation coefficients among gross domestic product (LNGDP), total export-import volume (LNXM), cargo throughput (LNTTL)and cargo throughput (LNYSL) are more than 0.9850, which means there are significantly correlated with each other.

3.3 The test of data stability

Through the test of correlation from data dependence analysis, it can be concluded that there are a significantly positive correlation among the 4 variables in some degree. However, there is another kind of possibility in the factual economic situation, that is there may be no correlation in theory between two or among several variables, just because some kind of economic phenomenon leading to the same correlation among 4 variables. Therefore, the illusion of the correlation has shown up, which is so called spurious regression. So in order to discuss the relationship between the international trade and the international logistics, this paper will give experimental variables stationary test and then the Granger Causality Test.

This paper will make use of ADF unit-root test to implement the stationary test, that is by using Eviews8.0 to take the Augmented Dickey-Fuller test on the gross domestic product (GDP), total export-import volume (XM),cargo throughput (TTL) and cargo throughput (YSL)in Shandong province and difference sequence. The results are as follows in the Table Ⅲ.

Table 3. ADF test results

Variables	ADF statistical value	Data test type (c, t, q)	Critical Value			Prob.*
			1%	*5%*	*10%*	
Ln(DGP)	-3.007936	(c, t, 1)	-4.571559	-3.690814	-3.286909	0.1568
D(lnDGP)	-4.411935	(c, t, 3)	-4.616209	-3.710482	-3.297799	0.0145
Ln(XM)	-1.617651	(c, t, 1)	-4.532598	-3.673616	-3.277364	0.7469
D(lnXM)	-3.687115	(c, 0, 0)	-3.857386	-3.040391	-2.660551	0.0141
Ln(TTL)	-2.849741	(c, t, 1)	-4.571559	-3.690814	-3.286909	0.1995
D(lnTTL)	-3.815301	(c, t, 0)	-4.616209	-3.710482	-3.297799	0.0417
Ln(YSL)	-1.500834	(c, t, 1)	-4.571559	-3.690814	-3.286909	0.7906
D(lnYSL)	-4.533857	(c, t, 0)	-4.616209	-3.710482	-3.297799	0.0116

D: first difference; c: intercept; t: time trend; q: lag order.

By results of ADF test, when significance level is 0.05, the ln(GDP),ln(XM),ln(TTL) and ln(YSL) are all non stationary series, but after first difference, they are all stationary series under the same significance level, which fully indicate that ln(GDP),ln(XM),ln(TTL) and ln(YSL) are all integrated of order one.

3.4 The test of Co-integration relationship

Granger and Engel once put it forward in the 1987 that if there were at least two time series which were non stationary but their linear combination was stationary, it mean that there was a co-integration relationship among the sequences. Through the ADF test results from data stability test showing the conditions that the four time sequences ln(GDP), ln(XM), ln(TTL) and ln(YSL) meet. This paper will take VAR co-integration relationship test on the sequences above. Johansen trace approach is used to test the number of integration relationship, and the results are in the Table Ⅳ.

Table 4. Johansen trace test

Unrestricted Co-integration Rank Test (Trace)				
Hypothesized		Trace	0.05	
No. of CE(s)	Eigenvalue	Statistic	Critical	Prob.**
None *	0.957950	79.10161	47.85613	0.0000
At most 1	0.591453	28.39934	29.79707	0.0718
At most 2	0.470111	14.07696	15.49471	0.0808
At most 3*	0.217078	3.915553	3.841466	0.0478
Trace test indicates 1 co-integrating eqn(s) at the 0.05 level				
* denotes rejection of the hypothesis at the 0.05 level				
**MacKinnon-Haug-Michelis (1999) p-values				

Table Ⅳ shows that there are two Co-integration equation among the four variables, ln(GDP), ln(XM), ln(TTL) and ln(YSL), the conclusion is that there are co-integration relationship among the variables.

3.5 The test of Granger causality

This research employs a lagged variable into other equation, and the Granger relationship exists when the variable in equation is affected by the lagged one. Using command Granger causality test from Eviews8.0, the model tests the causality of ln(GDP), ln(XM), ln(TTL) and ln(YSL) to see if the Granger causality exist.

Table 5. Granger causality test.

Null Hypothesis:	Obs	F-Statistic	Prob.	Conclusion
①LNTTL does not Granger Cause LNGDP	19	4.63062	0.0470	Reject
	18	0.00535	0.9426	Accept
	17	1.06977	0.4051	Accept
②LNTTL does not Granger Cause LNXM	19	0.02311	0.8811	Accept
	18	0.93444	0.4177	Accept
	17	0.91445	0.4684	Accept
③LNXM does not Granger Cause LNGDP	19	10.6389	0.0049	Reject
	18	0.31014	0.7386	Accept
	17	0.60058	0.6292	Accept
④LNGDP does not Granger Cause LNXM	19	4.7E-06	0.9983	Accept
	18	0.03301	0.9676	Accept
	17	0.17485	0.9110	Accept
⑤LNXM does not Granger Cause LNTTL	19	7.78327	0.0131	Reject
	18	1.13336	0.3518	Accept
	17	1.56248	0.2590	Accept
⑥LNXM does not Granger Cause LNYSL	19	9.33627	0.0076	Reject
	18	3.03615	0.0828	Reject
	17	3.54048	0.0560	Reject
⑦LNYSL does not Granger Cause LNXM	19	0.00535	0.9426	Accept
	18	0.01526	0.9849	Accept
	17	0.51472	0.6813	Accept
⑧LNGDP does not Granger Cause LNYSL	19	2.62531	0.1247	Accept
	18	3.39637	0.0651	Reject
	17	2.85079	0.0912	Reject
②LNTTL does not Granger Cause LNXM	19	0.02311	0.8811	Accept
	18	0.93444	0.4177	Accept
	17	0.91445	0.4684	Accept

Results show in TableⅥ:on the assumption of 10% significance level, Empirical results of hypothesis include ①LNTTL does not Granger Cause LNGDP and②LNXM does not Granger Cause and the LNGDP indicate that with a lag phase the total amount of import and export cargo throughput is the Granger reason of GDP. But in the lags of two and three phases this Granger causality does not exist.

In addition, the null hypothesis can be accepted under the 10% significant level that the cargo throughput and the gross domestic product is not the Granger reason of the total amount of imports. Conclusion can be tested from one and three phase lagged hypothesis including ② LNTTL does not Granger Cause LNXM and ④LNGDP does not Granger Cause LNXM.

Through the comparison of⑤LNXM does not Granger Cause LNTTL and ⑥LNXM does not Granger Cause LNYSL, concluding that import and export are always the Granger reason of freight volume growth, it is also the lagged Granger reason of cargo throughput growth, which shows a one-way causal relationship between the amount of cargo transport, cargo throughput and the total amount of import and export. Namely, the development of the total import and export together with import and export growth will accelerate cargo transportation volume and cargo throughput of Shandong province in the long term. However, this promoting effect is not obvious between the amount of cargo transport, cargo throughput and the total import and export trade growth in Shandong Province.

Finally, test results of hypothesis ③LNXM does not Granger Cause LNGDP and ⑧LNGDP does not Granger Cause LNYSL show that in one phase lag, the total amount of import and export is the Granger reason of GDP, and in the two period lag, domestic product is the Granger reason of freight volume of Shandong Province. Therefore, causality of the total amount of import and export in addition to the domestic product and the total freight volume both have certain time-lag effect and even longer effect of the latter.

4 Conclusions and policy suggestions

4.1 Conclusions

Relying on the results of above empirical analysis, this paper summarized relevant main conclusions as follows.

Firstly, seen from a long-term, there is a long-term equilibrium relationship between economic development and international trade, international logistics in Shandong province. The respective time series of gross regional production, total export-import volume, freight traffic volume and cargo handling capacity are all non-stationary series. But after the first order difference, these four variables are stationary time series, thus indicating they are all one-order integrated series essentially that the long-term co-integration relationship exists among all data.

Secondly, the total export-import volume has a one-way causal relation to freight traffic volume and cargo handling capacity that shows the increase of total amount of international trade in Shandong province will corresponding promote the development of international logistics in the province and improve the benefits of international logistics. However, the development of international logistics in Shandong province has a relatively weak impact on international trade.

Thirdly, the analysis results show that total export-import volume is the Granger-cause of gross regional production in the condition of lag phase I, and the gross regional production in Shandong province is the Granger-cause of freight traffic volume in lag phase II. Therefore, the effect of gross regional production on total export-import volume and the impact of gross regional production on freight traffic volume have certain hysteresis effect, and the latter's lagging effect longer. In other words, in the long-term, the increase of international trade in Shandong province will drive growth of gross regional production which in turn drives growth of international logistics yield in Shandong province. This well confirmed that there is a permanently interconnected guide mechanism between international trade and international logistics in Shandong province.

4.2 Policy suggestions

Based on the conclusion above and integrated Shandong's own development status, the paper raised three policy suggestions:

Firstly, as to macro-policy, Shandong province should make full use of congenital advantage such as geographical location, resource endowment and sufficient labor. Then, based on the strategic background of the flourish of Shandong Peninsula Blue Economic Zone, Shandong province should strengthen the link and interact with international trade and international logistics, and enact relevant policies to promote the development of international trade and international logistics. In order to standardize the order of the international trade market and construct the international logistics development environment, the concrete method including support policy of international logistics enterprise, establish and perfect the international logistics infrastructure and international logistics system, improve efficiency, reduce the intermediate process, and strengthen departments' coordination. Meanwhile, Shandong province should grasp the development opportunity with Japan and South Korea, with its unique geographical location and good open condition to vigorously attract foreign investment, optimize investment environment, introduce foreign advanced logistics management and technology experience, and expand domestic and foreign logistics strategic alliance, driving the development of international trade and international logistics to the largest extent in Shandong province.

Secondly, from the micro-construction aspect, Shandong province has superior location since it located in the northeast Asia ocean transport hub. Therefore, it should strengthen the international logistics infrastructure construction in bonded zones to ensure the easy access to highway, railway and maritime transportation, and realize regional economic integration to ensure the international multimodal transport. Meanwhile, Shandong should enhance the modernization of distribution park management process by making innovative reconstruction based on the existing infrastructure and setting up the internationalization of public information platform, as well as make full use of information technology and high-tech products improving the internationalization and efficiency of logistics operations in order to provide good hardware guarantee for the international trade in bonded area. About port construction, Shandong province should take the port of Qingdao for the center, coordinate with the port of Yantai, Weihai and Rizhao for common development to promote the integration of resources and realize the port area linkage. In the meantime, Shandong should increase port construction, expand radiating and driving function of ports, give full play to its own characteristics to avoid the "convergence" of port development. Finally, drive the construction of international logistics through the expansion of port economic hinterland, and promote the development of international trade in Shandong province.

There are lots of colleges, universities, scientific research institutions in Shandong province, and the scholars majoring in trade engineering have a firm background and professional. Therefore, to train more and better professional trade worker, Shandong province should aim to create a mode for integration of training and researching, learn advanced philosophy and training experience from abroad, and cooperate with international trade enterprises. To stimulate the development of international trade induced by Shandong province, Shandong province should develop a comprehensive, effective, creative policy based on the future development direction of Shandong Province free trade zone and by integrating of experienced advanced and fresh innovated logistics workers.

References

1. Q.C. Liu, "Discussing on the impact of international logistics on international trade", Jiangsu. Comm. Rev., vol.12. pp.60-62, 2011.
2. T.Jose, N.Owen, China's economic rise and its implications for logistics: The Australian case ,Trans. Pol., vol.16, pp. 224-231, 2009.
3. N.Owen, T.Jose, Causal Nexus between the Transport and Logistics Sector and Trade: The Case of Australia, Trans. Pol., vol.5, pp. 135-146, 2009.
4. C.C.Yang, The relationship between international trade and international logisticsc, Int'l Trade., vol.10, pp. 28-31, 2007.
5. Y.Y.Zhang,: A study of the relation develpopment between Sino-Thai mutual trade and logistics, 2011.
6. Y.Yue: Analysis of China-Japan international logistics impact on the bilateral trade, 2013.
7. M.L.Wang, B.L.Liu, Analysis on the influence factors of international logistics development in Tianjin Binhai New Area, Mod. Mgmt. Sci, vol.11, pp. 63-65, 2014.
8. L.Wang, Empirical Study on Relationship between Foreign Trade and Modern Logistics:Based on Shanghai Data from 1978 to 2008, Int'l Trade J., vol.1, pp. 59-65, 2010.
9. Y.L.Wang, J.Wang, Empirical Analysis on the Relationship between AquaticProduct Trade and Logistics in Taiwan, J. F N Univ: Soc. Sci. Ed., vol.5, pp. 50-55, 2011.

10. X.L.Gao, F.R.Meng, Empirical Analysis on Relationship between GuangdongForeign Trade and International Logistics, Tech. Mgmt. Res., vol. 21, pp. 86-90, 2012.
11. Y.Kong, Empirical study on relationship between international logistics and international trade, J S Univ Technol: Soc. Sci. Ed., vol.4, pp. 335-341, 2010.
12. J.Devlin, P.Yee. Global Links to regional networks: Trade logistics in MENA countries, MENA Trade & Investment in the New Economy, 2002.
13. M.P. Keane, S.E. Feinberg, Advances in Logistics and the Growth of Intra—Firm Trade:The Case of Canadian Affiliates of U.S. Multinationals, J. Ind. Econ., vol.4, pp. 571-632, 2007.
14. H.K. Nordås, E. Pinali, Logistics and Time as a Trade Barrier, OECD Trade Policy Papers, 2006.
15. P.Li, Research on the influence of logistics competitiveness on trade, Know. Econ., vol.8, pp. 129-130, 2010.
16. R.J.Chen, An empirical analysis of Shanghai port logistics and international trade development, Chin. Col. Econ., vol.1, pp. 124-125, 2010.
17. J.Y.Li, L.T.Sun, Empirical Analysis of Interaction between Yunnan Logistics Industryand International Trade Based on Granger-causal Test, Log. Tech., vol.13, pp. 242-244, 2013.

A Game Analysis on Information Sharing of Military Procurement and Logistics Standardization

Darong Ling[1], Fei Zhang[1], Ziyi Wang[1] and Yukun Li[1]

[1]*Military Economics Academy, Wuhan, China*

Abstract. Military procurement and logistics standardization is the base of the efficient and reasonable operation of modern military procurement &logistics. It's significant to study logistics standardization and the standardization of logistics information management. This paper constructs a game model which aims at comparing costs and benefits on condition of absolute sharing, partial sharing or not sharing information based on the different information sharing rate and logistics institutes' expectation, furthermore, it discusses how to build a relationship of mutual trust, in order to make the logistics institutes develop activeness to fasten the pace of logistics standardization and the standardization of logistics information management.

With the continuous development of military procurement & logistics standardization construction, military procurement & logistics standardization is seen as the foundation of the modern military procurement & logistics, at logistics information flow has become the key logistics activities, at the same time, more attention should be paid to logistics standardization research, and exerting efforts to expand standardized data exchange and application between logistics operation organizations, speeding up the transformation of military procurement & logistics information management into scientific, standardized in order to improve the efficiency of the military procurement & logistics and promote the construction of military procurement & logistics standardization, which is of great significance to improve the efficiency of military procurement & logistics support and efficiency.

1 Introduction

Logistics standardization and normalization could be seen as the foundation of the realization of logistics rationalization and efficiency, also, the standardization of logistics information management is viewed as a watershed in modern military procurement & logistics and traditional military procurement & logistics, unified information data exchange and sharing is very important for every member in military procurement & logistics system. However, in the process of logistics standardization, because of having no specific standard or some standards has not got recognition and promotion, what is more, some institutional barriers of information resources management, and some related departments care about their own interests, which concerned about their logistics information becomes public would make them lose their own advantages in the field of information possession. Consequently, they have no initiative to share logistics information, which result out slowly information communication and inefficiencies in logistics operation.

Information exchange and sharing between logistics nodes and standardization organizations is a process of mutual benefit. It can realize the bilateral benefit and facilitate the military procurement & logistics system coordinating and efficient operation, otherwise, military procurement & logistics system will be blocked. Therefore, it is the key to improve the overall efficiency and benefit of military procurement & logistics that dedicated to information sharing and standardization of information management.

Taking military procurement & logistics information flow as the research object, exploring the motivation that push on the standardization of military procurement & logistics information management, and promoting the standardization of military procurement & logistics management information construction, how to participate in both sides achieve this kind of information sharing and increase their benefit through the game, and improve the overall efficiency should be understood, thus in a higher level and deeper extent promoting the military procurement & logistics standardization construction. In view of this, cooperation game model is constructed to analyze the evolution mechanism of this kind of a win-win relationship.

2 The establishment of the model

Generally speaking, in the advance of military procurement & logistics standardization should be led by the military procurement & logistics standardization committee which working with the military procurement

& logistics operation organizations and logistics nodes that usually more than one independent agency or organization. In order to simplify the research, here only consider the one-to-one game mode.

2.1 Bilateral earnings analysis of the payoff function

The effective exchange of information between the standardization committee and military procurement & logistics operation organizations would make the military procurement & logistics operation smoothly. From the perspective of standardization committee, on the one hand, it can get information feedback from military procurement & logistics organizations to get firsthand material to set standards, promote military procurement & logistics information standardization and management standardization easily, and improve the performance. On the other hand, bilateral accelerating construction progress together that can save the special funds of military procurement & logistics standardization construction. For military procurement & logistics operation organizations, implement the logistics standardization could reduce logistics operation cost, improve efficiency, and improve business performance. As a result, this paper considers this behavior as a specific payoff function of decision-making. Since the decision-making and strategy game with similar characteristics, thus assumed that the military procurement & logistics standardization committee's benefits paid function is(π_A^i),the military procurement & logistics operation organizations' benefits paid function is(π_B^i).

2.2 Construction of basic model

Game bilateral are rational and willing to pursuit of self-interest maximization and risk-averse. In military procurement & logistics operation process, regional military procurement & logistics standardization organizations and institutions may retain a lot of important exclusive information, which result out the game is carried under the condition of asymmetric information. If logistics standardization committee obtain more information, but do not share it with military procurement & logistics operation organizations, what is more, the committee thinks it would be the dominant in the military procurement & logistics standardization and information management standardization; on the other hand, the military procurement & logistics operation organizations also could retain their information and do not involved in information sharing; otherwise, both sides share information with each other to realize win-win situation, improve business efficiency and reduce cost. The specific game situation and the basic model is shown in table 1:

military procurement & logistics standardization committee

Table 1. Statics game model of military procurement & logistics information sharing

military procurement and logistics operation organizations

	sharing	not sharing
sharing	a , a	b , c
not sharing	c , b	d, d

Assuming that bilateral side choose information sharing strategy, the earnings is respectively a and a; unilaterally choosing information sharing strategy, sharing the benefits for b, information alone is protected well income to c; Bilateral choice alone retain strategy, revenue d and d respectively, what is more, for general, $c > a > d > b > 0$.

3 Static game under information sharing

It assumes that logistics information sharing rate of military procurement & logistics standardization organization is ρ $(0 < \rho < 1)$ in reality, while, logistics information sharing rate of military procurement & logistics organization is ψ $(0 < \psi < 1)$, then, the game model is as follows:

military procurement & logistics standardization organization

Table 2. Information sharing game model of military procurement & logistics

	sharing rate (ρ)	not sharing rate ($1-\rho$)
Sharing rate (ψ)	ψa, ρa	ψb, $c-\rho c$
not sharing rate ($1-\psi$)	$c-\psi c$, ρb	$d-\psi d$, $d-\rho d$

(1) Assuming that military procurement & logistics standardization organization choose complete information sharing, logistics operation organization made the same decision as well , namely $\rho=1$, $\psi=1$, thus, it forms a static game model.

The specific process of a single stage game is shown in figure 1, which illustrates that the standardization organization adopts information sharing strategy, logistics operation organization would compare cost and gains($c > a$) ,and then selects information retaining strategy. On the other hand, if standardization organization selects information retaining strategy, logistics operation organization may yield profits d ($d>b$) and select information retaining strategy. Consequently, logistics standardization organization and logistics operation organization would both choose information preserving strategy under those conditions, also, the benefits is d respectively, while the earnings $d < a$, therefore, both of them would eventually prefer the dominant strategy choice to share information and get profits a. According to the theory of Nash equilibrium, if the operation of military procurement & logistics is just only a one time static game, then the gains group (*a*, *a*) would be seen as the Nash equilibrium. However, it is a long-term process of cooperative game that the logistics operation and information sharing for the standardization organization and operation organization as well, accordingly, this strategy choice is not proper for both in reality.

In table 2, the military procurement & logistics standardization committee and the logistics operation organizations yield profits are:

$$\pi_A{}^1 = n\rho a - \frac{\delta\rho a}{1-\delta}$$

$$\pi_B{}^1 = n\psi a - \frac{\delta\psi a}{1-\delta}$$

Then, the bilateral total revenue is:

$$\pi_t = n(\rho+\psi)a - \frac{\delta a}{1-\delta}(\rho+\psi)$$

Since, ρ=1, ψ=1, so,

$$\pi_{t1} = 2na - \frac{2\delta a}{1-\delta}$$

(2) Assuming that both of them share part of information, namely preserve some logistics information respectively, might as well set the possibility ρ=0.5, ψ=0.5, then the total revenue is:

$$\pi_{t2} = \frac{1}{2}(2na - \frac{2\delta a}{1-\delta})$$

(3) Assuming that both of them don't share any information and preserve all the information by themselves, obviously, the possibility ρ=0, ψ=0, the profits of military procurement & logistics standardization committee is as follows:

$$\pi_A^3 = (1-\rho)c + (n-1)(1-\rho)d - \frac{\delta(1-\rho)}{1-\delta}(c+\delta d-\delta c)$$

The earnings of logistics operation organization is:

$$\pi_B^3 = (1-\psi)c + (n-1)(1-\psi)d - \frac{\delta(1-\psi)}{1-\delta}(c+\delta d-\delta c)$$

The total earnings of bilateral organization is:

$$\pi_{t3} = 2c - 2(n-1)d - \frac{2\delta}{1-\delta}(c+\delta d-\delta c)$$

Through comprehensive comparison of the earnings in formulations above, it could conclude that:

$$\pi_{t1} = 2\pi_{t2} = (2na - \frac{2\delta a}{1-\delta})$$

Therefore, the total earnings gap of standardized information sharing and bilateral sharing with sharing rate is as follows:

$$\Delta\pi = (\rho+\psi)\left[na + (n-1)d + (1-\delta)c - \frac{\delta a}{1-\delta}\right] + 2\left[\frac{\delta^2 d}{1-\delta} - (1-\delta)c - (n-1)d\right]$$

Calculating the integral derivation on parameters $(\rho+\psi)$, then,

$$\frac{\partial\Delta\pi}{\partial(\rho+\psi)} = na + (n-1)d + (1-\delta)c - \frac{\delta a}{1-\delta}$$

Since 0 <δ< 1, thus, $\frac{\partial\Delta\pi}{\partial(\rho+\psi)} > 0$,this formulation also indicates that along with the increase of $\rho+\psi$, the total earnings gap of complete standardized information sharing and bilateral information sharing with sharing rate becomes greater and bigger, which means the greater sharing ratio, the greater total earnings could get. Influencing by this game result, the military procurement & logistics operation organizations are willing to cooperate and share information with military procurement & logistics standardization committee, and constantly improve their additional value of information resources they owned, what is more, they would try to make the information become standardized and unified information management standard in order to accumulate its own operation basis and achieve the target of "win-win" in the military procurement & logistics information sharing.

If bilateral organizations both adopt information preserving strategy completely, then, the total earnings gap between complete information sharing is:

$$\Delta\pi_1 = (1-\delta)\Delta\pi = 2\delta^2(d-c) + 2\delta\left[2c + (n-1)d\right] - 2\left[c + (n-d)d\right]$$

Calculating the partial derivation on parameter δ, then,

$$\frac{d\Delta\pi_1}{d\delta} = 4\delta(c-d) - 4c - 2(n-1)d < 0$$

Formulation above indicates that the total earnings gap between complete information sharing would become smaller and smaller under the condition of bilateral organizations both adopt information preserving strategy, namely the total earnings would decrease along with the incensement of preserving ratio. Accordingly, the operation cost is on the rise and the yield decreases when bilateral information don't share with each other, which means the logistics system lies in a state of mutual consumption and stalemate that both two sides are not willing to accept. Consequently, both of them hope to gaining more earnings from the process of information sharing, promoting the construction of logistics information standardization management process. In other words, the analysis result indicates that the relationship between logistics information sharing and logistics operating income as well as information management standardization is 'win-win' effect.

4 Repeated game and subjetive expectation analysis

4.1 The infinite repeated game of military procurement & logistics information sharing

Trigger strategy: assuming game is infinitely repeated, military procurement & logistics standardization committee take the strategy of logistics information sharing, also, the logistics operation organizations adopt information sharing strategy in the early stage continued to stage t. If the earnings of bilateral organizations are (a, a) until stage (t – 1), then, they would continue to adopt the strategy of logistics information sharing, however, once they found that opposite side take preserving strategy alone, they would choose information retaining strategy without any doubt.

Assuming that logistics standardization committee takes trigger strategy, then, logistics operation organizations would also take the trigger strategy, if the coefficient of strategy selection costs and earnings is δ(0 <δ<1), the earnings of military procurement & logistics operation organizations is:

If the military procurement & logistics operation organizations do not take the trigger strategy, on the contrary, adopt retaining strategy in the first stage, thus,

$$\pi_B{}' = \lim_{t\to\infty}\left[ta-(\delta a+\delta^2 a+\ldots+\delta^{t-1}a+\delta^t a)\right]$$
$$= ta-\frac{\delta a}{1-\delta}$$

they would receive a revenge and they have to adopt trigger strategy from the second phase. The earnings for the military procurement & logistics operation organizations is:

$$\pi_B{}^i=\left(c-\delta c, d-\delta^2 d, d-\delta^3 d,\ldots, d-\delta^n d\right)$$

$$\pi_B{}''=\lim_{n\to\infty}\left[c+(n-1)d-\delta c-\frac{\delta^2 d}{1-\delta}\right]$$

Thus, as long as the $\pi_B{}' < \pi_B{}''$, namely $\frac{c-a}{c-d}<\delta<1$, trigger strategy is dominant strategy for military procurement & logistics operation organizations, at the same time, trigger strategy is also dominant strategy for military procurement & logistics standardization committee as well, accordingly, the trigger strategies construct a Nash equilibrium in the game of the military procurement & logistics information sharing standardization. Pareto optimality policy (sharing , and sharing) constructs the equilibrium result of each stage, because of the unlimited times game this result would also be the final ending of long-standing game relation, which means the result and structure of any sub-game would be same no matter starting from any stage, therefore, the trigger strategy constructs a refining Nash equilibrium to the sub-game of infinite times repeated games.

Scripps and others have shown that, as long as the number of game is large enough, trigger strategy would also be positive to solve the "prisoner's dilemma" problem. In specific, information sharing game is repeated for infinite times in military procurement & logistics operation because of its long term running, what is more, military procurement & logistics operation organizations and military procurement & logistics standardization committee both have patience and confidence to punish the opportunism behavior, thus, any earnings of short-term opportunism behavior appears insignificant, as well as both two sides have enthusiasm to build the reputation of cooperation willing. Through this way, military procurement & logistics military procurement & logistics operation organizations and military procurement & logistics standardization committee both and military procurement & logistics standardization committee could both avoid dilemma, what is more, the earnings from the cooperation would make both sides continue to cooperate and establish a mutual trust relationship in long-term cooperation, which could also further make them actively promote logistics information management standardization.

4.2 Analysis based on of military procurement & logistics operation organizations subjective expectations

The discussion above suggests that military procurement & logistics operation organizations and military procurement & logistics standardization committee both have enthusiasm of sharing logistics information with other members. Consequently, assuming that military procurement & logistics operation organizations pursuit sustainable long-term expectations on future earnings in addition to the earnings at this stage in the game process of standardization of logistics information sharing. Assuming that the earnings at this stage is *a*, its probability is 1 - μ, *E* for long-term expectations, its probability is μ ($0<\mu<1$), therefore, the earnings is:

$$\pi=(1-\mu)\rho a+\mu\rho E-(1-\mu)\rho\delta_a>\lambda a,(\lambda>1)$$

then,

$$E>\frac{\lambda-(1-\mu)\rho}{\rho\mu}+1-\delta$$

Thus, μ is larger, the greater the chance of inequality could be established, which illustrates that military procurement & logistics operation organizations would be more motivated to participate in logistics information sharing based on the higher earnings expectations in the future. At the same time, the greater the value of ρ , which represents the standardized logistics information sharing, the more obvious the positive correlation between λ and *E*, which also means that the higher enthusiasm of sharing logistics information.

5 Conclusion

The game analysis above shows that it is benefited to both military procurement & logistics standardization committee and military procurement & logistics organizations for sharing standardization information with each other. Military procurement & logistics organizations have the higher expectations of future earnings and long-term cooperation, the higher enthusiasm of military procurement & logistics information sharing. When this kind of enthusiasm into standardized information sharing actions, namely taking the sharing policy, contributing to the behavior of the information management standardization, and promoting the construction of logistics standardization more actively, their earnings and bilateral total revenue will be growing. Organizations for standardization play a key role of the implementation of the logistics information management of military procurement & logistics organizations. The application and exchange of standardized information between the military procurement & logistics organizations in order to promote the informatization construction of military procurement & logistics and standardization of information management construction.

This paper insists that in the process of promoting the military procurement & logistics information management standardization, standardization committee first should focus on strengthening of the standardization promotion, making more military procurement & logistics organizations more comprehensive understanding of military procurement & logistics information management standardization. Second, it should establish a standardization of military procurement & logistics information management incentive mechanism. Incentive in the process of information exchange and sharing of military procurement & logistics organizations can affect each logistics organization goals or needs, and affect the logistics organization action motivation. Third, it can also cultivate strategic partnership, and promote the continuous development of military procurement & logistics information management standardization. This is the important foundation of improving the support efficiency and the important condition of civil-military integration support. So it can obtain a more competitive advantage, achieve their own goals and interests of the logistics organizations, and eliminate the differences of military procurement & logistics management information system nodes, bring out economies of scale. Standardization committee is committed to improving the incentive of logistics information sharing, making each member of the military procurement & logistics system enhance the consciousness of sharing standardize information, cooperation and resources integration, which in order to achieve a win-win situation, and improve the support capability of military procurement & logistics .

References

1. Li ling, Wang xiang. Strategic thought in game theory. Shanghai Journal of Economics, 2010, vol.1, pp.35-41.
2. Schliekelman, P, and M. Slatkin. "A Game Analysis of Standardization of Aquatic Production Enterprises and Cold Chain Logistics Enterprises".LISS 2014, Springer Berlin Heidelberg, 2015, pp.3-7.
3. Sun, Shusheng, and G. Gui. "A Game Analysis of Standardization of Regional Modern Logistics." Logistics Technology. 2013, vol.64, pp. 4965-4972.
4. Wang, Cai Xia. "Analysis of the Value of Logistics Standardization to the Management and Decision-making of Logistics Corporation." Logistics Sci-Tech, 2009, vol.5, pp. 756-767.
5. Xiao, Rong Na, and Q. W. Yin. "Mechanism for Promoting Logistics Information Sharing System Based on Game Theory." Transportation Standardization, 2012, pp.231-237.
6. Yuan, Qing. "Analysis on Fresh Agriculture Foods Logistics Mode in China and Cool Logistics Standardization Construction." Journal of Anhui Agricultural Sciences, 2013, vol.4, pp. 73-77.
7. Yang weilong, Lin jian. Study on the enterprise strategic management based on the game theory. Science of science and management, 2007, vol.8, pp.136-139.
8. Zhang, Yin Cong, and G. Y. Cui. "The Application Analysis of Logistics Standardization in Enterprise and the Model of Informational Share Game." Sci-Tech, 2009, vol.3, pp. 453-460.
9. Yang, Zou. "Analysis of China's Logistics Standardization Based on Economic Globalization." Storage Transportation & Preservation of Commodities, 2008, pp.254-261.
10. Zheng mingchun. Research of baseding on the game for service overlay networks resources providing questions. Shangdong normal university, 2011, vol.8, pp.136-139.
11. Zhang, Ai Hua,. A Game Analysis of New Technical Equipment Procurement. The 19th International Conference on Industrial Engineering and Engineering Management. Springer Berlin Heidelberg, 2013, pp.703-708.
12. Zhang yincong, Gui gengyu. The application analysis of logistics standardization in enterprise and the model of information share game. Logistics science and technology, 2009, vol.5, pp.69-75.

Optimizing the Distribution System of Fast Moving Consumer Goods Company Using Sweeping Methods with Discrete Event Simulation Approach

Arry Rahmawan D. [1], Bhagas Aufa H. [1], Fairuz Qalbi A. [1], Satrio Nur K. [1], Bagas Brahmantyo[1]

[1]*Department of Industrial Engineering, Universitas Indonesia, Depok, West Java, Indonesi*

Abstract. FMCG is considered as one of the industry with the highest competition in the world. To win this industry, customer satisfaction becomes the main thing. In this paper, authors will simulate a system of distribution of goods from PT X TBK which has 3 categories, namely M1 (biscuit), M2 (powder) and M3 (liquid). The distribution system has a problem on time delivery of goods which have a wide variety of obstacles, mainly on routing. There are 8 trucks with different routes and different utilization. There's trucks with over hour work and trucks with long idle time. The routing also causes problem for unused truck weight and volume capacity, where cost could possibly be reduced. Through Discrete event simulation and sweeping method, authors try to solve the problem by modelling the existing system and then find the solution to improve the distribution system and reaching the model objective.

1 Introduction

Logistics management is that part of supply chain management that plans, implements, and controls the efficient, effective forward and reverse flow and storage of goods, services and related information between the point of origin and the point of consumption in order to meet customers' requirements. (Council of Supply Chain Management, 2014)

PT X TBK is one of the consumer products company in Indonesia, focused in packaged foods. Packaged foods industry is one of the industries with biggest market in Indonesia. The transaction volume of the packaged food in 2015 reached 6 million Tons and according to Indonesia-Investment, in Quarter 1 year 2016, packaged foods sales already reached IDR 400 trillion. With the growing GDP and population in Indonesia (about 250 million people, the 4th biggest in the world), Indonesia is without a doubt an interesting market for packaged food Industry. (Euromonitor International, 2016)

To win this industry, one of the key factors is the capability to reach and keep the customer. Hereby, supply chain and logistic will be one of the determinant factor for PT X to achieve them. The purpose of the logistic system is how to deliver the products to consumer with the lowest cost possible. Logistic function in PT X is involving several third parties. One of the partners is PT Y Distribution. This company has a partnership with PT X to distribute PT X's product in Depok, West Java areas. The planned process by PT Y Distribution is they distribute the products based on the category that they've made, consist of M1, M2, and M3 (M1 for snack products, M2 for powder products, and M3 for liquid products). But, this distribution system is facing several problems that are mainly caused by the routing, where the usage of resources is not balanced. There's a route where a truck could finish their delivery far before its shift ends while still having unused space in it's box, wasting its weight and volume capacity. In the same day, there's a route where a truck must work overtime to finish its delivery, arriving very late at the headquarter warehouse and causing the worker to went home late, feeling more tired than others. The same route also causes some of the customer stores receive the delivery late.

In this paper, authors are not only making a model about the actual condition, but also displays methodology using Discrete Event Model and Sweeping Method that will improve the distribution routes, where truck as the main resources is utilized more efficiently by usage time, volume, and weight capacity to create a balanced work load for each worker, enhance delivery on time, and reducing the total cost for PT Y Distribution.

2 Literature review

This project is closely related to the Time-Dependent Dynamic Vehicle Routing Problem that reflects varying traffic conditions and also considers carbon emission, fuel and time efficiency of vehicles. The Time-Dependent Vehicle Routing Problem has been introduced to solve problem in a traffic network. A mixed integer formulation of the time-dependent vehicle routing problem where the travel time on each arc is a step function of the departure time was proposed (Daskin, 1992). Several years later, this routing problem improved by anticipating the potential

 DOI:10.1051/matecconf/201712405006

value by considering path changing in the Dynamic Vehicle Routing Problem.

Research of Dynamic Vehicle Routing Problem has been an interesting subject for many years until now. The Green Time-Dependent Routing, which consider vehicle routing problem to minimize emissions (Jan Fabian Ehmke, 2015). They investigated how to simplify the pre-computation of emission-minimizing paths by identifying the optimal paths that are independent of the vehicle load and used a tabu search method to solve the routing problem. Their test instances are based on a large real road network and extensive experiments are tested for analysis. On the same year, used a fuzzy ant colony system to solve the dynamic vehicle routing problem with uncertain service time. A fuzzy ant colony system uses the cluster insertion algorithm to divide customers. This algorithm groups the customers into several small clusters based on the distance between customers. The routes between customers within a cluster have higher pheromone levels than routes between customers in two different clusters. Besides, a vehicle tends to serve customers within the same cluster. Newly arrived customers are assigned to the nearest cluster. (Kuo, 2016)

From this literature, authors make four main sections to be reviewed. The first one is about what is the literature about, and then what the problem is, how the authors save it, and the conclusion about the problem and all things about the literature. This literature is all about comparing two modeling approaches to help people solve about logistics and supply chain management (LSCM) problems. Discrete event simulation (DES) is one out of a two modeling approaches used by this literature. It's widely used as decision support tools in LSCM. DES is used at an operational/tactical level. This paper explores the application of DES as decision support systems (DSS) for LSCM by looking at the nature and level of issues modeled. The authors findings suggest that DES has been used more frequently to model supply chains than the system dynamics one (modeling approaches that used in this literature).

This research is about PT Y Distribution's Distribution and logistic systems route around Depok, West Java Area. The solution can be gained by using DES, because the problem is more of operational/tactical level issues. Authors could use SD to model it or give some solution to solve the problem, but the DES could give more choices of solution.

(Nurcahyo, 2002) used as the main basis to solve routing problem. Sweep Algorithm consist of two stages. Stage 1 is called clustering stage, it illustrates a clustering process of several nodes and sweep hand that moves counter-clockwise. Clustering performed by joining nodes, beginning from the closest node to any node that is chosen to be the first node. The closest node is the one that has smallest angle. This process is repeated by joining the rest of the nodes until it satisfies the constraints. Stage 2 is called route generation. The objective of this stage is to link all nodes in every cluster of path network.

From all of the papers mentioned above, authors found novelty as it combines the Discrete Event Simulation with Sweep Algorithm. The real situation could be modeled with Discrete Event Model and use Sweep Algorithm as the main basis to find a better routing. Then, the Discrete Event Simulation is used to check the result and find other possible solutions.

3 Research method

There are some steps to doing the research and modeling the simulation for getting a great implementation in the end.

3.1 Research background

PT Y Distribution applies the take-order system, where there's salesman that visit customer stores using motorcycle. They take orders from the customers and bring them back to headquarter. The Headquarter then would plan a delivery system for the next day.

In the field, product distribution activities are always facing various problems that mainly caused by the routing, where the usage of resources is not balanced. There's a route where a truck could finish their delivery far before its shift ends while still having unused space in its box, and there's a route where a truck must work overtime to finish its delivery, arriving very late at the headquarter warehouse. The same route also causes some of the customer stores receive the delivery late. The biggest sales is happening on Wednesday, causing the number of products to be delivered reach its peak on Thursday. This day also has the largest distribution cost compared to other days. So, the research would focus on Thursday because it would give the biggest Impact for the stakeholder.

Table.1 describes the product sales PT X in Depok for a week starting from 28 November 2016 to 3 December 2016. The table describes the number of stores visited by the delivery trucks to deliver products PT X. According to distribution mechanism by PT Y Distribution, the orders taken today will be delivered on the next day.

Table 1. Product sales, number of stores, and delivery cost

Day	*M1*	*M2*	*M3*	*Total Product Per Day*	*Number of Stores*	*Delivery Cost (Rp)*
Monday	834	365	1615	2814	114	1.865.009
Tuesday	751	551	1755	3027	75	1.672.778
Wednesday	1008	563	1931	3502	92	1.896.566
Thursday	975	422	1584	2981	134	2.521.327
Friday	789	332	1446	2567	67	1.924.553
Saturday	896	447	1540	2883	34	1.108.896

It also describes the costs incurred by PT Y Distribution PT X in trucking the products to the stores according to order during the week from 28 November 2016 until December 3, 2016.

3.2 Problem definition

From the background that has been explained, there are 3 main problems:

1) The weight and volume capacity of the truck is not optimal yet, there's still trucks with many unused space in weight and volume capacity in their distribution
2) There's still truck that has to do overwork to finish their delivery, causing the worker to come late to the headquarter and feel more tired than others
3) There's still stores that didn't get the delivery on time (by 5PM according to the salesman deal).

3.3 Objectives

Based on the observations made directly in the field, as well as supporting data from the company, there are four objectives which authors will achieve through the solution from the model. The Model Objectives are:
1) Reducing total unused trucks volume and weight capacity by 50%
2) Ensure that 100% truck finished their delivery without doing overwork and could get back to headquarter on time
3) Ensure that 100% stores could get the delivery on time (by 5PM according to the salesman deal)
4) Reducing total cost by 20%:

4 The model

The Model was constructed by using actual observation on the day that the delivery actually happened. Table 2. Shows the components in constructing our model.

4.1 Model construction

Table 2. Model Components

Location	Store 1-134, warehouse
Entitiy	The Products (M1,M2,M3)
Resource	Truck 1-8 (Ankle and Double)
Simulation Time	09.00 – 17.00 (one work day)
Shift	09.00 – 17.00 (Break in 12.00 – 13.00 for lunch)
Path Network	Truck Route 1-8

4.2 Model equations

There are some key equations to help authors achieve the model's goals:

1) Moving Time

TW = Truck Weight
TW = Standard Truck Weigh (5500)

$$MT = \frac{\sum Distance}{Speed} \tag{1}$$

2) Volume Capacity

$$0{,}036M1 + 0{,}033M2 + 0{,}01M3 \leq 11{,}628 \tag{2}$$
$(Ankle\ truck)$

$$0{,}036M1 + 0{,}033M2 + 0{,}01M3 \leq 21{,}5 \tag{3}$$
$(Double\ truck)$

i. Truck Utilization

$$TU = (Time\ Used)/\ (Time\ Available) \tag{4}$$

ii. Weight Capacity

$$2M1 + 4M2 + 10M3 \leq 4000 \tag{5}$$
$(Ankle\ truck)$
$$2M1 + 4M2 + 10M3 \leq 6000 \tag{6}$$
$(Double\ truck)$

iii. Total Cost

$$TC = Driver\ Cost + FC \tag{7}$$

iv. Fuel Cost

3) Fuel Consumption (100 Km)

$$\frac{TW-\overline{TW}}{100}\ x\ 0{,}5 + 25 \tag{8}$$

A recent study found that for every 100-kg reduction, the combined city/highway fuel consumption could decrease by about 0.4 L/100 km for cars and about 0.5 L/100 km for light trucks (MIT 2008).

4.3 Model scenarios

This model is run with 2 kinds of scenarios. The first scenario is based on the real data, while the second scenario is modeling the distribution system in a heavy rain condition.

4.4 Verification and validation

Validation techniques has two step. First is model conceptualization validation and the second is model validation. In model conceptualization validation, the research team use face validity by asking feedbacks from our stakeholder, which is the distribution manager in PT. Y Distribution. In model validation, the research team used extreme condition test and sensitivity test. The model is also verified by face verification, asking for feedbacks one more time after the model is done. The feedbacks are positive and the stakeholder is pleased.

5 Results and discussion

5.1. Model construction

Table 3. Scenario Result

	Utiliz – ation	*Work Overti me*	*Late Deli v – eries*	*Unuse d Volum e (m3)*	*Unuse d Weigh t (kg)*	*Total Cost (Rp)*

Scenario 1	61.53 %	1	0	55.9	16422	2.481.012
Scenario 2	84.48 %	5	16	55.9	16422	2.481.012

Table 3. shows the result of model on the real data scenario. The utilization is achieved from the equation that mentioned before. The work overtime defined as a truck that hasn't been back to the warehouse in 5 PM. The late deliveries defined as number of stores that hasn't get their deliveries in 5 PM. The 5 PM parameter is from the deal that salesman make with the stores when they sales the products. Authors also got the unused volume, weight, and total cost from the equation that mentioned before. The result found that there is a room for improvement, where the unused volume and weight is really high in number while the utilization is not really high. It is also seen there's still a truck that has to work overtime to finish their job while the others don't. This result can be used as main base to later generate alternative solutions.

5.2 The heavy rain scenario result

On the heavy rain scenario, the speed of truck is dropped and the pickup and deposit time is doubled. Compared to the real data scenario, it is seen that there's a significant increase in the number of trucks that has to work overtime and the late deliveries.

Table 4. Solution Implementation

	Result in Scenario 2
Utilization	86.89%
Work Overtime	3
Late Deliveries	23
Unused Volume (m3)	22.8
Unused Weight(kg)	6422
Total Cost (Rp)	1.762.067

5.3 Generating solution

From the result that mentioned above, authors did a brainstorming with the help of fishbone tool. This tool is used to find the root cause in this distribution problems so that authors could generate alternative solutions. It is decided that the root cause is in the methods. There's 2 methods that could be improved. First, the routing. The current routing didn't consider the truck capacity, resulting in great number of unused truck capacity. The second one is the products layout method in the truck box. The products is put randomly as fast as possible when the truck is doing pickup at the warehouse, resulting in long searching time when they did the delivery in each stores, they have to match the specific demand with the randomly-put products in the truck. From those root causes, authors propose 3 alternative solutions:

1) Use the sweeping method in order to determine a better distribution route that has considered truck capacity
2) Spend more time in pickup at the warehouse, where the workers put the products following the stores demand that has been ordered according to the route.
3) The combination of solution 1 and 2

5.3.1 Solutions Implementation at the real data scenario model

Table 5. Heavy Rain Scenario result

	Real	*Solution 1*	*Solution 2*	*Solution 3*
Utilization	61.53%	66.63%	62.72%	64.43%
Work Overtime	1	0	0	0
Late Deliveries	0	0	0	0
Unused Volume (m3)	55.9	22.8	55.9	22.8
Unused Weight (kg)	16422	6442	16422	6442
Total Cost (Rp)	2.481.012	1.762.067	2.481.012	1.762.067

In Solution 1, the sweeping method is used to create a new route, authors did a counter clockwise stores grouping based on shortest distance between stores and the truck capacity. The new truck only used if the demand in the stores group couldn't fit in anymore with the truck volume or weight capacity. From the Sweeping Method, authors create a new path network for each of the trucks in the model. The result is the trucks could accomplish zero work overtime and late deliveries by only using 6 trucks. Then, authors eliminate one Ankle and one Double because according to our stakeholders, both of them actually a backup trucks that only used if the deliveries in the day couldn't be accomplished by normal amount of trucks. By this model, it is proved that instead of 8 trucks, 6 trucks could maintain the utilization, accomplish zero work overtime and late deliveries, reducing unused volume and weight, and reducing total cost.

In solution 2, it is assumed that the pickup time at the warehouse would increase 30 minutes because the workers have to put the products in order according to the stores demand that has been ordered along the route. But, the benefit is assumed that the deposit time in each store would decrease 25% because the reduced searching time. The result is there's zero work overtime and late deliveries, but, the unused truck capacity remains the same because there's no change in the route and the amount of trucks.

In solution 3, authors combine both of solution 1 and 2. The new route is used using sweeping method and gave the better product layout within the trucks. The result shows that this solution is the best one, because it gives the same improvement with solution 1 with lower utilization, meaning that in solution 3, the trucks could accomplish the task faster than in solution 1. So, it is recommended that solution 3 could improve the system.

5.3.2 Solutions Implementation at the Heavy Rain Scenario

Table 6. Solution implementation on heavy rain scenario

	Real	*Solution 1*	*Solution 2*	*Solution 3*
Utilization	86.89%	74.4%	79.01%	64.08%
Work Overtime	3	2	3	0
Late Deliveries	23	0	6	0
Unused Volume (m3)	22.8	22.8	22.8	22.8
Unused Weight (kg)	6442	6442	6442	6442
Total Cost (Rp)	1.762.067	2.587.067	1.962.067	2.787.067

After choose the best solution, authors applied the solution into heavy rain scenario model to see if the goals could also be accomplished in the worst condition or not. The result is shown at the table above. The result is pretty disappointing because even though the decrease of unused volume, weight, and total cost meet the same accomplishment, the percentage of work overtime and the amount of late deliveries is increased.

So, authors went back to brainstorming phase to find new alternative solutions that could be applied in the new model using solution 3 in heavy rain scenario. It is found that the root cause of the long-deposit time is that the workers has to wait until the rain is slowing down and then use an umbrella when doing the delivery in each store because they afraid that if the products get wet, the cardboard would be damaged and the stores would reject the products, causing lost profits for the firm. So, based on that root cause, our team propose 3 alternative solutions to improve the distribution system of PT Y Distribution in heavy rain scenario:

1) Hire one more worker to help with the distribution to fasten the deposit time

2) Buys a raincoat for the workers and a large plastic to cover the products when they did the deliveries

3) The combination of solution one and two

The table below (Table. 7) highlights the result of each alternative solutions.

In solution 1, the company hire one more worker to help fasten the deliveries. The deposit system that the workers applied is that one person stack the products in the truck box so that one more person could bring the stack in one go. It is assumed that by adding one more person to bring the stack could reduce the deposit time by 40% because the time needed to stack the product is far lower than the time needed to bring the stack. So, by doing this, the idle time could be reduced for the worker that only stack the products. The result is pretty satisfying because the number of late deliveries is reduced to zero. But, there's still trucks that has to work overtime to accomplish that. The cost is rather high because the driver (worker) cost is increased 150% by adding one more person.

In solution 2, the company would buy a raincoat for all of the workers and a very large plastic to cover the stacked products when it's being delivered in each stores. It is assumed by doing this, the deposit time could be reduced by 20% because they didn't have to wait for the rain becomes lighter and they didn't have to use umbrella anymore where the products that could be brought in one go is smaller in amount than our alternative solution. It is seen that the result is not as good as solution one because there's still 6 late deliveries and the number of work overtime is also bigger. But, it's cheaper because buying raincoat and plastic only cost Rp.200.000 and it last for a long time, could be used for a lot of deliveries.

In solution 3, authors combined both solution 1 and 2, which are add one more workers and bought raincoat and plastic to help fasten the deliveries. By doing this, the company could accomplish zero late deliveries and work overtime. The threat off is this solution has the biggest cost among all of the solutions. In the end, it is recommended that solution 3 could be implemented because the utilization is the lowest among all, meaning that it's the fastest to finish the deliveries. The "bonus time" could be used as anticipation if the rain is really heavy and the delivery time becomes longer than our scenario. To show our overall research goal achievement, authors put all of the improvement results in scenario 1 and scenario 2 in the table below. It is seen that all of the research goals are achieved in both scenarios, except reducing 20% cost in scenario 2. From the research, it is decided that in order to achieve other important goals, an extra cost would be needed.

DOI:10.1051/matecconf/201712405006

Table 7. Result of Each Alternative Solutions

	Scenario 1			Scenario 2			
	Before	*Final Improvement*	*Research Goal*	*Before*	*First Improvement*	*Final Improvement*	*Research Goal*
Utilization	61.53%	64.43%		84.48%	86.89%	64.08%	
Work Overtime	1	0	v	5	3	0	v
Late Deliveries	0	0	v	16	23	0	v
Unused Volume (m3)	55.9	22.8	v	55.9	22.8	22.8	v
Unused Weight(kg)	16422	6442	v	16422	6442	6442	v
Total Cost (Rp)	2481.012	1.762.067	v	2.481.012	1.762.067	2.787.067	x

This research shows that Discrete-Event Simulation (DES) performed preferable solutions than System Dynamic (SD). It is because DES Modeling solutions provide greater number of calculations at the individual level while SD has minimal data requirements to build a model. Dynamic Vehicle Routing Problem (DVRP) is one of System Dynamic example. In different point of solutions, Discrete-Event Simulation supports Sweep Algorithm Method to generate a new alternative solution by represent the result of Algorithm in a modeling simulation.

Sweep Algorithm have divided locations equally rather than fuzzy ant colony in Dynamic Vehicle Routing because each cluster has a large different number of locations. It also has more detailed variables of observation. Authors are not only observe the use of the transportation but also count the total cost for the solution and the net profit gained from the model. It is also count the cost of the driver, the fuel, the products production, the weight, the volume and else, so that the company could choose the better solution. Nonetheless, real-life situations may not be totally performed as the model assumes several constant rates and uncertain distributions.

6. Conclusion and Suggestion

After doing research, it can be concluded that the unbalance use of resources which caused overtime for many workers can be solved by the use of Sweep Algorithm combined with Discrete-Event Simulation and put the products following the stores demand that has been ordered according to the route to match the resource utilization and improve the customer's satisfaction. Our research provides 2 types of scenario, the first one shows the original condition and the second one shows the first scenario that face the rainy weather disturbances.

Having performed at Pro-Model simulation using the sweeping method, the use of 6 trucks can still be utilized better because capacity utilization is still low. If used properly, it will increase sales and profit targets obtained will also be more. For the first scenario, the most suitable solution is build the new route using sweeping method and make a better product layout within the trucks.. Then, for the second scenario the most suitable solution is add more workers and bought the raincoat and plastic to fasten the delivery. These solutions can reduce the total unused trucks capacity by 50%, ensure that trucks finished their total delivery without being late in every stores by 100%, and reduce the total cost by 20% in scenario 1. In scenario 2, the firm have to pay more to add workers and buy the plastic and raincoat as its best solution. However, the usage of assumptions and limitations for Discrete-Event Modeling in Pro-model need to be minimized to get the clearer data. The calculation of distances between locations in Pro-model can also be developed to obtain more specific scale

References

1. Nurcahyo, Gunadi W et all. 2002. *Sweep Algorithm in Vehicle Routing Problem For Public Transport.* Jurnal Antarabangsa (Teknologi Maklumat) **2**(2002): 51-64
2. Malandraki, C and Daskin, M. 1992. *Time Dependent Vehicle Routing Problems: Formulations, Properties and Heuristic Algorithms.* Transportation Science 26(3):185-200
3. A.A Tako, S. Robinson. 2002. *The application of discrete event simulation and system dynamics in the logistics and supply chain context.* Decision Support Systems, Vol **52**, pp. 802-815
4. Ahmadi, Mansour. 2012. *The application of system dynamics and discrete event simulation in supply chain management of Swedish manufacturing industries.* Product and Process Development Innovative Product
5. B. Angerhofer and M. Angelides. 2000. 2000 Winter Simulation Conference, pp. 342-351.
6. L. F. Cheng and M. A. Duran. 2004. *Logistics for world-wide crude oil transportation using discrete event simulation and optimal control.* Computers & Chemical Engineering, vol. **28**, pp. 897-911
7. Huang, Yixiao. 2016. *Time-dependent vehicle routing problem with path flexibility.* Transportation Research Part B **95** (2017) 169–195
8. Kuo, R.J. 2016. Application of a fuzzy ant colony system to solve the dynamic vehicle routing problem

with uncertain service time. Applied Mathematical Modelling **40** (2016) 9990–10 0 01

9. Arifin, Aridhanyati. 2014. *Model transportasi untuk masalah pendistribusian air minum (studi kasus pdam surakarta)*. TEKNOMATIKA Vol. **7**, No. 1, JULI 2014
10. Wassan, Niaz A et al. *Vehicle Routing Problem with Deliveries and Pickups: Modelling Issues and Meta-heuristics Solution Approaches.* International Journal of Transportation
11. Council of Supply Chain Management. *Cscmp.org*. Web. 10 Nov 2016
12. Euromonitor International. *Euromonitor.com*. Web. 10 Nov 2016

DOI:10.1051/matecconf/201712405007

Research on InlandPort Facilities Adaptability Evaluation Indicator System

Yaofeng Xie[1], Chenglin Liu[2], Suyang Gao[3], Yan Chen[4] and Ning Jiang[5]

[1-5]*School of Transportation, Southeast University, Si Pai Lou #2, Nanjing 210096, China*

Abstract: Upgrading engineering of inland simple backward port facilities is urgent and of great significance in the aspect of economy and society. Port facilities adaptability evaluation is the foundation work of the research on upgrading strategy. Based on Anhui inland simple backward port facilities upgrading strategy research, AHP is introduced in the analysis of port facilities adaptability evaluation indicator system. The weights of 40 evaluation indicators are obtained through establishing an AHP structure model, constructing judgment matrixes, checking consistency, and calculating combination weights. By combining the analysis of relevant relationships between the indicators, 16 indicators are selected as indicators of port facilities adaptability evaluation system. The research results lay the foundation for the upgrading strategy research on Anhui inland simple backward port facilities, which can be an essential reference for similar projects.

1 Introduction

Inland water transportation is an important part of comprehensive transport system and water resource utilization and management system, and it is an essential part of basin economy because of large freight volume, low transportation cost, less ground demand, less energy consumption, and less pollution. Yangtze River, Huai'he River and Xin'an River cut through AnhuiProvince, forming a natural advantage of water transportation. However, some ports lack systematic plan and scientific management, which results in the inefficient utilization of some inland river shoreline resources. Especially for the 600 simple backward berths with 500-ton level or below 500-ton level along the branches of three large drainage, the development level fall behind the development status of local economy and Anhui economy. These simple backward ports have already become negative factor blocking the transformation of Anhui economy.

Based on the development status of Anhui inland simple backward port facilities, port facilities adaptability evaluation system is established to evaluate the ability of port facilities to adapt the change of external environment factors and to keep pace with port development needs. In order to confirm the main indicators of port facilities adaptability evaluation, AHP is introduced to determine the weights of all the evaluation indicators, which is in support of selecting several relatively important indicators from many indicators proposed. AHP is an effective multi-target decision method combining qualitative analysis with quantitative analysis, which is a common method with easy calculation and clear results [1-2].

A lot of researches have been done with AHP in port management,green port development, port logisticand some other relevant areas. Mabrouki C et al analyzed the problem of operational risk management in port terminal with AHP method[3], P. Zhang et al studied the evaluation indexed system of dangerous good logistics port with AHP[4],Y. Xie et al developed the port ecological suitability evaluation model with AHP[5].

AHP is also widely adopted in indicator management system and decision-making system [6-8] and it is often combined with other methods for instance, expert consults and fuzzy set theory.Chiu.R revealed top five priority attributes of green port operation with Fuzzy AHP model[9].F. Zhu et al evaluated the function of the port logistics with AHP- FCE model[10]. Nevertheless, fewer literatures have been published on the port facilities adaptability evaluation system.

In this paper, MATLAB software is adopted to programmatically deal with judgment, analysis and calculation process of AHP in order to conveniently confirm the weights of port facilities adaptability evaluation indicators and to lay the foundation for the analysis of indicator selection results. The research results by AHP can be a valid reference for establishing port facilities adaptability evaluation system.

2 Basic rules of AHP

AHP is a weight decision method proposed by an American scholar named Saaty, T. L in the 1970s. AHP is characterized by making the highest goal hierarchy and resolving the goal hierarchy into small targets in the bottom hierarchy according to the influence indicators, which makes the decision method more convenient. The method combines quantitative analysis with qualitative analysis, using the experience of the decision makers to judge the relative importance of each target, which can be effectively used in some problems difficult to adopt quantitative method[11-13]. The basic ideas of AHP are as follows:

(1)According to the properties of the problem and the total target expected to reach, the influence indicators of the decision-making problem are listed. Moreover, an ordered hierarchy is established on the basis of the connection, interaction and membership between the indicators to obtain a multilevel analysis model, which methodizes the decision-making problem.

(2)The subjective judgments of indicator importance are reflected in quantitative way, based on which, judgment matrix is set up by comparing each indicator in the same hierarchy.

(3)The weights of indicators in each hierarchy are calculated by mathematical method.

(4)Through the comprehensive analysis of the hierarchies from low to high, the weights of indicators in the bottom hierarchy which are relative to the highest goal are calculated. The weight of the indicators represents the importance of each indicator, determining the important indicators which influence the highest goal.

3 Basic steps of AHP

3.1 AHP structure model establishment

The premise of establishing an AHP structure model is to fully understand the problem to be solved, grasp the scope, the indicators of the problem and the relationship and membership of these indicators. On this basis, the involved indicators of the problem are classified, and an AHP structure model figure is shown as Figure. 1.

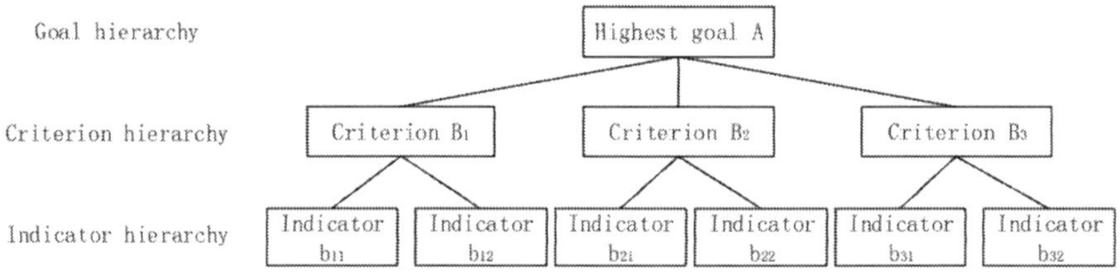

Figure 1. AHP structure model.

3.2 Judgment matrix construction

In general, the relative importance of each indicator in indicator hierarchy to criterion hierarchy is different, and the relative importance of each indicator in criterion hierarchy to the highest goal is different. The relative importance of two indicators is difficult to determine, therefore, through comparing these indicators in pairs, judgment matrix can be constructed according to the scale which describes the importance. The judgment matrix reflects the relative importance of indicators in the hierarchy to the specific indicator in the upper hierarchy, which is the basic ground for making decision.

Supposing that the indicator A_k of the upper hierarchy is relevant with n indicators ($B= \{B_1, B_2, \ldots\ldots, Bn\}$) in the lower hierarchy, and bij represents the importance of indicator B_i to B_j when considering indicator A_k, all comparison results can be shown as matrix $B=(bij)n\times n$ in Table 1. According to the characteristic of judgment matrix B, $bij>0$, $bii=1$, $bij=bji(i,j=1,2,\ldots,n)$.

In order to quantify the comparison, the scale method of 1~9 is introduced as shown in Table 2 on the basis of research conclusion proposed by psychologists that the ultimate ability of people in distinguishing the information level is 7±2. The value of bij can be 2, 4, 6 or 8 if the relative importance is between the levels. The reciprocal is the result of two indicators in reverse comparison.

Table 1. Matrix $B=(bij)n\times n$

A_k	B_1	B_2	……	B_j	……	B_n
B_1	b_{11}	b_{12}	……	b_{1j}	……	b_{1n}
B_2	b_{21}	……	……	……		
……	……	……	……	……	……	……
B_n	b_{n1}	……	……	B_{nj}	……	b_{nn}

Table 2. Scale method of 1~9

Comparing B_i with B_j	Equally important	Weakly important	Obviously more important	Demonstratively important	Absolutely important

b_{ij}	1	3	5	7	9

3.3 Consistency check of Hierarchical single arrangement

The hierarchical single arrangement weight is the order weight of indicators in this hierarchy to an indicator in the upper hierarchy. Firstly, the characteristic vector solution ($W= [W_1, W_2, W_3, \ldots\ldots, Wn]^T$) of the judgment matrix ($B= [bij]n\times n$) is calculated. Then, the weight ($C= [C_1, C2, C_3, \ldots\ldots, Cn]^T$) is obtained by normalizing the characteristic root solution.

$$BW = \lambda_{max} W (1)$$

For the problem with many indicators and large scale, due to the complexity of the thing and the one-sidedness of people's cognition, the judgment matrix is not necessarily completely consistent. If the judgment matrix is not completely consistent, mark judgment B as B' and the maximum characteristic root (λ'_{max}) obtained from characteristic equation ($B'W = \lambda'_{max} W$) is bigger than n, and the bigger $\lambda'_{max}$is, the more serious the inconsistent degree of B' is. In this situation, the characteristic vector corresponding to λ'_{max} is not a true reflection of the influence weight of indicators ($B= \{B_1, B_2, \ldots\ldots, Bn\}$) in this hierarchy to the indicator in the upper hierarchy. The following indicators need to be calculated in order to check the consistency ofHierarchical single arrangement:

Consistency indicator:

$$CI = \frac{\lambda_{max} - n}{n-1} (2)$$

If $CI \neq 0$, the maximum characteristic root is expected to be slightly bigger than n and the other characteristic roots are expected to be close to 0, then, the judgment matrix is considered to be consistent. In order to provide specific measure indicator, the mean random consistency indicator (RI) is introduced to judge whether the judgment matrix is consistent.

Mean random consistency indicator:

$$CR = \frac{CI}{RI} (3)$$

The values of mean random consistency indicators can be found in Table 3[14].

Table 3. Mean random consistency indicator.

n	1	2	3	4	5	6	7	8	9
RI	0	0	0.58	0.90	1.12	1.24	1.32	1.41	1.45

When $CR<0.10$, the hierarchical single arrangement is considered to be consistent, otherwise, adjustments should be made for the judgment matrix. In the case of $n=1$ or 2, CR is only formal and it needs no checking.

3.4 Consistency check of Hierarchical total arrangement

For obtaining the combination weights of indicators in a certain hierarchy to the highest goal and the interaction between these indicators and the indicator in the upper hierarchy, the weights of indicators in this hierarchy are calculated by using the arrangement results and the combination weights of the indicators in the upper hierarchy. In other words, the weights of all indicators in a certain hierarchy to the highest goal are calculated, which is called hierarchical total arrangement.

There are m indicators ($A_1, A_2, \ldots\ldots, Am$) in hierarchy A, and the weights are $a_1, a_2, \cdots\cdots, a_m$ respectively. There are n indicators ($B_1, B_2, \ldots\ldots, Bn$) in the lower hierarchy B, the hierarchical single arrangement weights of them are known as $b_{1j}, b_{2j}, \ldots\ldots, bnj$. If B_k is not relevant with A_j, then $b_{kj}=0$. The hierarchical total arrangement weights can be obtained by following the hierarchical total arrangement matrix shown in Table 4.

Table 4. Hierarchical total arrangement weights of hierarchy B

B \ A	A_1	A_2	......	A_j	......	A_m	Weight of hierarchy
	a_1	a_2	......	a_j	......	a_m	
B_1	b_{11}	b_{12}	......	b_{1j}	......	b_{1m}	$\sum_{j=1}^{m} a_j b_{1j}$
B_2	b_{21}	b_{22}	......	b_{2j}	......	b_{2m}	$\sum_{j=1}^{m} a_j b_{2j}$

......	......	......	......	......	......	......	......
B_n	b_{n1}	b_{n2}	......	b_{nj}	......	b_{nm}	$\sum_{j=1}^{m} a_j b_{nj}$

Hierarchical total arrangement consistency indicator of hierarchy *B* is shown as Equation 4, where *CIj* is hierarchical single arrangement consistency indicator of hierarchy *B* to indicator *Aj*, and *RIj* is mean random consistency indicator of hierarchy B to indicator *Aj*.

$$CR = \frac{\sum_{j=1}^{m} a_j CI_j}{\sum_{j=1}^{m} a_j RI_j} (4)$$

Similarly, hierarchical total arrangement is consistent if $CR<0.10$. Otherwise, judgment matrix needs adjustments.

3.5 Indicator combination weight calculation

The geometric method[15]can be adopted in the calculation of matrix weight. The steps are as follows:

(1) Calculate the product of all elements in every row in the judgment matrix:

$$m_i = \prod_{i=1}^{n} b_{ij} (i = 1,2 \ldots .n) (5)$$

(2) Calculate n-th root of mi ($\overline{\omega_i} = \sqrt[n]{m_i}$):

$$\omega_i = \overline{\omega_i} / \sum_{j=1}^{n} \overline{\omega_j} (6)$$

(3) Calculate the maximum characteristic root (λ_{max}) of matrix B:

$$\lambda_{max} = \frac{1}{n} \sum_{i=1}^{n} (B\omega)_i / \omega_i (7)$$

The weight of indicator hierarchy can be calculated with the weight of criterion hierarchy *B*. Assuming that the relative weight of criterion hierarchy *B* to goal hierarchy *A* is $\bar{\omega} = (\overline{\omega_1}, \overline{\omega_2}, \ldots . . \overline{\omega_k})^T$, and the relative weight of indicator hierarchy *C* to criterion hierarchy *Bi* is $\bar{\omega} = (\omega_{1i}, \omega_{2i}, \ldots \ldots, \omega_{ni})^T$, then the relative weight (*Wi*) of indicator hierarchy *C* to goal hierarchy *A* can be obtained:

$W_1 = \sum_{j=1}^{k} \omega_j \omega_{1j}$, $W_2 = \sum_{j=1}^{k} \omega_j \omega_{2j}$,......, $W_n = \sum_{j=1}^{k} \omega_j \omega_{nj}$.

4 Determination of Anhui inland port facilities adaptability evaluation indicator weight

In the process of researching upgrading strategy of Anhui inland simple backward port facilities, port facilities adaptability evaluation system should be established to evaluate the adaptability for all the port facilities, and judge their development level to lay the theoretical foundation for proposing the batch-upgrading scheme. Several representative indicators are chosen from those initial indicators when establishing port facilities adaptability evaluation system. These representative indicators should cover every criterion hierarchy, and the starting point is "keep to the point" and "from point to area", which is significant in choosing relatively important and representative indicators.

AHP is adopted to deal with expert scoring data, and MATLAB functions and matrix calculation are used in the input and creation of judgment matrix with loop structure to checking the consistency of judgment matrix. The weights of all indicators in indicator hierarchy to the goal hierarchy are calculated when the judgment matrix is consistent. The program flow chart of AHP is shown as Figure.2. The weights of indicators of criterion hierarchy (port classification throughput and capacity) and the checking result of single arrangement consistency are shown in Table 5, and the final weights of all the indicators are shown in Table 6.

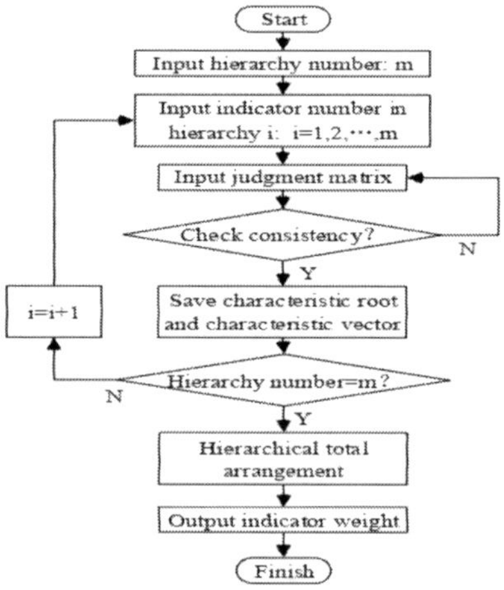

Figure 2. Program flow chart of AHP.

According to the calculation results listed in Table 6, in criterion hierarchy, port economy, port classification throughput and capacity, port coastline efficiency and port energy consumption are the 4 indicators that have

DOI:10.1051/matecconf/201712405007

most influence on port facilities adaptability evaluation, which also means they determine port development level. In indicator hierarchy, profit margin of the cost and expense, port throughput, net assets return ratio and energy consumption of every 10000-ton goods mark the highest weight, and these 4 indicators reflect the essence of port facilities adaptability evaluation directly. Among these 4 indicators, profit margin of the cost and expense, port throughput and net assets return ratio represent port economy development level, and energy consumption of every 10000-ton goods represents port energy consumption efficiency.

According to the weight values of indicators to the goal hierarchy listed in Table 6, weight contribution threshold value of these indicators to the goal hierarchy is taken as 85%, and the weight contribution sum of 16 indicators whose weights are relatively larger is 86.16%.(0.1995+0.1176+0.1155+0.0560+0.0546+0.0510 +0.0459+0.0350+0.0320+0.0252+0.0243+0.0221+0.0221 +0.0221+0.0221+0.0166=86.16%). Considering the correlation analysis of these indicators, 16 evaluation indicators are chosen as follows: port safety, port usability, channel level, energy consumption of every 10000-ton goods, energy consumption of every 10000-yuan income, operation capacity, checked capacity, port throughput, throughput of unit shoreline length, shoreline length of every 10000-ton capacity, berth capacity utilization, berth operation rate, berth utilization, productivity of staffs, profit margin of the cost and expense and net assets return ratio.

Compare the 16 selected main valuation indicators with indicators in Table 6, it can be found that the main evaluation indicators selected by AHP have a wide coverage: they involve 6 of the all 8 criterion hierarchies. Moreover, these selected indicators contain main valuation contents such as port safety, port throughput, port shoreline efficiency and port economy, which means they consist of those major problems in port facilities adaptability evaluation.

The development reality of Anhui ports and quantification valuation operability of these indicators should be taken into consideration to make some fine-adjustments in the determination of port facilities adaptability valuation indicators. The fine-adjustments are helpful for the perfection of port facilities adaptability evaluation system, which makes its results provide scientific guidance for port facilities batch-upgrading scheme in the future.

5 Conclusions and suggestions

(1) AHP is a concise and practical decision-making method. On the basis of expert scoring, the weights of port facilities adaptability evaluation indicators can be determined. AHP is used to combine qualitative judgment with quantitative judgment and transform a multi-criteria and multi-objective problem into a multi-hierarchy and single objective problem, whose calculation is simplex and results are clear.

(2) AHP is adopted in selecting port facilities adaptability evaluation indicators. With MATLAB programming calculation and analysis, 16 relatively important and representative evaluation indicators are chosen, which lay the foundation for establishing port facilities adaptability evaluation system in the future.

(3) The development reality of Anhui ports and quantification valuation operability of these indicators should be taken into consideration in the determination of port facilities adaptability valuation indicators. On the basis of expert scoring data, the weight of port facilities adaptability evaluation indicators are determined and port facilities adaptability evaluation indicators are selected with AHP. The research method and research results can be an essential reference for decision making of port navigation management department.

(4) A cmprehensive model such as fuzzy-AHP method are supposed to be conducted in the future to improve the quality of the evaluation system.

Table 5. Port classification throughput and capacity indicators.

Port classification throughput and capacity	Designed capacity	Checked capacity	Operation capacity	Port throughput	Weight: ω_i
Designed capacity	1	1/3	1/5	1/7	0.06
Checked capacity	3	1	1/3	1/5	0.12

Operation capacity	5	3	1	1/3	0.26
Port throughput	7	5	3	1	0.56
λ_{max}=4.12; CR=0.0438, matrix is consistent ω_i —Normalization weight					

Table 6. Final weights of indicators.

Goal	Criterion	Indicator	Weight
Port facilities adaptability valuation	Hydraulic structure 0.090	Berth structure and level	0.0117
		Berth depth	0.0027
		Port technical state	0.0054
		Port safety	0.0459
		Port usability	0.0243
	Channel and ship type 0.022	Channel level	0.0166
		Ship type	0.0034
	Port classification throughput and capacity 0.208	Designed capacity	0.0126
		Checked capacity	0.0252
		Operation capacity	0.0546
		Port throughput	0.1176
	Handling operation 0.046	Classification loading machinery allocation	0.0020
		Classification loading machinery capacity and production	0.0030
		Loading machinery property and technical state	0.0050
		Loading efficiency	0.0100
		Ship average efficiency	0.0050
		Ship average amount	0.0050
		Time of loading a designed ship	0.0100
		Time of ship in port when loading every 100-ton goods	0.0100
	Port energy consumption 0.096	Energy consumption in loading goods	0.0120
		Energy consumption of every 10000-yuan income	0.0320
		Energy consumption of every 10000-ton goods	0.0560
	Port coastline efficiency 0.167	Port shoreline length	0.0034
		Shoreline condition and level	0.0085
		Berth number and length	0.0085
		Berth utilization	0.0221
		Berth operation rate	0.0510
		Throughput of unit shoreline length	0.0221
		Shoreline length of every 10000-ton capacity	0.0221
		Real value of unit coastal resource	0.0085
		Berth capacity utilization	0.0221
	Port warehouse and yard	Capacity	0.0010
		Effective area	0.0024

DOI:10.1051/matecconf/201712405007

	0.023	Storage quota of unit area	0.0064
		Average storage capacity	0.0026
		Average storage amount	0.0064
		Average storage period	0.0012
	Port economy 0.348	Productivity of staffs	0.0350
		Profit margin of the cost and expense	0.1955
		Net assets return ratio	0.1112

References

1. Z. Chen. Port & Waterway Engineering**07**,70-73(2009).
2. J. Guo, Z.Zhang, Q. Sun.China Safety Science Journal.**05**,148-153(2008).
3. Mabrouki C, Bentaleb F, Mousrij A. Safety Sci63(3):124-132(2014).
4. P. Zhang, J. Mao, L. Yang. Software Engineering and Knowledge Engineering: Theory and Practice.943-949(2012).
5. Y. Xie, Y.Liu, X. Xiao. Frontiers of Structural and Civil Engineering.**9**(1), 65-70(2015).
6. G.Chen, Y. Zheao.Journal of Shanghai Jiaotong University.**40**(4), 701-704(2006).
7. Moghadam M K, Bonsall S, Wang J, et al. Mar. Technol. Soc. J, **43**(3), 34-50(2009).
8. Dai J S L L J. Transportation Planning & Technology, **35**(4), 509-524(2012).
9. Chiu R, Lin L, Ting S. Math ProblEng, **37**(3), 413-418(2014).
10. F. Zhu. International Conference on Management & Service Science, 1-4 (2011).
11. X.Zhou, H. Lin. China Water Transport,**12**,68-70(2008).
12. Z.Wang, X.Li.. Port & Waterway Engineering, **02**, 81-85(2001).
13. X.Wang, L.Qin, Y.Chen.Journal of Chongqing Jiaotong University(Natural Science), **04**,628-631(2010).
14. X. WANG, Q. Kang, J.Qin, Q. Zhang, S. Wang. Journal of Central South University (Science and Technology), **06**, 2455-2462(2013).
15. Saaty, T L, Wong M M. J. MathSociol, **9**,181-209(1993).

Dynamic Energy Consumption and Emission Modelling of Container Terminal based on Multi Agents

Jue Hou

China Waterborne Transport Research Institute, Ministry of Transport of the People's Republic of China, Haidian District, Beijing, China

Abstract. Environmental protection and energy saving pressure press the increasing attention of container terminal operators. In order to comply with the more and more strict environmental regulation, reducing energy consumption and air pollution emissions, meanwhile, optimizing the operation efficiency, which, is an urgent problem to container terminal operator of China. This paper based on the characteristic of Container Terminal Operation System (CTOS), which includes several sections of container product processes, consist of berth allocation problem, truck dispatching problem, yard allocation problem and auxiliary process. Dynamic energy consumption and emissions characteristic of each equipment and process is modelled, this paper presents the architecture of CTOS based on the multi agent system with early-warning model, which is based on multi-class support vector machines (SVM). A simulation on container terminal is built on the JADE platform to support the decision-making of container terminal, which can reduce energy consumption and air pollution emissions, allows the container terminal operator to be more flexible in their decision to meet the Emission Control Area regulation and Green Port Plan of China.

1 Introduction

In the last more than a decade years, with China's domestic and foreign trade continues to expand, container transport is increasingly valued, environmental protection and energy saving pressure press the increasing attention of government and container terminal operators. China has started the Green Port Plan from 2011, which is total life cycle and closed-loop in operation of port, includes several evaluation indicators, to evaluate operation management level of environment and energy efficiency of port. Four years later, China has established the three Emission Control Areas (ECAs) in 2015, Three ECAs include the Pearl River Delta, the Yangtze River Delta, and in the Bohai Sea. They have been created to reduce the levels of ship-generated air pollution and mainly focus on the sulphur content of fuels.

In order to improve energy efficiency and reduce emission, and get dynamic information in advance, this paper based on the characteristic of Container Terminal Operation System (CTOS), which includes several sections of container product processes, and consist of berth allocation problem, truck dispatching problem, yard allocation problem and auxiliary process. Dynamic energy consumption and emissions characteristic of each equipment and process is modelled, this paper presents the architecture of CTOS based on the multi agent system with early-warning model, which is based on multi-class Support Vector Machines (SVM) algorithm. A simulation on container terminal is built on the JADE platform to support the decision-making of container terminal, which can reduce energy consumption and air pollution emissions, and allows the container terminal operator to be more flexible in their decision to meet the environmental protection regulation and green port plan of China.

2 Modelling

In order to reflect the energy consumption and emission of container terminal comprehensively, this paper presents a model, considering the dynamic energy consumption and mission of process of container terminal, which based on Container Terminal Operation System (CTOS) with early warning model.

2.1 Container Terminal Operation System

CTOS is a complex discrete dynamic system which reflect a series of operation process in container terminal. The energy consumption and emission problem based on CTOS is mainly for four problems in this study, when container ships arrive at terminal through anchorage, which berth should be allocated to the mooring ship, called berth allocation problem (BAP) [1~2], includes ship, quay crane and shore power system (SPS). Then, ships are handled by quay cranes, and containers of ships are transhipped by trucks between quay and yard, which called the truck allocation problem (TDP) [3~6]. In yard stack of container terminal, yard crane and truck need cooperate with each other, to transport and storage

 DOI:10.1051/matecconf/201712406001

container, called the yard allocation problem (YAP) [7], as shown in Figure 1. The auxiliary process of operation includes lighting, refrigeration and heating, cleaning and so on.

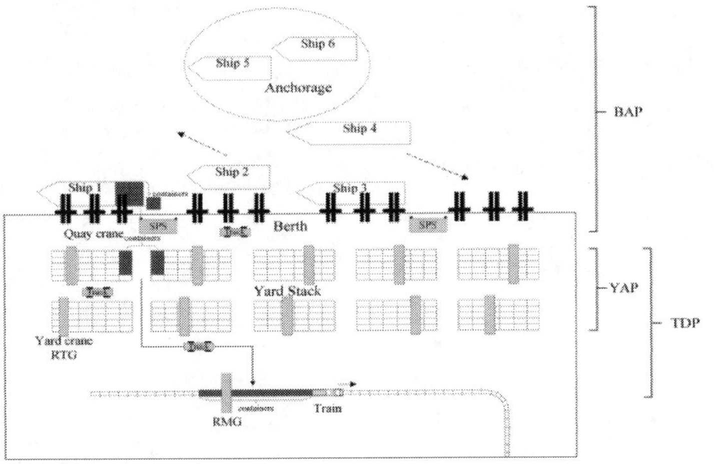

Figure 1. A typical container terminal and its operation process.

The subsystems structure of CTOS in this paper as shown in Figure 2, including location subsystem and equipment subsystem. Each subsystem represents the transhipment of location and handle equipment of container flow in CTOS, which could reflect dynamic characteristic of energy consumption and emission. CTOS will be modelled in the next section.

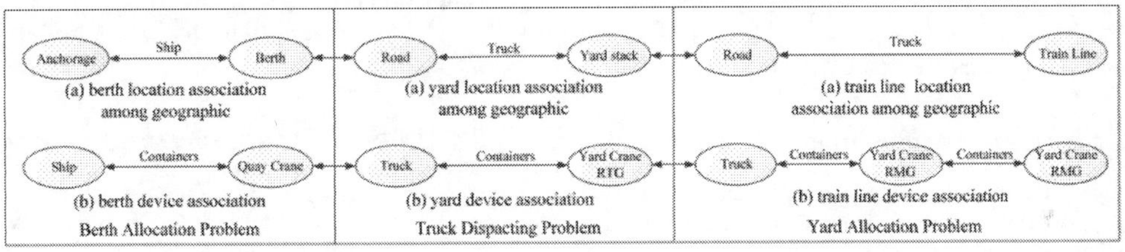

Figure 2. Flowchart of Container Terminal Operation System

2.2 Dynamic Energy Consumption and Emission Model

Consider the multi variables of energy consumption (electric, diesel, LNG, etc.). The model as shown in Figure 3. Based on the dynamic energy consumption and emission, container operator analysis the process which is high energy consumption and emission, and take actions to reduce.

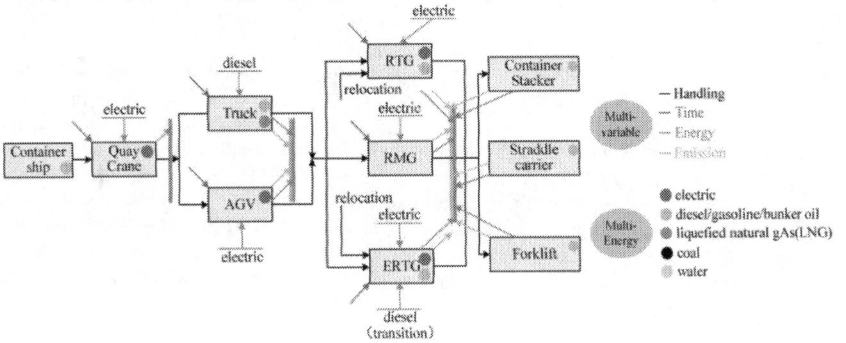

Figure 3. Multi variable and energy of container terminal.

2.3 Early-warning model

To evaluate the level of terminals in energy efficiency and emission, the early warning model is built, which could help the container operator to obtain the information about energy consumption and emission. An early warning model dealing with terminal energy consumption and pollutant emission is built based on three levels, including normal, early warning and warning, Multi-Class Support Vector Machine (M-SVM) is used to solve the model.

The evaluation of energy consumption and emission of container terminal can be regarded as a machine learning problem which based on empirical data, and the solution steps as follows:

- **Step 1** selects training data set

Collect energy consumption and emission training data set.

- **Step 2** identifies training models

Expert experience is used to assign weights and determine training data set.

- **Step 3** selects the kernel function

The training data is analysed and judged to be linear separable or non-separable, and different kernel functions are designed to solve the model.

- **Step 4** forecast early warning level

Training data and cross validation and scoring the three SVM models, warning levels, classification hyperplane and support vector are obtained.

- **Step 5** accuracy verification

Model accuracy verification.

2.4 Approach of early-warning model

Support Vector Machine (SVM) is based on statistical learning theory which is a kind of machine learning method, considering the training error and the generalization ability, to search for the best compromise solution. SVM is originally proposed for two classification problem, but it is often proposed for multi classification problem. If the input space of the training set is linearly separable, there is a hyperplane as:

$$y_i\left(\mathbf{w}^T x_i + b\right) \geq \pm 1 \quad i = 1,2,\mathrm{L}\ ,n. \tag{1}$$

The interval between the two classes can be calculated as $2/\|\mathbf{w}\|$ the minimum value of the interval between the two classes:

$$\begin{cases} \min\limits_{w,b} \dfrac{1}{2}\mathbf{w}^T\mathbf{w} \\ \text{s.t. } y_i\left(\mathbf{w}^T x_i + b\right) \geq 1 \quad i = 1,2,\cdots,n. \end{cases} \tag{2}$$

Where the $\mathbf{w}$ is the direction vector of the hyperplane, b is the position offset of the classification hyperplane, n is the number of support vectors, then the optimal hyperplane f is determined, and the decision function is s. For a test sample x, if s=1, then the x belongs to the positive class, otherwise x belongs to the negative class, so as to achieve the classification of samples.

Using the Lagrange method, the Eq. (2) can be transformed into two programming problems:

$$\begin{cases} \min\limits_{w,b}\max\limits_{\alpha} L(\mathbf{w},b,\alpha) = \dfrac{1}{2}\mathbf{w}^T\mathbf{w} - \sum\limits_{i=1}^{n}\left[y_i\left(\mathbf{w}^T x_i + b\right) - 1\right] \\ \text{s.t. } \sum\limits_{i=1}^{n} y_i\alpha_i = 0 \qquad \alpha_i \geq 0, i = 1,2,\cdots,n. \end{cases} \tag{3}$$

If the input space of the training set is linearly non-separable, a relaxation term is added to the constraint condition Eq. (2), and then the two optimization problem is transformed into the following:

$$\begin{cases} \min\limits_{w,b_j,\xi} \dfrac{1}{2}\mathbf{w}^T \cdot \mathbf{w} + C\sum\limits_{i=1}^{n}\xi_i \\ \text{s.t. } y_i\left(\mathbf{w}^T x_i + b\right) \geq 1 - \xi_i, \quad \xi_i \geq 0 \quad i = 1,2,\mathrm{L}\ ,n. \end{cases} \tag{4}$$

Where C is a constant, used to control the trade-off between the maximum interval and the minimum training error, called the penalty factor, and the two programming problem can be transformed into the following dual forms:

$$\begin{cases} \max\limits_{\alpha} L(\alpha) = \sum\limits_{i=1}^{n}\alpha_i - \dfrac{1}{2}\sum\limits_{i=1}^{n}\sum\limits_{j=1}^{n} y_i y_i \alpha_i \alpha_j K\left(x_i, x_j\right) \\ \text{s.t. } \sum\limits_{i=1}^{n} y_i\alpha_i = 0 \qquad \alpha_i \geq 0, i = 1,2,\mathrm{L}\ ,n. \end{cases} \tag{5}$$

In order to solve the problem of multiple classifiers, this paper presents the optimal decision function as [8~9]:

$$f_{st}(x) = \mathbf{w}_{st} \cdot \phi(x) + b_{st} = \sum_{i=sv}\alpha_i^{st} y_i K\left(x_i, x\right) + b_{st} \tag{6}$$

3 Multi Agents System using for Modelling

3.1 Multi Agents System

CTOS could solve the conflicts between operation efficiency, energy consumption and emission in reasonable strategy of container terminal. Multi-Agents System (MAS) can solve some complex task and target.

Simulation is not so strongly which is limited by the size and complexity of the CTOS and can usually cover all required subsystems in a single model, respecting their mutual relations as well [10~12]. Multi-Agents System (MAS) is a complex system that consists of agents which are the characteristics of autonomy, cooperation, and sociality [13~16], and it can reflect container terminal operating process. MAS is typical discrete, distribute and dynamic system, could meet functional requirements of the CTOS. This paper study the CTOS based on MAS, using the below characteristics: the autonomy, when the CTOS sends request, the agent do not accept request until the task is finished. The cooperation: when the container transport or operation task could not be completed by single agent, the task by coordination and negotiation with other agents. The sociality: the agents bring about communication, collaboration, coordination and negotiation with the others. This study has a sequence

DOI:10.1051/matecconf/201712406001

of discrete time steps (1 sec), at each time step, the agent receives a state from the environment. At the next time step, as a consequence of its action, the agent receives a new task and moves to a new state.

This paper with the CTOS consists of two subsystems, the location subsystem and equipment subsystem. Location subsystem is composed of anchorage agents, berth agents, stack yard agents, road agents and train loading and unloading line agents. They are responsible for handling locations of the CTOS. The interaction sequence between these location agents as shown in Figure 4. Equipment subsystem is composed of ship agents, quay crane agents, truck agents, yard crane agents and train agents. They are responsible for handling operation of the CTOS. The flow of containers is bi-direction, the import and export, the interaction sequence of import container between these equipment agents as shown in Figure 5.

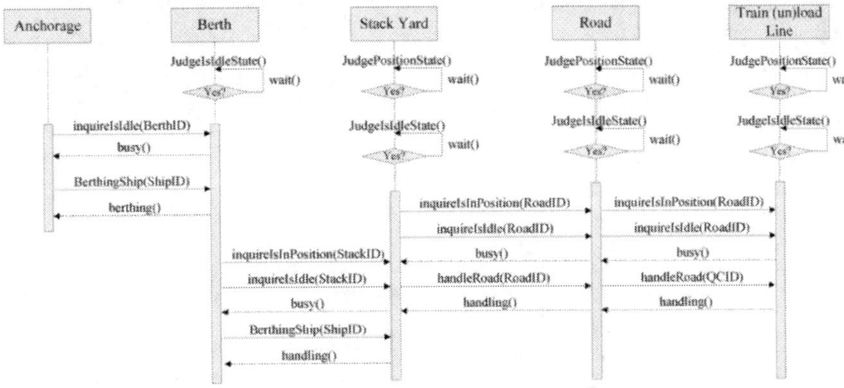

Figure 4. Location subsystem interaction sequence diagram of container terminal operation system

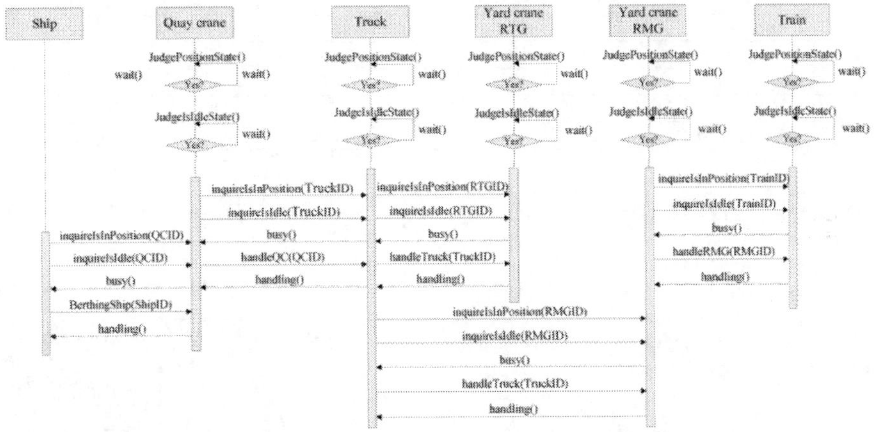

Figure 5. Equipment subsystem interaction sequence diagram of container terminal operation system

3.2 Platform

In this paper, all the container terminal agents are developed over the Java Agent Development framework (JADE) platform, which is a programming and simulation environment based on Java that matches with Unified Modelling Language (UML), and conforms to the Foundation for Intelligent Physical Agents (FIPA) specifications. MAS based on JADE platform allows the user to combine different techniques and approaches such as differential equations and discrete events with container terminal operation. This paper build a CTOS simulation model based on JADE platform, the framework of which is divided into four layers.

The first layer is Java platform, which ensure the flexibility and portability of the MAS system. The second layer is JADE platform, located in the upper Java platform, JADE platform is the API based on Java, completely, which has good compatibility with the Java platform, including Agent Manage System (AMS), Directory Facilitator (DF) and Agent Communication Language (ACL) services. The third layer is the simulation control system that provided AMS, DF and ACL services, mainly includes three important Agent. Simulation Agent (SA), which is the design and implementation of a simulation for all common Agent class. Simulation Controller Agent (SCA) the manager of SA, including start, termination, and synchronization. Environment Agent (ENVA) is the main work of storage simulation using data. The fourth layer is application layer, which is extended through the simulation control layer, to control location and equipment agent.

4 Empirical study

This study performs a simulation research on a container terminal in Yangtze ECA of China, which has a throughput capacity of 1,000,000 TEU per year. In order to evaluate possible impacts both on environmental and energy consumption. The container terminal design parameters as in Table 1.

Due to the transport distance of container is variable, which is calculated by distance divide to truck and train move speed, as shown in Table 2.

Table 1. The container terminal parameters

Agent	Item	Parameter
Quay	shoreline	1300 meter
	berth	4
	shore power	2
Stack Yard	yard area	30 kilo square
	stacks	8
	stack yard length	100 meter
	stack yard width	30 meter
	stack yard capacity	85 TEU
	each length	100~168 meter
	each width	61.2~83.5 meter
Road	north-south road	6
	east-west road	3
	road width	22 meter

Table 2. The equipment data in container terminal

Agent	Power *(kW)*	Size L×W×H *(meter)*	Operation Time *(secs/TEU)*	Move Speed *(meters/sec)*	Number	Energy
Quay Crane	900	65×30×10	50	0.40	12	electric
RTG	450	30×31×19	30	0.30	5	diesel
ERTG	480	30×31×19	35	0.25	28	electric
RMG	450	30×31×19	35	0.23	2	diesel
Truck	100	12.5×2×3	variable	1.10	50	diesel
Empty container stacker	100	12.5×2×3	50	1.50	5	diesel
Straddle carrier	100	12.5×2×3	100	0.40	1	diesel
Forklift	100	12.5×2×3	100	1.00	4	diesel
Train(locomotive)	2500	21×3×4	variable	variable	2	diesel

According to the ECA regulation, when the ship without on-board shore power system installing in berthing period, they need to take the 0.5% sulphur MGO fuel, and 0.1% sulphur diesel fuel for truck.

In exprientmat study, design two days to contrast. The throughput of container terminal is 12245 TEU in day 1, without optimization strategy, the energy consumption and emission as shown in Table 3.

Table 3. The energy consumption and emission of container terminal in day 1.

Process	No.	Index	Value
Energy and Emission	1	Diesel/ kg	1115.25
	2	Power/ kWh	5688.94
	3	CO_2/kg	3714.187
	4	SOx/kg	2398.52
	5	NOx/kg	595.93
	6	PM/kg	155.45
Equipment	7	QC kWh/TEU	0.221
	8	Truck kg/TEU	0.045
	9	ERTG kWh/TEU	0.046
	10	RTG kg/TEU	0.059
	11	RMG kg/TEU	0.054
	12	Empty stacker kg/TEU	0.078
	13	Straddle carrier kg/TEU	0.069
	14	Forklift kg/TEU	0.077

Figure 6 shows result of a early warning model in day 1 by using MAS simulation model of the CTOS, there are nine warning hour points in operation of container terminal, the Aux reprensents the auxiliary operation.

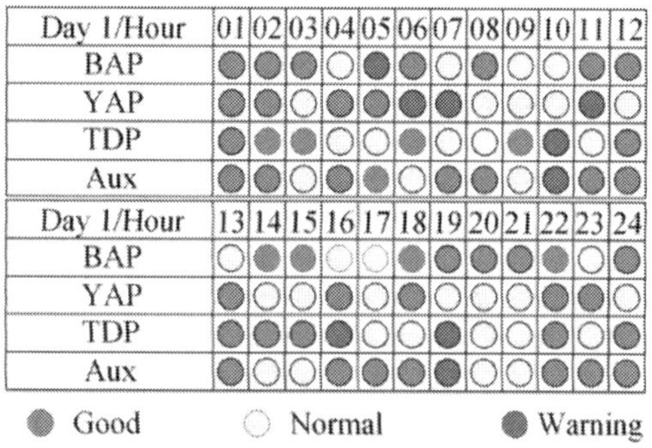

Figure 6. Diagram of early warning model in day 1.

As contrast, container terminal operator adopts optimization strategy to improve efficiency, throughput of container terminal is 11305 TEU in day 2, energy consumption and emission as shown in Table 4.

Table 4. The energy consumption and emission of container terminal in day 2.

Process	No.	Index	Value
Energy and Emission	1	Diesel/ kg	984.54
	2	Power/ kWh	5356.90
	3	CO_2/kg	3111.28
	4	SOx/kg	2198.52
	5	NOx/kg	595.93
	6	PM/kg	155.45
Equipment	7	QC kWh/TEU	0.201
	8	Truck kg/TEU	0.041
	9	ERTG kWh/TEU	0.040
	10	RTG kg/TEU	0.056
	11	RMG kg/TEU	0.050
	12	Empty stacker kg/TEU	0.063
	13	Straddle carrier kg/TEU	0.066
	14	Forklift kg/TEU	0.073

After the early warning model, contianer terminal operator adopts optimization strategy to improve efficiency, reduce energy consumption and emissions, Figure 7 shows result of a early warning model in day 2 by using MAS simulation model of the CTOS, only three warning hour points in operation of container terminal.

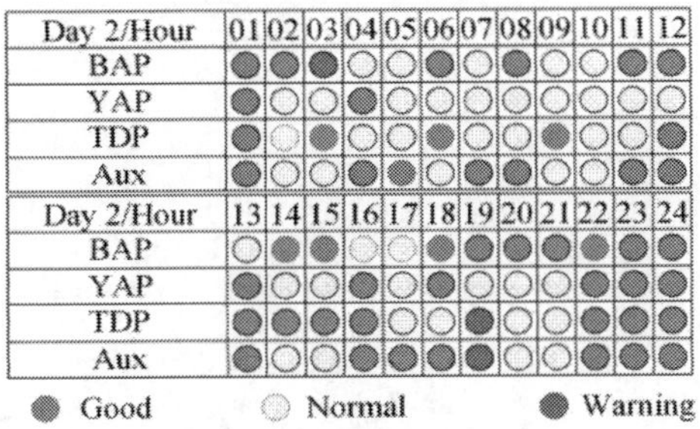

Figure 7. Diagram of early warning model in day 2.

Empirical study shows the dynamic energy consumption and emission model based on CTOS using MAS with early-warning model that can improve efficiency, reduce energy consumption and emissions.

5 Conclusions

In order to reflect the energy consumption and emission of container terminal comprehensively, this paper presents a model, considering the dynamic energy consumption and mission of process of container terminal, which based on Container Terminal Operation System (CTOS) with early warning model. This paper based on the characteristic of CTOS, dynamic energy consumption and emissions characteristic of each equipment and process is modelled, this paper presents the architecture of CTOS based on MAS with early-warning model with M-SVM. A simulation on container terminal is built on the JADE platform to support the decision-making of container terminal, which can reduce energy consumption and air pollution emissions, allows the container terminal operator to be more flexible in their decision to meet the Green Port Plan of China.

Acknowledgements

This research was partly supported carried out with financial support from the Conserve Energy and Reduce Emissions Grant (CEREG) No.2014-JNJP-008-037 and No.2015-JNJP-010-056, from the Ministry of Transport of the People's Republic of China.

References

1. Imai, A., Sun, X., Nishimura, E., & Papadimitriou, S. (2005). Berth allocation in a container port: using a continuous location space approach. Transportation Research Part B Methodological, 39(3), 199-221.
2. Golias, M.M., Saharidis, G.K., Boile, M., Theofanis, S., Ierapetritou, M.G., 2009. The berth allocation problem: optimizing vessel arrival time. Marit. Econom.Logist. 11 (4), 358–377.
3. Yan S, Lai W, Chen M. Production scheduling and truck dispatching of ready mixed concrete[J]. Transportation Research Part E, 2008, 44(1):164-179.
4. Liao C H, Tseng P H, Lu C S. Comparing carbon dioxide emissions of trucking and intermodal container transport in Taiwan[J]. Transportation Research Part D, 2009, 14(7):493-496.
5. Yan S, Lai W, Chen M. Production scheduling and truck dispatching of ready mixed concrete[J]. Transportation Research Part E, 2008, 44(1):164-179.
6. Liao C H, Tseng P H, Lu C S. Comparing carbon dioxide emissions of trucking and intermodal container transport in Taiwan[J]. Transportation Research Part D, 2009, 14(7):493-496.
7. Ya X, Qiushuang C, Xiongwen Q. A robust integrated approach to yard space allocation and crane scheduling in container terminals[C]// International Conference on Service Systems and Service Management. IEEE Xplore, 2010:1-6.
8. Li B, Wang Q W, HU J L. Multi-SVM classifier system with piecewise interpolation[J]. IEEJTrans on Electrical and Electronic Engineering, 2013, 8(2):132-138.
9. Li Boyang, Wang Qiangwei, Hu Jinglu. Multi-SVM classifier system with piecewise interpolation[J]. IEEJ Trans on Electrical and Electronic Engineering，2013，8(2) : 132-138.
10. Sun Z, Lee L H, Chew E P, et al. MicroPort: A general simulation platform for seaport container terminals[J]. Advanced Engineering Informatics, 2012, 26(1):80-89.
11. Chhetri P, Jayatilleke G B, Gekara V O, et al. Container terminal operations simulator (CTOS) – Simulating the impact of extreme weather events on port operation[J]. European Journal of Transport & Infrastructure Research, 2016.
12. Kocifaj M, Adamko N. Modelling of container terminals using two-layer agent architecture[C]// IEEE, International Symposium on Applied Machine Intelligence and Informatics. IEEE, 2014:251-256

Estimates the Emission of Passenger Cars Based on the First Developed Driving Cycle in Nakhon Ratchasima Municipality, Thailand

Noppatsorn Muangnak[1], and Nares Chuersuwan[2]

[1]*Graduate Program in Environmental Pollution and Safety, Suranaree University of Technology, Nakhon Ratchasima, Thailand*
[2]*School of Environmental Health, Suranaree University of Technology, Nakhon Ratchasima, Thailand*

Abstract. This study aims to estimate emissions of passenger cars traveled on the highways in Nakhon Ratchasima Municipality (NRM) based on recently developed driving cycles. A mobile phone application (SafeMate) was used to record speed-time of passenger car on the highways both weekday and weekend. The results showed that the driving cycles of the passenger cars on highway 224 and 304 were quite different. More traffic congestion on highway 224 impacted the average speeds, ranging from 10 to 34 km/h with longer idling time. The average speeds on highway 304 ranged from 49 to 69 km/h with more cruise time. The combined driving cycles on both highways showed the average speeds of 20-40 km/h depending upon the time of days. The traffic density on weekday was higher than the weekend. Peak-hour weekday showed lower traffic flow than off-peak hour which was the opposite of the weekend. Based on the derived driving cycles, the passenger car traveled on the highways in NRM emitted CO, NOx and HC approximately 90.7-129.8, 25.5-34.5 and 6.9-11.7 tons/year, respectively. The passenger cars were estimated to emit about 13,927.7-20,419.3 tons/year of carbon dioxide on the highways. The highway 2 accounted more of the emissions due to the traffic volume and coverage distance in NRM.

1 Introduction

Road transport is a major source of air pollution in Thailand. A number of vehicles has been increased on roadway and caused traffic congestion in several areas, especially in large cities, such as Bangkok, Chiang Mai, Nakhon Ratchasima. Traffic density correlates with high levels of vehicular exhaust emission [1, 2] and has potential to cause health effects in urban residents.

Estimating exhaust emissions are conveniently used emission factors and economically attractive. However, the emission inventory requires decent inputs, such as, emission factor of vehicles (EFs), travel activity, etc. Emissions from vehicle depend on several factors, including driving cycle, fuel type, emission control technology, vehicle weight, engine capacity, environment conditions road gradient and the level of vehicle maintenance [3]. Driving cycles have been developed to represent conditions on roads and used in the emission estimates for specific city. The emissions from vehicles are also related to the driving mode: idle, steady-state cruise, acceleration and deceleration.

Driving cycles provide the variations of the vehicle speed with time for a certain period of travel for a specific route. Vehicle emission factors are commonly derived from chassis dynamometer tests programmed with standard driving cycle, such as U.S. Federal Test Procedure (FTP75), Economic Commission for Europe (ECE) and Japanese 10 mode cycle [4]. In turn, the driving cycles used in the U.S. and Europe are not represented the traffic conditions in other regions. Many driving cycles developed elsewhere under specific driving characteristic are clearly differ from one area to another, even within the same city [5] such as Thailand [6] Turkey [7] and China [8]. Consequently, researchers should set priority to specific of driving characteristics in each area to achieve a reliable estimate of the emissions.

This study aims to estimate the emissions of passenger cars traveled on the highways in Nakhon Ratchasima Municipality (NRM) based on recently developed driving cycles with the help of a mobile phone application to record speed-time of the passenger cars.

2 Methodology

2.1 Route selection

Nakhon Ratchasima Municipality (NRM) is the fifth largest municipality in Thailand that experienced increasing volume of vehicle and traffic congestion due to the subsidy of the First Car Program by the previous government and economic improvement. Four highways, highway number 2, 205, 224, and 304, pass through the municipality. The road sections of those highways were selected to develop the driving cycles of route between Rajamangala University of Technology Isan (RMUTI) to

DOI:10.1051/matecconf/201712406002

Save One Market, approximately 10.17 kilometers trip (Figure 1).

2.2 Speed data collection

Since the passenger car shared the largest portion on the highways in NRM, the driving cycle of the passenger car was our focus. In-vehicle method was used to collect speed-time data with the assistant of the mobile application, SafeMate. Data were collected from Monday to Sunday during the morning peak hour (7.00 a.m. - 9.00 a.m.), off-peak hour (9.00 a.m. - 3.00 p.m.), and the evening peak hour (3.00 p.m. - 7.00 p.m.). The speed-time data were collected for both inbound and outbound directions.

2.3 Speed-time analysis and driving cycle development

The graph data of the speed-time were analyzed by transforming to micro trips for every 7.3 seconds. The speed-time data were separated into weekday and weekend according to peak hour and off-peak hour. Descriptive statistics were used to estimate mean, difference mean, and standard deviation. Difference mean is an accepted mean different at less than 10%. The data were tested for the differences in the dataset using inferential statistic (T-test, Anova test, Mann-Whitney U-test or Kruskal-Wallis). The derived speed data were selected to represent the real driving characteristics of the route. Then, the speed-time data were used to develop driving cycles of the trip.

The micro trips of each highway section were used to construct the driving cycles of the route. Driving characteristic were described by the parameters including average speed (V_{avg}), average running speed (V_{avg1}), average acceleration of all acceleration phase (Acc), average deceleration of all deceleration phases (Dec), idling mode (T_{idle}), acceleration mode (T_{acc}), deceleration mode (T_{dec}), and cruising mode (T_{cruise}).

2.4 Estimating passenger car emissions on the highways

Mobile source emission factors in Thailand has developed on a standard chassis dynamometer, which the exhaust from the vehicles were measured under different situation such as fuel type, emission control technology and driving speed. The air pollutants and greenhouse gas of passenger car were measured including hydrocarbon (HC), carbon monoxide (CO), oxide of nitrogen (NO_x) and carbon dioxide (CO_2) and classified according to speed used in the test [9]. Data from the dynamometer tests were extracted and used to calculate passenger car emissions under the developed driving cycles in NRM. Traffic volume was counted and estimated on all highways according to the Department of Highway, Thailand. The vehicle emission rate was obtained by multiplying vehicle activity with the derived EFs of the passenger car (Eq. 1).

$$E_i = EF_i \times D \times N \quad (1)$$

Where EF_i (g/km) was an emission factor of i pollutant from vehicle, D (km) was the travel distance and N (vehicles/day) was average daily traffic. The calculations of exhaust emission under specific driving cycle were estimated using (Eq. 2).

$$E_i = \sum_{i=1}^{a} \sum_{j=1}^{b} \left[EF_{ij} \times T \times n_j \times j \right] \times N \quad (2)$$

Where EF_{ij} (g/km) was an emission factor of i pollutant at j travel speed range, n_j was the number of micro-trip at j speed range of vehicle on driving cycle, T=7.31 (s) was the time scale of micro-trip, j (km/h) was a travel speed range of vehicle (including 0-10, 11-20, 21-30, 31-40, 41-50, 51-60, 61-70, 71-80, 81-90 km/h).

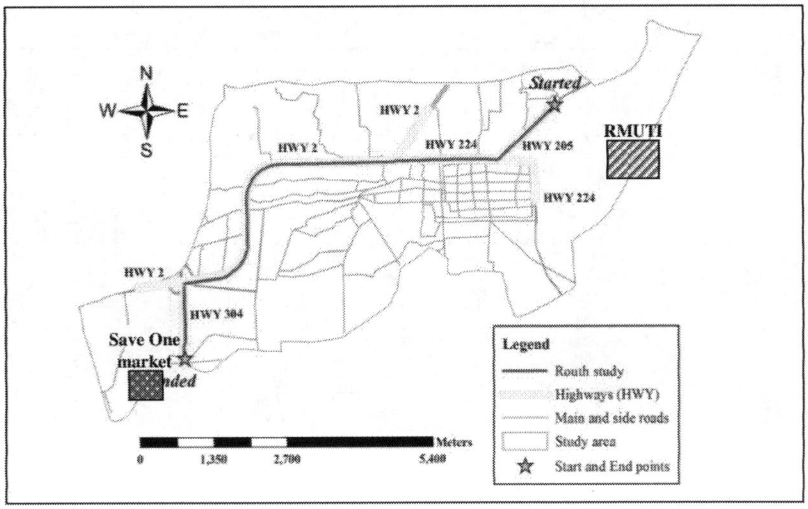

Figure 1. The study area in Nakhon Ratchasima Municipality (NRM) and selected highway route

3 Results and discussion

3.1. Deriving the highway driving cycles

Collected speed-time data showed less than 10 percent differences ($p<0.05$) in the dataset. The driving parameters of all sections of the highways are presented in Table 1. The results showed that the driving cycles on highway 224 and highway 304 differed from the rest, caused by congestion on highways. The average speed of highway 224 was ranged from 10 to 34 km/h with longer idling time. The evening peak hour on weekday experienced the congestion. With the average speed of 10 km/h, the proportion of idle time and cruise were approximately 50% and 9%, respectively.

Highway 304 experienced less traffic congestion with the average speed ranging from 49 to 69 km/h and relatively smooth traffic flow indicating by no idling time and high cruise time. The driving cycle in the morning during the weekend was rather smooth with 50% of cruise time

The driving cycles on highway 2 and highway 205 were quite fluctuated which the average speed ranged from 32 to 54 and 22 to 42 km/h, respectively. Weekday had high traffic congestion on both highways. During peak hour, the driving cycles showed high traffic intensity indicated by low speed and fluctuating traffic flow.

The micro trips were used to construct the driving cycles on the highways in NRM for the passenger car travelling about 10.17 km from the northeast to the southwest locations (Figure 2). With the driving cycles, the driving parameters were derived (Table 2). The results showed that the average speed on weekday and weekend were approximately 23.6 and 34.7 km/h, respectively. The weekday had more fluctuation than the weekend, indicated by the average idle mode on weekday with the highest proportion of 39%, while the average cruise mode was only 9.7%. Traffic intensity was high on weekday during peak hour while the weekend occurred during off-peak hour. Morning peak and evening peak on weekday showed serious traffic congestion, with the average speed of 22.1 and 20.3 km/h, respectively. The traffic flow, however, was smooth on morning peak during the weekend with twice the average speed typically found weekday peak hour. The morning peak hour during the weekend had the lowest proportion of idling, about 11.6%. The driving characteristics varied during the day and different times of day and daily. The important factors affected the driving variations in NRM included road characteristics, driving behaviour, and traffic volume, etc. Several studies used speed-time data in morning peak to represent the driving cycles [6], [10], and [11]. In NRM, however, the traffic volume was high during the evening peak hour.

Table 1. The driving parameters on highways in NRM

Road sections	Days of week	Times of day	V_{avg} (km/h)	V_{avg1} (km/h)	Acc (m/s^2)	Dec (m/s^2)	T_{idle} (%)	T_{acc} (%)	T_{dec} (%)	T_{cruise} (%)	Time (s)	Length (km)
Highways 224	Weekdays	Morning peak	11.7	24.2	0.1	-0.3	49.3	17.4	21.0	12.3	1009	3.20
		off peak	20.4	35.2	0.2	-0.6	36.8	33.8	25.0	4.4	497	
		Evening peak	10.0	13.2	0.1	-0.4	49.3	20.8	19.4	10.4	1053	
		Average	14.0	24.2	0.1	-0.5	45.1	24.0	21.8	9.0	853	
	Weekend	Morning peak	28.7	35.9	0.3	-0.5	15.3	39.0	33.9	11.9	431	
		off peak	34.2	38.7	0.3	-0.5	11.9	47.5	37.3	3.4	431	
		Evening peak	19.0	40.5	0.1	-0.6	50.0	23.1	17.9	9.0	570	
		Average	27.3	38.4	0.2	-0.5	25.7	36.5	29.7	8.1	477	
Highways 2	Weekdays	Morning peak	44.0	58.3	0.2	-0.4	22.5	33.7	31.5	12.4	651	7.66
		off peak	39.6	50.9	0.1	-0.3	21.5	33.6	25.2	19.6	782	
		Evening peak	28.9	40.1	0.2	-0.3	25.8	31.5	30.6	12.1	906	
		Average	37.5	49.8	0.2	-0.3	23.3	32.9	29.1	14.7	780	
	Weekend	Morning peak	53.5	59.4	0.2	-0.4	7.1	32.9	32.9	27.1	512	
		off peak	31.7	45.6	0.2	-0.4	28.2	32.5	29.1	10.3	855	
		Evening peak	46.4	54.5	0.3	-0.5	11.3	43.8	31.3	13.8	585	
		Average	43.9	53.2	0.2	-0.5	15.5	36.4	31.1	17.0	651	
Highways 205	Weekdays	Morning peak	22.3	27.6	0.2	-0.4	16.7	43.3	30.0	10.0	219	1.45
		off peak	29.8	42.9	0.3	-0.6	27.3	36.4	31.8	4.5	181	
		Evening peak	30.2	30.2	0.2	-0.3	0.0	50.0	40.9	9.1	181	
		Average	27.4	33.6	0.2	-0.5	14.6	43.2	34.2	7.9	194	
	Weekend	Morning peak	26.5	37.1	0.3	-0.7	18.5	44.4	33.3	3.7	197	
		off peak	42.3	42.3	0.2	-0.5	0.0	56.3	31.3	12.5	117	
		Evening peak	32.7	37.9	0.2	-0.5	9.5	47.6	28.6	14.3	154	
		Average	33.8	39.1	0.2	-0.6	9.3	49.4	31.1	10.2	156	
Highways 304	Weekdays	Morning peak	48.8	48.8	0.1	-0.3	0.0	40.0	40.0	20.0	110	1.63
		off peak	54.0	54.0	0.3	-0.2	0.0	53.8	38.5	7.7	95	
		Evening peak	53.9	53.9	0.2	-0.3	0.0	53.3	40.0	6.7	110	
		Average	52.2	52.2	0.2	-0.3	0.0	49.1	39.5	11.5	105	
	Weekend	Morning peak	68.7	68.7	0.0	-0.2	0.0	0.0	50.0	50.0	73	
		off peak	58.4	58.4	0.1	-0.2	0.0	33.3	40.0	26.7	110	
		Evening peak	62.1	66.8	0.1	-0.6	0.0	30.8	38.5	30.8	95	
		Average	63.1	64.7	0.1	-0.3	0.0	21.4	42.8	35.8	93	

Note: Acc* = Acceleration; Dec** = Deceleration

Table 2. The driving parameters of passenger car on highway route in NRM

Driving cycle	Days of week	Times of days	V_{avg} (km/h)	V_{avg1} (km/h)	Acc (m/s^2)	Dec (m/s^2)	T_{idle} (%)	T_{acc} (%)	T_{dec} (%)	T_{cruise} (%)	Time (s)	Length (km)
NRMDC (RMUTI to Save one market)	Weekdays	Morning peak	22.1	44.2	0.1	-0.4	46.5	24.3	21.6	7.6	1352	10.3
		Off peak	28.4	48.0	0.1	-0.5	37.0	29.5	23.3	10.3	1067	
		Evening peak	20.3	33.0	0.1	-0.4	33.8	28.4	26.5	11.3	1491	
		Average	23.6	41.7	0.1	-0.4	39.1	27.4	23.8	9.7	1303	
	Weekend	Morning peak	39.9	49.5	0.2	-0.5	11.6	37.5	33.0	17.9	826	
		Off peak	31.6	42.4	0.2	-0.5	23.1	37.1	30.1	9.8	1052	
		Evening peak	32.5	51.1	0.2	-0.6	31.5	32.3	24.6	11.5	957	
		Average	34.7	47.6	0.2	-0.5	22.1	35.6	29.2	13.1	945	

Table 3. The comparison of developed driving cycle of passenger car in NRM with international cycles

Driving cycle	Days of week	Times of days	V_{avg} (km/h)	V_{avg1} (km/h)	Acc (m/s^2)	Dec (m/s^2)	T_{idle} (%)	T_{acc} (%)	T_{dec} (%)	T_{cruise} (%)	Time (s)	Length (km)	Description
NRMDC	Weekdays	Morning peak	22.1	44.2	0.1	-0.4	46.5	24.3	21.6	7.6	1352	10.3	Highway
	Weekend	Morning peak	39.9	49.5	0.2	-0.5	11.6	37.5	33.0	17.9	826		
	Week	Morning peak	31.0	46.9	0.2	-0.5	29.1	30.9	27.3	12.8	1089		
MIDC	Weekdays	Morning peak	26.9	37.5	0.6	-0.9	28.4	20.7	15.8	35.2	1744	13.0	Highway [10]
HKHDC	Week	Morning peak	38.3	41.8	0.4	-0.4	8.4	37.5	36.2	17.2	1401	14.9	Highway [11]
BDC	Week	Morning peak	17.7	28.8	0.7	-0.7	37.7	15.3	23.2	23.8	1160	5.7	Arterial and highway [6]

Note: NRMDC = Nakhon Ratchasima Municipality Driving Cycle, MIDC = Modified Indian Driving Cycle, HKHDC = Hong Kong Highway Driving Cycle and BDC = Bangkok Driving Cycle.

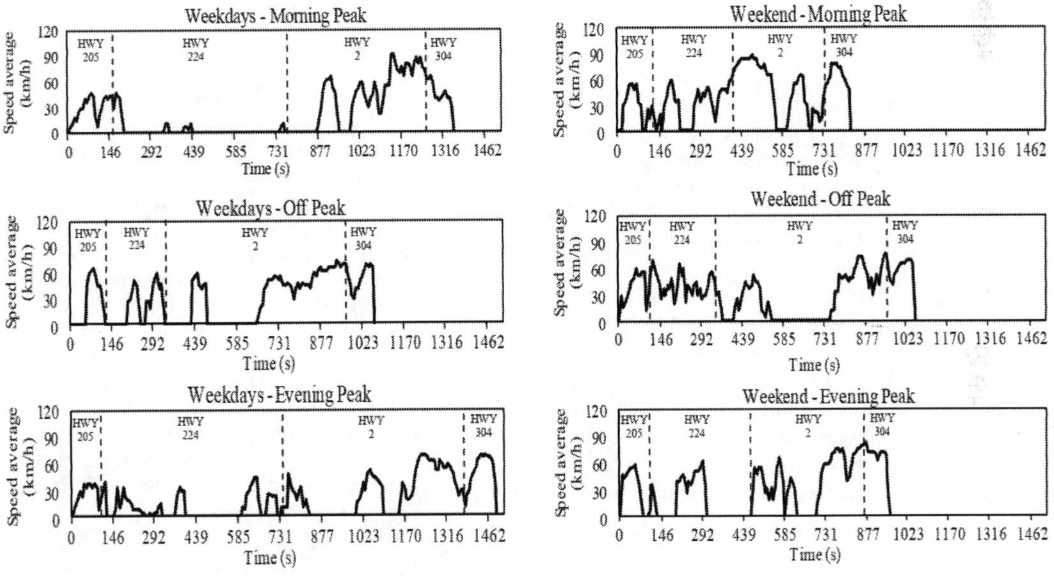

Figure 2. Driving cycles on highway (HWY) route in NRM.

The morning peak was selected as the driving cycles of passenger car in NRM. The comparison with other driving cycles in other countries is in Table 3. The average speed of NRMDC on weekday morning peak was lower than MIDC due to the high traffic congestion in NRM. NRMDC had high portion of idle time, approximately 46.5%. The driving cycle on the morning peak of HKHDC and BDC were obviously differed from NRMDC. The average speed of HKHDC, NRMDC and BDC were about 38.3, 31.0 and 17.7 km/h, respectively.

HKHDC showed higher fluctuation than other driving cycles. High portion of acceleration time and deceleration was approximately 37.5% and 36.2%, respectively. Values of the driving parameters between BDC and NRMDC were different. BDC had the lowest average speed and running speed comparing to others driving cycles. The reason that BDC was developed based on traffic conditions of both highway and arterial road, where the arterial roads in Bangkok had serious traffic congestion [6]. Therefore, it was necessary to develop the specific driving cycles in NRM with more appropriate speed-time data for the estimation of the vehicular emissions.

3.2 Estimated passenger car emissions on highways in NRM

Estimation of mobile emission in NRM was obtained from multiplying average annual daily traffic (AADT) with vehicle kilometer travel and the emission factors.

The estimation of exhaust emissions from passenger car on highways in NRM is shown in Table 4. Based on 2015 data, the main air pollutant was carbon monoxide, about 90.7-129.8 tons annually. As a greenhouse gas, annual CO_2 emission was at high as 20,419.3 tons. Highway 2 accounted a large portion of the emissions due to the travel distance and traffic volume. The estimated annual emissions of CO, NO_x, and HC from highway 2 were approximately 67.4-90.2, 19.2-24.3 and 4.9-7.4 tons, respectively, whereas highway 304 had the smallest portion of the emissions from passenger cars.

Table 4. Estimated total emissions from passenger cars on the highways in NRM, 2015

Section road	Distance (km)	AADT (vehicle/day)	Emissions (tons/year)							
			HC		CO		NO_x		CO_2	
			Low	High	Low	High	Low	High	Low	High
Highway 2	7.66	30839	4.9	7.4	67.4	90.2	19.2	24.3	10242.1	13883.8
Highway 205	1.46	22908	0.7	1.2	9.4	13.4	2.5	3.5	1456.1	2132.7
Highway 224	3.20	12285	1.1	2.8	11.4	23.2	3.1	5.9	1858.3	3964.0
Highway 304	1.69	5743	0.2	0.2	2.5	2.9	0.7	0.8	371.2	438.8
Total emission (tons/year)			6.9	11.7	90.7	129.8	25.5	34.5	13927.7	20419.3
Percent emission (%)			0.03		0.32		0.09		49.57	

4 Conclusions

Specific driving cycle is important for reasonable estimate of air emissions in area of interest like NRM. The driving cycles can be constructed based on the information of micro trips from speed-time data. Current technology allows a mobile application to assist data gathering of the traffic related data virtually in any city. Along with the traffic volume and appropriate emission factors, vehicle emissions can be routinely estimate. The passenger cars in NRM had different driving patterns depending on time of the day and day of the week. Long idling time of passenger car on highway in NRM was affected by the traffic activities. The average speed on weekday was about 23.6 km/h, and 34.7 km/h during weekend. The weekday had more speed fluctuation than the weekend, indicated by the average idle mode on weekday with the highest proportion of 39%, while the average cruise mode was only 9.7%. In 2015, annual carbon monoxide was emitted about 91-130 tons, 26-35 tons of NO_x, and 7-12 tons of HC. The future research should further investigate other types of roads and vehicles to quantify the whole fleet and to provide the overall vehicle emission inventory for the comprehensive air quality planning.

Acknowledgement

The authors would like to thank the Automotive Emission Laboratory of Pollution Control Department, Thailand for supporting the vehicle emission factors in Thailand.

References

1. S. T. Leong, S. Muttamara and P. Laortanakul, Air Pollution and Traffic Measurements in Bangkok Streets, Asian Journal on Energy and Environment, Vol. 3, pp. 185-213, (2003).
2. S. Pandian, S. Gokhale, A. K. Ghoshal, Evaluating effects of traffic and vehicle characteristics on vehicular emissions near traffic intersections, Transportation Research Part D: Transport and Environment, Vol. 14, pp. 180-196, (2009).
3. A. Faiz, C. S. Weaver and M. P. Walsh, Air pollution from motor vehicles: Standards and technologies for controlling emissions. 1st ed. Washington DC, USA: The World Bank, ch. 1, pp. 1-24, (1996).
4. US. EPA. Vehicle and Fuel Emissions Testing. [Online]. Available: https://www.epa.gov/vehicle-and-fuel-emissions -testing/dynamo meter-drive-schedules, (January, 2017).
5. H.Y. Tong, W.T. Hung and C.S. Cheung, Development of a driving cycle for Hong Kong, Atmospheric Environment, Vol. 33, pp. 2323-2335, (1999).
6. N. Tamsanya and S. Chungpaibulpatana, Influence of driving cycles on exhaust emissions and fuel consumption of gasoline passenger car in Bangkok, Journal of Environmental Sciences, Vol. 21, pp. 604-611, (2009).
7. J. Lin and D. A. Niemeier, Regional driving characteristics, regional driving cycles, Transportation Research Part D: Transport and Environment, Vol. 8, pp. 330-346, (2003).
8. Q. Wang, H. Huo, Kebin He, Z. Yao and Q. Zhang, Characterization of vehicle driving patterns and development of driving cycles in Chinese cities, Transportation Research Part D: Transport and Environment, Vol. 13, pp. 289-297, (2008).
9. I. Pawarmart, T. Petplangsri and N. Pala-en, Update mobile source emission factors in Thailand, presented at 5th International Conference on Sastainable Energy and Environment, Bangkok, Thailand, (2014).
10. P. Adak, R. Sahu and S. P. Elumalai, Development of emission factors for motorcycles and shared auto-rickshaws using real world driving cycle for a typical Indian city, Science of the Total Environment, vol. 544, pp. 299-308, (2016).
11. W. T. Hung, H. Y. Tong, C. P. Lee, K. Ha and L. Y. Pao, Development of a practical driving cycle construction methodology: A case study in Hong Kong, Transportation Research Part D, vol. 12, pp. 115-128, (2007).

MATEC Web of Conferences **124**, 06003 (2017) DOI:10.1051/matecconf/201712406003
ICTTE 2017

The Effective Spatial Qualities on Urban Tourism in Iran Focusing on Social Factors

Kambiz Razavi[1], Filipe Themudo Barta[2] and Elham Amini [3]

[1]*Member of the Board of Directors, and the chairman of Professional Group of Traffic in Construction Engineering Organization of Tehran Province; Student of Urban Planning in Universidade Lusofona, Lisbon*
[2]*Chair Holder of the University of Evora UNESCO Chair*
[3] *Associate Professor, Islamic Azad University, Pardis Branch, Tehran, Iran*

Abstract. Considering the increasing challenges of the current cities, the urban tourism has been proposed as a solution for solving or minimizing different problems. Including economic, social and environmental aspects, this approach can lead to the fulfillment of sustainable development theory. Accordingly, there is no doubt about the successful results of the urban tourism development and all urban managers and authorities agree on this point. Meanwhile, what threatens the success of the urban tourism development plans is to ignore the all effective factors that make effect on the quality of travel experience. In their best, the cities are projected in their public spaces, and any tourist spent a considerable part of his travel in such spaces. Thus the quality of urban spaces is an important factor on the successfulness of urban tourism development plans and hence their environmental desirability must be studied and evaluated more precise under a comprehensive approach. The main objective of this research is to answer the main question of how the effective spatial qualities on tourism are explain with regard to the social factors. Since the man and his activities are the base of the urban studies, this article uses an urban-strategic approach based on the descriptive-analytical method using library resources to study the social factors that are effective on the qualities of urban spaces. The analytical model of the research will be presented based on the theoretical studies. The results show that some qualities such as flexibility, functional diversity, physical diversity, vegetation, safety, lightening, landscape attractiveness, identity, and memorability in relation to the urban space, and social and arbitrary activities based on Gehl theory can make effect as the complementary factors of the urban tourism attraction and make effect on social factors such as vitality, sense of security, arbitrary activity, social activity and tourists sense of peace. Emphasizing on the social exchange theory, such factors can pave the way for the sustainability of urban tourism industry. The results of this research can be used in identifying and reinforcing the spatial qualities relating to the social factors that are effective on tourism in Iran. Consequently, they can be useful in the tourism development. The essential value of this article is its simultaneous focus on both spatial qualities (physical dimension) and social factors (human dimension) in relation to the tourism development. In this regard, the unidimensional attitude is rejected and the adaptation of a holistic approach to the urban tourism is highlighted.

1 Introduction

Urban tourism is a growing approach. Although it has a long history, but due to its positive effects on the urban life, today it has attracted more attention. Apart from its effects on the tourists, urban tourism can make considerable effects on the quality of the residents' of tourism destination. Global experiences show that tourism is not based only on the natural and historical attractions but nowadays, in many cases, the man-made environmental desirability is preferred over the old heritage of the touristic host societies. Thus it is necessary to improve the attitude toward urban tourism, and the role of urban spaces as the most important setting of the social interactions between the host and guest societies on one hand and the meaningfulness of the urban space and time for tourists on the other hand. This article first reviews the concepts, theories and opinions on the research keywords with emphasis on case of Iran and then we will present our theoretical framework in order to re-emphasize the role of the quality of urban spaces in urban tourism focusing on the social factors.

2 The role of social factors in urban tourism

Paying attention to the sociocultural factors in tourism-related studies is mainly directed toward the mutual effects of the host society and tourists on each other. In other words, the mentioned factors have the effects that are the result of the touristic activities in the social interactions of the host society and tourists. Emphasizing on the cultural aspect, the cultural effects are the changes occurring in the rituals, arts, architecture and spiritual and intellectual values (Saeidnia and Mehdizadeh, 2012: 66). This type of attitude in turn is very important. Orbasl (2000) believes that the sociocultural aspect of the tourists' movement in urban spaces is a very important point in the field of urban tourism because the encounter of the host and guest is the result of cultural differences in using the urban space or the multiplicity of the urban function and diversity of the urban users (Orbasl, 200: 56). In this regard, the fundamental subjects in the urban spaces management that are involved in identification and application of such spaces include the spatial analysis of urban functions and activities, value of the urban uses, transportation models, citizens' mobility and the urban networks that in sum, form the logical order of the urban behaviors (Rabbani and Vahida, 2007: 45).

It is obvious that in touristic activities, different types of cultures and behaviors of the two societies, i.e. the host and guest societies are involved. But the host society has its own goals, values and interests that are expected to be met in the development of touristic activities. Undoubtedly, the mutual reactions between the two cultures and their tacit motivations make different effects and changes. Such effects and changes sometimes are in behalf of the host society and sometimes they act against the host society. Today, new approaches, particularly the approaches toward sustainable tourism development attempt to reinforce the positive values and interests of tourism and reduce its negative effects (Saeidnia and Mehdizadeh, 2012: 67-8). Table 1 shows a part of the most important positive and negative impacts of tourism focusing on the social and cultural aspects.

Since the social context includes culture, religion and ethnicity, particularly in Iran with its considerable diversity of mentioned categories, it seems necessary to pay attention to cultural tourism, religious tourism and ethnic tourism.

3 Tourism in Iran with emphasis on the social factors

Iran is a country with many natural, historical, cultural, religious and ethnic attractions and hence the tourism in Iran is potentially ideal. In other words, material and nonmaterial potentials (both natural and man-made ones) are very high in this country. Focusing on the social factors, Iranians are known as warm and hospitable people. Compared to cultural, religious and ethnic tourism, there are little studies on other social factors in this country. Accordingly, in this study we mainly focus on the three mentioned types of tourism.

3.1 Cultural tourism in Iran

The recourse of the cultural tourism to both old and contemporary cultural symbols implies the exclusive conditions of Iran in urban tourism. With its old history and civilization, Iran has provided a wide extent for tourists who want to visit and enjoy the glorious history of the country. Cultural masterpieces from the old history of Iran have made Iranian cities such as Isfahan, Yazd, Shiraz, Kermanshah and Susa famous worldwide, such that the identity of Iran and its cities are tied with the elements and spaces such as Chogha Zanbil in Susa, Naqsh-e Jahan Square in Isfahan, Persepolis and Pasargadae historical sites near Shiraz, Takht-e Soleiman historical site in Takab in West Azerbaijan, Falak-ol-Aflak Castle in Khorramabad, Sultanieh in Zanjan, Bistoun Site in Kermanshah, Traditional Bazaar of Tabriz, Fin Garden in Kashan, Golestan Palace in Tehran, Delgosha Garden in Yazd, etc.

Table 1. Negative and positive sociocultural effects of tourism

Effective factors	Positive impacts	Negative impacts
Using culture as an attraction	• Increase of supporting the traditional culture • Introducing the ethnic identity • Revival the traditional arts and rites • Revival the local language and literature	• Changes in the traditional arts due to the tourists' tastes • Influencing the privacy of traditional areas • Crowdedness in the traditional place • Commercialization of the cultural products
Direct contact between tourists and local people	• Reduction of the negative behavioral representations • Increase of the social opportunities	• Reinforcement of the negative behavioral representations • Outbreak of new diseases • Increase of the population density and crowdedness
Changes of social roles and jobs	• Emergence of new jobs • Reduction of social inequities	• Increase of social conflicts • Increase of social inequities • Destruction of local languages and accents
Development of tourism bodies	• Increase of recreational facilities • Increase of the accessibility facilities	• Limitedness of the local people's accesses • Increase of the bureaucracy
Population increase due to the tourism development	• Increase of the quality and quantity of welfare facilities • Increase of the quality of life	• Increase of the crowdedness and population mix • Increase of the crimes

Source: World Tourism Organization, 1999: 328

On the other hand, focusing on Iranian contemporary culture, the cultural diversity of the country is considerable while the diversity of its subcultures are more important among the other aspects of cultural diversity. In this regard one can refer to the rites and rituals in the subcultures in different regions of the country, specifically in especial events such as Yalda Night.

3.2 Religious tourism in Iran

Due to the diversity of ethnics believing in Islam in Iran, the religious tourism is very common in this country. Some holly places such as the Imam Reza Shrine (the shrine of 8th Shiite Imam) in Mashhad and Fatima Masumeh Shrine in Qom as the religious centers of the country along with the tombs of many Imamzadehs all around the country has made Iran a destination for religious tourists all seasons of the year. Religious ceremonies, particularly the Ramadan, Sha'ban and Muharram months in cities like Zanjan and Niasar are attractive both for religious people and other tourists.

The positive factors of the Iranian religious tourism can be classified as follows (Aqajani and Farahanifard, 2015: 56):

- Very attractive rituals and ceremonies in different regions: different months have different feasts and mourning days when the Iranian of different regions take those ceremonies differently from each other.
- Accommodations: considering the social, cultural and religious diversities, there are diverse hospitality and accommodation facilities in Iran. This factor besides the experience of living in traditional environment is considered as the tourist attraction factors.
- Islamic holy sites: apart from the spiritual characteristics of the Islamic holy sites, these places are indeed the exhibition of arts and symbols of the Islamic culture and civilization that normally attract non-Muslim tourists
- Offering Halal Foods and respecting the prayer times

By contrast, the negative and deterrent factors are as follows (Aqajani and Farahanifard, 2015: 56-8):

- Undesirable information and promotion: promotion and advertisement is a modern way of informing. In order to persuade the target society, the promotions and advertisement must be based on different tools such as written, audio and visual communication tools to make changes in the values, beliefs, attitudes and behaviors of the people. This aspect is weak in Islam world in general and Iran in particular, and there is no precise information about the centers, rituals and plans of religious tourism.
- The weakness in providing the security and financial, physical and mental comfort of the tourists: even a very light sense of insecurity can make irreparable damages on tourism industry (including the religious tourism) of any country. A major obstacle to religious tourism in Islam and Iran is the real or unreal descriptions of conflicts and insecurities at international level. The undesirable and negative effects of Iran-Iraq war is still alive in the minds of the people.
- Undesirability of the amenities: one of the most primary wants of any tourist is to benefit from the accommodation and amenities according the most updated standards of the world. Unfortunately, since the religious tourists mainly root in the poor classes of the society and since they are always willing to pilgrim despite all problems, the attitude toward the quality and quantity of the accommodation and amenities of such tourists are usually weak and neglected.

3.3 Ethnic tourism in Iran

The wide span and diversity of the ethnicities in Iran promotes the ethnic tourism in this country. The dispersal of Lor, Kurd, Turk, Balouch, Persian, Turkman and other ethnicities around Iran along with the long roots of their civilization can be really attractive for both Iranian and foreign tourists. This specific type of tourism not only can make those ethnicities closer to each other, but it can revive their historical and traditional rites on one hand and reinforce their social and ethnic structures to support their collective culture and heritage on the other hand. As an example, the traditional rites of Lori and Ilami Chamar Mourning event is one of the most important rituals of this ethnicity with a very long history. The people of this 10-thousand year old ethnicity has not been attracted to modernity during all the past years, and hence their social and traditional ceremonies, rites and rituals are still alive. Ilam Province people still believe in the values of such rituals of their own land. Chamar is the traditional, objective and theatrical mourning being held in Ilam province of Iran.

4 Conclusion and Discussion

Encountering the challenges and problems of the current cities has led to several theoretical and practical attempts. One of the most recent and most comprehensive attempts in this regard is the sustainable development theory focusing on the social, economic and environmental aspects. This theory almost covers all dimensions of the human collective life and creates a logical cycle between the physical, human, and behavioral aspects. Indeed the durability of the cities is not possible without having a comprehensive view to all the mentioned three aspects. Meanwhile, urban tourism is an industry that makes effect on and is effected by the society, economy and environment in the framework on urban sustainability. In terms of the positive impacts of tourism as an environmentally clean industry, there is no doubt about the economic effects (less costs) and social interactions between the host and guest societies. Such factors can lead to considerable positive results if there is a proper planning for them. Considering the social exchange theory, in facing the natural and artificial phenomena (including the cities) human beings always compare their inputs and outputs. Indeed they always do cost-benefit calculations and make their further decisions based on the result of those calculations. In focusing on the subject of urban tourism we accept the social exchange theory and recognize that the satisfactory experiences of tourists depends on several criteria and factors. Although these factors can be general or individual (varying from one person to another or be based on the personal traits) but they play a key role in the durability and sustainability of the urban tourism. This is while the urban spaces are the main point of interaction between the physical and social aspects of the host city and the tourists. The quality of such spaces can contribute in the fulfilment of guests' social exchange analysis considerably.

On the other hand, the quality of urban spaces (as the best representation of collective life) cannot be defined as an inherent physical feature independent from the visitor, or as a completely mental and subjective category regardless of the relationship with the physical aspect. Adopting a phenomenological approach we can state that the urban spaces are phenomena that are formed in the interaction of social and physical aspects and are highly effected by the man and his activities. Such a definition requires the simultaneous attention to three components (i.e. the form, meaning and function) of the urban spaces that is the setting for interactions between man and place on one hand and man and other men on the other hand. The environmental desirability of this phenomena encompasses several factors. Table 2 shows a part of attempts being done by some theorists and researchers in relation to the quality of urban public spaces.

Table 2. Quality of urban public space according to different researchers

Researcher	Key components implying the quality of urban space
Qalenoei and Bahrami (2014)	Climatic comfort, comfort and peace, mixed land uses, social mix, security, pause, high density, physical diversity, enclosure, access, permeability
Handy (2006)	Density, mixed land uses, availability and connectedness to communication networks, regional structure, climatic comfort, shadow of the environment, aesthetic qualities, place attractiveness
Mohammadi & Changlavaei (2012)	Mental perception environment, objective perception environment, physical quality environment, security, safety, and behavioral qualities
Project of Public Space (PPS)	Accessibility and connectedness, comfort and mental image, application, activities and sociability
Gehl (2007)	Priority of social and optional behaviors over the compulsory activities
London Planning Advisory Committee (LPAC)	Human scale and intensity of the urban fabric, suitable structure, readability and identity, cleanness and safety, desirable urban management, visual richness, mixed land uses and activities, existence of public spaces and especial spaces, and finally the easiness of walking and driving movements
Jacobs (1961) cited in Yoke Lai et. al (2013)	Social interactions, vitality, day and night function, urban security, functional diversity and attractiveness of the place.

As the table shows, most theories have emphasized on the role of social factors in the quality of urban spaces. Professional literature of the urban planning shows that such factors can be clarified with the combination of factors like the vitality, security, behavioral patterns and human activities. A significant point is that the space the context and ground of the activities without which no activity is imaginable. Thus the emphasis on the social factors in the discussion of quality desirability of the urban spaces doesn't mean to ignore the physical aspect but it is merely an emphasis on the human centrality in the urban planning and urban tourism development planning. Accordingly, urban tourism is formed in the urban spaces and hence it is affected by the quality of such spaces directly or indirectly. On one hand, the urban tourism is affected by the social factors that can act as attraction, and on the other hand, they can make effect on the social and cultural structures of both host and guest societies. Of course these two types of effects are mutual and it is merely possible to identify the weakness and strength of interactional relationships. The relationships between urban tourism and social factors are illustrated in Fig. 1.

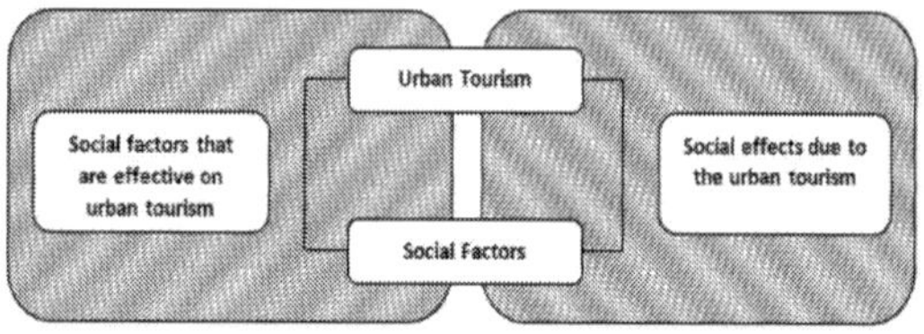

Fig. 1. Mutual relationship between urban tourism and social factors

The humane approach in the urban tourism states that the social and cultural attitude is more important than the economic aspect of tourism. In this approach, the sociocultural factors can be studied at two levels: first, the sociocultural factors, which are indeed a type of social capital of the host society and facilitate this function of urban tourism. Examples of these factors include the social participation, social trust, hospitality, intimate relationship, social solidarity, identity and national (local) sense of pride. Such traits not only can improve the sociocultural conditions of the society, but they play role in the ways of interaction with tourists and creating their satisfaction. For example, the identity and national pride will not let the shortages remain in the tourists' mind as the main features of that place, but the citizens always attempt to create a desirable image of their city for the tourists and make them leave the host city with good memories.

The second level that is proposed in relation to sociocultural factors in the urban tourism is the function of these factors as the destination attractions. As coordinated with the cultural, social and historical heritage, these factors can be investigated under cultural tourism or historical tourism as well. Some of the most important sociocultural factors that are recognized as the urban tourist attractions of Iran are as follows: ritual ceremonies, religious and national ceremonies, unofficial events such as weekly bazaars with emphasis on charities and cultural products, cultural diversity, ethnical diversity, religious diversity, music, local dances and signs, local languages and cloths, and identity symbols with emphasis on social, cultural and historical symbols.

Urban tourism with emphasis on the social factors cannot be fulfilled without paying attention to the cultural, religious and ethnical aspects. In fact the mentioned concept are so inseparably interconnected and form the totality of the social context hand in hand. The professional view to this issue leads us to the

 DOI:10.1051/matecconf/201712406003

subcategories of urban tourism including the cultural tourism, religious tourism and ethnical tourism. In light of the quality of urban spaces, the cultural and religious activities and ethnical rites must be classified in terms of Gehl's threefold classification of the feasible activities in urban spaces (Gehl, 1987). Indeed all mentioned aspects (i.e. cultural, religious and ethnical aspects) are parts of social and optional activities and they somehow confirm the fact that undesirable urban spaces cannot be host the tourism events. Fig 2 shows the graphical representation of this claim.

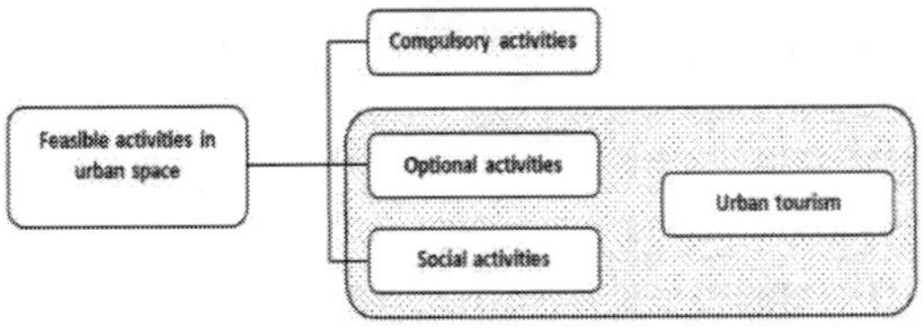

Fig. 2. Feasible activities in urban spaces for tourism

Tourist attraction of each region or city can be man-made or natural. The latter is divided into old and contemporary, and material and intellectual. Iran not only has diverse climates and diverse natural elements (from mountains to deserts, plains, seas and forests) it has very long cultural history, considerable diversity of ethnics, etc. and thus Iran is a desirable potential for the urban tourism development. Indeed with its unique historical, cultural and religious monuments in its different urban regions and old roots of civilization and history on one hand and the modern buildings of its metropolises on the other hand, Iran can introduces many attractions as the old and contemporary man-made elements. This claim is true for the material and intellectual classification of urban tourist attractions as well, while the rites and rituals of different urban subcultures and the diverse religions and ethnics around the country can confirm the mentioned claim.

Focusing on the optional and compulsory activities and the levels of cultural, religious and ethnical tourism in the context of urban spaces, we can propose a model as the proposed analytical model of the research (according to table 3) with the aim of studying the spatial qualities that are effective on the urban tourism focusing on the social factors in Iran.

The proposed framework in combination to the urban tourist attractions can make more effects on the development of tourism industry and on the realization of urban tourism potentials in Iran. In other words, paying attention to the spatial qualities in urban design and planning in combination to the social and cultural features can lead to the sustainability of urban tourism processes in a more efficient way.

Table 3. Proposed analytical framework of the research

Urban space quality	Social factors				
	Vitality	Optional activity	Security	Social activity	Comfort
Flexibility	*			*	
Functional diversity	*	*	*	*	
Physical diversity	*		*	*	
Vegetation	*				*
Security		*	*	*	*
Lighting	*	*	*	*	*
Landscape attractiveness	*		*	*	
Identity	*				*
Memorability	*				*

References

1. Aqajani, Masumeh and Farahani Fard, Saeid, "Religious tourism and its effective factors (case study: Iran)", *Strategic Policies Quarterly*, No. 9, pp. 43-66 (2015) [in Persian].
2. Rabbani, Rasoul and Vahida, Fereidoon, *Urban Sociology*, Isfahan: Isfahan University Press and SAMT Pub (2007) [in Persian].
3. Rahimpur, Ali and Seyyed Hosseini, Davood, *Landscape of Islamic Tourism*, Tehran: Mahya (2000) [in Persian].
4. World Tourism Organization, *International Tourism, A Global Perspective*, translated by Ali Parsaeian and Mohammad A'rabi, Tehran: Cultural Research Office (1998) [in Persian].
5. Saeidnia, Ahmad and Mehdizadeh, Javad, *Urban Tourism (Third Book of Municipality Green Books)*, Tehran: National Organization of Municipalities (2012) [in Persian].
6. Qalenoei, Mahmood and Bahrami, Maryam, "Evaluating the effective qualities on the urban tourism in Shariati St. of Kerman", *Research and Scientific Journal of Iranian Society of Architecture and Urban Planning*, No. 7, pp. 23-33 (2014) [in Persian].
7. Mohammadi, Mahmood and Changlavaei, Yunes, "Evaluating the components of urban space quality on the desirability level of touristic walking pathways (case study: prioritization of walking pathways in Isfahan)", *Research and Scientific Journal of Iranian Society of Architecture and Urban Planning*, No. 5, pp. 15-32 (2012) [in Persian].
8. Handy, S., Self-selection in the relationship between the built environment and walking, *Journal of the American planning association*, 5 (72), 55- 74. (2006).
9. Gehl, I., *Life between buildings*, translated by J. Koch, New York (1987).
10. LPAC (London Planning Advisory Committee), *London's Urban Environmental Quality*, Tibbalds Colbourne Karaski Williams Monro, Remford (1993).
11. Orbaşl, A. Is Tourism Governing Conservation in Historic Towns? *Journal of Architectural Conservation* 6(3), 7-19 (2000).

DOI:10.1051/matecconf/201712406003

12. Project of Public Space (PPS), How To Turn a Place Around: A Handbook for Creating Successful Public Spaces, Project For Public Spaces, Inc., New York (2001).
13. Yoke Lai, L., Said, I. & Kubota, A. The Roles of Cultural Spaces in Malaysia's Historic Towns: The case of Kuala Dungun and Taiping, Social and Behavioral Sciences, 85, 602 –25 (2013).

MATEC Web of Conferences **124**, 07001 (2017) DOI:10.1051/matecconf/201712407001
ICTTE 2017

Experimental Analysis of HC Emission Behavior from a Lean Burn LPG Engine during Transient Operating Conditions

Zhang Zhongbo[1], Li Lifu[1] and Ou Jiajie[2]

[1] *School of Mechanical and Automotive Engineering, South China University of Technology, Guangzhou, 510641, China*
[2] *Institute of Mechanical and Electrical Engineering, Guangzhou University, Guangzhou, 510006, China*

Abstract. Based on engine bench test system, the HC emission behavior of a lean burn LPG engine during constant speed and changing (increasing and decreasing) torque transient conditions was investigated. Results show that HC is significantly deteriorated under the experimental operating conditions. HC concentration value is higher than the steady-state value by 2.3% and 14.2%, respectively. The response of the actual torque lags behind the desired torque by 2.047s and 2.109s, respectively, and leads to the actual speed deviates from the desired value by 4.5% and 3.7%. The deviation of the actual speed causes the fluctuations of IAA, which is the key reason for the deterioration of HC under increasing torque operation. The control of fuel valve cannot meet the demand of the actual LPG mass during the transient operating conditions, leads to the actual LPG mass value deviate from the target value by 4.7% and 8.8%, respectively. And this is also the main responsibility for the actual AFR deviate from the target value by 5.1% and 8.2%, respectively. During the decreasing torque operation, caused by the deviation of the actual AFR, the mixture is too rich, which results in the deterioration of HC.

1 Introduction

LPG (Liquefied Petroleum Gas) has been widely used as an alternative fuel for vehicle engine due to its good fuel properties and environmental protection[1]. However, with gradually stricter emission regulations, to further reduce the emissions of LPG engine is a vital issue for its manufacturers.

Scholars have done many researches on the subject of LPG engine emissions[2-5]. However, the researches mainly have focused on the steady-state emissions behavior of LPG engine, with much less attention paid to the emissions behavior under transient operating conditions, especially acceleration and deceleration. As we all know, the vehicle engines mainly operate under transient conditions during daily driving. Under the transient conditions, the rapid change of the parameters (such as fuelling and air-supply) and the unstable combustion boundaries (such as thermodynamic state and in-cylinder mixture state) can easily cause the deterioration of combustion in-cylinder and lead to the deterioration of emissions eventually [6].

HC (Hydrocarbon) is one of the main emissions of LPG engine and is mainly produced from the unburned or incomplete burned LPG. Due to its vital contribution for the formation of photochemical smog, most countries and regions put forward strict limits on it.

To investigate the HC behavior of LPG engine during the transient conditions, this study takes a premixed spark ignition lean burn LPG engine as its research object. Based on engine bench test system, the HC behavior of LPG engine under constant speed and changing torque transient conditions was experimented. Also, together with the formation mechanism of HC, the related factors were analyzed.

2 Experimental system and conditions

2.1 Experimental system and principle

The engine used in this study is an YC6112ZLQE in-line 6-cylinder, 4-stroke, turbocharged, after-cooled, closed loop control of AFR (air-fuel ratio), spark ignition, premixed lean burn LPG engine, and is widely used to power city buses. Its basic technical data are given in Table 1.

Table 1. LPG engine specifications

Item	Parameter
Compression ratio	8.5
Rated power(kW)	162
Rated speed($r \cdot min^{-1}$)	2300
Maximum torque ($N \cdot m$)	800
Maximum torque speed($r \cdot min^{-1}$)	1300~1400
Idling Speed($r \cdot min^{-1}$)	650±25
Exhaust temperature after the turbine (℃)	≤640

DOI:10.1051/matecconf/201712407001

The principle of LPG engine bench test system is illustrated in Fig.1. The main instrumentations include: Xiangyi GW250 eddy current dynamometer, Xiangyi FC2012 MCI(Measurement and Control Instrument), Xiangyi FC2110 drive, Guangzhou Fuli FLB-501 exhaust analyzer and Advantech 610MB IPC (Industrial Personal Computer). With IPC used as the carrier, engine experimental management software based on Lab VIEW is developed.

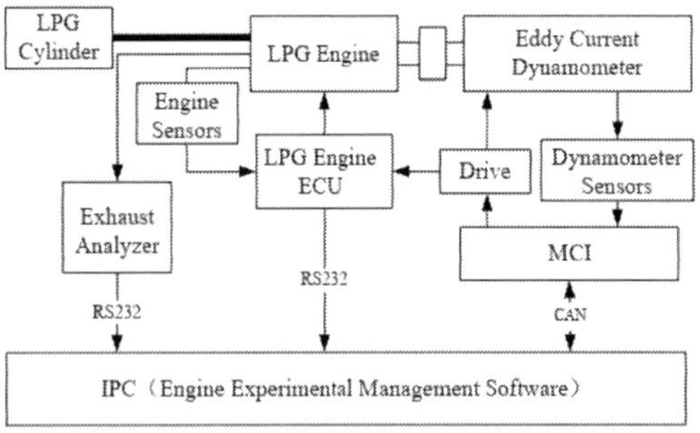

Figure 1. Schematic diagram of LPG engine bench test system

During the test, LPG engine experimental operations (desired speed and torque) are preset in IPC who controls the drive. The drive sends electronic accelerator pedal signals to LPG engine ECU (Electronic Control Unit) to control the brake power while changes the excitation current to adjust the brake torque of eddy current dynamometer. Collecting the actual engine operations (actual speed and torque) in real time, the dynamometer sensors continuously feed it back to MCI who adjusts the engine operations again, so that LPG engine can be operated following the preset conditions. Experimental data are recorded in real-time by IPC, include: pressure before throttle, MAP (manifold absolute pressure), fuel valve opening, AFR, IAA (ignition advance angle), engine torque, speed and HC emission.

2.2 Experimental operating conditions

The engine transient operating conditions mainly include: starting, constant speed and changing torque, constant torque and changing speed, etc[7]. However, constant speed and changing torque operation is in a large proportion in all types of test cycle, e.g. the ratio that this condition in the European Transient Cycle is 61.13% [8]. As a result, constant speed and changing torque operation is usually used as a typical transient condition [9] and is widely used in engine transient performance experiments.

Therefore, the transient HC behavior of LPG engine was experimented under constant speed and changing torque operation in this test. According to the characteristic of this LPG engine, the transient operation was based on constant speed 1100 r•min^{-1} and changing torque from 110N•m to 320 N•m over a 3s transition time. The detailed test operation is shown in Fig.2. And the constant speed and changing torque condition was divided into two operations: increasing torque operation and decreasing torque condition, respectively.

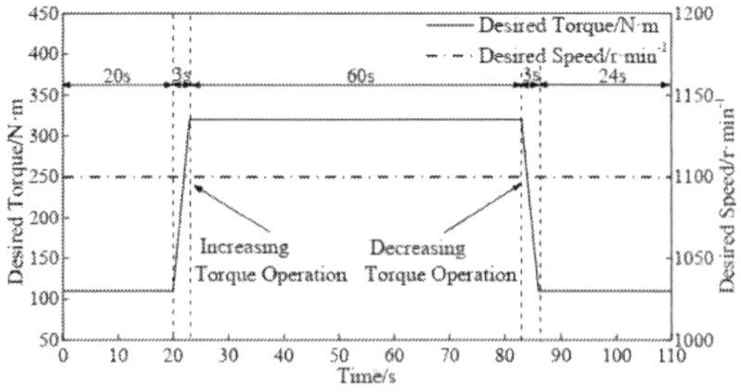

Figure 2. Experimental operation of LPG engine

3 Experimental results and analysis

3.1 Transient HC emission behavior

Under the experimental conditions, the respond of HC concentration (volume concentration) in the exhaust is shown in Fig.3.

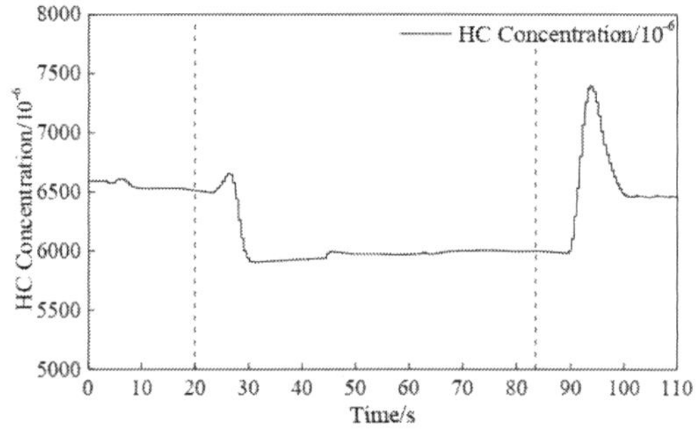

Figure 3. The respond of HC concentration in the exhaust

Fig.3 presents that during the increasing torque operation, HC concentration increases from 6511$\times$10^{-6} (23.672s) to a peak of 6659 $\times$ 10^{-6}(26.172s) and subsequently decreases to 5906$\times$10^{-6}(30.625s) with a small spike formed, and it is higher than its steady-state value by 2.3%(26.172s), indicates that HC is slightly deteriorated. During the decreasing torque operation, HC concentration increases from 6000$\times$10^{-6} (90.062s) to a maximum of 7396$\times$10^{-6} (94.015s) and then decreases to 6474$\times$10^{-6} (100.078s) and forms a larger peak, and it is higher than its steady-state value by 14.2% (94.015s), shows that HC is seriously deteriorated.

3.2 Analysis of actual experimental conditions

According to the engine principle, AFR and temperature in-cylinder is different among various operating conditions, thus the emissions are different [10]. For the transient HC behavior test of LPG engine, the experimental system changes the engine brake power by controlling electronic accelerator pedal signal, and the load of engine is changed by controlling the excitation current of eddy current dynamometer (that is to say, to change the engine brake torque). However, due to the slow response of dynamometer, the engine brake torque will lag behind. According to the definition of engine brake power

$$P = Mn \tag{1}$$

where M is the brake torque and n is the rotational speed. When the brake power of LPG engine changes, the lag of brake torque will cause the fluctuations of engine speed. Therefore, the actual conditions of engine are difficult to operate in accordance with the experimental conditions. For this reason, the influence of actual experimental conditions must be taken into account in the analysis of HC behavior of LPG engine under transient operating conditions. The respond of LPG engine torque and rotational speed under experimental operations is given in Fig.4 and Fig.5, respectively.

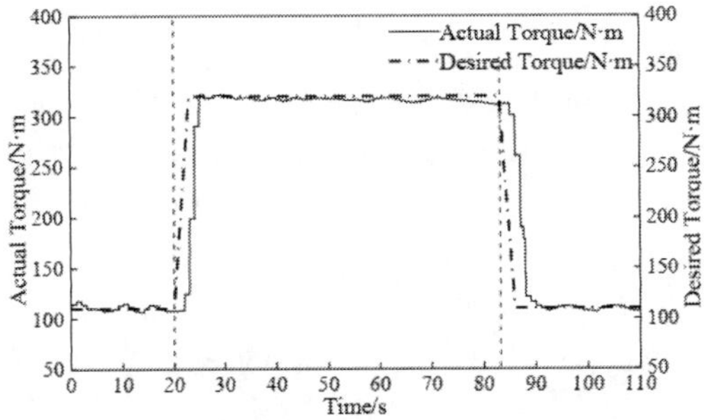

Figure 4. The respond of torque

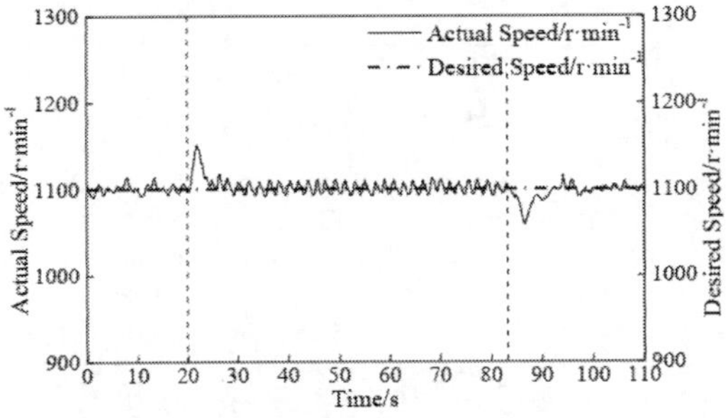

Figure 5. The respond of speed

Fig.4 presents the respond of the actual torque lags behind the desired torque by 2.047s and 2.109s, respectively. Fig.5 indicates, due to the lags of actual torque, the actual speed deviates from the desired speed by 4.5%(21.843s) and 3.7%(86.531s), respectively.

3.3 Analysis of control parameters

According to the formation mechanisms of engine emissions, AFR and IAA is the main control parameter that affects the generation of spark ignition engine emissions [10].

As shown in Fig.6 and Fig.7, the respond of AFR and IAA under experimental conditions is given, respectively. Fig.6 presents that the target AFR is stable without obvious fluctuations under the experimental conditions. However, the actual AFR fluctuates greatly and deviates from the target AFR by 5.1% (22.765s) and 8.2% (87.218s), respectively. Fig.7 shows that the IAA also fluctuates greatly during the experimental conditions.

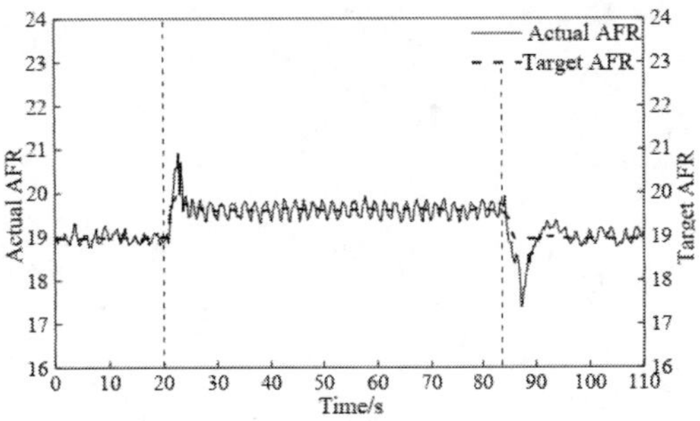

Figure 6. The respond of AFR

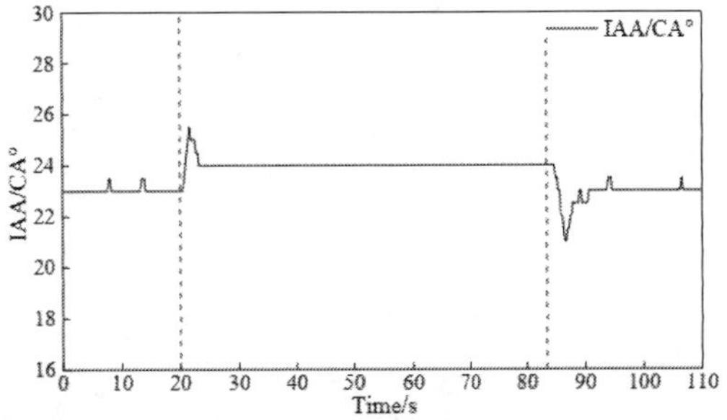

Figure 7. The respond of IAA

(1) Analysis of AFR

According to the basic principle of premixed mixture formation of LPG engine, as shown in Fig.8, with regard to the respond of the actual AFR under the experimental conditions, AFR is defined by

$$\alpha_{AFR} = \frac{m_{air}}{m_{LPG}} \tag{2}$$

where m_{air} is the air inflow and m_{LPG} is the LPG mass.

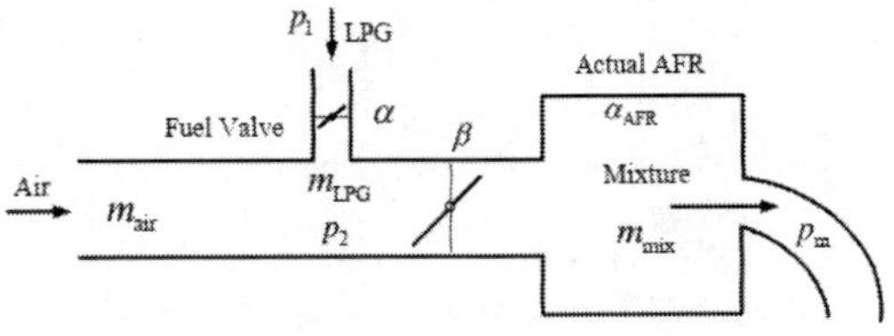

Figure 8. Schematic diagram of the premixed mixture formation of LPG engine

The speed-density equation is usually used to calculate the air inflow

$$m_{air} = (P_m / (RT_m)) N \bullet n \bullet V_h \eta_v \tag{3}$$

where P_m is the MAP; R is the gas constant; T_m is the manifold air temperature(MAT); N is the number of engine cylinders; V_h is the displacement and η_v is the volumetric efficiency.

For the Eq. (3), V_h is fixed, N is 6, the experimental data of n is shown in Fig.5 and the respond of MAP and

MAT during the experimental conditions is given in Fig. 9 and Fig. 10, respectively.

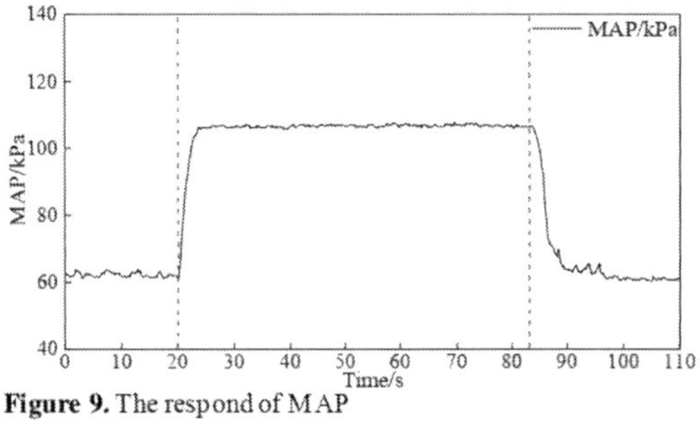

Figure 9. The respond of MAP

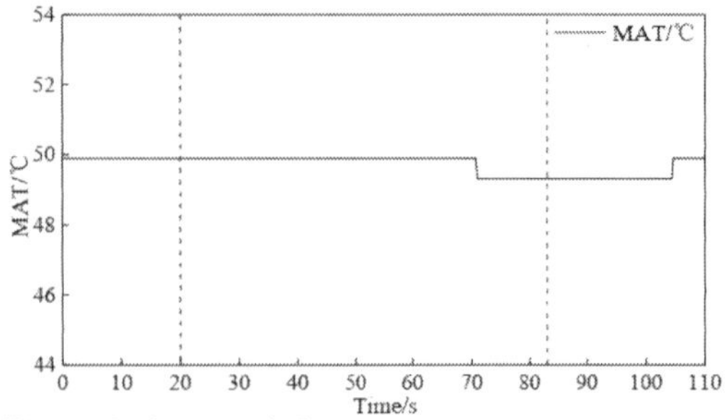

Figure 10. The respond of MAT

Fig.9 presents that the transition of the MAP during the experimental conditions is relatively stable without obviously fluctuations. It is shown in Fig.10, under the experimental operations, the variation of MAT is very small, only 0.7℃.

For the volumetric efficiency, it can be calculated by the relation [11]

$$\eta_v = \frac{T_0}{P_0} \bullet \frac{\varepsilon}{\varepsilon - 1} \bullet (\frac{P_a}{T_a} - \frac{P_0}{T_0}) \tag{4}$$

where T_0 is the atmospheric temperature; P_0 is the atmospheric pressure; ε is the compression ratio; P_a and T_a is the pressure and temperature of the mixture at the end of intake and is approximate equal to MAP and MAT, respectively.

Therefore, with Eqs.(3), (4) and the experimental data of MAP (Fig.9), MAT (Fig.10) and rotational speed (Fig.5), the respond of air inflow during the experimental conditions is calculated, which is given in Fig.11.

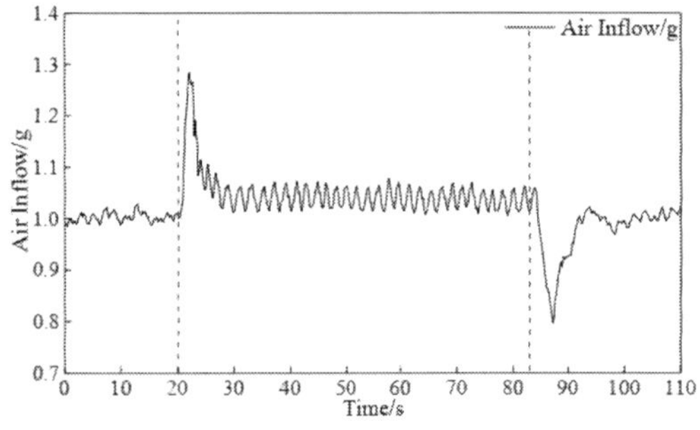

Figure 11. The respond of air inflow

Fig.11 indicates that, with the increase of MAP, air inflow increases rapidly during the increasing torque condition. However, the increase of the speed further results in the increase of air inflow, which leads to the air inflow forms a large peak. Under the decreasing torque condition, the respond of MAP and speed is opposite to that during the increasing torque condition, which leads to the air inflow forms a large valley. Therefore, the fluctuations of engine speed (Fig.5) lead to the fluctuations of air inflow during the actual experimental conditions.

For the LPG mass, based on the calculation of air inflow and the determining of the target AFR by look-up table, ECU calculates the LPG mass under the current operating condition.

Therefore, by Eq.(2) and the experimental data of AFR (Fig.6) and air inflow(Fig.11), the respond of the actual and target LPG mass during the experimental operations is calculated, as shown in Fig.12.

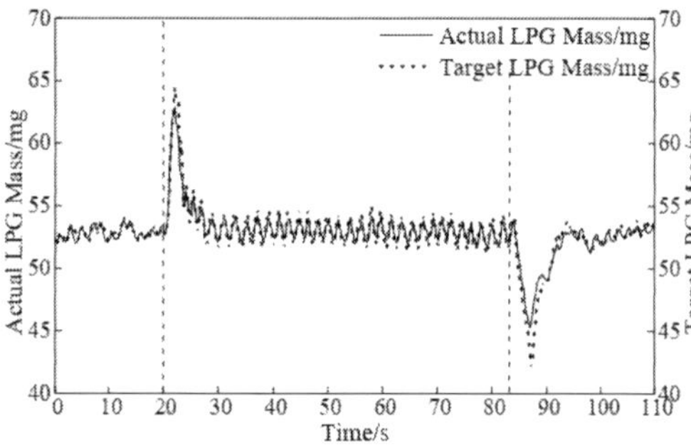

Figure 12. The respond of LPG mass

Fig.12 presents, under the steady-state operations, the deviation from actual LPG mass to target value is small and is less than 1.85%. However, during the transient operations, the deviation is much larger. Under the increasing torque condition, the actual LPG mass is smaller than the target value, and the maximal deviation is 4.7% (22.765s). During the decreasing torque condition, the actual LPG mass is much larger than the target one, and the maximal deviation is 8.8% (87.218s).

For the respond of LPG mass, according to the basic principle of premixed mixture formation of LPG engine in Fig.8, the LPG mass can be expressed by

$$m_{LPG} = C_v A_v \psi \frac{P_1^2}{P_2 \sqrt{RT_v}} \tag{5}$$

where T_v is the LPG temperature; P_1 is the upstream pressure of the fuel valve; P_2 is the downstream pressure of the fuel valve that is equal to the pressure before throttle; C_v is the flow coefficient of the fuel valve which is related to the physical behavior of LPG and the difference of fuel valve upstream/downstream pressure, and can be calculated by the experimental method and the empirical equation Eq.(6)

$$C_v = a_1 + a_2\alpha + a_3\alpha^2 \tag{6}$$

where α is the fuel valve opening; a_1, a_2 and a_3 is the coefficient can be acquired through experiment.

A_v is the effective cross-sectional area of the fuel valve and is a function of fuel valve opening, its expression is

$$A_v = \frac{\pi d^2}{4}(1-\cos^2\alpha) \quad (7)$$

where d is the fuel valve diameter.

ψ is the flow function that is related to the difference of the fuel valve upstream/downstream pressure, and its expression is

$$\psi(\frac{P_1}{P_2}) = \begin{cases} \sqrt{\gamma[\frac{2}{\gamma+1}]^{\frac{\gamma+1}{\gamma-1}}} & \frac{P_2}{P_1} \le [\frac{2}{\gamma+1}]^{\frac{\gamma}{\gamma-1}} \\ (\frac{P_2(t)}{P_1(t)})^{\frac{1}{\gamma}} \sqrt{\frac{2\gamma}{\gamma-1}[1-(\frac{P_2}{P_1})^{\frac{\gamma-1}{\gamma}}]} & \frac{P_2}{P_1} > [\frac{2}{\gamma+1}]^{\frac{\gamma}{\gamma-1}} \end{cases} \quad (8)$$

where γ is the specific heat ratio of LPG.

Therefore, according to Eqs.(5) ~ (8), the LPG mass is a function of α, T_v, d, P_1, P_2 and γ, i.e.

$$m_{LPG} = f(\alpha, T_v, d, P_1, P_2, \gamma) \quad (9)$$

However, in Eqs. (9) d and γ is constant, and due to the regulator, T_v and P_1 changed slightly can be seen as constant, so m_{LPG} is mainly affected by α and P_2, i.e.:

$$m_{LPG} = f(\alpha, P_2) \quad (10)$$

The respond of fuel valve opening and pressure before throttle during the experimental operations is shown in Fig.13 and Fig.14, respectively.

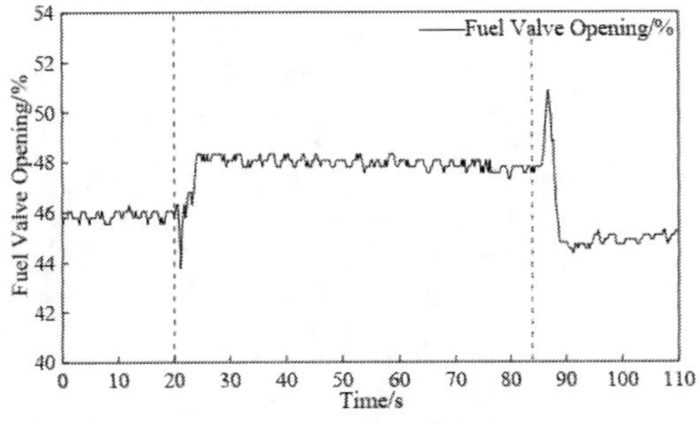

Figure 13. The respond of fuel valve opening

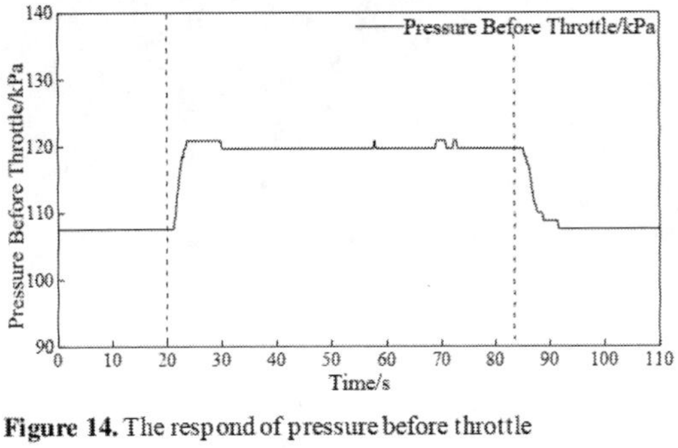

Figure 14. The respond of pressure before throttle

Fig.13 shows that the fuel valve opening under the experimental operations fluctuates obviously. Fig.14 presents that the pressure before throttle under experimental operations has a smooth transition without noted fluctuations. According to the principle of LPG engine control system, fuel valve is the actuator of LPG engine control system. Therefore, the control of fuel valve cannot meet the needs of the actual LPG mass is the reason for the actual LPG mass deviation from the target value under transient operating conditions.

Hence, by the Eq.(2) and the analysis of the respond of air inflow and LPG mass under experimental condition, the control of fuel valve cannot meet the demand of actual LPG mass is the root cause of the LPG engine actual AFR deviates from the target value.

(2) Analysis of IAA

For the respond of IAA during the experimental conditions, by the control principle of the engine IAA, during the working process of engine, according to the MAP and the speed, the basic spark angle is obtained by look-up table and interpolation method. Based on the basic spark angle, the actual spark angle is obtained by a series of amendment. Therefore, the respond of IAA is mainly affected by MAP and speed.

According to the respond of speed (Fig.5), MAP (Fig.9) and IAA (Fig.7), during the increasing torque condition, with the increase of MAP, the IAA increases rapidly. However, the fluctuations of engine speed (increased first and then decreased) lead to the decrease of IAA after a further increase. Under the decreasing torque operation, the respond of the influencing factors of IAA are opposite to that during the increasing torque condition, IAA increases after a rapid decrease.

3.4 Analysis of transient HC emission behavior

For LPG engines, HC is the main product of incomplete combusted of LPG. The formation reaction of HC is complex in-cylinder and influenced by AFR and IAA obviously [10]. If the AFR of mixture is too small, LPG cannot be fully burned, which results in the increase of HC. The increase of AFR is beneficial to the complete combustion of LPG to reduce HC. However, a larger AFR will lead to the deterioration of combustion in-cylinder and even result in the phenomenon of misfire, which cause the increase of HC finally. Reducing the IAA is a favourable way to increase the temperature of engine exhaust, so that HC can be oxidized in exhaust systemand reduced consequently.

With the respond of HC (Fig.3), AFR (Fig.6) and IAA (Fig.7), during the increasing torque condition, the rapid increase of the actual AFR is beneficial to the complete combustion of LPG, which helps out with lighter HC concentration. Although the mixture concentration in this process is too thin, it can still be stable burn [12], so the dilution of mixture do not lead to the deterioration of HC. However, whilst the IAA increases rapidly, which leads to the short of after combustion period and the decrease of exhaust temperature, weakens the oxidization of HC in exhaust stroke and exhaust system, results in the increase of HC eventually. Therefore, a small HC concentration peak is formed during the process. During the decreasing torque condition, the actual AFR decreases rapidly, and results in the unburned partial LPG. At the same time, the excessive rich of the mixture in this process further leads to the increase of the proportion of incomplete burned LPG. Therefore, HC concentration increases rapidly. Although the reduction of IAA is beneficial to reduce HC, AFR is the dominant factors of HC form in this whole process, so a large HC concentration peak is formed.

4 Conclusions

To study HC emission behaviour of LPG engine during the transient conditions, based on engine bench test system, under constant speed and changing (increasing and decreasing) torque transient conditions, a lean burn LPG engine was tested and analysed.

(1)Under the experimental conditions, HC is significantly deteriorated. HC concentration value is higher than the steady-state value by 2.3% and 14.2%, respectively.

(2)The respond of the actual torque lags behind the desired value by 2.047s and 2.109s, respectively, leads to the actual speed deviates from the desired value with 4.5% and 3.7%.

(3)The deviation of the actual speed causes the fluctuations of IAA, which is the key reason for the deterioration of HC under the increasing torque operation.

(4)During the experimental conditions, the control of fuel valve cannot meet the demand of actual LPG mass leads to the actual LPG mass value deviates from the target value by 4.7% and 8.8%, respectively, and this is also the main responsibility for the actual AFR deviates from the target value (the deviation is 5.1% and 8.2%, respectively).

(5)During the decreasing torque operation, caused by the deviation of actual AFR, the mixture is too rich, which results in the deterioration of HC, eventually.

In order to decrease HC emission of LPG engine under transient operating conditions, new control methods of fuel valve and IAA need to be studied in the future.

Acknowledgements

This project is supported by National Natural Science Foundation of China (Grant No. 51605104).

References

1. Mitukiewicz G, Dychto R, Leyko J. Relationship between LPG fuel and gasoline injection duration for gasoline direct injection engines[J]. Fuel, 2015, 153: 526-534.
2. Erkus Baris, Karamangil M. Ihsan, Surmen Ali. Enhancing the heavy load performance of a gasoline engine converted for LPG use by modifying the ignition timings[J]. Applied Thermal Engineering, 2015, 85: 188-194.
3. Oh Seungmook. Effects of intake and exhaust valve timing on combustion and emission characteristics of lean-burn direct-injection LPG engine[J]. Transactions of the KSME. B, 2015, 1(39): 45-51.
4. Can I, Oner C, Sugozu I. Experimental investigation of effect of stratified charging system adopted to gasoline engine running with LPG on engine performance[J]. Journal of the Energy Institute, 2013, 2(86): 113-118.
5. Pradeep V, Ramesh A. A novel twin injector strategy for direct cylinder barrel injection of LPG in a 2S-SI engine[J]. Fuel, 2015, 162: 128-137.
6. C.D.Rakopoulos, A.M.Dimaratos, E.G.Giakoumis, et al. Evaluation of the effect of engine, load and turbocharger parameters on transient emissions of diesel engine[J]. Energy Conversion and Management, 2009, 50: 2381-2393.
7. Fu Jianqin. Continuous detecting on the operating and performance parameters of automotive engine under transient conditions and study of the heat-work conversion process[D]. Changsha:Hunan University, 2014.
8. Zhang Longping. Investigation of performance deterioration and control strategy of automotive diesel engine under transient operation conditions [D]. Changchun: Jilin University, 2015.
9. Zhang Longping, Liu Zhongchang, Tian Jing, et al. Investigation of the combustion deterioration of an automotive diesel engine under transient operation [J]. Science China Technological Sciences, 2014, 57(3): 480-488.
10. Heywood JB. Internal combustion engine fundamentals[M]. New York: McGrawHill, 1988.
11. Chen Gong. The analysis and test study on the influence of the engine combustion process to the exhaust temperature[D]. Harbin: Northeast Forestry University, 2013.
12. Wang Zhensuo, Deng Baoqing, Li Liguang, et al. Judgement of lean burn limit for SI engine using LPG[J].Journal of Jilin University (Engineering and Technology Edition), 2002, 32(04): 7-11.

 DOI:10.1051/matecconf/201712407002

Research on Loading System of the Rail Vehicle Bogie Test Bed

Ning Lu[1], Li Na Ren[1] and Zhong Wei Chen[1]

[1]School of Mechanical-electronic and Automobile Engineering, BUCEA, China

Abstract. The bogie is one of the most important parts of metro vehicles, to provide traction, support and guidance of the vehicle in the process of the vehicle. To a great extent, the rationality of the structure of the bogie can directly affect the dynamic performance of the metro vehicles and the safety and comfort of the ride. This paper researches on the detection equipment of the Metro Bogie, mainly researches on the loading system of the metro static load test bed.

1 Introduction

With the rapid development of China's economy and the increase of urban population, the pressure on urban public transport is becoming larger. The metro is an important part of urban public transport and plays an important role in easing traffic congestion and other issues. The bogie is one of the most important parts of metro vehicles, to provide traction, support and guidance of the vehicle in the process of the vehicle. To a great extent, the rationality of the structure of the bogie can directly affect the dynamic performance of the metro vehicles and the safety and comfort of the ride[1]. This paper research on the detection equipment of the Metro Bogie, mainly research on the loading system of the metro static load test bed, has the actual application value, can improve the vehicle running safety and comfort, and will promote the development of metro maintenance in our country.

2 Composition of the test bed

The test bed consists of the following components: mechanical system, hydraulic systems, electrical control systems.

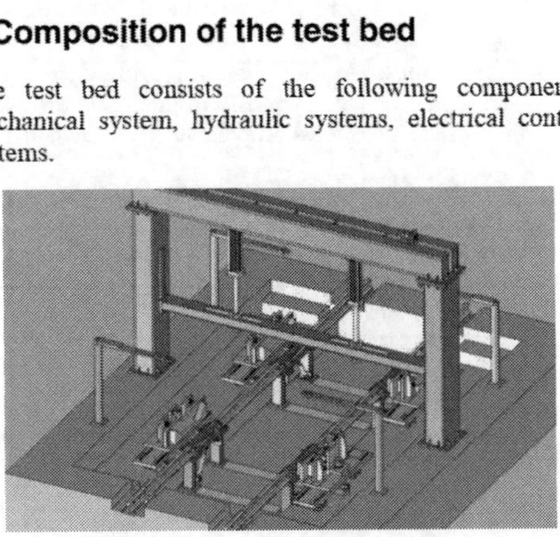

Figure 1. Electro-hydraulic servo double actuators rail vehicle bogie test bed.

2.1 Mechanical system

Mechanical system is composed of fixed transom door frame, bogie centering device, weight measurement units. Fixed transom door frame is a structural steel frame. Bogie centering device is set in one side of the standard track. Weight measurement units are core part of the test bed, and the units are the introduction of China's railway technical authorities newly developed system weighing units, which can entirely monitor the wheel weight during the bogie is loaded, weighing data bus transferring to the computer.

2.2 Hydraulic systems

Hydraulic station is the hydraulic power source of the entire system. It is installed in the vicinity of the gantry, the tubing along the column and beam receiving cylinder[2]. The hydraulic pressure source is provided with a cooling cycle apparatus. Hydraulic station is in accordance with the relevant provisions in order to meet the requirements of the servo system for cleaning, installation, to ensure the normal operation of the various parts of the hydraulic systems.

The device is installed two ram cylinders, cylinder stroke 400mm, and single cylinder is greater than the maximum output pressure 200KN. Import displacement sensors are in the cylinder, and import force sensor is connected to the piston rod end. Pressing cylinder uses the method of electro-hydraulic servo intelligent control to achieve load control and displacement control, which can simulate vehicle coordination loaded in the case of partial load.

2.3 Electrical control systems

The control of the test bed is composed of a remote computer cabinets and site electrical control cabinet, including the hardware of the remote computer cabinet consists of: industrial control computers, monitors, UPS uninterruptible power supply, A4 printer and the corresponding buttons and lights; by means of configuration software operation cabinet to complete the test bed data storage, graphics, data reporting, query and printing. Site electrical control cabinet consists of a DC power supply, PLC modules, data acquisition modules scene, analog data output modules and other components, which mainly achieve the function of site control actuator, sensor signal detection and processing, fault information collection and more. Operation platform and control cabinet have a reliable ground, zero connection protection, terminal blocks using high-quality insulating material. Control systems have the functions of overload, overcurrent, overvoltage, overheating, under voltage, short circuit, lack of equal protection. Operation platform and control cabinets are equipped with an emergency stop button.

3 Loading system composition

The loading system adopts hydraulic loading. Basis of hydraulic loading method is that the motor provides the power source, then the hydraulic pump to convert mechanical energy into pressure energy to drive the hydraulic oil, by changing the valve opening direction or the size of the opening to control the flow of hydraulic oil and flow, makes the hydraulic cylinder can make different reaction route and action to complete the system needed to travel or movement[3].

The loading system of the static load test platform of the Metro Bogie is composed of 2 sets of hydraulic cylinders, 2 servo valves, 2 pressure sensors and 2 displacement sensors. The hydraulic cylinder is connected to the movable beam by the front flange, the hydraulic cylinder is 500mm, and the built-in displacement sensor is adopted to reduce the influence of environmental factors. The hydraulic cylinder can be moved automatically on the movable beam, and the adjustment range is 500~2000mm. The displacement control or the force loading is set by the industrial computer, and the loading results are displayed in the monitoring interface in real time, the equipment can be loaded synchronously, and can be loaded separately.

4 Loading system working principle

The working principle of this system is that the loading system is a closed-loop system. First, the required displacement is provided by the industrial computer, hydraulic pump provides power for hydraulic cylinder and servo valve, the input signal of the displacement sensor is compared with the input signal provided by the industrial computer. Then the signal is transmitted to the servo valve to control the opening size of the valve core. Thus, the oil flow rate of the hydraulic cylinder is controlled, so as to achieve the control of the displacement, and the actual displacement size is fed back to the industrial computer through a displacement sensor arranged on the hydraulic cylinder.

There are two kinds of synchronization control strategies of the system, the same way and the master-slave mode. Master-slave mode refers to multiple hydraulic cylinder synchronous control, one hydraulic cylinder is selected as the active cylinder, output active cylinder set as output, other hydraulic cylinder for the slave cylinder, common active cylinder electro-hydraulic valve, cylinder driven selection of electro-hydraulic servo valve, cylinder driven with active cylinder movement the same way refers to the synchronization control; pre setting an ideal input, a plurality of hydraulic cylinders and tracking operation to achieve synchronous control. In the same way, it can eliminate the interaction between the hydraulic cylinders, and do not need to adjust the running of the hydraulic cylinder, so the time needed to realize the synchronization is short, and the dynamic performance of the system is better. Therefore, the system uses the same control strategy

The same type of displacement synchronous control system structure as shown in Figure 4-8, control the use of two controller, while tracking the same amount of input, and then the two loop respectively according to the difference between the actual input and output adjustments, two circuits do not interfere with each other, their control, effectively eliminate the influence of two loops. The system can be in a very short period of time to make adjustments and achieve synchronization output.

The main technical indexes of the system are:

1) Single channel maximum loading capacity: 15 tons;
2) Servo cylinder displacement measurement range: 0-500mm, system accuracy 0.1%F.S;
3) The lifting speed of the cylinder: 60~200 mm / min, adjustable at any time, can work continuously;
4) Load measurement accuracy: 0.1%F.S;
5) Displacement measurement accuracy: 0.1%F.S;
6) The working environment temperature: -20 to +45 DEG C.

According to the working principle and work requirements, the system design proposal, select the appropriate hydraulic components. As shown in Table 1.

Table 1 Components models

Name	Model	Quantity
Servo valve	R02KO1M0NSP2	2
Hydraulic cylinder	USYWF10ϕ125× 500-25LHA	2
Displacement sensor	LWF-500-A1	2
Force sensor	DBSL20T	2
Hydraulic pump	A3H16-FRO1KK-10	1
Electric machinery	QA132S4A	1

One-way valve	S8A3.0/2	1
Relief valve	DBDS6P10/31.5	1
Tank	SS304	1
Accumulator	S50250S-A	1

5 Simulation experiment

This paper will also take trying to determine the parameters of PID: the process of trying in the parameters according to the variation tendency of the system to adjust, first with proportional coefficient adjustment, and then adjusting the integral coefficient, finally by differential coefficient adjustment. First adjust the proportion coefficient, integral time adjustment to infinity, the differential time is adjusted to 0, the proportion coefficient from adjusting observation curve, if serious oscillation increases proportional coefficient, if serious overshoot decreases proportional coefficient, 4:1 curve to stop; then by adjusting the integral coefficient, if only use ratio can not eliminate static error coefficient regulation requires the use of integral coefficient adjustment, first set a larger integration time value, adjusted from large to small, if volatility increases the integral time value, if the deviation from the given value is reduced, the integral time value. Adjust the proportion coefficient during May, in order to achieve better control effect; finally, differential coefficient adjustment, if the proportion coefficient and integral coefficient after repeated adjustment has not reached the desired effect, the need for differential adjustment, the differential time is set to 0, if the system has high overshoot and long time is required to reach a steady state on the contrary, increase the differential time, decrease the differential time during this period can also adjust the proportion coefficient and integral coefficient, the ideal curve, to achieve the control effect[4].

The fuzzy PID controller selection fuzzy controller displacement control system and its control principle is: according to the deviation and deviation change displacement input displacement value rate, according to the fuzzy rules set for correction of PID parameters.

Displacement control system, fuzzy PID controller, parameter tuning principle: Based on the repeated tests of the displacement control system, the deviations of the system, the variation rate of the deviation, the coefficient of proportionality, the integral coefficient and the coefficient of differentiation are obtained, and summarized as follows: When the displacement control system of displacement deviation of |e| is large, should take a larger value to the system response speed, but should not be too large, otherwise it will make the system unstable; should take a smaller value to prevent the system out of control at first, if the overshoot can be removed more points. When the displacement deviation of displacement control system is equal to |e|, smaller values should be taken to obtain smaller overshoot, and the values should be selected appropriately to speed up the response, while the value should be increased in an appropriate range. When the displacement deviation of displacement control system is less than |e|, in order to make the system stable, we should select the larger sum and select the appropriate one to reduce oscillation[5].

The fuzzy subsets of input and output are defined as: {NB, NM, NS, Z, PS, PM, PB}, respectively, representing negative, large, negative, middle, negative, small, zero, positive, small, middle, and large.

The domain of definition of input and output for input: e, domain EC: {-3, -2, -1, 0, 1, 2, 3}; domain output: {-3, -2, -1, 0, 1, 2, 3}; domain: {-0.6, -0.4, -0.2,0,0.2,0.4,0.6}; domain: {-0.06, -0.04, -0.02, 0, 0.02, 0.04, 0.06};

The deviation of the basic domain of [-10, 10]mm, basic theory, deviation rate of change EC domain [-0.5, 0.5]mm/s, quantization factor calculated deviation E and deviation change rate of EC, respectively, the output, the scale factor, respectively, K1=1, K2=1/5, K3=1/50.

Co simulation of displacement control system is used in AMESim and Simulink, establish the model of hydraulic system in hydraulic system simulation has advantages in AMESim, with the establishment of control system model of the advantages of the Simulink in control system simulation.

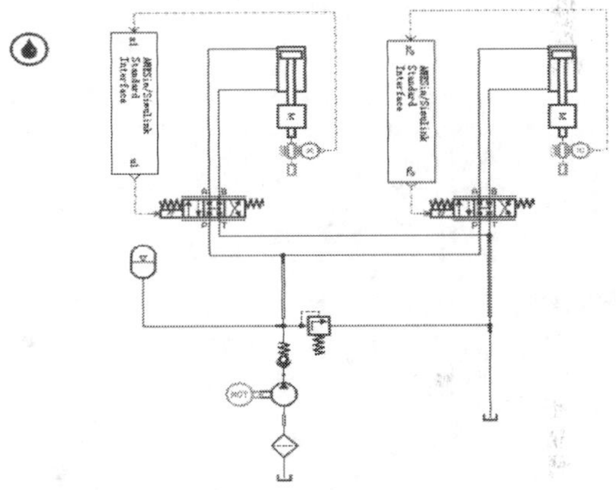

Figure 2. Displacement control hydraulic system model built in AMESim

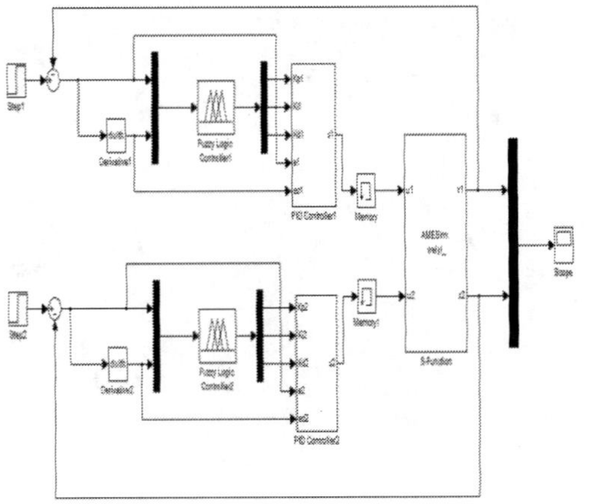

Figure 3. Fuzzy PID control system model built in Simulink

DOI:10.1051/matecconf/201712407002

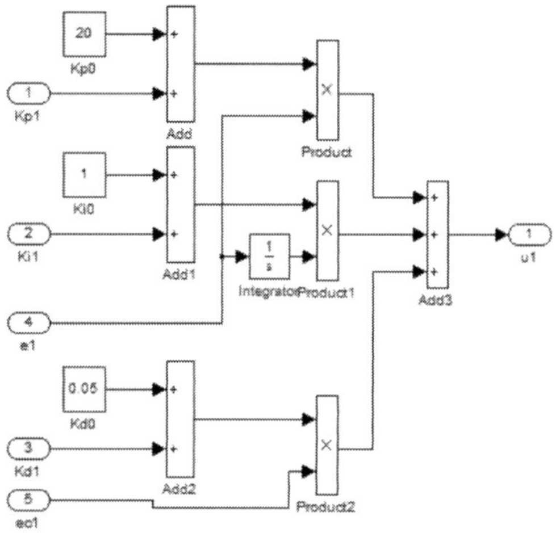

Figure 4. PID controller structure

The initial parameters of the PID control structure of fuzzy PID controller are set up $K_p = 20$, $K_i = 1$, $K_d = 0.05$, and simulation is carried out to get the step response curve of the fuzzy PID control, as shown in figure5, the displacement error curves of the two hydraulic cylinders are shown in figure6.

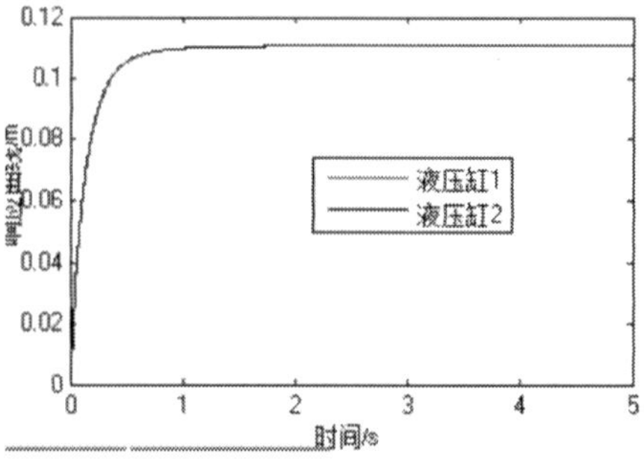

Figure 5. Fuzzy PID control step response curve of displacement control system

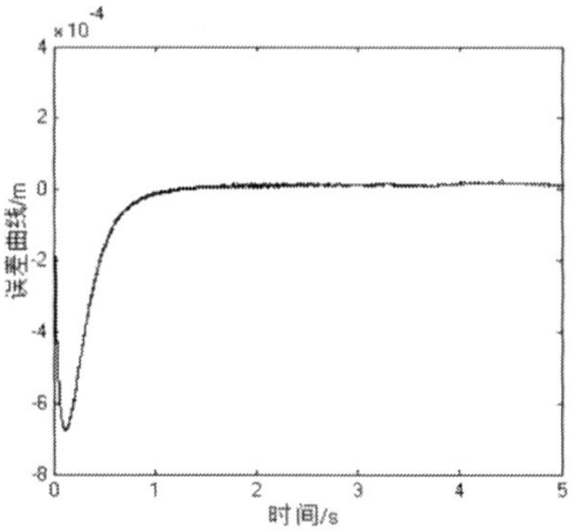

Figure 6. Displacement error curve of two hydraulic cylinders

From the above two figure can be seen, the use of fuzzy PID synchronous control, hydraulic cylinder 1 and the hydraulic cylinder 2 displacement can reach steady state in about 1s, fast response speed, and according to the error curve can also be seen to achieve good synchronization effect.

References

1. Chr Angeli, An online expert system for fault diagnosis in hydraulic systems J. Expert System. vol. 2, pp. 115-119, 1999.
2. M Lizell, Dynamic leveling a low power active suspension with adaptative control J. Vehicle System Dynamics. vol. 3, 1991.
3. Marquis-Favre, W.Bideaux, E.Scavarda. A planar mechanical library in the AMESim simulation software[J]. Formulation of dynamics equations, Simulation modeling practice and theory.2006,14(1):25-46.
4. Edik Arakeljan, Mark Panko, Vasili Usenko. Comparative Analysis of Classical and Fuzzy PID Algorithms[J]. Advances in Soft Computing.2000,6: 206-212.
5. X.C. Liu, Q.T Huang & D.C. Cong. New modeling and analysis of three-stage electro-hydraulic servo valve. 2008 International Workshop on Modeling[J]. Simulation and Optimization.2008,8:146-150.

Dry Friction Clutch Disc of an Automobile under Transient Thermal Load: A Comparison of Friction Lining Materials

Anosh Ali[1], Liaqat Ali[2], Samiur Rahman Shah[2], Mushtaq Khan[2], Syed Husain Imran[2] and Shahid Ikramullah Butt[2]

[1]*Department of Mechatronics Engineering, College of Electrical and Mechanical Engineering, National University of Sciences and Technology (NUST), Islamabad 44000, Pakistan*
[2]*School of Mechanical and Manufacturing Engineering (SMME), National University of Sciences and Technology (NUST), Islamabad 44000, Pakistan*

Abstract. This paper shows the comparison of temperatures produced in a dry friction clutch disc with different materials during a single engagement to assist in clutch plate design and analysis. A study of usage of different materials for friction lining of clutch disc is required, which will provide improved performance and enhanced life. This investigation is modelled mathematically and solved numerically using finite element method. ANSYS® 15.0 is a dedicated finite element package used for determining the temperature distribution across a clutch disc. In the present work, an investigation of a conventionally used harmful friction lining material asbestos is compared with carbon-carbon composite, S2-glass fibre and aluminium metal matrix composite. The transient thermal analysis of a clutch disc with different materials is performed and the temperature distribution on the clutch system is compared. Simulation results indicate that all the values of the temperature obtained from the analysis of aluminium metal matrix are less than those of asbestos based lining material, therefore clutch disc made up of aluminium metal matrix composite will assure the extended service life and the longer stability due to the fact that the temperature responsible for the wear and tear has been reduced. Furthermore, the slipping time is also considered in this investigation.

1 Introduction

A clutch is commonly used in automotive vehicles for power transfer from the engine to the transmission system. Since a clutch has to be replaced many times before the car reaches end of its lifetime, so it becomes necessary to enhance its functional life. The lifetime of a clutch is highly dependent on the life and performance of the clutch plate. An automobile dry friction clutch system consists of a flywheel, clutch plate and a pressure plate.

Clutch disc comes in direct contact with the flywheel of the engine for power transmission. It consists of frictional material on both sides and an axial cushion is sandwiched between them. Friction lining is used to prevent the slippage between the two contacting surfaces. The engine side face of the friction lining couples with the flywheel while the transmission side couples with the pressure plate. When a clutch is engaged, the relative motion due to slippage between the contacting surfaces of a clutch system results in high surface temperatures which may lead to the surface cracks, enhanced wear rate as well as permanent distortions which can cause premature failure of the contacting surfaces [1-2].

Slippage occurring between the two contacting surfaces results in heat generation and consequently, increase in temperature. Increased temperature in clutch disc results in elastic deformations and thermal cracking which is responsible for the failure of the friction material before the expected lifespan of the clutch [3]. The generated temperature will be conducted between the clutch system as well as dissipated to the atmosphere [4].

Recently, thermal problems due to sliding of dry friction clutches and brakes are being extensively studied. It is found that the temperatures fields of the brake disc are highly effected by the contact pressure with specific material properties [5].

For a clutch disc, a study of using new friction materials is needed for improved performance and enhanced life of plate as well as clutch itself. Conventional materials are being replaced by the composites in both the clutch and brake applications. Aluminium composite reinforced with natural fibre is a good option in the application of automotive brake pads [6-7]. Behaviour of Aluminium metal matrix composites and high strength glass fibres and towards friction and wear is proved to be promising. S2 glass fibre is proved to be a suitable material for braking operation as a result of transient thermos-elastic analysis of brake disc for various materials in frequent braking operation [8]. Thermal resistance of glass fibres is more as compared to conventionally used asbestos [9]. There is a lot of research still to be done in the area of composite friction materials [4].

The study of thermo-mechanical characteristics reveals the dependence of heat transfer on slippage time and contact area ratio. Since, the pressure is maximum in constant pressure, hence, heat distribution is also maximum in this case. The maximum temperature appear at the half of the slippage time. [1, 10].

During clutch engagement, temperature distribution under constant pressure is higher than pressures of other type and the uniform temperature in radial direction due to continuous contact surface [11]. The one dimensional time variant heat transfer is dependent on friction in two sliding contact brake disk [12]. Expressions for determining the heat dissipation rate during single clutch engagement are a function of thermal conductivity, elastic modulus, and coefficient of thermal expansion [13].

In this study, the clutch lining temperature during contact between clutch disc and flywheel and also between clutch disc and pressure plate during single engagement is evaluated by finite element method. Transient thermal module in ANSYS® 15.0 is used to determine the temperatures produced in the friction lining of a clutch disc for different materials and the results obtained are compared.

2 Problem formulation

During slipping period (t_s), a lot of kinetic energy produced is transformed into thermal energy. In the present work, it is assumed that because of slipping, the energy due to friction between two sliding surfaces is converted into heat energy at the interface [14]. During slipping, the total heat produced is given as

$$Q(r,t) = \begin{cases} \mu p V_s & ; \quad 0 \le t \le t_s \\ 0 & ; \quad t > t_s \end{cases} \tag{1}$$

$$V_s = \omega_s r \tag{2}$$

Here sliding velocity is denoted by V_s whereas ω_s is used for the sliding angular velocity, and is assumed to decrease linearly with time as,

$$\omega_s(t) = \omega_0\left(1 - \frac{t}{t_s}\right) \quad ; \quad 0 \le t \le t_s \tag{3}$$

The initial sliding angular velocity is ω_0 (at $t=0$) at the start of the slippage. The heat produced at any time of slipping on the clutch surface is

$$Q_c(r,t) = f_c \mu p r \omega_0\left(1 - \frac{t}{t_s}\right) \quad ; \quad 0 \le t \le t_s \tag{4}$$

The heat dissipation between friction lining and pressure plate/flywheel is given by f_c which is the heat partition ratio and is calculated as

$$f_c = \frac{\sqrt{k_c\rho_c c_c}}{\sqrt{k_c\rho_c c_c} + \sqrt{k_f\rho_f c_f}} = \frac{\sqrt{k_c\rho_c c_c}}{\sqrt{k_c\rho_c c_c} + \sqrt{k_p\rho_p c_p}} \tag{5}$$

where the indexes c, f and p denotes friction lining, flywheel and pressure plate [11].

During start of clutch engagement process, there will be a difference of velocities between friction lining and flywheel i.e. between driven and the driving shaft and as a result, slippage occurs between the contacting surfaces and it will be responsible for the heat generation [15].

This heat produced is dissipated through the process of conduction between components of the clutch system and also through the process of convection to the surrounding environment [13] as shown in Fig.1.

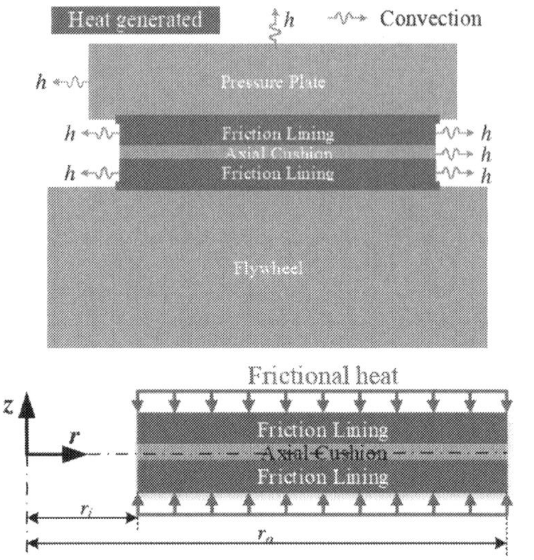

Figure 1. Heat Transfer.

Clutch parameters taken for the analysis are shown in Table 1 [16] and the three-dimensional model for the clutch system developed with the help of these parameters is shown in Fig. 2.

Table 1. Clutch Parameters

Parameters	Values
Inner radius of friction lining and axial cushion, r_i [m]	0.06298
Outer radius of friction lining and axial cushion, r_o [m]	0.08721
Inner radius of pressure plate [m]	0.05814
Outer radius of pressure plate [m]	0.09205
Inner radius of flywheel [m]	0.04845
Outer radius of flywheel [m]	0.0969
Thickness of friction lining, [m]	0.003
Thickness of axial cushion [m]	0.0015
Thickness of pressure plate [m]	0.00969
Thickness of flywheel [m]	0.01938
Slippage time, t_s [s]	0.4
Maximum pressure, p_o [Pa]	1×10^6
Initial relative angular velocity, ω_{ro} [rad s^{-1}]	200
Ambient temperature, T_a [K]	295.15

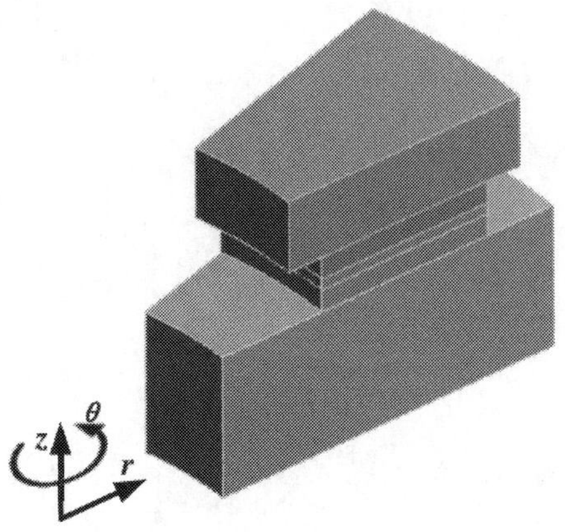

Figure 2. Coordinate system of analysis.

For the temperature field analysis of the clutch system, the starting point is the parabolic heat conduction equation in the cylindrical coordinate system, (r-radial coordinate (m), θ-circumferential coordinate (rad), z-axial coordinate (m)), which is centred in the axis of the disc and z refers to its thickness [1, 17].

$$\frac{\partial^2 T}{\partial r^2}+\frac{1}{r}\frac{\partial T}{\partial r}+\frac{1}{r^2}\frac{\partial^2 T}{\partial \theta^2}+\frac{\partial^2 T}{\partial z^2}=\frac{1}{\alpha}\frac{\partial T}{\partial t} \quad ; \qquad (6)$$
$$r_i \le r \le r_o, \quad 0 \le \theta \le 2\pi, \quad 0 \le z \le \delta, \quad t > 0$$

Here α represents the thermal diffusivity.

In the present work, the temperature in z-axis is symmetric due to same applied load value and also same heat transfer coefficient on friction clutch surfaces. Convection takes place from the exposed surfaces of the clutch model with the heat transfer coefficient taken as independent of the temperature. The material properties of the friction material, flywheel, pressure plate and axial cushion are assumed to be isotropic and independent of the temperature [18]. The temperature fields in the clutch are found with the help of the below mentioned three-dimensional boundary-value problems.

$$K_{cu}\left.\frac{\partial T}{\partial r}\right|_{r=r_0}=h\left[T(r_0,z,t)-T_a\right] \quad ; \qquad (7)$$
$$0 \le \theta \le 2\pi, \quad 0 \le z \le \frac{t_{cu}}{2}, \quad t \ge 0$$

Where h is the convection heat transfer and T_a is the ambient temperature.

$$K_c\left.\frac{\partial T}{\partial z}\right|_{z=\frac{t_{cu}}{2}+t_c}=Q_c(r,t) \quad ; \qquad (8)$$
$$r_i \le r \le r_0, \quad 0 \le \theta \le 2\pi, \quad 0 \le t \le t_s$$

$$K_c\left.\frac{\partial T}{\partial r}\right|_{r=r_i}=h\left[T(r_0,z,t)-T_a\right] \quad ; \qquad (9)$$
$$0 \le \theta \le 2\pi, \quad \frac{t_{cu}}{2} \le z \le \frac{t_{cu}}{2}+t_c, \quad t \ge 0$$

The initial temperature is given as

$$T(r,\theta,z,0)=T_i \quad ; \qquad (10)$$
$$r_i \le r \le r_o \quad 0 \le \theta \le 2\pi, \quad 0 \le z \le t_c+\frac{t_{cu}}{2}$$

$$\left.\frac{\partial T}{\partial r}\right|_{r=r_i}=0 \quad ; \qquad (11)$$
$$0 \le z \le \frac{t_{cu}}{2}, \quad t \ge 0$$

$$\left.\frac{\partial T}{\partial z}\right|_{z=0}=0 \quad ; \qquad (12)$$
$$r_i \le r \le r_o, \quad 0 \le \theta \le 2\pi, \quad t \ge 0$$

Rate of convection heat transfer $Q_{conv.}$ is

$$Q_{conv.}=h[T(r,z,t)-T_a]; \quad t \ge 0 \qquad (13)$$

During launch, the total clutch energy Q produced as shown in Fig. 1 can be calculated by

$$Q(t)=\int_0^{t_s}\tau\omega_r(t)\cdot dt \qquad (14)$$

Initially pressure is uniform for new plates. Wear at outer radius will be more and as a result, the shape will be changed and pressure will be redistributed .i.e. less at outer radius and more at inner radius. Torque on friction lining is

$$\tau=r\mu F=2\pi\mu p_o r_i\int_{r_i}^{r_o} r\cdot dr \qquad (15)$$

Heat generated between flywheel/pressure plate and friction lining due to relative angular velocity is

$$Q_c(r,t)=\tau\omega_r(t)=2\pi\mu p_o r_i\omega_r(t)\int_{r_i}^{r_o} r\cdot dr \qquad (16)$$

$$Q(r,t)=2\pi\mu p_o r_i\omega_{ro}\left(1-\frac{t}{t_s}\right)\int_{r_i}^{r_o} r\cdot dr \qquad (17)$$

$$Q(t)=\pi\mu p_o r_i\omega_{ro}\left(r_o^2-r_i^2\right)\left(1-\frac{t}{t_s}\right) \qquad (18)$$

$$Q(t)=\begin{cases}\pi\mu p_o r_i\omega_{ro}\left(r_o^2-r_i^2\right)\left(1-\frac{t}{t_s}\right) & ; \quad 0 \le t \le t_s \\ 0 & ; \quad t > t_s\end{cases} \qquad (19)$$

The heat generation with respect to time on the clutch plate lining and pressure plate/flywheel interaction is calculated as above.

3 Finite element formulation

The finite element (FE) modelling is carried out with ANSYS Meshing as shown in Fig. 3 using 3-D 20-node thermal solid element SOLID90, whereas, the contact between friction lining and pressure plate/flywheel is modelled with 3-D target surface TARGE170 and 3-D 8-

node surface-to-surface contact element CONTA174. The convection and heat generation surfaces are modelled with 3-D thermal surface effect element SURF152. The model is discretised into 127,650 nodes and 335,494 elements to assure seamless transfer of heat throughout the model. The mesh is kept uniform by mapped face meshing and relevance between the contacts of friction lining and pressure plate/flywheel is set to 100% to avoid any discontinuity in heat transfer. This relevance is not affecting the contacts between friction lining and axial cushion, since the heat is dissipated through flywheel and pressure plate, not through axial cushion.

Figure 3. Finite Element Model.

4 Results

The slippage time taken in the simulation is 0.4 second hence the maximum temperature i.e. 361.83 K is occurring at half of the slippage time. The same trend is observed if the friction lining of asbestos is replaced by carbon-carbon composite, S2-glass fibre and aluminium metal matrix composite as shown in Fig. 4.

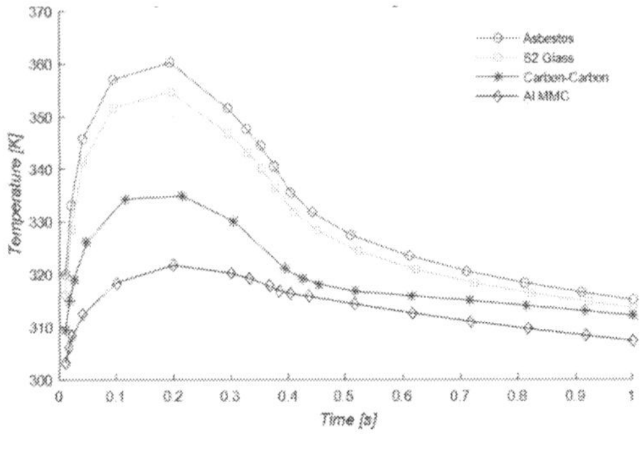

(a)

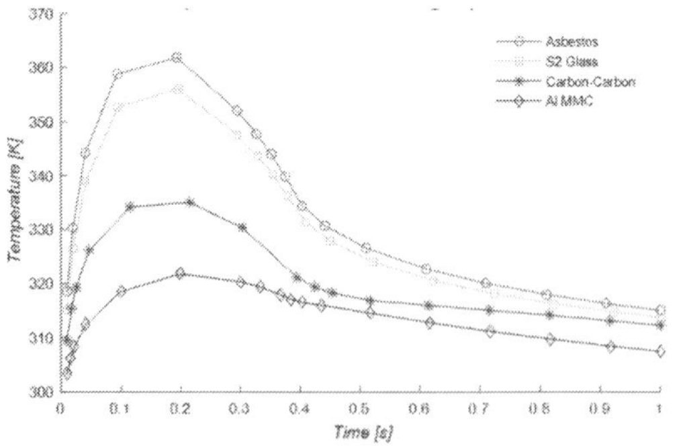

(b)

Figure 4. Temperature Profile of: (a) Friction Lining/Pressure Plate side and (b) Friction Lining/Flywheel side

A comparison of the maximum temperature of various materials as a result of the transient thermal analysis is given in the Table 2.

Table 2. Maximum temperature for the tested friction materials

Material	Temperature at Pressure Plate side (K)	Temperature at Flywheel side (K)
Asbestos	360.39	361.83
S2 Glass Fibre	354.73	356.04
Carbon-Carbon Composite	334. 64	335.12
Aluminium Metal Matrix Composite	321. 6	321.79

The temperature is found maximum in conventional friction lining material asbestos and minimum in aluminium metal matrix composite as shown in Fig. 5. The heat dissipation trend is established in the opposite manner, i.e. maximum for aluminium metal matrix composite and minimum for asbestos.

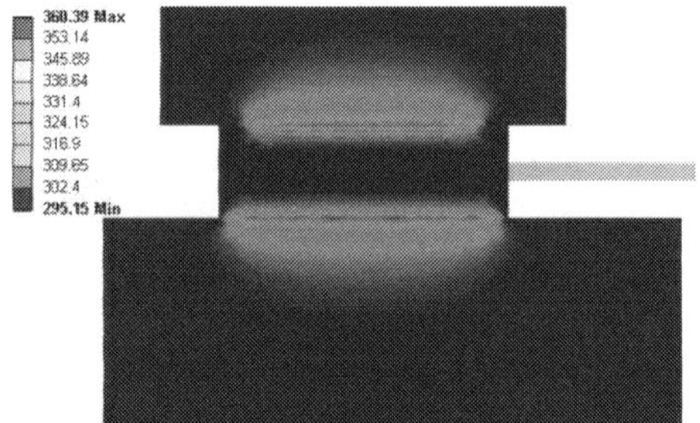

(a)

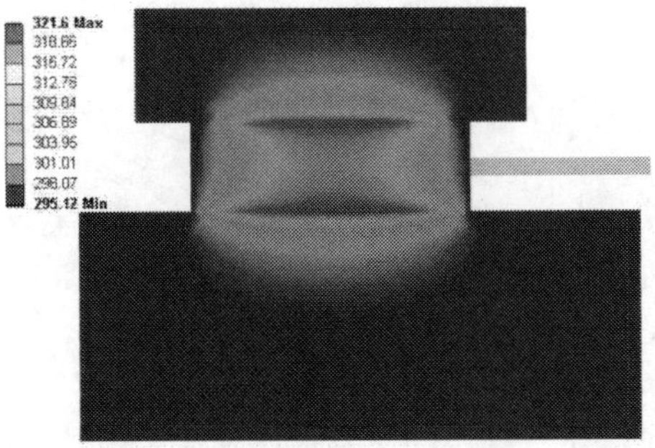

(b)

Figure 5. Temperature Distribution at clutch system with a friction lining of: (a) Asbestos and (b) Aluminium Metal Matrix Composite

5 Conclusions and future scope

In this paper, transient thermal analysis of three different materials for the replacement of Asbestos based clutch lining is performed. The analysis is done to assure the extended functional life and longer stability of the clutch disc and ultimately the clutch itself. Simulation results successfully indicate that Aluminium Metal Matrix composite is better material selection for friction lining in order to reduce rise in temperature and to maximize heat dissipation as compared to Asbestos based friction lining.

Future investigation will attempt to develop a system that will take initial design parameters and material properties of friction lining as an input, analyze clutch disc (structurally and thermally) and will give optimized design parameters and friction lining material for specific purpose. Also, other non-asbestos materials can also be compared for the particular application.

References

1. Oday I. Abdullah, Josef Schlattmann, "Thermal behavior of friction clutch disc based on uniform pressure and uniform wear assumptions", Friction, **4**, pp 228-237, (2016)
2. Y. H. Jang and S.-h. Ahn, "Frictionally-excited thermoelastic instability in functionally graded material," Wear, **262**, pp. 1102–1112, (2006)
3. O. I. Abdullah and J. Schlattmann, "The Effect of Disc Radius on Heat Flux and Temperature Distribution in Friction Clutches," Journal of Advanced Materials Research, **505**, pp. 154-164, (2012)
4. R. Biczó, G. Kalácska, "Composite Friction Materials for Brakes and Clutches", Scientific Bulletin, Tribology, **2016,** pp 21-26, (2016)
5. Piotr GRZEŚ, "Finite Element Analysis of disc temperature during braking process", acta mechanica et automatica, **3,** (2009)
6. M.A. Maleque, A. Atiqah, "New natural fibre reinforced aluminium composite for automotive brake pad", International journal of mechanical and materials engineering, **7**, pp 166-170, (2012)
7. Priyanka S. Bankar, S. N. Khan, "A Review Paper on Composite Brake Friction Lining for lining Applications on Band Brake", IJEERT, **2**, pp 116-122, (2014)
8. Yathish K.O, Arun L.R, "Performance Analysis And Material Optimization Of Disc Brake using Mmc", IJIRSET, **2,** pp 4101-4108, (2013)
9. M. G. Jacko, S. K. Rhee, Brake Linings and Clutch Facings, Kirk-Othmer Encyclopedia of Chemical Technology, (2000)
10. E. Mouffak, M. Bouchetara, "Transient thermal behavior of automotive dry clutch discs by using Ansys software ", Mechanika, **22 (6),** pp 562–570, (2016)
11. Oday I. Abdullah, J. Schlattmann, "Effect of Band Contact on the Temperature Distribution for Dry Friction Clutch", International Journal of Mechanical, Aerospace, Industrial, Mechatronic and Manufacturing Engineering, **6**, (2012)
12. T. C. Kennedy and S. Traiviratana, "Transient Effects on Heat Conduction in Sliding Bodies," IJCM, **47**, pp. 57-77, (2007)
13. Faramarz Talati, Salman Jalalifar, "Analysis of heat conduction in a disk brake system",Heat Mass Transfer, Vol. **45**, pp 1047–1059, (2009)
14. O.I. Abdullah, J. Schlattmann, "Stresses and Deformations Analysis of a Dry Friction Clutch System", Tribology in Industry, **35**, pp155 - 162, (2013)
15. O.I. Abdullah, J. Schlattmann, "Thermo-mechanical Analysis of the Dry Clutches under different Boundary Conditions", Tribology in Industry, **36**, pp 172-180, (2014)
16. O.I. Abdullah, J. Schlattmann, "Contact Analysis of a Dry Friction Clutch System", ISRN, **2013**, ID 495918, (2013)
17. Nowacki W., Nowacki W. "Thermoelasticity", Oxford (UK), Pergamon Press, (1962)
18. Oday I. Abdullah, Josef Schlattmann, "An Investigation Into the Thermal Behavior of the Grooved Dry Friction Clutch", Journal of Tribology, **136**, (2014)

Experimental Investigation of Performance and Emissions of a Stratified Charge CNG Direct Injection Engine with Turbocharger

Hilmi Amiruddin[1,2], Wan Mohd Faizal Wan Mahmood[1], Shahrir Abdullah[1], Mohd Radzi Mansor[1] and Mohd Fadzli Abdollah[2]

[1]*Faculty of Engineering and Built Environment, Universiti Kebangsaan Malaysia, 43600 UKM Bangi, Selangor, Malaysia*
[2]*Faculty of Mechanical Engineering, Universiti Teknikal Malaysia Melaka, Hang Tuah Jaya, 76100 Durian Tunggal, Melaka, Malaysia*

Abstract. This paper presents the results from 1.6 litre, 4 cylinders stratified charge compressed natural gas (CNG) direct injection engine with boosting device. A turbocharger with compressor trim of 40 was used to increase engine output. The engine was tested at wide open throttle (WOT) and speed ranging from 1000 to 5000 rpm. Engine performance and emissions data were recorded under steady state condition. Results show turbocharged CNG engine produced an average of 26% increment in brake power and 24% additional maximum brake torque as compared with natural aspirated (NA) CNG engine. Turbocharged CNG engine improved brake specific fuel consumption (BSFC) and yielded higher fuel conversion efficiency (FCE). Relatively turbocharged CNG engine showed lower emission of hydrocarbon (HC) and carbon monoxide (CO) throughout tested engine speed. Conversely, the carbon dioxide (CO_2) and nitrogen oxide (NO_x) emission produced were slightly higher compared with NA CNG engine.

1 Introduction

The exhaust gases of vehicles are one of the main contributors to the world's greenhouse gases problem. Quest for alternative fuel for automotive engine has gained more attentions mainly due to stringent emission limit and instability of world liquid fossil fuel price. The use of natural gas as an alternative fuel in spark ignition (SI) direct injection (DI) engine have been studied expansively and utilized in vehicles [1-2].

Higher octane number of natural gas gives several advantages as a fuel for SI engine including suitable for high compression ratio engine operation, higher thermal efficiency and less knocking problem.

Stratified charge engines have rich mixture around spark plug and leaner mixture for the rest of the combustion chamber.

Currently, boosting device technology has been widely used in automotive engine due to increasing demand of engine power output, better fuel economy and reduces emission level [3]. The use of turbocharging technology in gasoline SI engine is limited due to knocking and premature combustion [4]. Nowadays, interest in application of boosting device operating on natural gas engine produced promising results. This is due to higher knocking resistance properties of natural gas and suitable for high compression ratio SI engine [5].

The main objective of this study is to experimentally investigate the stratified charged CNG DI engine with turbocharger output and emission performance.

2 Experimental setup and procedures

Tests were conducted on a 1600 cm^3, 4 cylinders, spark ignition and direct injection CNG engine. The arrangement of the engine experimental setup is shown in Figure 1 and the specifications of the engine are listed in Table 1. An eddy current dynamometer (Apicom Model FR250) and KRONOS 4 software were used to program the engine test and recorded all the performance results. The engine was tested at steady state conditions with wide open throttle (WOT) at constant speed ranging from 1000 rpm to 4500 rpm with 500 rpm increment. A T25/T28 turbocharger with compressor trim of 40 was installed to increase the CNG engine output performance. The inducer and exducer size of the turbocharger are 37.8 and 59.7 mm respectively. In this test, the CNG engine installed with stratified piston. Figure 2 shows the stratified piston with bowl shape geometry on its crown.

The CNG was stored at 3000 psi pressure in cascade tank and its pressure was reduced by pressure regulator before injected into the combustion chamber. The properties of the gasoline and CNG fuels are shown in Table 2 and composition of CNG used in this test is listed in Table 3. The mass flow rate of CNG was measured using gas flow meter. A pressure sensor (Kistler type 6125B) was installed in cylinder number 1 to record the inside cylinder pressure. Exhaust emissions were measured using EMS Model 5002 portable exhaust gas analyser.

DOI:10.1051/matecconf/201712407004

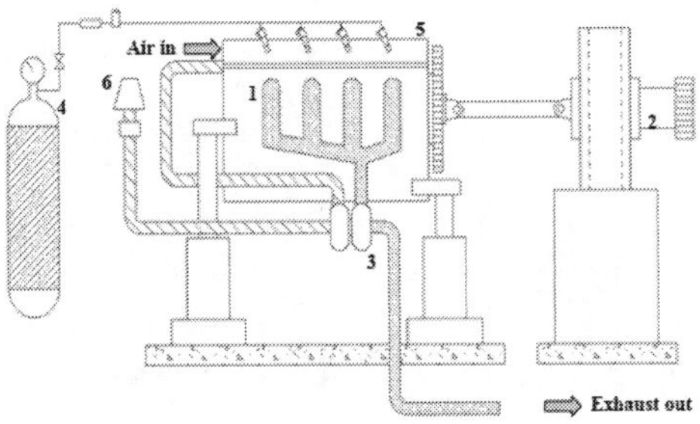

1. CNGDI engine	4. CNG Tank
2. Eddy current dynamometer	5. Fuel injection rail
3. Turbocharger	6. Air filter

Figure 1. Schematic of experiment setup.

Table 1. Specification of CNGDI engine.

Engine Parameters	Value
Number of cylinder	4
Displacement volume (cm^3)	1596
Bore (mm)	76
Stroke (mm)	88
Compression ratio	14:1
Connecting rod length (mm)	131

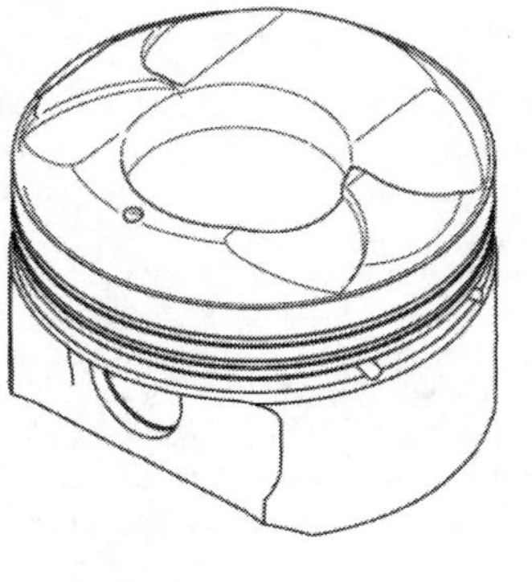

Figure 2. Stratified piston crown.

Table 2. Properties of gasoline and CNG fuels [6].

Properties	Gasoline	CNG
Specific gravity (kg/m^3)	0.72-0.78	0.72
Heat of vaporization (KJ/kg)	305	509
Laminar burning velocity (m/s)	0.43	0.50
Higher heating value (MJ/kg)	47.3	55.5
Lower heating value (MJ/kg)	44.0	50.0
Stoichiometric AFR	14.6	17.23
Research octane number	92-98	120

Table 3. Typical composition (vol. %) of CNG [1].

Component	Symbol	Volumetric (%)
Methane	CH_4	94.42
Ethane	C_2H_6	2.29
Propane	C_3H_8	0.03
Butane	C_4H_{10}	0.25
Carbon Dioxide	CO_2	0.57
Nitrogen	N_2	0.44
Others	H_2O+	2.00

3 Results and discussions

The performance of CNG engine with respect to brake power, brake torque, brake mean effective pressure, BSFC, FCE and exhaust emissions were examined for NA and force induction under several steady state conditions. Figure 3 and 4 present the results of brake power and brake torque at WOT for both NA and turbocharger CNG engine. On average, the increment of 26% in brake power can be seen throughout the speed range. The maximum brake power obtained by NA and turbocharged were 50 kW and 59 kW respectively both at 4500 pm. In the case of brake torque, maximum torque obtained by NA and turbocharged were 119 Nm and 148 Nm respectively both at 3000 rpm. There is 24% of additional output at maximum torque. This is mainly due to the excess of oxygen available to convert the fuel energy to useful work.

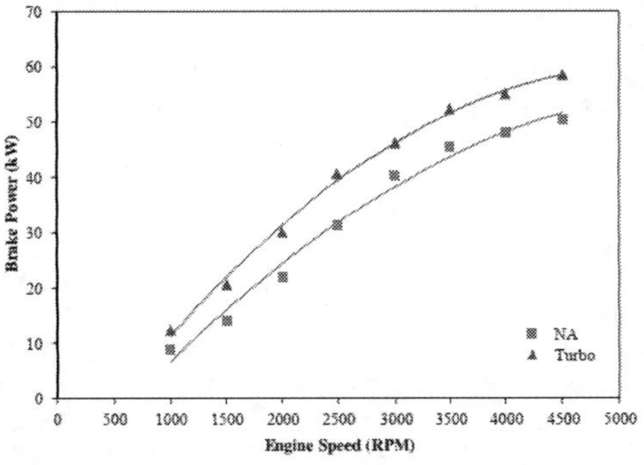

Figure 3. Brake power versus engine speeds.

DOI: 10.1051/matecconf/201712407004

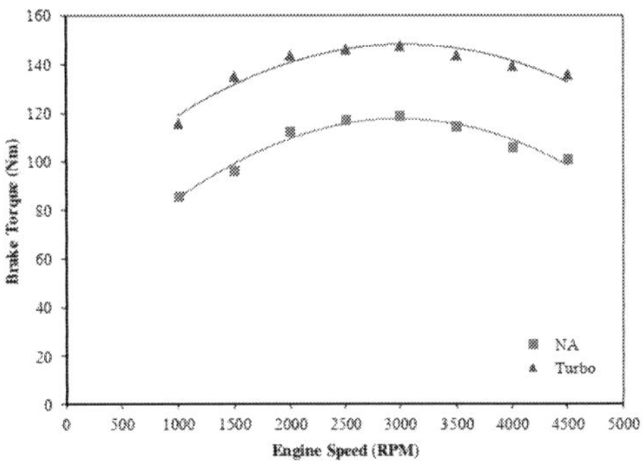

Figure 4. Brake torque versus engine speeds.

As shown in Figure 5, the BMEP of turbocharged is 8 – 18% higher than NA CNG engine. This is because higher intake pressure created by turbocharger increased peak pressure during compression stroke. Volumetric efficiency measured the maximum amount or air into the engine and higher volumetric efficiency increase the power output. There are 7 – 53% volumetric efficiency rise with turbocharged operation compared to NA as shown in Figure 6.

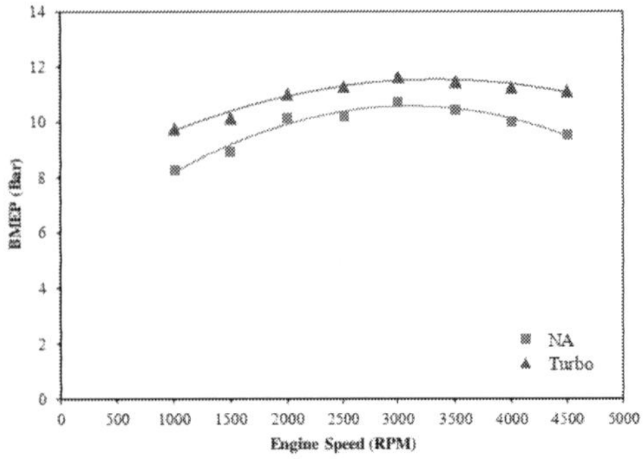

Figure 5. BMEP versus engine speeds.

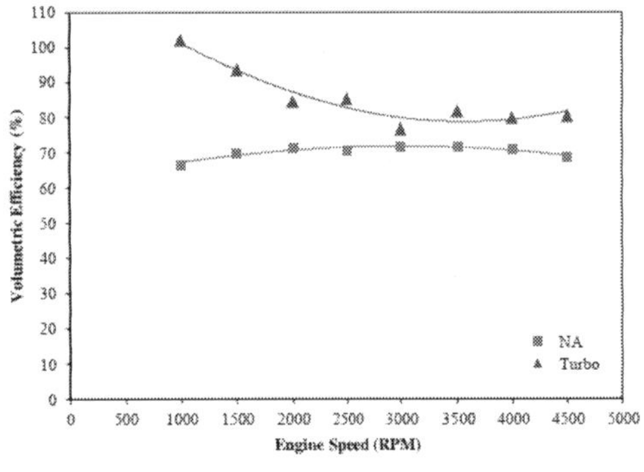

Figure 6. Volumetric efficiency versus engine speeds.

The BSFC curve in Figure 7 shows turbocharged results produced remarkably 3 - 20% lower fuel consumption compared to NA. The minimum BSFC of NA and turbocharged are 157 g/kW·h and 130 g/kW·h respectively both at 3000 rpm. Because of lower BSFC, turbocharged CNG engine achieves 3 – 22% higher FCE compare to NA as shown in Figure 8.

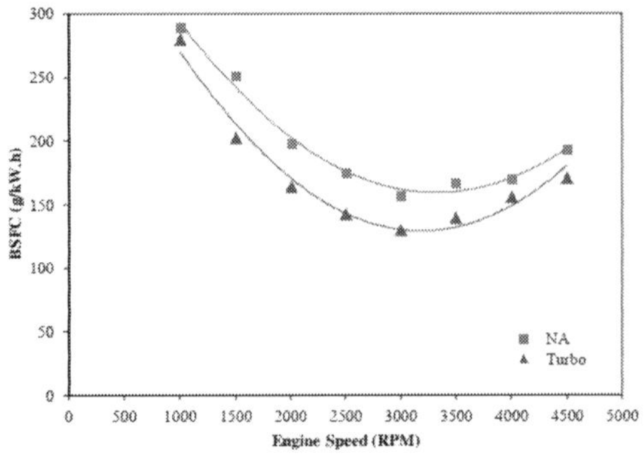

Figure 7. BSFC versus engine speeds.

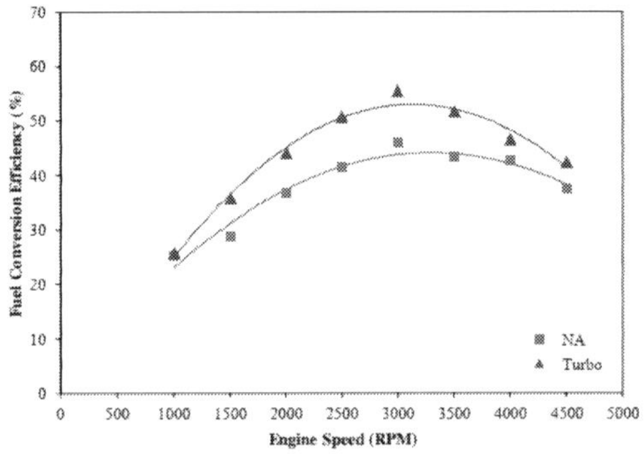

Figure 8. Fuel conversion efficiency versus engine speeds.

Figure 9 shows the Lambda value both NA and turbocharged. Generally, throughout the tested speed, both engine condition operated under lean mixture. The exhaust emission of HC for both NA and turbocharged are presented in Figure 10. It shows that CNG engine with turbocharger lower the unburned hydrocarbon emission compare to NA. The emission of HC is reduced by 3 – 15% with turbocharger due to a more complete combustion of CNG. Furthermore, it was found that CNG engine with turbocharger produced less CO, 14 – 25% in reduction compare to NA as shown in Figure 11.

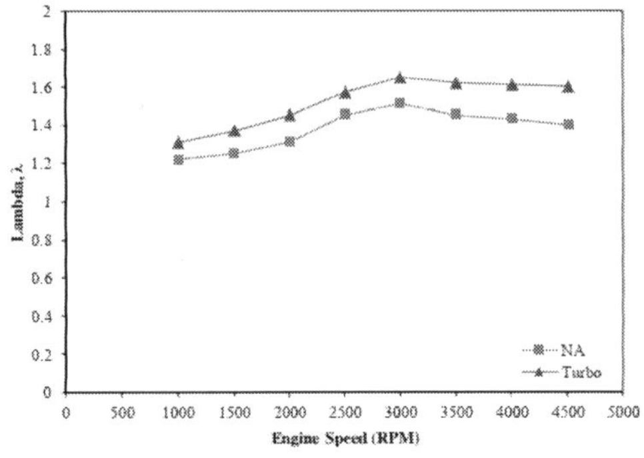

Figure 9. Lambda versus engine speeds.

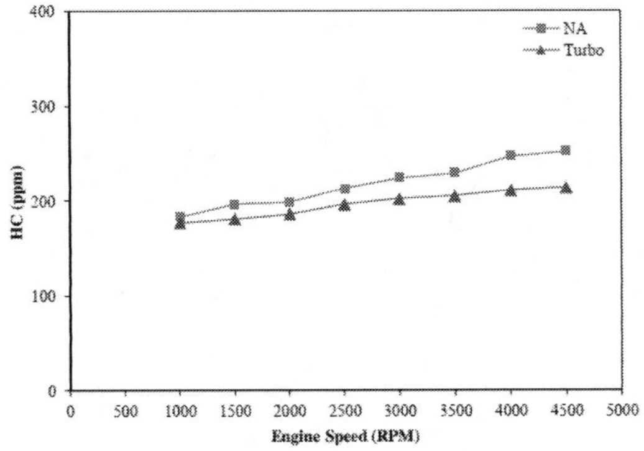

Figure 10. Hydrocarbon emission versus engine speeds.

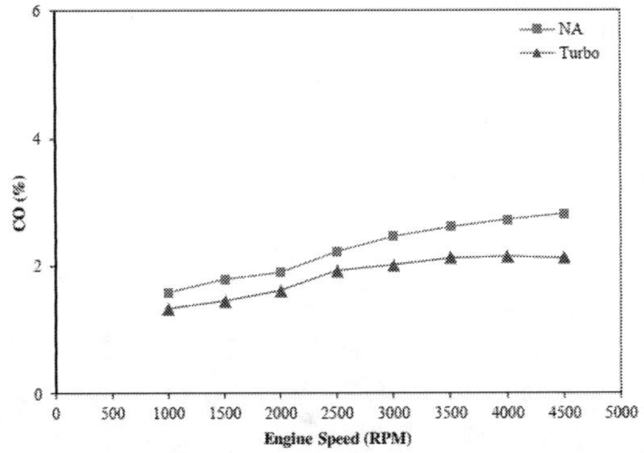

Figure 11. Carbon monoxide emission versus engine speeds.

Figure 12 shows the formation of CO_2 emission on turbocharged is 5 – 17% higher compare to NA. This trend also true on NO_x emission where turbocharged produced 7 – 29% higher compare to NA as shown in Figure 13. This higher formation of NO_x in turbocharged engine is primarily cause by dissociation on N_2 due to high combustion temperature, pressure and leaner mixture.

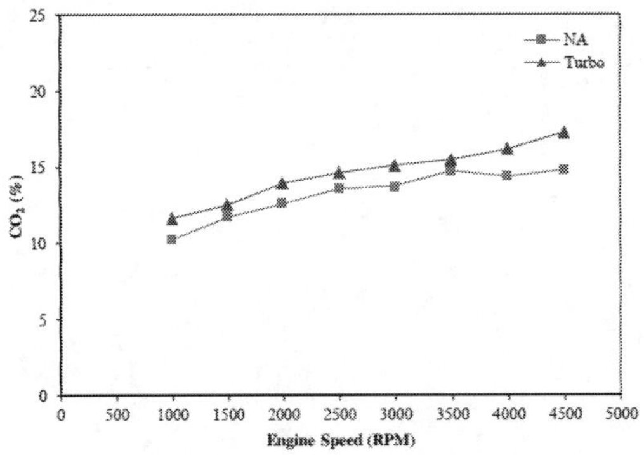

Figure 12. Carbon dioxide emission versus engine speeds.

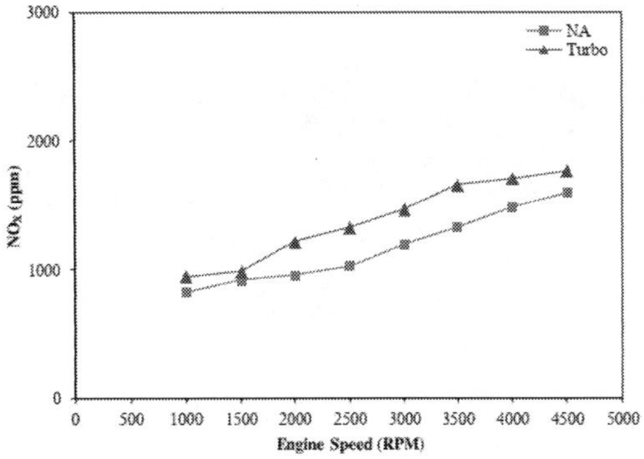

Figure 13. Nitrogen oxides emission versus engine speeds.

4 Conclusion

This study demonstrates that stratified charge CNG DI with turbocharger has a potential for higher engine output and improved fuel economy. The conclusion of these test are list as given below.

1. On average, CNG engine with turbocharger results 26% higher in brake power and 31% increment in brake torque.
2. Turbocharged CNG engine produced minimum BSFC of 130 g/kW·h and 3 - 22% higher FCE compare to NA CNG engine.
3. Turbocharger improve volumetric efficiency of CNG engine by 7 – 53%.
4. Emission of pollutant gaseous from NA CNG engine significantly reduced by application turbocharger with 3 – 15% reduction of unburned HC and 14 – 25% of CO.

Acknowledgement

The authors would like to acknowledge to Ministry of Science, Technology and Innovation (MOSTI), Ministry of Higher Education Malaysia and Universiti Teknikal Malaysia Melaka (UTeM) for sponsoring the Ph.D study and research work under project 03-01-02 SF0995.

References

[1] M. A. Kalam and H. H. Masjuki, "An experimental investigation of high performance natural gas engine with direct injection" Energy **36**, 3563-3571, (2011)

[2] S. Aljamali, S. Abdullah, W. M. F. Wan Mahmood and Y. Ali, "Effect of fuel injection timing on performance and emissions of stratified combustion CNGDI engine" Applied Thermal Engineering **109**, 619-629, (2016)

[3] J. E. Kirwan, M. Shost, G. Rothn and J. Zizelman, "3-Cylinder turbocharged gasoline direct injection: A high value solution for Low CO_2 and NO_x emissions" SAE Technical Paper 2010-01-0590, (2010)

[4] H. Zhao, *Advanced direct injection combustion engine technologies and development Volume 1: Gasoline and gas engine* (CRC Press, UK, 2010)
[5] G. T. Chala, A. R. A. Aziz and F. Y. Hagos, "*Combined effect of boost pressure and injection timing on the performance and combustion of CNG in a DI spark ignition engine*" Int. J. of Automotive Technology **18**, No. 1, 85-96, (2016)
[6] J. B. Heywood, *Internal combustion engine fundamental* (McGraw-Hill, New York, 1988)

On Optimizing Steering Performance of Multi-axle Vehicle Based on Driving Force Control

Zhicheng Wu[1], Cong Chen

[1]*Beijing Institute of Technology, School of Mechanical Engineering, Beijing, China*

Abstract. The steering performance of multi-axle vehicle with independent driving system is affected by the distribution of the wheel driving force. A nonlinear vehicle dynamics model including magic formula tire model for describing 11 DoF four-axle vehicle with dual-front-axle-steering (DFAS) system was presented. The influence of different driving force distribution scheme on the steering performance of the vehicle was analyzed. A control strategy for improving the steady response and transient response of the vehicle steering is proposed. The results show: For the steady response, setting different drive force for internal and external wheels according to the actual steering characteristics of the vehicle can effectively improve its steering characteristics; For the transient response, adopting the zero sideslip angle control strategy and using the PID control algorithm to control the driving force of the outside wheel of tear-two-axle, under angle step input, the vehicle sideslip angle can quickly stabilize to 0 and yaw rate also significantly decreases.

1 Introduction

With the development of industrial, the application of multi-axle vehicle more widely. Multi-axle vehicles have large weight and moment of inertia, the handling stability of these vehicles is poor when steering[1]. So there is a strong practical significance to improve the steering performance of multi-axle vehicles.

At present, the worldwide researchers mainly focus on the control and distribution of steering angle of multi-axle vehicles. Hakima A found a relationship of steering angle between front and rear wheel to ensure the stability of the vehicle in different speed and different steering angle by simulation analysis, and designed a fuzzy controller to improve the steering stability of the vehicle[2]. Hamzah N used 2-DOF vehicle model as a reference to design a sliding mode controller, adjusted the four wheel steering angles in real time to stabilize the vehicle, and proved the stability of the controller[3]. Williams D E derived the degree of understeer and equivalent wheelbase expression of multi-axis vehicle, where the axle can contain any number of steering axes, resulting in more general conclusions.[4] Tabbache B taked a vehicle which driven by two electric wheels as research object, and proposed a method based on adaptive control, which used direct torque control and adaptive flux-speed observer to control the each wheel of the vehicle, to improve the reliability of the vehicle.[5]

This paper researches to optimize the steering performance of four-axle vehicle with DFAS system by reasonably distributing the driving force of different wheels. In this paper, an 11-DOF nonlinear vehicle dynamics model with wheel motion is established using the magic formula tire model. In this paper, the control strategy was researched using Matlab/SIMULINK. Based on the zero centroid side angle principle, the PID control algorithm is used to distribute the wheel driving force reasonably to optimize the steering performance of independent driving multi-axle vehicle.

2 Nonlinear Vehicle Dynamics Model

This paper establishes a nonlinear vehicle model with 11-DOF, which is longitudinal, lateral, yaw motion and eight-wheel rotary motion. The model can simulate the linear drive and steering drive and other complex conditions[6], the vehicle dynamics model is shown in Figure.1.

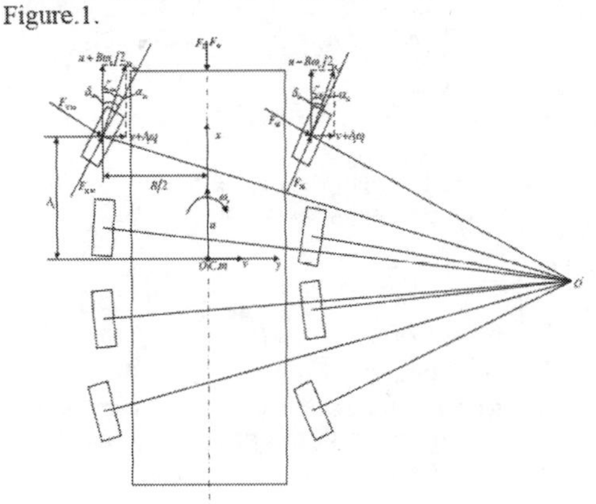

Figure 1. Vehicle dynamics model

Where B is the tread of vehicle, F_w is air resistance, F_f is rolling resistance, F_{Yii} are the lateral reaction force of the ground to the inner wheels, F_{Yio} are the lateral reaction force of the ground to the outer wheels, F_{Xii} are the tire longitudinal force of the inner wheels, F_{Xio} are the tire longitudinal force of the outer wheels, δ_{ii}, δ_{io} are steering angles of inside and outside wheels respectively, specifies that the wheel steering angle is positive when the car turns right, α_{ii}, α_{io} are sideslip angle of inside and outside wheels respectively.

To facilitate the research, the following assumptions are made in this paper in the modeling process of vehicle dynamics:

(1) Ignoring the impact of the car steering system, regard the wheel steering angle as input;

(2) Ignoring the role of the car suspension system, regard that the car only move parallel to the ground, which is the displacement of the car along the Z axis is 0;

(3) The roll angle of the car around the Y axis and the pitch angle around the X axis are both 0. The characteristics of each tire are same.

2.1 Vehicle dynamics model

The motion equation of the rigid body established by Newtonian mechanics can be described as[7]:

Equation of longitudinal motion:

$$\sum F_{Xii}\cos\delta_{ii}+\sum F_{Xio}\cos\delta_{io}+\frac{|i-2.5|}{i-2.5}\left(\sum F_{Yii}\sin\delta_{ii}+\sum F_{Yio}\sin\delta_{io}\right)-F_w-F_f=m(\dot{u}-v\omega_r) \quad (1)$$

Equation of lateral motion:

$$\sum F_{Yii}\cos\delta_{ii}+\sum F_{Yio}\cos\delta_{io}-\frac{|i-2.5|}{i-2.5}\left(\sum F_{Xii}\sin\delta_{ii}+\sum F_{Xio}\sin\delta_{io}\right)=m(\dot{v}+u\omega_r) \quad (2)$$

Equation of yaw motion:

$$\begin{aligned}&\sum A_i F_{Yii}\cos\delta_{ii}+\sum A_i F_{Yio}\cos\delta_{io}-\frac{|i-2.5|}{i-2.5}\\&\left(\sum A_i F_{Xii}\sin\delta_{ii}+\sum A_i F_{Xio}\sin\delta_{io}\right)+\\&\frac{B}{2}\left(\sum F_{Xio}\cos\delta_{io}-\sum F_{Xii}\cos\delta_{ii}+\right.\\&\left.\frac{|i-2.5|}{i-2.5}\left(\sum F_{Yii}\sin\delta_{ii}-\sum F_{Yio}\sin\delta_{io}\right)\right)=I_Z\dot{\omega}_r\end{aligned} \quad (3)$$

The eight wheels were subjected to force analysis to obtain the wheel motion equation:

Inner wheels:

$$J_{\omega ii}\dot{\omega}_{ii}=T_{dii}-T_{bii}-T_{fii}-F_{Xii}R_{ii} \quad (4)$$

Outer wheels:

$$J_{\omega io}\dot{\omega}_{io}=T_{dio}-T_{bio}-T_{fio}-F_{Xio}R_{io} \quad (5)$$

Where $J_{\omega ii}$, $J_{\omega io}$ are the moment of inertia of the inner and outer wheels respectively, ω_{ii}, ω_{io} are the rotation angular velocity of the inner and outer wheels respectively, T_{dii}, T_{dio} are the driving torque of the inner and outer wheels respectively, T_{bii}, T_{bio} are the braking torque of the inner and outer wheels respectively, T_{fii}, T_{fio} are the rolling resistance torque of the inner and outer wheels respectively, R_{ii}, R_{io} are rotation radius the of the inner and outer wheels respectively.

2.2 Nonlinear dynamics model

In this paper, the Pacejka89 magic formula is used to establish the nonlinear vehicle dynamics model[8]. The magic formula tire model is described as:

$$y(x)=D\sin\left\{C\arctan\left[BX-E\left(BX-\arctan BX\right)\right]\right\} \quad (6)$$

$$Y(X)=y(x)+S_v \quad (7)$$

$$X=x+S_h \quad (8)$$

Where Y is longitudinal force, lateral force or aligning torque, x is longitudinal slip rate or sideslip angle, the rest are parameters of magic formula tire model.

The tire vertical load, sideslip angle and longitudinal slip rate are important input parameters in the magic formula tire model. In this model, the effect of aligning torque is not considered.

According to the coordinate system, the sideslip angle of each wheel is:

$$\alpha_{ii}=\frac{\beta+\frac{A_i}{u}\omega_r}{1-\frac{B}{2u}\omega_r}-\delta_{ii} \quad (9)$$

$$\alpha_{io}=\frac{\beta+\frac{A_i}{u}\omega_r}{1+\frac{B}{2u}\omega_r}-\delta_{io} \quad (10)$$

The moving speed of each wheel center in the direction of its deflection is:

$$u_{ii}=\left(u-\frac{B}{2}\omega_r\right)\cos\delta_{ii}-\frac{|i-2.5|}{i-2.5}(v+A_i\omega_r)\sin\delta_{ii} \quad (11)$$

$$u_{io}=\left(u+\frac{B}{2}\omega_r\right)\cos\delta_{io}-\frac{|i-2.5|}{i-2.5}(v+A_i\omega_r)\sin\delta_{io} \quad (12)$$

Let s_{ii}, s_{io} be the slip rate of the inner and outer wheels respectively, the value is:

$$s_{ii}=\frac{R_{ii}\omega_{ii}}{u_{ii}}-1=\frac{R_{ii}\omega_{ii}}{\left(u-\frac{B}{2}\omega_r\right)\cos\delta_{ii}-\frac{|i-2.5|}{i-2.5}(v+A_i\omega_r)\sin\delta_{ii}}-1 \quad (13)$$

$$s_{io}=\frac{R_{io}\omega_{io}}{u_{io}}-1=\frac{R_{io}\omega_{io}}{\left(u+\frac{B}{2}\omega_r\right)\cos\delta_{io}-\frac{|i-2.5|}{i-2.5}(v+A_i\omega_r)\sin\delta_{io}}-1 \quad (14)$$

The tire vertical load consists of static load and dynamic load, in which the dynamic load, ignoring the effects of lateral winds and road slopes, is caused by the longitudinal acceleration and lateral acceleration of vehicle. This paper uses the solution method in [9].

In this paper, the study of the four-axle vehicle with DFAS system, the steering angle distribution follows the Ackerman principle.

DOI:10.1051/matecconf/201712407005

3 Control Scheme Design

In order to improve the steering performance of the four-axle vehicle with independent driving system, in this paper, the steady response and transient response of steering are analyzed and studied. Through establishing the vehicle dynamics simulation model in Matlab/SIMULINK, study the influence of different driving force distribution on the steering performance of the vehicle, and design a control scheme suitable for four-axle vehicle with independent driving system. The parameters of a four-axle vehicle are shown in Table 1.

Table 1. Main Parameters of Vehicle

Symbol	Value	Unit
m	2500	kg
I_z	4295	kg.m^2
B	1.4	m
$L_1/L_2/L_3$	0.75/1.2/0.75	m
$A_1/A_2/A_3/A_4$	1.35/0.6/-0.6/-1.35	m
h_g	0.5	m
R	0.36	m
J_w	2.5	kg.m^2
C_d	0.8	
A	1.4	m^2
f	0.015	

3.1 The effect on steady response of different wheel driving force

Establish the steady steering response model of four-axle vehicle in Matlab / SIMULINK and simulate. The steering angle of the outer wheel of the first axle is 0.008rad, and steering angles of other wheels are set according to Ackerman principle. Set different driving force of inside and outside wheels, front-two-axle and rear-two-axle wheels, different-axle wheels respectively. The simulation results of steady steering characteristics of the vehicle are shown in Figure 2. Where C and C 'are positive constants.

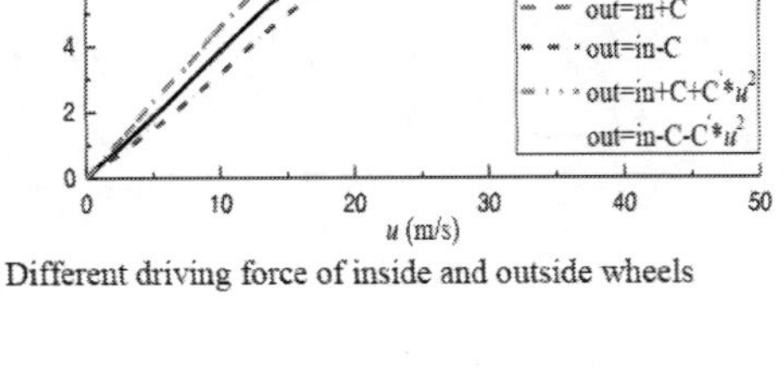

(a) Different driving force of inside and outside wheels

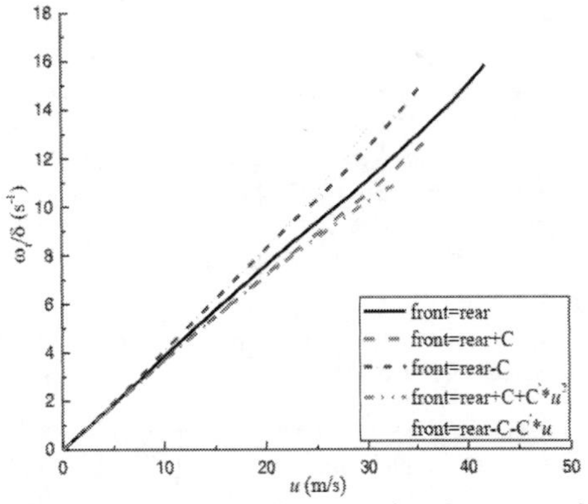

(b) Different driving force of front-two-axle and rear-two-axle wheels

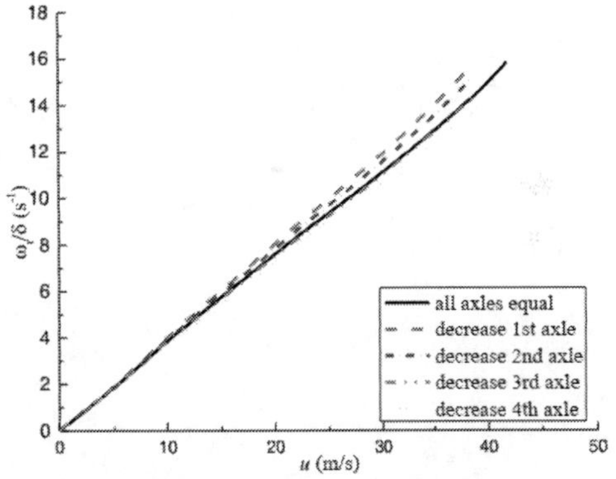

(c) Different driving force of different-axle wheels

Figure. 2 The effect on steady response of different wheel driving force

As seen as in the Figure 2, the steering characteristic of the four-axle vehicle is neutral steering. The Figure 2.(a) (b) indicates that when the driving force of the outer wheel is larger than the driving force of the inner wheel, the yaw rate gain of the vehicle increases, otherwise the yaw rate gain decreases; when the driving force of the front-two-axle wheels is greater than the driving force of the rear-two-axle wheels, the yaw rate gain of the vehicle decreases; otherwise the yaw rate gain increases. With the increase of speed, increase the difference between the driving forces of two parts wheels, the steering characteristics of the vehicle will change, become oversteering or understeering, and it can be seen that the difference between inside and outside wheels can change the steering characteristics of the vehicle more effectively. The Figure 2.(c) indicates that reducing the wheel driving force of first or second axle will cause the yaw rate gain increase, reducing the wheel driving force of forth axle will cause the yaw rate gain decrease, changing the wheel driving force of the third axle has little effect on the yaw rate gain; wherein changing the wheel driving force of the first axle has the greatest effect on the yaw rate gain.

So when the steering characteristic of vehicles change from neutral steering to oversteering or severe understeering due to the changes of vehicle parameters in the actual use, the steering characteristics can be improved by changing the driving force of the different wheels. As the difference of driving force between inside and outside wheels can greatly change the steering characteristics, it is possible to improve the steady response by setting different driving force between inside and outside wheel according to the actual steering characteristics of the vehicle when steering.

3.2 The effect on transient response of different wheel driving force

Establish the transient steering response model of four-axle vehicle in Matlab / SIMULINK and simulate under the steering wheel angle step input. The step value of the steering angle of the outer wheel of the first axle is 0.02rad, and steering angles of other wheels are set according to Ackerman principle. The initial speed of vehicle is 72km/h. Will be 8 wheels of the four-axle vehicle are divided into 4 groups, namely, outside of front-two-axle(OF), inside of front-two-axle(IF), outside of rear-two-axle(OR), inside of rear-two-axle(IR). Reduce 28% of the driving force, simulate and analyze the change of sideslip angle and yaw rate, the results are shown in Figure 3.

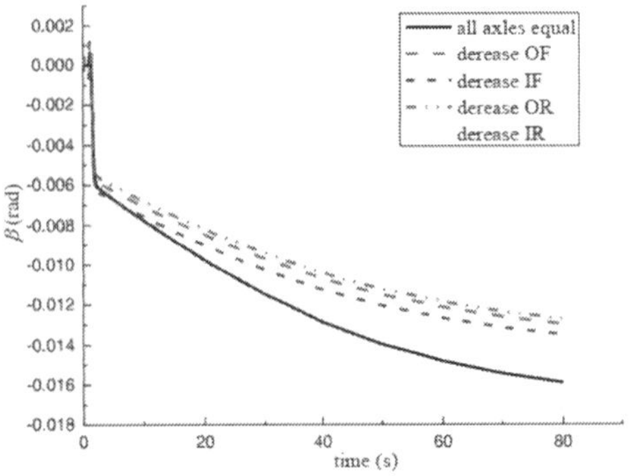

(a)Changes of sideslip angle under different driving force

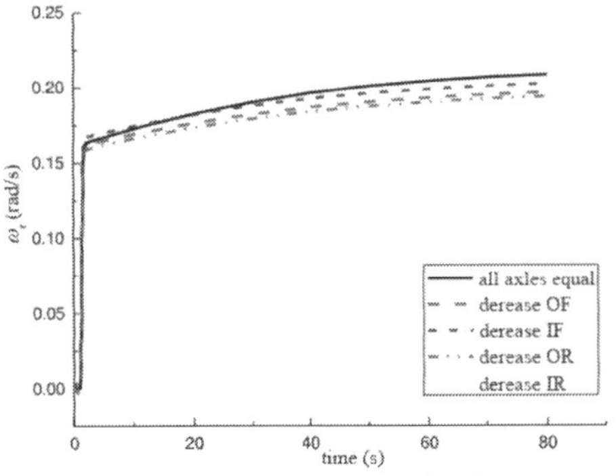

(b) Changes of yaw rate under different driving force

Figure 3. The effect on transient response of different wheel driving force

As seen as in the Figure 3, reduce the driving force of each group can make the value of sideslip angle and yaw rate decrease. When the steering angle is positive, reducing the driving force of outside wheels of rear-two-axle is most effective for improving the transient response, and the value decreases by 20%. If the steering angle is negative, it should reduce the driving force of outside wheels of rear-two-axle. (Corresponding outside wheel is opposite when the steering angle is positive or negative)

Establish the control strategy of transient response of the four-axle vehicle according to the above analysis. Select wheels to control driving force according to the positive and negative of steering angle. In this paper, we use the zero sideslip angle control strategy[10]. When steering, the corresponding wheels are selected by the driving force distribution module, and the driving force is controlled by the PID control algorithm to prevent the vehicle from entering the unsteady state. The structure of the control system is shown in Figure 4.

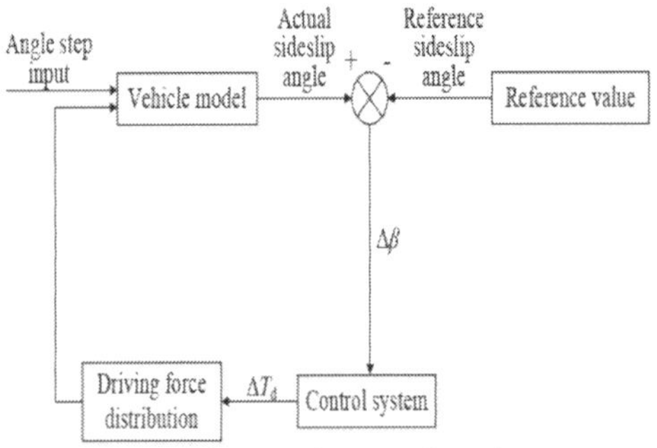

Figure 4. Structure of the control system of transient response

4 Simulation And Analysis Of Zero Sideslip Angle Control

Establish the transient response control model based on the control strategy described in 3.2 to optimize sideslip angle by feedback control.

4.1 Simulation under angle step input

The step value of the steering angle of the outer wheel of the first axle is 0.02rad, and steering angles of other wheels are set according to Ackerman principle. The initial speed of vehicle is 72km/h. The simulation results are shown in Figure 5.

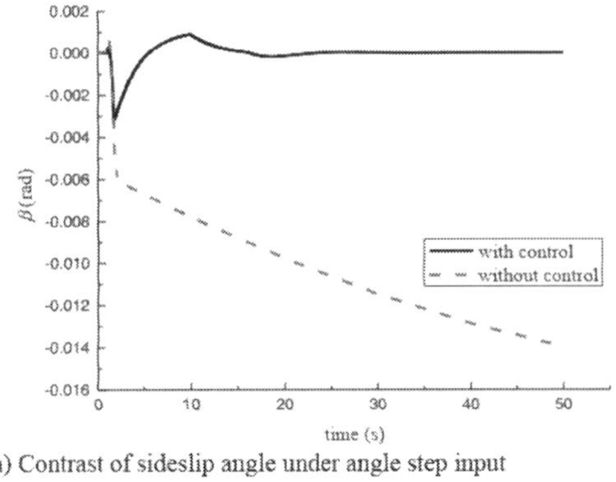

(a) Contrast of sideslip angle under angle step input

DOI:10.1051/matecconf/201712407005

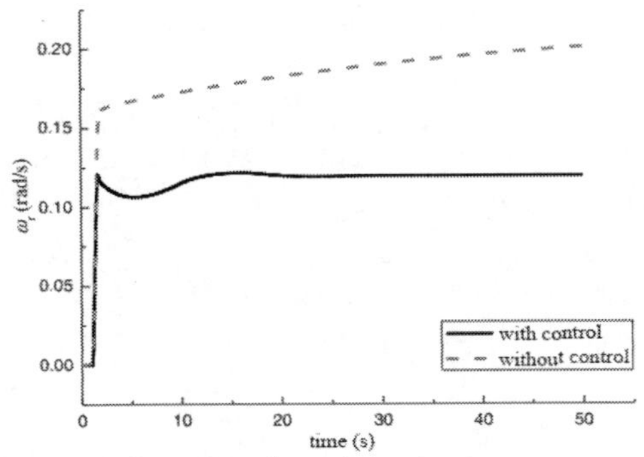

(b) Contrast of yaw rate under angle step input

Figure 5. Contrast of transient response under angle step input

As seen as in the Figure 5. under the control of PID based on the control strategy of zero sideslip angle, the sideslip angle can be quickly stabilized at around 0 by controlling the driving force of the outside wheels of rear-two-axle, but in the case of no control, the value of sideslip angle is larger and there is a trend of instability. And the yaw rate of the vehicle also greatly decreases relative to the situation of no control.

4.2 Simulation under sine input

The amplitude of sine input of the steering angle of the outer wheel of the first axle is 0.02rad, the frequency is $1/\pi$ Hz, and steering angles of other wheels are set according to Ackerman principle. The initial speed of vehicle is 72km/h. The simulation results are shown in Figure 6.

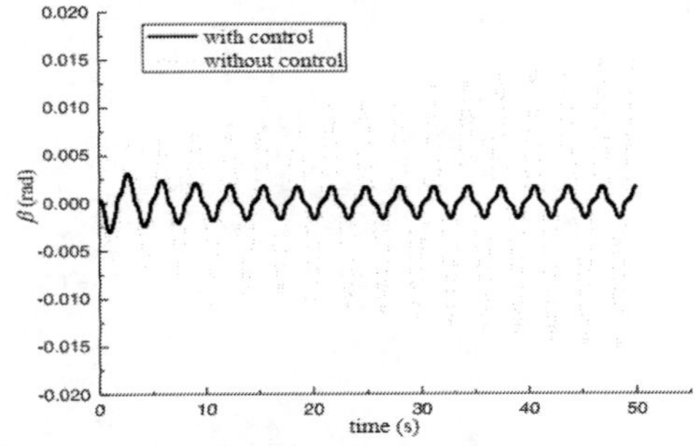

(a) Contrast of sideslip angle under sine input

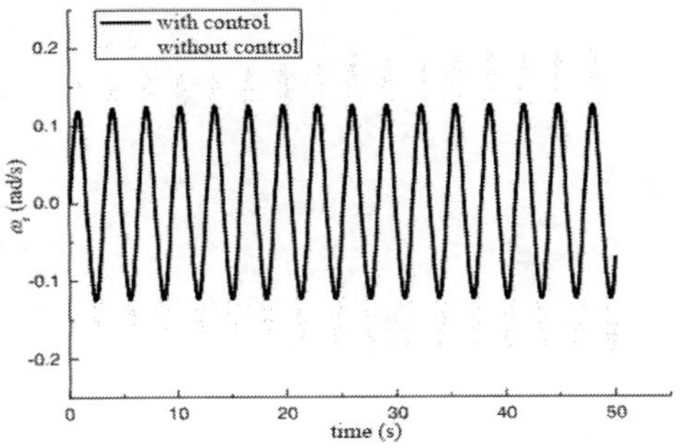

(b) Contrast of yaw rate under sine input

Figure 6. Contrast of transient response under sine input

As seen as in the Figure 6. relativing to the situation of no control, the PID control based on the zero sideslip angle can significantly reduce the sideslip angle of the vehicle under sine input, and the yaw rate also greatly decreases.

According to the simulation and analysis, for four-axle vehicle with independent driving system, controlling the driving force of outside wheels of rear-two-axle by using the PID control based on control strategy of zero sideslip angle can improve the transient response characteristics well to make vehicle maintain a steady state during steering.

5 Conclusion

(1) In this paper, the influence of different driving force distribution scheme on the steering performance of the vehicle is analyzed by establishing the nonlinear multi-axis vehicle dynamics model. According to the above method, the control strategy of improving the steering response and transient response of the vehicle can be obtained.

(2) For steady steering characteristics, set different driving force between inside and outside wheels or front-two-axle and rear-two-axle wheels of the vehicle can both change its characteristics, wherein, the difference of driving force between inside and outside wheels is more effective. For transient steering characteristics, reducing the driving force of outside wheels of rear-two-axle is most effective for reducing the sideslip angle, and the value decreases by 20% when the driving force is reduced by 28%.

(3) Using PID control algorithm based on control strategy of zero sideslip angle to control the driving force of outside wheels of rear-two-axle when steering, under angle step input the sideslip angle can be quickly stabilized at around 0 and the yaw rate also greatly decreases, and under sine input the sideslip angle significantly reduces and the yaw rate also greatly decreases, which prove that the control method is valid.

References

1. Jin Shu, Sizhong Chen, Lin Yang. Design of multi-wheel independent steering control system for multi-axle vehicle [J]. Traffic and Computer, (06): 57-59, (2003).
2. A. Hakima, S. Ameli, Designing a fuzzy logic controller to adjust the angle of tires in four wheel steering vehicles[C]//Control Automation Robotics & Vision (ICARCV), 2010 11th International Conference on. IEEE, 2010: 2208-2213.
3. N. Hamzah, M.K. Aripin, Y.M. Sam, *et al.* Yaw stability improvement for four-wheel active steering vehicle using sliding mode control[C]//Signal Processing and its Applications (CSPA), 2012 IEEE 8th International Colloquium on. IEEE, 127-132, (2012).
4. D.E. Williams, Generalised multi-axle vehicle handling [J]. Vehicle System Dynamics, 50(1): 149-166, (2012).

5. B. Tabbache, A. Kheloui, M.E.H. Benbouzid, An adaptive electric differential for electric vehicles motion stabilization[J]. IEEE Transactions on Vehicular Technology, 60(1): 104-110, (2011).
6. Lei Qiu. Simulation and Research on Vehicle Electronic Stability Program Control System [D]: Chongqing University, (2013).
7. Zhisheng Yu. The theory of automobile [M]. Beijing: Mechanical Industry Press, (2009).
8. Feng Yang, Parameters Identification for common steady-state tyre models based on excel solver [D]: Jilin University, (2011).
9. Xiao Jiang. Research on AWS System Dynamics of Multi-axle Research on AWS System Dynamics of Multi-axle Vehicle [D]: Beijing Institute of Technology,(2017).
10. Yunchao Wang, Research on steering performance of multi-axle steering vehicle[D]: Jilin University, (2007).

DOI:10.1051/matecconf/201712407006

Design of a suspension system and determining suspension parameters of a medium downforce small Formula type car

Sadjyot Biswal[1], Aravind Prasanth[1], Dr. R Udayakumar[2], Shobhit Deva[1], Aman Gupta

[1]*Automotive, Mechanical Department, BITS Pilani Dubai Campus, DIAC, 345055*
[2] *Mechanical Department, BITS Pilani Dubai Campus, DIAC, 345055*

Abstract. The focus of the paper is on designing a suspension system for a medium downforce small Formula type race car. The paper not only focusses on step by step design for a double wishbone type suspension but will also show the use and role of kinematics software in determining the optimized suspension of the car. The paper will also focus on the use of tire data in determining suspension parameters and the design of the double wishbone suspension. Various parameters, their design importance and the process to optimize them according to suspension design goals will be covered. The easiest and best ways to change the suspension parameters to get the best results will also be covered.

1 Introduction

The suspension system is one of the most important components of a race car because it not only connects the sprung mass to the chassis but is important for maintaining maximum tire contact patch at all times. Along with the tires and steering geometry, it plays a very crucial role in determining the vehicle dynamics of a race car. The objective of this paper is to present a step-by-step design for a good suspension system of a low downforce small formula type race car such as those used in Formula student. The paper will also talk about the parameters, which will affect suspension design; their optimization and the use of kinematic software to further improve the designs.

2 Requirements

Formula student competition like F1 uses wishbone suspension system due to its high structural strength and its ease of track tuning. The main requirements for the suspension system are:

- It should have a minimum travel of 25.4 mm in both jounce and travel.
- The system must be responsive and also stable enough to give best performance in cornering and straight line.
- The system should be such that it should help maintain maximum tire contact patch for better friction.
- It must be structurally strong but also lightweight to maintain low unsprung mass.
- It such be such that major suspension design parameters can be tuned for on track testing.

3 Design

Certain parameters need to be decided while designing a suspension geometry. The first things to be decided are the trackwidth and wheelbase. The wheelbase will affect primarily two things:

- The longitudinal weight transfers
- Steering angle required

Trackwidth can be decided also based on a few parameters. Trackwidth affects lateral weight transfer too according to the following formula:

$$F_{zr} = 0.5\rho mg + \frac{\theta_{roll} K_f}{t_f}$$

$$F_{zl} = 0.5\rho mg - \frac{\theta_{roll} K_f}{t_f}$$

According to the formula increasing track width will reduce lateral weight transfer but having a bigger track width might cause trouble in tight cornering. The difference in front and rear track width values is due to the fact to give the driver more freedom in not hitting the cones in the race events during Formula student with the rear of the car.

Other important factors to be considered are inertia and weight. Larger the wheelbase and trackwidth higher is the inertia and also weight of the car. Both higher inertia and weight are undesirable characteristics and the aim is to keep both low

 DOI:10.1051/matecconf/201712407006

Combining all the above factors, the values have been chosen accordingly:
Wb= 1665mm
Front Track= 1204mm
Rear Track= 1150mm

3.1 Outboard suspension pickup points

Once the wheelbase and trackwidth has been decided, the outboard suspension pickup points need to be decided. The main criteria for deciding the outboard pickup points is the packaging. The rim diameter, it's thickness, the hub and upright, brake rotor and calliper all affect the suspension design.

Except this the distance between the two outer ball joints on the upright also affect roll centre migration. The graph showing the trend is shown below.

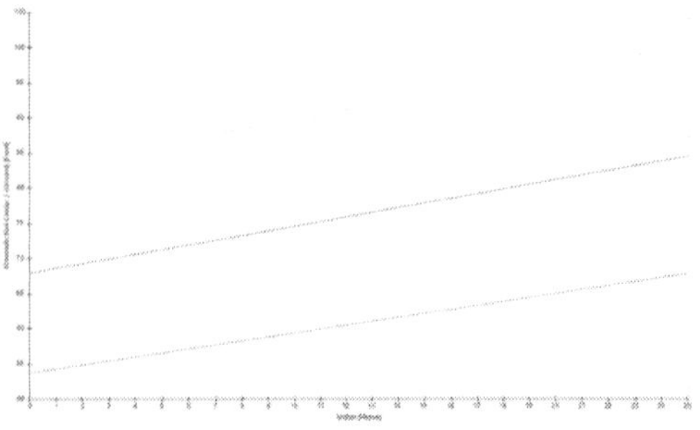

Figure 1. Roll Centre height migration in heave

From the graph, it can be inferred that as the distance between the ball joints increases the roll centre migration increases proportionally

The first step is to create the 2-d geometry both in front view and side view. The two outermost points are located keeping in mind all the above-mentioned packaging constraints. Hence the points are decided in front view. The points are kept as far away from each other to provide higher structural strength. The ratio between lower to upper wishbone length also plays an important part in determining the roll centre migration. However, if the kingpin angle and scrub radius get fixed the lengths can be altered only by changing the inboard pickup points.

The kingpin inclination and scrub radius also play an integral part in designing the front view. The focus is to maintain a scrub radius as low as possible to avoid scrubbing of tires in tight corners. The kingpin causes camber variation in steering so a kingpin value of 8 degrees can be chosen after analysis in Kinematics software.

3.2 Side view geometry and caster

In side view the points are not in a straight line to give the front wheels some caster. As a rule of thumb the steer moment should not be more than 10 Nm and should be near 5 Nm caster and mechanical trail affect the steer torque along with the pneumatic trail.

The steer torque is calculated as follows:

$$\text{Steer Torque} = \frac{\Sigma(M_z, M_y, F_y, F_x, F_z)}{\text{Pitman Arm Trail}} \times \text{Steering Rack trail}$$

⇒ Steering Rack trail = 0.025m
⇒ Pitman Arm trail = 0.2m

$$\Rightarrow \text{Steer Torque out} = \frac{35\ \text{N.m} + 1700 \times 0.025}{0.2} \times 0.025 \quad \text{(only } F_y \text{ \& } M_z \text{ is Considered)}$$

$$= 9.68\ \text{N.m}$$

$$\Rightarrow \text{Steer Torque in} = \frac{0.25\ \text{N.m} + 350 \times 0.025}{0.2} \times 0.025$$

$$= 1.125\ \text{N.m}$$

$$\text{Total} = \text{Steer Torque}_{out} + \text{Steer Torque}_{in}$$

$$= 9.68 + 1.125$$

$$= 10.805\ \text{Nm}$$

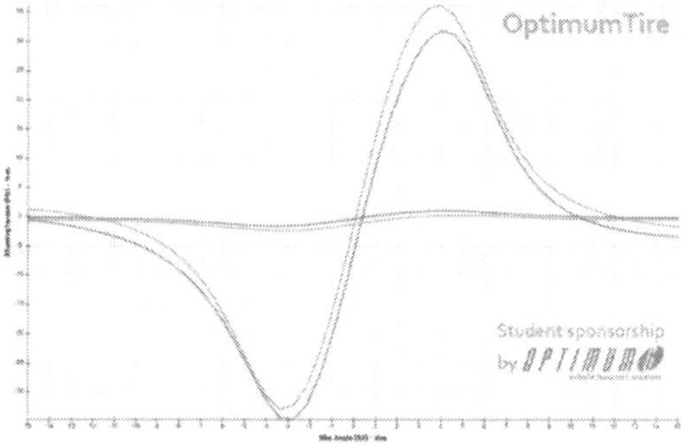

Figure 2. Self-aligning torque vs slip angle

Self aligning torque is a property of the tyre and can be found out using tyre data by creating a graph between self aligning torque and slip angle from the graph it can be seen that peak sat is 35Nm.

Therefore, a caster of degrees is kept and this will help determine the side view outboard points which will be further optimized in a kinematic software to give the final set of points. The side view geometry is as follows:

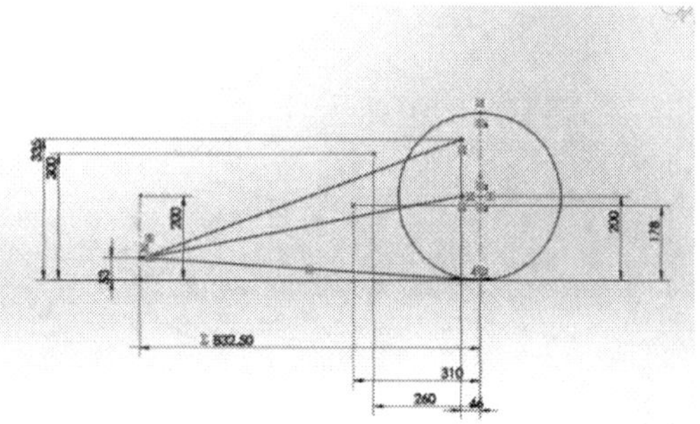

Figure 3. Side View Suspension Geometry

DOI:10.1051/matecconf/201712407006

3.3 Inboard suspension points

These are the factors that need to be considered to decide the inboard pick up points:

- Anti-Dive
- Anti-Squat
- Roll Centre Migration
- Chassis Design
- Roll centre static height

The goal is to keep chassis as narrow as possible to make it light weight. Once the chassis width is decided according to rules, it gives a ballpark to place the inner points. The inner points can be placed till they are further optimized by kinematics software.

3.4 Anti-Dive

Anti-features are to be added to the suspension to stop pitching of the vehicle. More anti-percentage simply means that more load is going to the A-arms rather than through the springs. This prevents vehicle pitching. Setting Anti features is a matter of experience and is generally kept between 0-20 percent. As the inboard pickup points affect the Side View Geometry the percent anti-dive also gives a fair idea about the placement of the inboard pickup points.

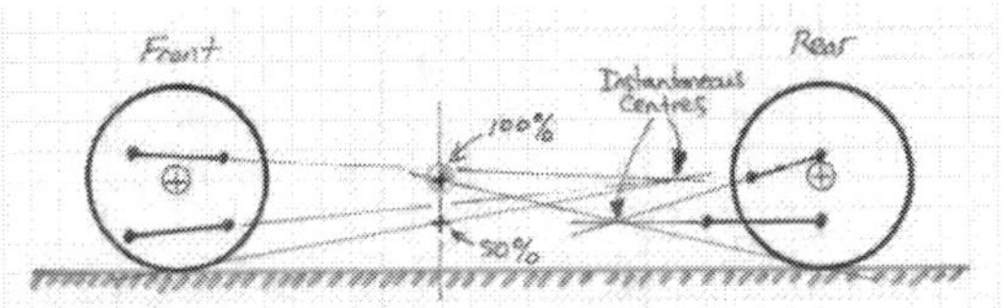

Figure 4. Anti-dive Geometry

3.5 Anti-Squat

The same logic goes for Anti Squat as Anti dive and some amount of it can be given to avoid vehicle pitching during acceleration.

The anti- squat percent affects the inboard points in the same way as anti-squat because it helps determine the side view geometry.

The percent anti-squat can be calculated by making use of the following formula:

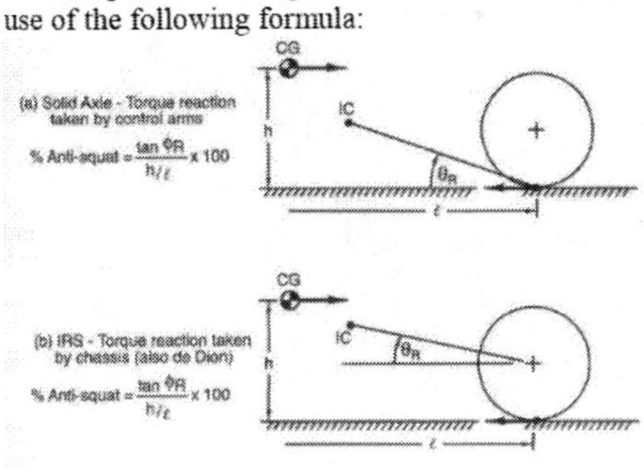

Figure 5. Anti-squat Geometry

3.6 Roll centre migration

Roll Centre migration also affects the inboard pickup points. The roll centre migration follows a trend where in the migration reduces drastically if the lower wishbone length is increased compared to the upper wishbone length. Thus, the inboard wishbone points can be kept under the chassis if space is available to make the lower wishbone longer while keeping the upper wishbone pick up points at their desired position relative to chassis design.

The graph showing this particular trend is shown later on.

3.7 Roll centre static height

As the front view IC is determined by the inboard pickup points, by changing the position of the inboard pick up points along the Z axis, the static roll centre height can be fixed. The static roll centre height at the front is generally kept greater than the rear to allow the rear of the car to catch up during cornering with the front. However too high a roll centre will cause jacking which is unwanted.

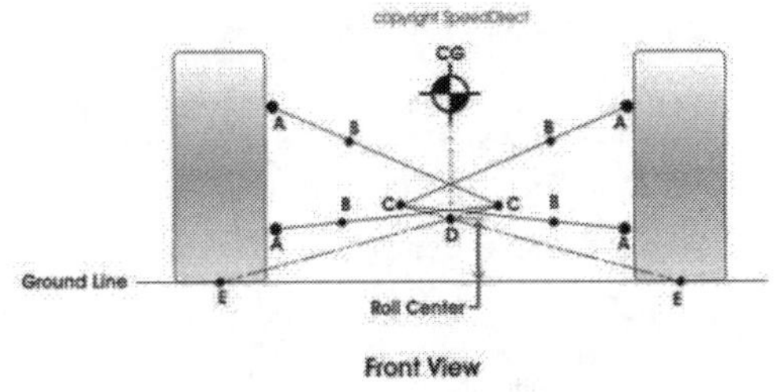

Figure 6. Static Roll centre

3.8 Camber variation in steering

The camber variation in steering is dependent on 3 variables:

- Steering Angle
- Kingpin angle
- Caster

The camber variation is directly proportional to the caster and increases with an increase in caster. Keeping Caster constant an increase in kPI will decrease camber variation with a positive steering angle and vice versa. The graph shown in Appendix shows the general trend:

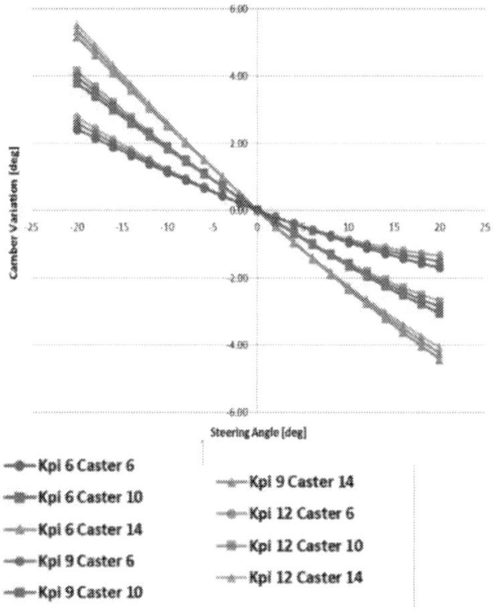

Figure 7. Camber variation vs Steering angle

3.9 Camber change in bump and roll

The camber change in bump and roll is an important factor to maintain maximum tire contact patch.

Camber change in Bump= $CB = \tan^{-1}\left(\frac{heave}{lat} * \frac{1}{VASL}\right)$

Camber change in Roll= $CR = \theta * 1 - \left(\frac{track}{2 * lat\ VASL}\right)$

Where:

- CB is camber change in Bump
- Heave is movement in z direction
- Lat VSAL is lateral Virtual Swing Arm Length
- CR is Camber change in roll
- θ is roll angle
- Track is trackwidth

Camber change in bump can also be analysed in kinematics software with varying VSAL for maximum calculated roll.

The total roll is calculated as 2.7°

3.10 Camber compliance

Every system has some kind of compliance and the compliance must be accounted for when going for any design. The camber compliance plays a crucial part in deciding static camber

The camber compliance for a 13-inch wheel is generally 1.7.

The following formula can be used to calculate actual camber taking compliance into account.

Camber=Static Camber + camber change-compliances

3.11 Roll centre and steering linkages

As opposed to popular belief the roll centre height static is affected by steering linkages. The effect is demonstrated by the graph.

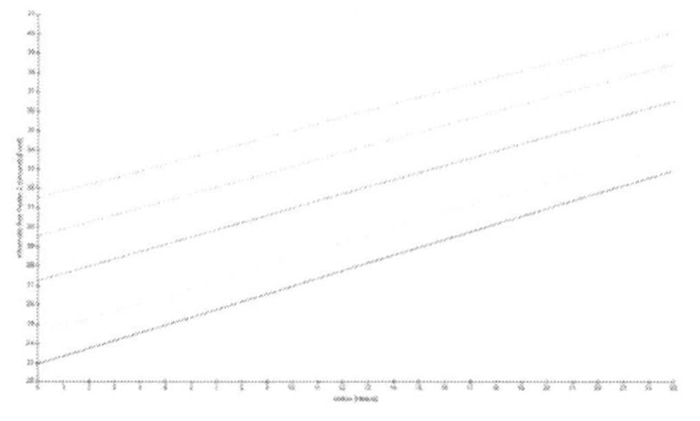

Figure 8. Roll centre height vs change in steering linkage height

as shown by the trends the roll centre height increases with decreases in z steering co ordinates and vice versa. The same trends as above can also be seen for changing the Y axis coordinates of the steering linkages.

3.12 Caster variation

Another important parameter to be discussed is caster variation. This depends on the side view swing arm length. With a long SVSAL with pitching there is a high caster change whereas the caster change is small in roll. With a short SVSAL the opposite applies.

3.13 Optimization in Kinematics software

The process to optimize in a kinematics software is tedious and involves a lot of trial and error. However certain trends were established regarding certain parameters in the software. These trends are seen later. Parameters like roll centre migration, static roll centre height, bump steer, camber variation in bump and roll can be analysed and the results of simulation can help determine final pick up points subject to vehicle testing. The bump steer is discussed in a parallel paper.

3.14 Outboard pick up points

As discussed above the outboard pickup points affect the roll centre migration and static height as they determine the A –arm lengths along with the upright length.

3.15 Inboard pick up points

The inboard pick up points also affect the same parameters and the results of analysis in Kinematics Software can be seen below.

DOI:10.1051/matecconf/201712407006

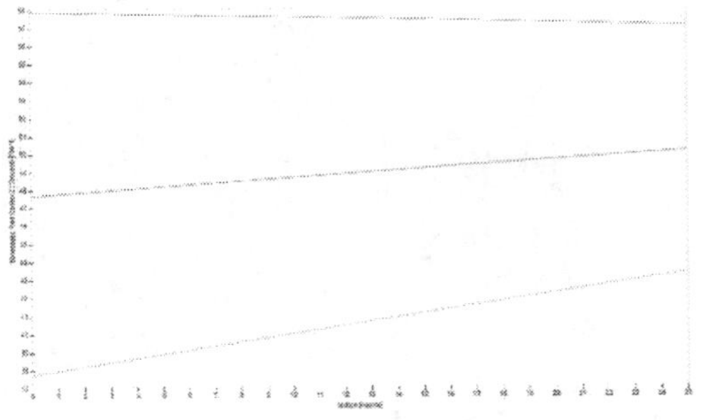

Figure 9. Roll centre migration vs inboard suspension pickup points

the graph shows that keeping the lower A arm constant and decreasing the upper A arm length the roll center migration goes down drastically

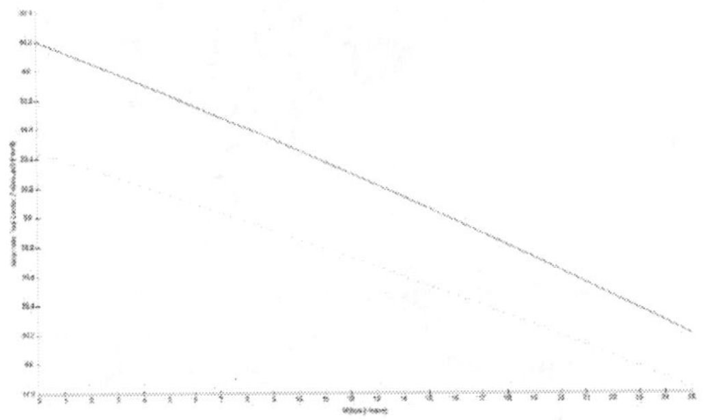

Figure 10. Roll centre migration vs inboard suspension pickup points

Similarly keeping the Upper A arm constant and increasing the lower A arm length also decreases RC migration.

3.16 Camber Variation in Roll

One of the main purpose of the suspension is to maintain maximum contact patch during all types of vehicle motion. To best use the kinematics software firstly the vehicle roll is calculated using and the roll motion is simulated. The aim is to maintain 0.4 degrees of camber at full roll to maximize contact patch. Thus, a static camber is chosen accordingly with trial and error to get 0.4 degrees of camber at full roll. Now the same static camber will be analysed in heave to see if it is close to the 0.4 degrees of camber needed. So, a static camber of -1.48 can be set.

The graph below shows the results of camber variation in roll.

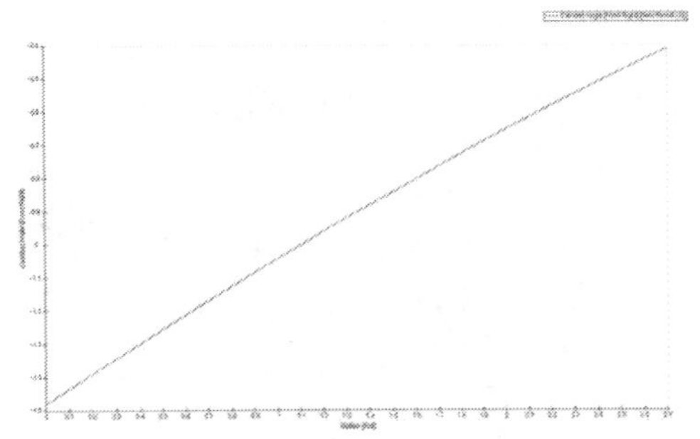

Figure 11. Camber variation in roll

as seen from the graph the camber changenges to -.4 deg at max roll

3.17 Camber Variation in Bump

Similarly, the camber variation in bump can be seen by simulating vehicle pitch motion. The pitch motion can be specified as according to the pitch expected. We have taken the vehicle to pitch 25 mm up and down. For the following motion, the camber changes in bump from -1.48 to -0.35. As it is close to the camber required at full pitch for maximum contact, the above particular value of static camber can be finalized.

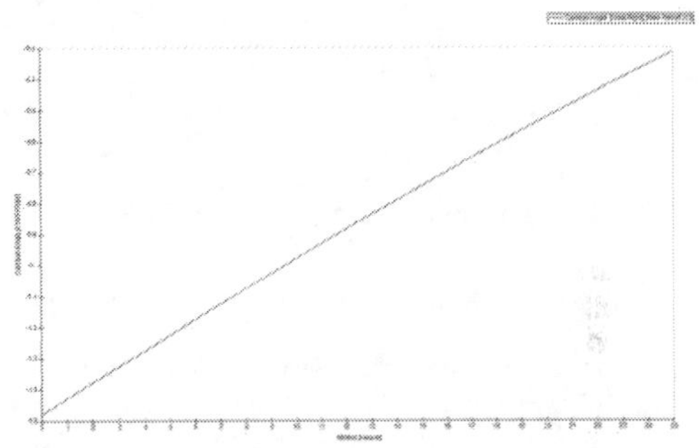

Figure 12. Camber variation in roll

as seen from the graph the camber changenges to -.35 deg at max bump

3.18 Forces on the system

The load on the suspension system can be calculated by taking moments of forces available at tire contact patch. The maximum normal load is calculated by taking into account weight transfers. The lateral maximum is known at optimum slip angle and longitudinal peak is known at optimum slip ratio by looking into the tire data.
The calculations are as follows:

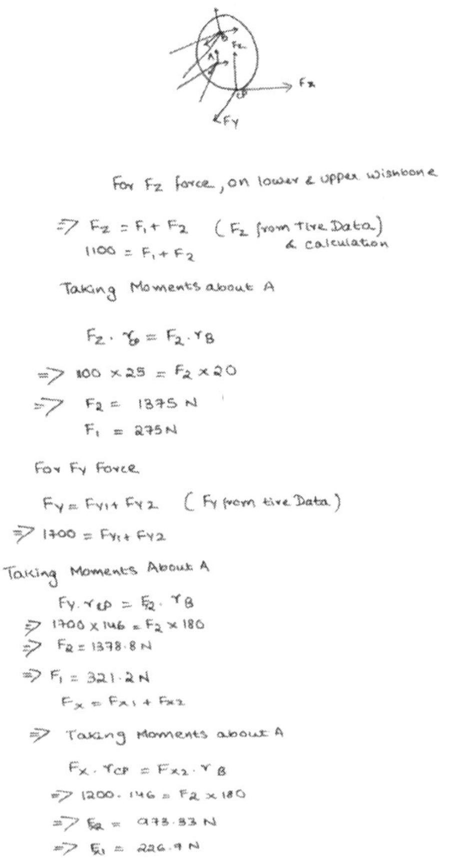

Figure 13. Calculation of forces on suspension arms

3.19 Load Analysis on carbon fibre rods

One of the main goals of suspension design is to reduce the unsprung mass. So, a decision was taken to use carbon fibre rods for the A-Arms. The process to be used was to use carbon fibre rods and attach it to the rod ends with the use of Aluminium inserts which will be stuck to the carbon fibre tubes with the help of high grade epoxy. The design of the aluminium inserts is of utmost importance as weight needs to be minimum while still keeping strength intact.

3.20 Al inserts

As mentioned above Al inserts are designed in Solidworks and analysed. The results of analysis are shown below.

As the displacement is very low the above design can be considered good for the application. The weight being 70 grams also is very light and can be used as part of the suspension system.

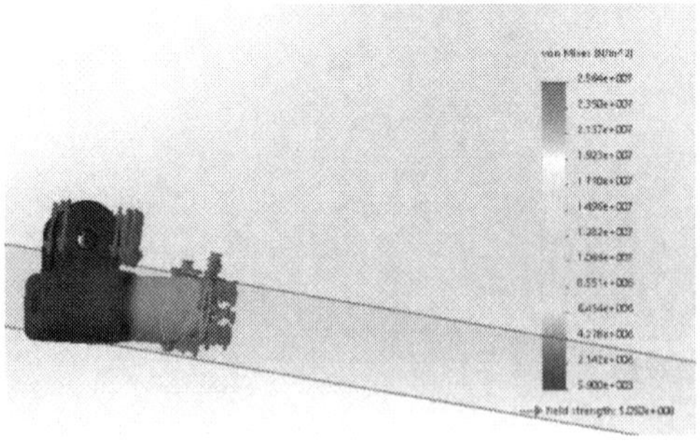

Figure 14. Stress on insert

Final Design in Solidworks

Figure 15. Final Design

References

1. www.iaeng.org/publication/WCE2016/WCE2016_pp1062-1066.pdf
2. Ganesh Mohan, Claude Rouelle, Emmanuel Hugon, OptimumG LLC, Denver, Colorado *A New Method to Evaluate Bump Steer and Steering Influence on Kinematic Roll and Pitch Axes for All Independent Suspension Types*
3. http://www.optimumg.com/docs/SteeringGeometry.pdf
4. William F. Milliken and Douglas L. Milliken, 1995, "Race Car Vehicle Dynamics", Society of Automotive Engineers, Inc.
5. Gaffney, Edmund F., and Anthony R. Salinas. Introduction to Formula SAE® Suspension and Frame Design. No. 971584. SAE Technical Paper, 1997.
6. Robertson, Dennis, and George J. Delagrammatikas. The suspension system of the 2009 cooper union FSAE vehicle: A comprehensive design review. No. 2010-01-0311. SAE Technical Paper, 2010.
7. Gillespie, Thomas D., Fundamentals of Vehicle Dynamics. Society of Automotive Engineers, Inc., Pennsylvania, 1992.
8. Smith, Carroll, Engineer to Win. MBI Publishing Company, Minnesota, 1984.

Design of steering system for a small Formula type vehicle using tire data and slip angles

Sadjyot Biswal[1], Aravind Prasanth[1], Dr. R Udayakumar[2], Dr. M. N. Sankaram[2], Dhruvil Patel[1]

[1]*Automotive, Mechanical Department, BITS Pilani Dubai Campus, DIAC, 345055*
[2] *Mechanical Department, BITS Pilani Dubai Campus, DIAC, 345055*

Abstract. The focus of this paper would be on the design of steering system for a small formula race type vehicle. The design would consist of determining steering geometry with the use of tire models and optimizing the deign to have the best possible design. Ackermann Geometry, Anti Ackermann, parallel steer will be discussed and the ways to obtain the desired geometries will also be focussed on. The effect of steering on other parameters like roll centre will also be discussed. The importance of reducing bump steer and the methods of doing it will also be discussed.

1 Introduction

The focus of this paper is to understand dynamics of vehicle related to design of steering geometry and how to design a steering geometry as per track requirement and use of tire data. The paper will focus both on static and dynamic design of the steering system and will take into consideration the parameters to be kept in mind while designing a fast responding system with enough steering feel. The paper will present a step by step analysis of the design phase and how a kinematic steering software can be used to determine correct geometry.

2 Requirements of steering system

The steering system needs to be such that the car can cover all turns on the track and also give accurate steering feel to the driver. The steering effort required by the driver should not be too much that it makes it uncomfortable and at the same time maintains enough steering feel.

3 Design

The most important concepts for designing a race car is to keep the tire operating at peak lateral force. The peak lateral force of the tire is a property of the tire and depends on the inclination angle and normal load on the tire. To determine the peak lateral force at a particular slip angle, at a normal load and slip angle is determined by the tire data which can be purchased from FSAE TTC. This will provide some knowledge as to what type of steering geometry is suited for the tires.

3.1 Three types of steering geometry

Ackermann: In Ackermann geometry, the inner tire in a corner turns more than the outer tire with the same input steering angle.

Anti-Ackermann: In Anti Ackermann geometry, the inner tire does not turn as much as the outer tire with the same input steering angle.

Parallel Steer: In parallel steer both the wheels will turn by the same amount with same input steering angle.

According to the weight transfers in the corners, the different peak lateral forces and the slip angles for it can be noted according to the Lateral force vs Slip angle graph. Then depending on it the steering geometry which is preferred by the tires to operate at peak lateral force can be found out

The weight transfer can be found out using the below Formulae.

$$M_{roll} = A_y m h$$

$$\theta_{roll} = \frac{M_{roll}}{K_f + K_r}$$

$$F_{zr} = 0.5\rho m g + \frac{\theta_{roll} K_f}{t_f}$$

$$F_{zl} = 0.5\rho m g - \frac{\theta_{roll} K_f}{t_f}$$

$$A_y = \frac{F_{yl} + F_{yr}}{\rho m}$$

Where

- Ay is the lateral acceleration
- Mroll is the roll moment
- θ roll is the roll angle
- Fzr and Fzl are the right and left tire vertical loads

 DOI:10.1051/matecconf/201712407007

•Fyr and Fyl are the right and left tire lateral forces

•g is the acceleration due to gravity

The normal load after weight transfers have been found out as 143lb and 260lb respectively.

Now once the weight transfer has been found out, we need to find the inclination angle of the tires for the cornering as well.

This can be found out using the below formula:

$$\gamma_r = \gamma_{r0} + \lambda(1 - \cos\delta_r) - \upsilon\sin\delta_r + \theta_{roll}C$$

$$\gamma_l = -\gamma_{r0} + \lambda(1 - \cos\delta_l) - \upsilon\sin\delta_l + \theta_{roll}C$$

Now once the inclination angle and normal load on both the front tires have been found out, we can use the tire data and use it to find out the slip angles at which peak lateral force can be achieved. As the tire data is a raw data we need a tire analysis software to get the F_y vs SA graphs. Optimum T has been used for the following case.

We notice from the graph that the lateral force peaks out at a particular slip angle and then falls down. The graph points can be exported into an excel sheet and the peak lateral force can be found out and noted.

For the particular set of tires, the normal loads and inclination angles calculated are:

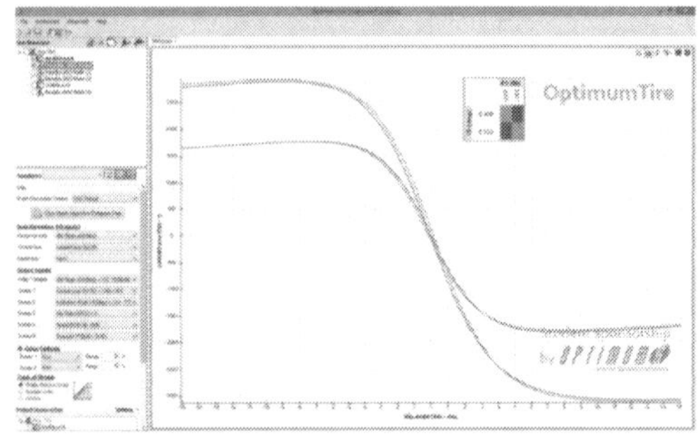

Figure 1. orce vs Slip angle

Fz Outer= -260 lb

Fz Inner=-143 lb

Inclination Angle Outer: - 0.4

Inclination Angle Inner: - 2.322

The corresponding slip angle values for the tires to operate at peak lateral force for the above-mentioned set of values are:

11.93 for the outer tire and 7.65 for inner tire.

The above results show that these particular set of tires prefer anti Ackermann steering geometry as the outer tire needs more slip angle and hence more steering input to obtain peak lateral force values.

4 Tire drag

For a formula student car and other low downforce formula type cars which run on tracks with tight corners, one more factor comes into play which determines what kind of steering geometry that should be used called tire drag.

Tyre drag is a effect of the lateral force often called as lateral force induced drag. It is a force in the opposite direction of vehicle movement analogous to aerodynamic drag but created due to the tire. Higher the slip angle at which the car operates more the drag produced by the tire. This tire drag has a huge effect on cornering of a vehicle.

Whenever a vehicle is needed to traverse a corner, it needs yaw moment. For a tight turn like a hairpin where a large lateral acceleration is needed in a small time, tire drag has a very negative impact on the turning of the car.

From the figure above it can be seen that the tire drag on the inner tire helps to traverse the corner but he drag on the outer side produces a moment out of the corner. If Anti Ackermann is preferred then the outer tire would have a large tire drag and hence the vehicle wont be able to take tight corners as found in Formula student circuits.

Therefore, based on one's track and requirement a particular steering geometry must be chosen. Due to the needs of FSAE track being tight corners Ackermann steering geometry is considered to be more suitable.

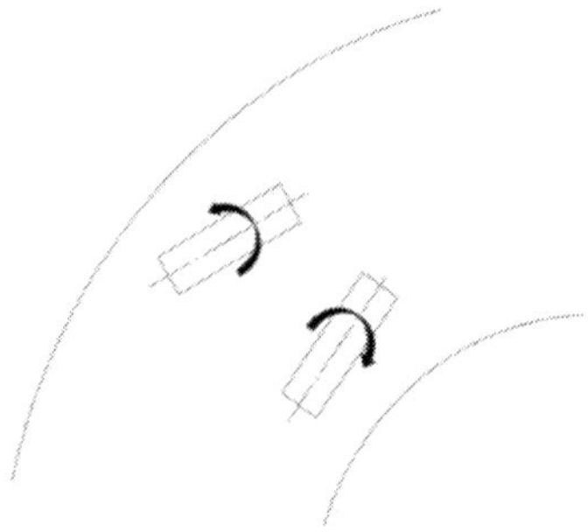

Figure 2. Tire Drag

4 How to determine steering curve

Now to determine the steering curve there were two things to be seen, the tires wanted a anti Ackermamn geometry whereas considering the tire drag Ackermann geometry would be more preferred. Therefore, a small Ackermann geometry was preferred so as to reduce the effects of tire drag and still not sacrifice a lot of the tire's peak lateral force.

The steering curve is basically a graph between the Tyre Inside-Tire Outside angle vs Ackermann Angle. To determine the difference between Inside and Outside angles the following set of formulas have been used.

$$\alpha_r^G = \frac{l}{R + \frac{t}{2}} + \beta$$

$$\alpha_l^G = \frac{l}{R - \frac{t}{2}} + \beta$$

$$\delta_r = \alpha_r^G - \alpha_r^i$$

$$\delta_l = \alpha_l^G - \alpha_l^i$$

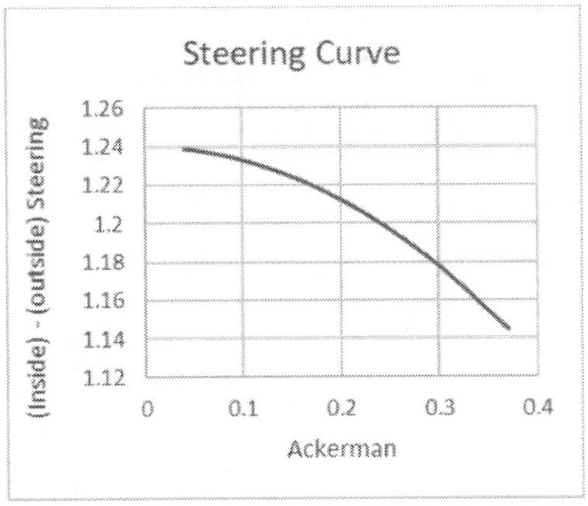

Figure 3. Steering Curve

The track radius can be determined form the Optimum L software which has inbuilt track co-ordinates of FSAE track. From the software, it can be seen that majority of the car travels at corners of radius and maximum radius required to go across is 25 m and minimum is 4.5m

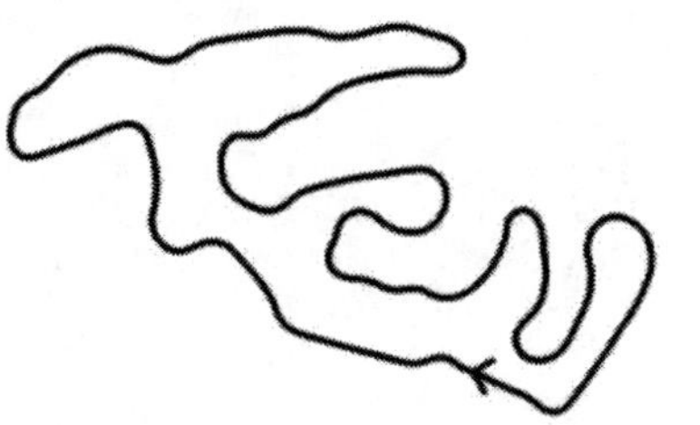

Figure 4. FSAE Germany Endurance track

Now it needs to be seen if the particular steering geometry can cover the lowest possible radius on the track. This can be determined using the following formulae.

Now once the geometry has been decided, a few more things like caster, bump steer and making the geometry into a real-world steering system need to be focused on.

5 Caster and mechanical trail

As mentioned before, the steering feel on the car is of utmost importance and the car should also have some self centring effect. However, if it is too much the driver will become uncomfortable.

The total aligning torque of the tires is determined by the mechanical trail and pneumatic trail. The pneumatic trail of the tires can be seen using tire data as it is a property of the tyre. The mechanical trail depends on the caster. The key in setting caster is to see how much camber change is occurring and what mechanical trail is needed.

The mechanical trail should be such that it does not over power pneumatic trail and it should not be too less also so that peak of self aligning torque can be achieved at higher slip angles. The self aligning torque peak is reached much before the Lateral force peak and so a higher peak of SAT will help driver to drive much more near the peak lateral force range and also have a good steering feel.

Looking at the tire data the pneumatic trail is around 30 mm. So a mechanical trail of 40 mm is chosen to keep the SAT peak higher without making steering force required by the driver too high. A provision can be kept on the wishbone mounts to keep caster adjustable, so that the system can be tested on track.

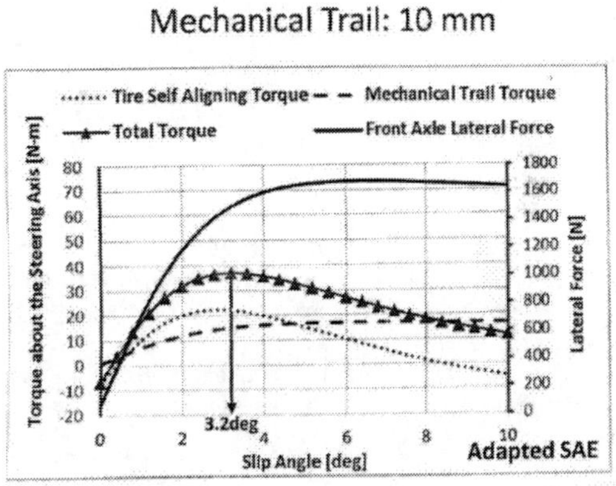

Figure 5. Steering torque vs slip angle (10 mm Mechanical trail)

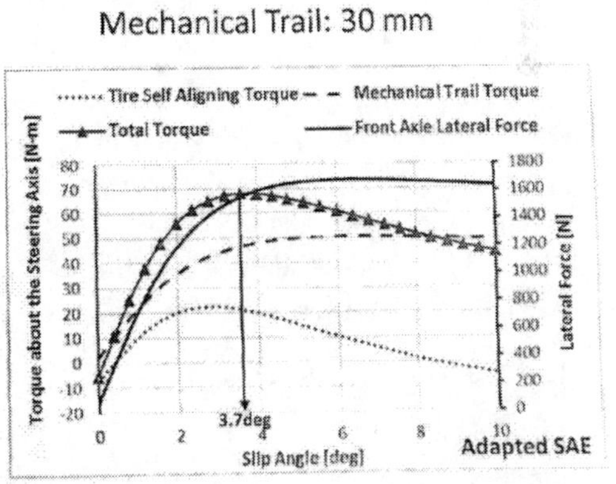

Figure 6. Steering torque vs slip angle (30 mm Mechanical trail)

6 Bump steer

Bump steer is a generally a negative characteristic and should be minimized. The two ways of doing it are first by using the right geometric design and then putting the points in the kinematic software and optimizing it. The geometrical ay to do it is shown below. The position of rack and pinion along the length of the car and the outboard toe link point affect bump steer. Toe base should be kept maximum and toe link should be perpendicular to chassis for best compliance related results

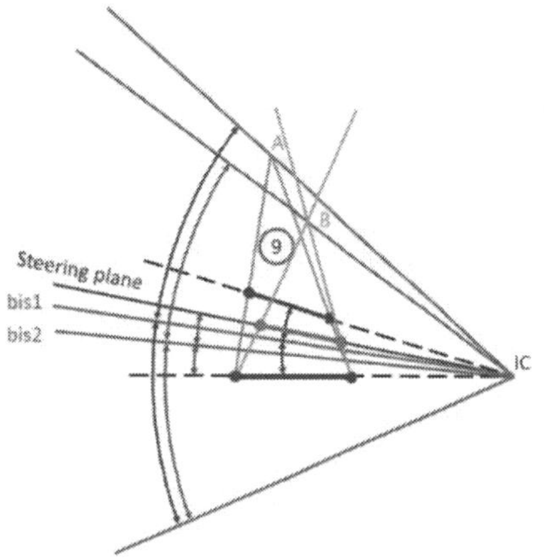

Figure 7. Geometry for minimum bump steer

The best results can be obtained if the toe link is facing the instantaneous centre. The below diagram gives a brief idea about how to go about creating negligible bump steer.

Once the diagram is done and points are known they can be entered into a kinematics software for simulation and further optimized.

7 Steering rack location

The location of the steering rack is very important in determining what type of geometry is used. The following table shows what each rack position and the outboard steering points with respect to outboard wishbone pickup points signifies.

Table 1. Effect of various parameters on steering geometry types

Steering Rack Location	Outboard Steering points with respect to outer A-arm points	Type of steering Geometry
Front Steer	Inside	anti-Ackerman
Front Steer	Inline	Parallel
Front Steer	Outboard	Ackerman
Rear Steer	Inside	Ackerman
Rear Steer	Inline	Parallel
Rear Steer	Outboard	anti-Ackerman

8 Roll centre and steering linkages

As opposed to popular belief the position of the roll centre is affected by the position of the toe link. This is because roll centre is not a 2-D concept but is in fact a 3-D concept. The method is based on the Reuleaux method, which is a geometrical method to locate the instant centre of rotation of a planar motion.

9 Conclusion

The focus of this paper is to give an idea on how to use tire data, kinematic software and use of other important parameters to design a competitive steering system which will give highest performance for the set of tyres. Important points like roll centre relation with steering linkage, use of peak lateral force along with the choice of Anti-Ackermann or pro Ackermann have been explained. It covers all important points needed to understand the designing of a steering system for providing peak performance. Trends have been generated to show the significance of change of RCH with change in steering linkage coordinates.

The influence of other parameters like bump steer, caster and mechanical trail have also been discussed. The significance of tire drag which is not generally given enough mention has also been discussed.

References

1. www.iaeng.org/publication/WCE2016/WCE2016_pp1062-1066.pdf
2. Ganesh Mohan, Claude Rouelle, Emmanuel Hugon, OptimumG LLC, Denver, Colorado *A New Method to Evaluate Bump Steer and Steering Influence on Kinematic Roll and Pitch Axes for All Independent Suspension Types*
3. http://www.optimumg.com/docs/SteeringGeometry.pdf
4. William F. Milliken and Douglas L. Milliken, 1995, "Race Car Vehicle Dynamics", Society of Automotive Engineers, Inc.
5. Gaffney, Edmund F., and Anthony R. Salinas. Introduction to Formula SAE® Suspension and Frame Design. No. 971584. SAE Technical Paper, 1997.
6. Robertson, Dennis, and George J. Delagrammatikas. The suspension system of the 2009 cooper union FSAE vehicle: A comprehensive design review. No. 2010-01-0311. SAE Technical Paper, 2010.
7. Gillespie, Thomas D., Fundamentals of Vehicle Dynamics. Society of Automotive Engineers, Inc., Pennsylvania, 1992.
8. Smith, Carroll, Engineer to Win. MBI Publishing Company, Minnesota, 1984.
9. Norton, R. Machine design. Pearson Prentice Hall, Upper Saddle River, NJ, 2006.
10. Heisler, H. (1989), Advanced Vehicle Technology, Arnold, 338 Euston Road, London NW1 3BH
11. Fenton, J. (1980), Vehicle Body Layout and Analysis, Mechanical Engineering Publications Ltd. London

DOI:10.1051/matecconf/201712407007

12. Smith, C. (1978), Tune to Win: The Art and Science of Race Car Development and Tuning, Aero Publishers, Inc. 329 West Aviation Road, Fallbrook, CA 29028
13. Vehicle Dynamics Terminology- SAEJ670, Revise 07-1976.
14. Pashley, T. (2008). Hoe to Build Motorcycle-engined racing cars, UK, Veloce Publishing

Identifying Unique Characteristics of Disassembly for Various Product Recovery Methods with A Focus on Remanufacturing

Anjar Priyono[1]

[1] Department of Management, Universitas Islam Indonesia

Abstract. This paper aims to identify differences between disassembly for three product recovery methods: remanufacturing, recycling, and repair. Existing studies focus on disassembly differences for the three recovery methods from technical perspective only; those studies focus on the level of disassembly and the method of disassembly. This study covers various aspects of disassembly such as knowledge of worker, value of the materials, and interplays with other processes in product recovery. As many as seven products were used as case studies were undertaken to identify differences of disassembly for three recovery methods. At the end of the paper, managerial implications are presented.

1. Introduction

Remanufacturing falls under the major theme of sustainable manufacturing, which has received considerable attention in recent years. There is a growing trend within government bodies, such as European Union Directives, Waste Electrical and Electronic Equipment (WEEE), Kyoto Protocol and End-of-Life Vehicles (ELV) in favour of adopting more environmentally friendly manufacturing practices and regulations. Economic analysis has shown remanufacturing to be a prospective business [1] and it is environmentally friendly from an ecological perspective [2]. This is because remanufacturing is not only viewed as an effective way to minimise discarded waste, but also as part of a company's manufacturing and marketing strategy [3].

In terms of product types, automotive products are the most common [3], but other product types, such as photocopiers [8, 11, 12], forklift trucks [7], toner cartridge [7] and telecommunication devices [7, 14] are all gaining in popularity. This evidence indicates that remanufacturing has spread widely across many countries and product types.

This significant increase in attention towards remanufacturing practices suggests that more studies in this area are badly needed. Such a study would contribute to a better understanding of remanufacturing, which would in turn lead to more economically feasible and environmentally friendly manufacturing practices. *This research aims to fill this particular need by investigating how disassembly for remanufacturing process is different from that of other recovery methods.*

Through the better understanding of the disassembly process, remanufacturing performance can be improved by minimising uncertainties [10], with disassembly being one of the main causes. Equally important are the uncertainties that result from disassembly and how these affect other processes in remanufacturing. As a result, better-managed disassembly will lead to more economically efficient remanufacturing operations.

Disassembly is a critical element of product recovery activities, as it is the key link that connects product return with product recovery, and a prerequisite for some processes [11]. Disassembly is a prerequisite for reprocessing and re-machining because the processes cannot be carried out whilst the components are still embedded in the products. In other processes, for example cleaning and testing, the components do not have to be entirely disassembled from the products; however, partial disassembly may still be necessary.

In this case, products must be disassembled to a certain level in order to give access to certain components so that cleaning and testing can be carried out. Without disassembly it may not be possible for other processes in remanufacturing to be undertaken, which would mean the products could not be remanufactured.

Disassembly is also the main gateway of information, where many data that is related to the remanufacturing operations originates [17, 18]. Information is valuable in order to minimise any uncertainties in activities that are related to remanufacture, such as purchasing new parts, inventory management and production planning, and scheduling [19–21]. The importance of information from disassembly increases when remanufacturers deal with complex products [14].

Uncertainties that arise on the disassembly shop floor can affect other processes in remanufacturing. During disassembly, not all of the parts can be recovered. It is rare to achieve full recovery rates, so there is an increase in the need for new parts [13]. As disassembly is cited as the main source of parts during reassembly [16] and availability of parts is the most costly operation in remanufacturing [17].

Based on the significance of successful disassembly demonstrated above, the overall remanufacturing process can be enhanced by improving disassembly. Hammond et al. [17] state in their survey report that disassembly is one of the most serious problems found in remanufacturing and, therefore, improving disassembly could improve overall remanufacturing processes. The ability to manage disassembly is the key to success for remanufacturers because it affects several other parameters, including lead times, price, delivery and quality [18].

There has been an abundance of research conducted regarding disassembly that concentrates on product disassembly as the focus of the investigation [19]. Many of the studies that investigate disassembly do not mention the ultimate purpose of product recovery from disassembly. The final purpose of disassembly is important because disassembly for remanufacturing is different from that of other recovery methods. The output of remanufacturing should be able to offer quality performance at the same level, or higher, than the new condition, whereas other recovery methods do not offer such a performance specification [22, 27]. Therefore, the relevant research question is as follow: *How does disassembly for remanufacturing differs from that of other recovery methods?*

The remainder of this paper is organised into 3 sections. Next section describes the method used to conduct this study. Section 3 discusses the findings and analysis. Existing studies in this topic are also discussed in the section. Section 4, last section, summarises the key findings of this paper.

2. Method

This study adopt inductive approach to analyse disassembly for different recovery methods. Also, multiple case study method was selected to allow researcher undertake within and cross case analysis [22] as well as direct and indirect replication logic [23]. Interview, observation, and ompany visits were conducted to collect data. In the company visits there were observations, interviews, document checks and discussion. Four companies that reproses various products participated in this study. In total there are 7 product types investigated in this study, including gearbox, clutches, jet engine, car engine, photocopier, truck engine, and heavy equipment.

3 Findings and Analysis

3.1 Existing concepts in disassembly

Several studies examine the differences between disassembly for remanufacturing and that for other recovery methods; these are presented in Table 1.

Table 1. Output and disassembly level differences across various end-of-life strategies

End-of-life strategies	Disassembly level	References
Remanufacturing	Total disassembly	[21]
Recycling	Total disassembly	[21]
Repairing	Partial disassembly	[22, 27, 28]
Reconditioning	Partial disassembly	[22, 27]

Source: adapted from [3]

From the perspective of the new process model of disassembly that use a comprehensive perspective [11], the above table has several drawbacks. First, the differences highlighted focus on level of disassembly only, which is related to 'hard' factors. By contrast, 'soft' factors, such as data and information about products, skills of employes, as well as an economic consideration of disassembly, are not discussed [25].

The differences above also ignore the suggestion by existing literature that disassembly can be improved using early information. For example, the development of embedded devices can provide data that can be used to support disassembly [29, 30]. Such technology can assist in reducing the amount of work involved in the process and, consequently, remanufacturers can carry out disassembly more efficiently [26]. In addition, from a strategic perspective, developing organisational relationship with OEMs would be helpful in obtaining product specifications [3], [28]. These specifications are useful as they assist remanufacturers in preparing a facility to carry out disassembly.

Second, the studies detailed above assume that disassembly is an independent activity, ignoring the fact that disassembly in remanufacturing is related to other processes. Coordination with other processes, such as material requirement planning (MRP), bill of material (BOM) and purchase of new parts should also be considered [29].

Further adding to the complexity of discussion is differences between the work content, warranty and quality of output in a remanufacturing context compared to disassembly for repair, reconditioning or recycling [24]. The differences between the three recovery methods, in terms of these parameters, are presented graphically in Figure 1. If highlighted, these differences can aid a better understanding of how disassembly for the three recovery methods is not identical.

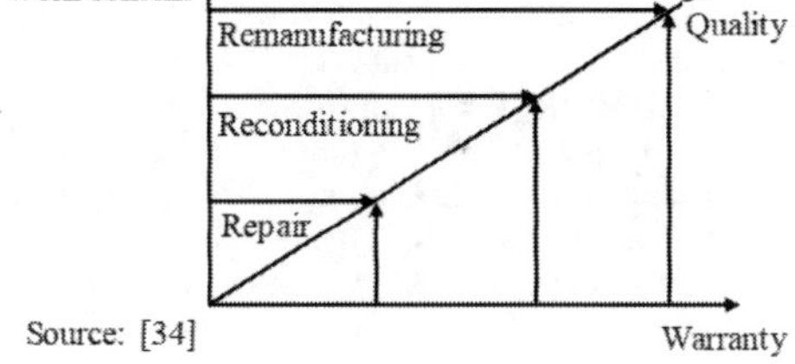

Source: [34]

Figure 1. Hierarchy of secondary production processes

3.2 Comparing disassembly for remanufacturing and recycling

When distinguishing the differences between disassembly for remanufacturing and disassembly for recycling, it is important to stress the sources of the recovered value of the cores. The recovered value of cores comes from two sources: (1) the materials used to make the components, and (2) the manufacturing process which made the components. These two sources of recovered value differ in importance between remanufacturing and recycling. In remanufacturing, both sources of recovered value are important, whilst in recycling, only the value that derives from materials is considered.

Table 2. Differences between disassembly for remanufacturing and recycling.

No.	Issue	Remanu-facturing	Recycling
1.	Method to disassemble.	Destructive disassembly is not allowed.	Both destructive and non-destructive methods are possible.
2.	Priority to recover.	Consider both the value that comes from the manufacturing process and the materials to make the products.	Only considers the value of materials to make the products.
3.	Coordination of disassembly with BOM and MRP	There needs to be coordination with BOM and MRP.	There is no need for any kind of coordination, as the disassembled components are going to be melted into materials, not to be reassembled.

In recycling, companies attempt to recover the value from materials, regardless of how high the potential value from the manufacturing process is. As the condition of the components is irrelevant, the destructive method is allowed in recycling. Components made from materials with higher value will be given priority. Two components that are made from the same materials would be treated in the same way during disassembly, even if the residual value of one component was higher than the other, for example if one of the components had a higher value due to a more complex manufacturing process, complex product structure or rare component.

In remanufacturing, however, both sources of recovered value are considered. For this reason, components made from less expensive materials might be given priority, rather than components that are made from more expensive materials. This is because the former has undergone complex production processes that make the residual value higher compared to the latter one. It is the shape of the components that makes the residual value high, not the materials used to make them. Furthermore, disassembly in recycling does not produce information about the recovery rates of components and it does not require the forecasting of new parts. The summary of all the differences is presented in Table 2.

3.3 Comparing disassembly for remanufacturing and repairing

As there is a need to ensure that the output of remanufacturing is equal to, or higher than, the new ones [22, 33, 34], compulsory disassembly has to be undertaken. This occurs because there are parts that must be replaced with new ones, regardless of the condition, in order to comply with legislation requirements, OEM standards or company policies. Parts are replaced because they have been identified as being typically worn out, a potential risk to the user, or have a limited amount of life remaining. In some circumstances, the parts are not necessarily replaced with new ones, but all the parts should be tested in order to ensure that they meet acceptable quality standards. In this case, the parts still have to be disassembled.

On the other hand, repairing that aims to bring back the functionality of products [22, 27] does not require any compulsory disassembly. Components are disassembled only if they are related to the fault of the products. There is no legislation, company policy or OEM standards that requires specific components to be replaced with new ones. Used components can be reinstalled in the products, provided that they function normally.

From an economic perspective, compulsory disassembly, which is found in remanufacturing, is not efficient, since remanufacturers have to carry out disassembles which are not always necessary [35, 36]. Even components whose conditions are almost equal to new ones have to undergo disassembly. As a result, there might be unnecessary activities conducted by remanufacturers. In contrast, when it comes to repairing, components are only disassembled when it is necessary, for example to enable employees to carry out work on the components and restore the functionality of the products. The summary of these differences is provided in Table 3.

Table 3. Differences between disassembly for remanufacturing and repairing.

No.	Issue	Remanu-facturing	Repairing
1.	Compulsory to disassemble.	There are compulsory disassembly.	There is no compulsory disassembly.
2.	Economics of disassembly.	Not always economically efficient.	More efficient economically than disassembly for remanufacturing.

3.4 Comparing disassembly for remanufacturing and reconditioning

Remanufacturing needs higher level of disassembly than reconditioning. This is because the resulting output of reconditioning is lower than the new products, while the output of remanufacturing is equal to or higher than the new products [22, 27, 37]. For this reason,

DOI:10.1051/matecconf/201712408001

remanufacturing requires more work and higher standards than reconditioning does [34] – see Figure 1. To permit more work in remanufacturing, there would have to be a higher level of disassembly than there is in reconditioning . More work in this regard refers to any remanufacturing processes, such as a more detailed inspection, more detailed testing, or a tighter criteria for sorting and inspection. On the other hand, the output of reconditioning has lower standards, which is not necessarily as strict as the criteria for new products. Disassembly for reconditioned products does not need to ensure that the quality of the resulting products is the same as new ones. There is no need to carry out disassembly, provided that the modules or parts reach acceptable quality standards.

Table 4. Differences between disassembly for remanufacturing and reconditioning.

No.	Issue	Remanu-facturing	Recondi-tioning
1.	Level of disassembly	Have higher level than reconditioning	Have lower level than remanu-facturing
2.	Compulsory disassembly	There are compulsory disassembly activities that should be carried out.	There is no compulsory disassembly activities that should be carried out.

As discussed previously, there are mandatory component replacements in remanufacturing. This causes compulsory disassembly, which is not always economically efficient. On the other hand, reconditioning does not need mandatory component replacements. As long as the components are still in acceptable condition, they are not disassembled. Thus, disassembly in reconditioning is only carried out whenever necessary to do. From economic perspective, this disassembly is more economically efficient than disassembly for mandatory component replacement. The summary of the difference is presented in Table 4.

4 Conclusion

This study offers insights regarding the differences of disassembly for three recovery methods : remanufacture, repair, and recycle. In this study, disassembly for the three recovery methods was analysed comprehensively. It is not only about the level, the method, and sequence of disassembly, but also about skills of employees as well as knowledge regarding the specification of new products. The knowledge is required to ensure that the results of remanufacturing process is equal or higher than the new one.

References

1. R. Subramoniam, D. Huisingh, and R. B. Chinnam, "Remanufacturing for the automotive aftermarket-strategic factors: literature review and future research needs," *J. Clean. Prod.*, vol. 17, no. 13, pp. 1163–1174, Sep. 2009.
2. G. D. Hatcher, W. L. Ijomah, and J. F. C. Windmill, "Design for remanufacture: a literature review and future research needs," *J. Clean. Prod.*, vol. 19, no. 17–18, pp. 2004–2014, Nov. 2011.
3. Y. M. B. Saavedra, A. P. B. Barquet, H. Rozenfeld, F. a. Forcellini, and A. R. Ometto, "Remanufacturing in Brazil: case studies on the automotive sector," *J. Clean. Prod.*, vol. 53, pp. 267–276, Aug. 2013.
4. B. Walsh, "An assessment of the remanufacture of photocopier equipment commissioned by Market Transformation Programme," *Cent. Remanufacturing Reuse*, no. May, pp. 1–14, 2009.
5. K. Mukherjee and S. Mondal, "Analysis of issues relating to remanufacturing technology – a case of an Indian company," *Technol. Anal. Strateg. Manag.*, vol. 21, no. 5, pp. 639–652, Jul. 2009.
6. A. King, J. Miemczyk, and D. Bufton, "Photocopier remanufacturing at Xerox UK," in *D. Brissaudet al. (eds.) Innovation in life cycle engineering and sustainable development.*, 2006, pp. 173–186.
7. J. Östlin, E. Sundin, and M. Björkman, "Importance of closed-loop supply chain relationships for product remanufacturing," *Int. J. Prod. Econ.*, vol. 115, no. 2, pp. 336–348, Oct. 2008.
8. C. Franke, B. Basdere, M. Ciupek, and S. Seliger, "Remanufacturing of mobile phones—capacity, program and facility adaptation planning," *Omega*, vol. 34, no. 6, pp. 562–570, Dec. 2006.
9. P. Rathore, S. Kota, and A. Chakrabarti, "Sustainability through remanufacturing in India: a case study on mobile handsets," *J. Clean. Prod.*, vol. 19, no. 15, pp. 1709–1722, Oct. 2011.
10. K. H. Aksoy and S. M. Gupta, "Near optimal buffer allocation in remanufacturing systems with N-policy," *Comput. Ind. Eng.*, vol. 59, no. 4, pp. 496–508, Nov. 2010.
11. A. Priyono, W. Ijomah, and U. Bititci, "Disassembly for remanufacturing – A systematic literature review, new model development and future research needs," *J. Ind. Eng. Manag. JIEM*, vol. 9, no. 4, pp. 2016–9, 2016.
12. M. L. Junior and M. G. Filho, "Production planning and control for remanufacturing : literature review and analysis," *Prod. Plan. Control*, vol. 23, no. 6, pp. 37–41, 2012.
13. V. Guide Jr., "Production planning and control for remanufacturing: Industry practice and research needs," *J. Oper. Manag.*, vol. 18, pp. 467–483, 2000.
14. G. Ferrer and M. E. Ketzenberg, "Value of information in remanufacturing complex products," *IIE Trans.*, vol. 36, no. 3, pp. 265–277, Mar. 2004.
15. G. Ferrer, "Yield information and supplier responsiveness in remanufacturing operations," *Eur. J. Oper. Res.*, vol. 149, no. 3, pp. 540–556, Sep. 2003.
16. G. Ferrer and D. C. Whybark, "Material planning for a remanufacturing facility," *Prod. Oper. Manag.*, vol. 10, no. 2, pp. 112–124, Jan. 2001.
17. R. Hammond, T. Amezquita, and B. Bras, "Issues in the automotive parts remanufacturing industry – A discussion of results from surveys performed among

remanufacturers," *Int. J. Eng. Des. Autom. – Spec. Issue Environ. Conscious Des. Manuf.*, vol. 4, no. 1, pp. 27–46, 1998.
18. J. R. Duflou, B. Willems, and W. Dewulf, "Towards self-disassembling products design solutions for economically feasible large-scale disassembly," in *D. Brissaud et al. (eds.), Innovation in Life Cycle Engineering and Sustainable Development*, 2006, pp. 87–110.
19. M. A. Ilgin and S. M. Gupta, "Environmentally conscious manufacturing and product recovery (ECMPRO): A review of the state of the art.," *J. Environ. Manage.*, vol. 91, no. 3, pp. 563–91, 2010.
20. W. Ijomah, S. Childe, and C. McMahon, "Remanufacturing - a key strategy for sustainable development.," in *Proceedings of the third international conference on design and manufacture for sustainable development, September 1-2, 2004 Loughborough, UK.*, 2004.
21. M. Thierry, M. Salomon, J. Van Nunen, and L. Van Wassenhove, "Strategic issues in product recovery management," *Calif. Manage. Rev.*, vol. 37, no. 2, pp. 114–135, 1995.
22. K. M. Eisenhardt, "Building theories from case study research," *Acad. Manag. Rev.*, vol. 14, no. 4, pp. 532–550, 1989.
23. R. K. Yin, *Case study research: Design and methods*, 4th ed. London, United Kingdom.: Sage Publications, Inc., 2009.
24. W. L. Ijomah, C. A. McMahon, G. P. Hammond, and S. T. Newman, "Development of robust design-for-remanufacturing guidelines to further the aims of sustainable development," *Int. J. Prod. Res.*, vol. 45, no. 18–19, pp. 4513–4536, Sep. 2007.
25. A. Priyono, W. L. Ijomah, and U. S. Bititci, "Strategic operations framework for disassembly in remanufacturing," *J. Remanufacturing*, vol. 5, no. 11, pp. 1–16, 2015.
26. M. A. Ilgin and S. M. Gupta, "Comparison of economic benefits of sensor embedded products and conventional products in a multi-product disassembly line," *Comput. Ind. Eng.*, vol. 59, no. 4, pp. 748–763, Nov. 2010.
27. O. Ondemir and S. M. Gupta, "Quality management in product recovery using the Internet of Things: An optimization approach," *Comput. Ind.*, vol. 65, no. 3, pp. 491–504, Apr. 2014.
28. S. Lind, D. Olsson, and E. Sundin, "Exploring inter-organizational relationships in automotive component remanufacturing," *J. Remanufacturing*, vol. 4, no. 1, pp. 1–14, 2014.
29. H. B. Lee, N. W. Cho, and Y. S. Hong, "A hierarchical end-of-life decision model for determining the economic levels of remanufacturing and disassembly under environmental regulations," *J. Clean. Prod.*, vol. 18, no. 13, pp. 1276–1283, Sep. 2010.
30. R. T. Lund, *Remanufacturing: the experience of the USA and implications for the developing countries*. Washington, D.C.: World Bank, 1984.
31. W. L. Ijomah, "A tool to improve training and operational effectiveness in remanufacturing," *Int. J. Comput. Integr. Manuf.*, vol. 21, no. 6, pp. 676–701, Sep. 2008.
32. A. Desai and A. Mital, "Evaluation of disassemblability to enable design for disassembly in mass production," *Int. J. Ind. Ergon.*, vol. 32, no. 4, pp. 265–281, Oct. 2003.
33. M. R. Johnson and M. H. Wang, "Planning product disassembly for material recovery opportunities," *Int. J. Prod. Res.*, vol. 33, no. 11, pp. 3119–3142., 1995.
34. W. L. Ijomah, C. A. McMahon, G. P. Hammond, and S. T. Newman, "Development of design for remanufacturing guidelines to support sustainable manufacturing," *Robot. Comput. Integr. Manuf.*, vol. 23, no. 6, pp. 712–719, Dec. 2007.

DOI:10.1051/matecconf/201712408002

Freezing issue on stability master production scheduling for supplier network: Decision making view

Azizah Aisyati[1,2], TMA Ari Samadhi[1], Anas Ma'ruf[1] and Andi Cakravastia[1]

[1]*Department of Industrial Engineering, Bandung Institute of Technology, 40116 Bandung, Indonesia*
[2]*Department of Industrial Engineering, Sebelas Maret University, 57126 Surakarta, Indonesia*

Abstract. In the daily operation, there are frequently changes in customer order requirement which will induce instability of the MPS. Moreover, the frequently adjustment of MPS can induce fluctuation of production and increasing of inventory cost as well as decreasing service level of customer. Most of studies about instability of MPS use freezing method and rolling procedure to adjust MPS periodically. Freezing is the proportion of planning horizon being frozen, whereas rolling procedure is a method replanning periodically of MPS using newly updated demand data. This study is focused on interval freezing length as an issue of decision making. In supply chain, a manufacturer is supported by suppliers to supply material requirement. Since a manufacturer plan production schedule on MPS the freezing interval is determined that will be informed to suppliers which supply the material requirement. In previous research, the freezing interval is decided by manufacturer as necessary decision maker. This decision must be followed by suppliers though it is not beneficial for them. It can be concluded that this condition is no win-win situation. Hence, this research proposes that suppliers will be involved as decision maker besides a manufacturer so the interval freezing is decided by two-side decision maker.

1 Introduction

In uncertain environment, master production scheduling often differs from actual demand. This is happened due to rolling effect and uncertain demand caused by forecast error [1]. Rolling schedule is replanning periodically the MPS depend on replanning interval, frozen interval and forecast window [2]. The MPS changes will have consequences to requirement of material planning (MRP) significantly called system nervousness [3]. MRP nervousness is defined as the significantly changes of MRP due to little changes on MPS [4]. The impact of MRP changes can increase inventory cost and disrupt production activity.

To maintain a stable of MPS is used freezing method that freezes some portion of horizon planning [3]. The portion of freezing can be using either interval freezing or order freezing of the MPS horizon planning. The MPS schedule in frozen interval is fixed whereas the uncertainty of demand as long as frozen interval would be anticipated by safety stock [2]. The variance of MPS planning and actual demand in frozen interval would be a newly data that determines the replanning of MPS for the next period schedule.

In the supply chain system, there are two types decision making namely hierarchical planning system and heterarchical planning system [5]. Hierarchical planning system or hierarchical coordination represent centralized decision making that is achieved by top-level of an organization decision process. Moreover, in order to organize and control the various operation processes the coordination of these activities is a common way in one hand. Conversely, heterarchical approach or decentralized decision making considers an agreement based on consensus between all parties involved in supply chain with different objectives, measures and rules. This approach is called collaborative planning which refers decision making process in decentralized perspective [6]. A decentralized approach will not guarantee global optimum solution. In addition, a decentralized approach can reduce complexity and flexibility of decision making [7].

The frameworks of the relationship between a manufacturer and suppliers are classified into transactional, strategic and collaborative relationship [8]. Moreover, suppliers included in collaboration relationship are strategic supplier. Collaboration proposes the better cooperation of supply chain members to achieve win-win situation in a long term relationship. In addition, collaboration refers coordination of all participants in the relationship to solve some problems together [9]. The other definition of collaboration is a relationship of independence participants to reach more successful [10]. Collaborative advantages is not only the individual partner but also the whole partnership in supply chain structure [6]. The benefits of collaboration are less characteristic variance of component, communication more effective and simple, fewer document, short lead time, low carrying cost, fixed schedule, low inventory level, faster design changes, and stability of price [11]. Based on stage of production process, there are two stages of collaboration consist of product development and production [8]. Due to this research discuss about production planning of MPS this study focuses on production stage.

2 Previous research

Most of researches solve stability Master Production Scheduling using freezing method can be classified into determining freezing parameters by simulation, freezing variables by mathematic model and planning-replanning model. Some researches determine freezing parameter using simulation for single and multi-level of product [3,4]. Meanwhile, mathematical model to find optimal freezing variables in uncertainty environment are develop to single level of product in single and multi-echelon of supply chain structure [2,10,13]. All of these studies were using rolling schedule planning.

Sridharan et. al. [3] discussed to solve instability MPS by determining freezing parameters using simulation for one item or one product. This study checked three parameters of freezing to the MPS stability such as the freezing method, the portion frozen of MPS and horizon planning length of MPS. The result of this research are freezing portion up to 50% reduce cost significantly and a long interval freezing more costly than freezing based on order.

Zhao and Lee [4] continued Sridharan et.al research included multi-level of product to find freezing parameter. The result of this result are the order based freezing method recommended than that of period based, and the increasing of freezing portion can reduce total inventory cost and instability of schedule, contradiction with prior research result.

Lin and Krajewski [2] defined freezing interval, replanning interval and forecast window based on the solution of mathematical model. Freezing interval is frozen period of each planning cycle which no schedule change are allowed. Replanning interval is an interval between two successive replanning of MPS that determines how often the replanning of MPS. This model is focused on single echelon or a manufacturer.

The study evaluated the effect of integration scheduling to reduce negative impact of schedule revision in two echelons supply chain between a manufacturer and suppliers is developed by Krajewski and Wei [12]. In this study stochastic cost model is developed able to analyze the value of scheduling integration by choosing the optimal variables of freezing interval and replanning interval. The model integrates manufacturer and supplier freezing model in two echelon supply chain.

The two echelon supply chain which consider uncertainty environment is also studied by Sundana and Samadhi [13]. Developing Krajewski and Wei research, this study considers safety level as decision variable than that of previous study as parameter model. Therefore the decision variables are freezing interval, replanning interval and safety stock to minimize total cost of system.

3 State of the art

All of the previous researches decide either freezing parameter or freezing interval in hierarchical decision making approach by dominance position member in supply chain. Based on previous researches about instability MPS in uncertainty environment with rolling schedule, they can be classified into one echelon level of supply chain model and two echelons model. For one echelon level the model can be divided to the model which is define the freezing parameters model using simulation and also the model which is determine the freezing decision variables using mathematical model. In two echelons level, the model determines the optimal solution of freezing decision variables using mathematical model.

Dolgui and Prodhon [14] define the technique commonly used to solve uncertainty environment based on recent studies consist of instability in MRP system and demand uncertainty. There are two actions to overcome instability in MRP system namely action on MPS and lot sizing. Meanwhile, the techniques accomplish the problem of demand uncertainty i.e. action on MPS, lot sizing, and safety stock. The state of the art research concerned about instability MPS can be illustrated in Figure 1.

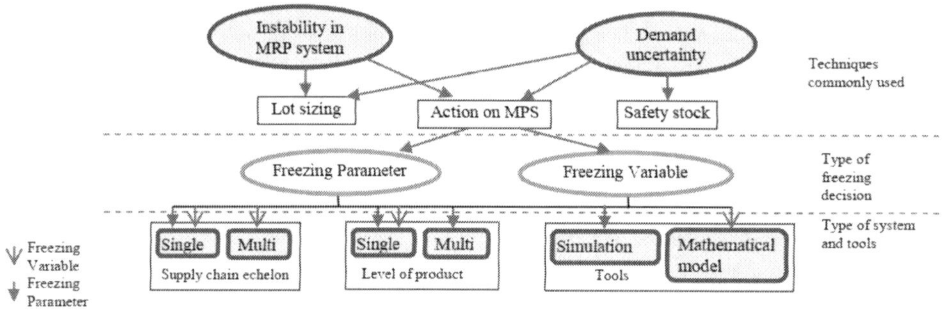

Figure 1. State of the art research on instability of MPS

DOI:10.1051/matecconf/201712408002

Table 1. Previous researches concern on instability of MPS using freezing issue

Paper	Type of system	Freezing parameters	Freezing variables	Research Result
Sridharan et. el. [3]	Single level of product, single echelon	Freezing method, portion of MPS frozen, planning horizon length		Order based is more preferable than period based, more portion of MPS frozen make MPS stable, too longer horizon planning length induce instability of MPS
Zhao and Lee [4]	Multi levels of product, single echelon	Freezing method, portion of MPS frozen, planning horizon length		The selection of freezing parameter is not influenced by product structure, cost parameter and lot sizing rule
Lin and Krajewski [2]	Single level of product, single echelon		Freezing interval, replanning interval, forecast window	The model can used estimated the expected cost by any combination of freezing variable
Krajeswki and Wei [12]	Single level of product, multi echelon		Freezing interval, replanning interval	Schedule integration can reduce total system cost, supplier with higher inventory cost and longer lead time may not get benefit from the integration
Wei and Krajewski [17]	Single level of product, multi echelon		Freezing interval, replanning interval	Increasing forecast effectiveness can reduce total system cost
Tang and Grubbstrom [15]	Single level of product, single echelon		Freezing interval, replanning interval	A model to find freezing variables
Vargas and Metters [16]	Single level of product, single echelon	Freezing method, portion of MPS frozen, planning horizon length		An optimal algorithm to solve stochastic MPS problems
Sundana and Samadhi [13]	Single level of product, multi echelon		Freezing interval, replanning interval, safety stock	Safety factor is optimized by considering shortage cost and limit of inventory space can result cost saving

4 Conclusion and the future research

From decision making perspective, all of the research concern on uncertainty demand solved by freezing method use hierarchical decision making [18]. In supply chain system, there are many parties involved in the system that it is common the dominance position member decide MPS planning which must be followed by all of the members. In uncertainty demand, the manufacturer must have flexibilities to respond customer requirement changing by replanning MPS. The changes of MPS effect some adjustments on MRP that leads reschedule cost of supplier excluded production cost. However, suppliers have no an authority decide their production planning to achieve satisfaction profit by giving a feedback about the planning. To obtain win-win situation, the members of supply chain should have equal right to respond the decision making about MPS planning. So the decision about MPS and freezing can be viewed from two side decision making perspective or decentralized decision making.

Decentralized decision making enable suppliers give a recommendation to manufacturer about the production schedule planning which profitable for them or arrange the counter proposal. Hence, the agreement about production schedule between manufacturer and suppliers should be in win-win situation that is needed negotiation. Negotiation is a tool to solve a complex situation caused the supply chain members have individual interest respectively which are conflicting. Negotiation enable each party make own decision production planning which will be communicated to the others in supply chain.

Previous researches concern about production planning negotiation between manufacturer and suppliers generally can group based on the negotiation decision maker. This grouping is summarized in Figure 2. The new issue is used in this research to accommodate negotiation to solve problem in an instability MPS that is freezing issue. This research will combine freezing method and negotiation of production planning which is a novelty of this research. The roadmap of this research is divided consist of three steps. Firstly, we develop coordination model between a manufacturer and a supplier then determine negotiation model between a manufacturer and two suppliers. Finally, we will construct architecture negotiation model which support negotiation automatically. The research roadmap and the research position in negotiation of production planning area are illustrated in Figure 3 and Table 2.

This research proposes an approach to solve instability of MPS from two-side decision maker in production planning problem between manufacturer and supplier. This gives equal rights for participants to make decision that satisfied them than that of one-side decision making. The one-side decision making is dominated by a participant having dominant position to decide profitable production planning which ignore the other party interests.

The previous researches about instability MPS uses hierarchical approach for decide production planning and freezing parameter by manufacturer. This decision perhaps not fair for supplier because the changes of MPS only gives beneficial effect for manufacturer than that of supplier so two-side decision making is propose to improve this condition. In supply chain system the increasing profit obtained by a party can affect the increasing expense for the others.

This research uses negotiation approach to decide an agreement about production planning and freezing that can satisfy both manufacturer and supplier. Firstly, we develop coordination model between a manufacturer and supplier to decide production planning and freezing because this step is only needed coordination mechanism without conflict. Furthermore, coordination between a manufacture and two-supplier can generate some conflict that will be solved using negotiation approach. Finally, it is needed to perform negotiation automatically so an architecture using multi agent system is designed to support it.

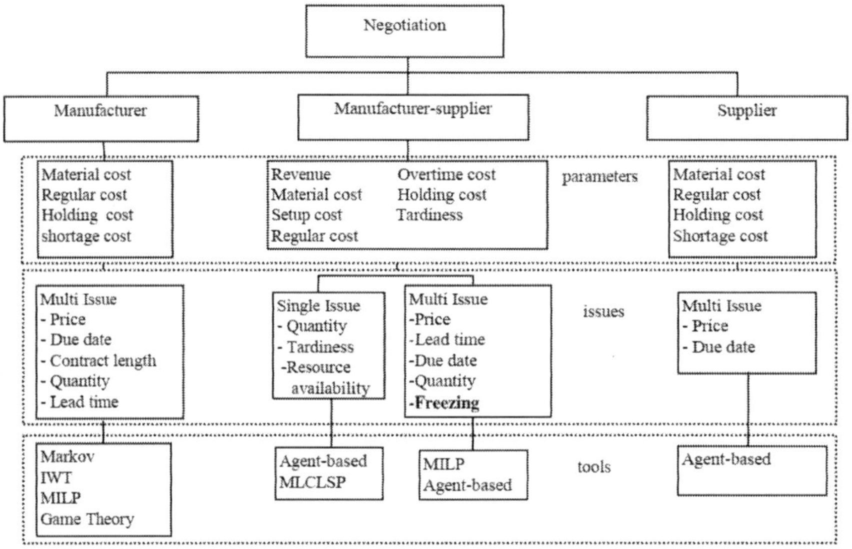

Figure 2. Negotiation research taxonomy

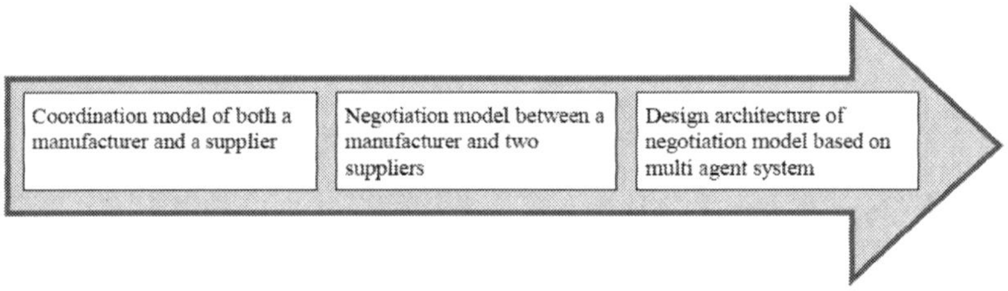

Figure 3. The research roadmap

Table 2. The research position

Paper	Decision Making		Issue	Tools
	One-side	*Two-side*		
Groosman et.al. [19]	√		price, quantity, contract length	Markov
Cakravastia and Nakamura [20]	√		price, due date	Interactive Weighted-Tchebychef
Calosso et.al. [21]		√	price, due date	Mixed Integer Linier Programming
Dudek and Stadtler [7]		√	Production planning	Multi-Level

				Capacitated Lot Sizing Problem
Dudek and Stadtler [22]		√	Production planning	Multi-Level Capacitated Lot Sizing Problem
Argoneto and Renna [23]	√		price, due date	Agent-based
Kim and Cho [24]		√	resource availability	Agent-based
Pan and Choi [25]		√	price, due date	Agent-based
Hernandez et. al. [6]		√	price, quantity	Agent-based
Yu et. al. [27]		√	price, quantity, leadtime	Game theory
This research		√	freezing, production planning	Agent-based

References

1. H. Tsubone and H. Furuta, Int. J. Prod. Ec. **44,** 53-61, (1996)
2. N. P. Lin and L. Krajewski, Dec. Sci. **23**, 839-61 (1992)
3. V. Sridharan, W. L. Berry, V. Udayabhanu, Man. Sci. **33**, 1137–1149 (1987)
4. X. Zhao and T. S. Lee, Prod. Plan. Con. **7,** 144-161 2 (1996)
5. G. Dudek, *Collaborative Planning in Supply Chains: A Negotiation-Based Approach*, (2nd Ed. Springer-Verlag Berlin Heidelberg, 2009).
6. J. E. Hernández, J Mula, R. Poler, A. C. Lyons, Group Dec. Neg. **23,** 235–269 (2014)
7. G. Dudek and H. Stadtler, Eur. J. Op. Res. **163**, 668-687 (2005)
8. J. Park, K. Shin, T. W. Chang, J. Park, Ind. Man. Dat. Sys. **110**, 4 495-515 (2010)
9. E. R. Lai, *Collaboration: A Literature Review*, (Research Report , 2011)
10. T. M. Simatupang and R. Sridharan, Int. J. Log. Man. **13**, 15 – 30 (2002)
11. M. Briers, S. Cuganesan, W. F. Chua, *Control in Strategic Supplier Relationship*, (Univ. New South Wales Press, 1999)
12. L. Krajewski and J. C. Wei, Dec. Sci. **32**, 601-633 (2001)
13. Sundana and TMA. A. Samadhi, *Proc. Seminar on Application and Research in Industrial Technology*, Dept. Mech. and Ind. Eng. UGM, E42-48 (2013)
14. A. Dolgui and C. Prodhon, An. Rev. Con. **31**, 269-79 (2007)
15. W. Tang and R. W. Grubbstrom, Int. J. Prod. Ec. **78**, 323-334 (2002)
16. V. Vargas and R. Metters, Int. J. Prod. Ec. **132**, 296-302 (2011)
17. J. C. Wei and L. Krajewski, Int. J. Prod. Res. **38**, 2099-2123 (2000)
18. H. Raiffa, J. Richardson, D. Metcalfe, *Negotiation Analysis: The Science and Art of Collaborative Decision Making*, (The Belknap Press of Harvard University Press, 2007).
19. T. A. Grossman, T. R. Rohleder, E. A. Silver, Int. J. Prod. Ec. **66**, 67-76 (2000)
20. A. Cakravastia and N. Nakamura, Int. J. Prod. Res. **40**, 3425-3440 2002
21. T. Calosso, M. Cantamessa, M. Gualano, Rob. Comp.Int. Man. **20**, 405–146 (2004)
22. G. Dudek and H. Stadtler, Int. J. Prod. Res. **45**, 465–484 (2007)
23. P. Argoneto and P. Renna, Rob. Comp.Int. Man. **26**, 1–12 (2010)
24. H. S. Kim and J. H. Cho, Dec. Sup. Sys. **49**, 77–90 (2010)
25. Pan and T. M. Choi, *An agent-based negotiation model on price and delivery date in a fashion supply chain*, (Annals of Operation Research, Springer Science+Business Media New York, 2013)
26. R. Neubert, O. Gorlitz, T. Teich, Int. J. Prod. Ec. **89**, 175–187 (2004)
27. F. Yu, T. Kaihara, N. Fuji, C. Sun, W. Yang, Int. J. of Prod. Res. **53,** 1-17 (2015)
28. J. E. Hernández, M. M. E. Alemany, F. C. Lario, R. Poler, *Innovar* **19**, 99-120 (2009)
29. C. Yu and T. N. Wong, Exp. Sys. App. **42**, 223-237 (2015)
30. I. W. Zartman, *Negotiation and Conflict Management: Essays on Theory and Practice*, (Routledge Taylor & Francis Group, 2008).
31. X.. Zhao and K. Lam, Int. J. Prod. Ec. **53**, 281-305 (1997)

AUTHOR INDEX

AUTHOR INDEX